Lecture Notes in Physics

Springer-Verlag Berlin Heidelberg GmbH

http://www.springer.de/phys/

The Editorial Policy for Edited Volumes

The series *Lecture Notes in Physics* (LNP), founded in 1969, reports new developments in physics research and teaching - quickly, informally but with a high degree of quality. Manuscripts to be considered for publication are topical volumes consisting of a limited number of contributions, carefully edited and closely related to each other. Each contribution should contain at least partly original and previously unpublished material, be written in a clear, pedagogical style and aimed at a broader readership, especially graduate students and nonspecialist researchers wishing to familiarize themselves with the topic concerned. For this reason, traditional proceedings cannot be considered for this series though volumes to appear in this series are often based on material presented at conferences, workshops and schools.

Acceptance

A project can only be accepted tentatively for publication, by both the editorial board and the publisher, following thorough examination of the material submitted. The book proposal sent to the publisher should consist at least of a preliminary table of contents outlining the structure of the book together with abstracts of all contributions to be included. Final acceptance is issued by the series editor in charge, in consultation with the publisher, only after receiving the complete manuscript. Final acceptance, possibly requiring minor corrections, usually follows the tentative acceptance unless the final manuscript differs significantly from expectations (project outline). In particular, the series editors are entitled to reject individual contributions if they do not meet the high quality standards of this series. The final manuscript must be ready to print, and should include both an informative introduction and a sufficiently detailed subject index.

Contractual Aspects

Publication in LNP is free of charge. There is no formal contract, no royalties are paid, and no bulk orders are required, although special discounts are offered in this case. The volume editors receive jointly 30 free copies for their personal use and are entitled, as are the contributing authors, to purchase Springer books at a reduced rate. The publisher secures the copyright for each volume. As a rule, no reprints of individual contributions can be supplied.

Manuscript Submission

The manuscript in its final and approved version must be submitted in ready to print form. The corresponding electronic source files are also required for the production process, in particular the online version. Technical assistance in compiling the final manuscript can be provided by the publisher's production editor(s), especially with regard to the publisher's own LaTeX macro package which has been specially designed for this series.

LNP Homepage (http://link.springer.de/series/lnp/)

On the LNP homepage you will find:

– The LNP online archive. It contains the full texts (PDF) of all volumes published since 2000. Abstracts, table of contents and prefaces are accessible free of charge to everyone. Information about the availability of printed volumes can be obtained.

– The subscription information. The online archive is free of charge to all subscribers of the printed volumes.

– The editorial contacts, with respect to both scientific and technical matters.

– The author's / editor's instructions.

G. Rangarajan M. Ding (Eds.)

Processes with Long-Range Correlations

Theory and Applications

Springer

Editors

Prof. Govindan Rangarajan
Dept. of Mathematics
Indian Institute of Science
Bangalore 560 012
India

Prof. Mingzhou Ding
Center for Complex Systems
and Brain Sciences
Florida Atlantic University
Boca Raton, FL 33431
USA

Cataloging-in-Publication Data applied for

A catalog record for this book is available from the Library of Congress.

Bibliographic information published by Die Deutsche Bibliothek

Die Deutsche Bibliothek lists this publication in the Deutsche Nationalbibliografie; detailed bibliographic data is available in the Internet at http://dnb.ddb.de

ISSN 0075-8450

DOI 10.1007/978-3-540-44832-7

http://www.springer.de

Originally published by Springer-Verlag Berlin Heidelberg New York in 2003
MyCopy version of the original edition 2003

Typesetting: Camera-ready by the authors/editor
Camera-data conversion by Steingraeber Satztechnik GmbH Heidelberg
Cover design: *design & production*, Heidelberg

Printed on acid-free paper
54/3141/du - 5 4 3 2 1 0
www.springer.com/mycopy

Preface

Processes with long range correlations (also called long range dependent processes) occur ubiquitously in nature. They are defined as random/stochastic processes whose autocorrelation function, decaying as a power law in the lag variable for large lag values, sums to infinity. Because of this slow decay (as opposed to an exponential decay), these processes are also said to have long memory. Since the sample paths of these processes exhibit self-similarity they are also referred to as fractal processes. Long range dependent processes were first introduced by Kolmogorov to model scaling behavior in turbulence. In parallel, Wiener considered extension of the Wiener process to more general diffusion processes, including fractional Brownian motion. More recently, Mandelbrot and colleagues defined the notion of fractional Gaussian noise as a model for the long memory observed by Hurst in the Nile river flood records. Subsequent to these developments, there have been many reports of discoveries of long memory properties in natural phenomena ranging from physics to biology. A partial list of problems involving long range dependence includes: Anomalous diffusion, potential energy fluctuations in small atomic clusters, Ethernet traffic, geophysical time series such as variation in temperature and rainfall records, electronic device noises in field effect and bipolar transistors, financial time series, amplitude and frequency variation in music, EEG signals etc.

Given the wide range of applications, research in this area has grown by leaps and bounds over the past several years. This growth has been aided by several Conferences and workshops devoted to this field. Recently, in an effort to bridge the gap between different disciplines in this inherently interdisciplinary area, an International Conference on long range dependent stochastic processes was held in January 2002 at the Indian Institute of Science, Bangalore, India. This volume is a natural outgrowth of this Conference. It brings together a collection of papers that review both the theory and applications of long range dependent stochastic processes. This book would be useful both to graduate students who wish to pursue research in this area and to researchers who need an overview of this subject.

The book is divided into two parts. The first part deals with theoretical developments in the area of long range dependent processes. The second part comprises chapters dealing primarily with three major areas of application: anomalous diffusion (in the context of statistical physics, chemical physics and porous media), economics and finance, and biology (especially neuroscience). Below we briefly review the contents of the papers appearing in this edited volume.

Bhansali and Kokoszka review different methods of predicting a long memory time series. They distinguish between two different approaches taken in the literature - one where the long memory time series itself is modeled using standard methods like ARMA etc and the other where the dth fractional difference of the long memory time series is studied using such methods. Different methods for evaluating multistep prediction constants are also investigated.

Fractional Brownian motion is a prototypical self similar Gaussian process with long memory. A full understanding of this process is essential since it (and fractional Gaussian noise) has been proposed as a model for myriad natural structures and processes. In his paper, Qian provides a simple introduction to fractional Brownian motion and its connection with renormalization group.

Abry provides a pedagogical introduction to scaling phenomena and self-similar processes with stationary increments. The paper describes in detail how wavelet transform can be used to detect, identify and estimate self similarity. Empirically observed scaling that can not be described by self-similar processes is sought to be explained using other models such as long range dependent processes, multifractal and multiplicative processes etc.

Hurst parameter is an important quantity characterizing time series data. Well known techniques exist for estimating this parameter in the finite variance case of self similar processes and for asymptotic self similarity or long range dependent processes. But these techniques do not work for heavy tailed infinite variance processes. Stoev and Taqqu review recent results for estimating the Hurst parameter using wavelets. Two estimators are investigated – one based on linear fractional stable motion and another based on FARIMA time series. These estimators are shown to be consistent and asymptotically normal under certain conditions.

The Lamperti transformation gives a one-to-one correspondence between stationary processes and self-similar processes. In their paper, Flandrin, Borgnat and Amblard review the basic theory of the transform. Then they investigate how classical concepts from stationary processes can be applied to self-similar processes using a proper "lampertization" and vice versa. These results are extended to weakened forms of stationarity and a stochastic extension to the notion of discrete scale invariance is given.

Partly random fractal sums of pulses (PFSP) form an interesting category of random functions. They are very useful in modeling phenomena that combine global dependence with long distribution tails. Mandelbrot investigates various properties of these functions in detail. Particular attention is paid to their global statistical dependence, lateral attractors etc.

The second part of the book deals with applications of long range dependent processes in various fields. The first set of papers deal with anomalous diffusion in its various manifestations. Shlesinger investigates the problem of producing supra-diffusion where the mean square displacement grows faster than linearly with time. Formulating the random walk as an integral master equation with a coupled space-time memory, conditions under which supra-diffusion emerges are obtained. Examples considered include random flights in a turbulent fluid, relativistic turbulent diffusion and accelerating random flights in a gravitational field.

Gorenflo and Mainardi study anomalous diffusion using the space-time fractional diffusion equation. This equation is obtained from the usual diffusion equation by replacing the second order space derivative by a Riesz-Feller derivative and the time derivative by a Caputo derivative of fractional orders. The solution to this equation is interpreted as a pdf of a fractional diffusion process. The connection with master equation approach for continuous-time random walks is also explored.

The paper by Rangarajan and Ding gives yet another perspective of anomalous diffusion and long memory processes. The first passage time problem for stochastic processes has applications in several areas of science and engineering. In this paper, first passage time distribution for a class of long memory processes (called Levy type anomalous diffusion) is investigated. Using fractional calculus and H-functions, the fractional Fokker-Planck equation corresponding to the first passage time problem is explicitly solved for the zero drift case.

Sen also investigates anomalous diffusion, but in the context of porous media. The main focus of the work is to show how geometric restrictions to diffusion and flow give rise to anomalous behavior. Potential applications of this work are numerous (contamination, polymer networks, nutrient transport etc).

Denisov, Klafter and Urbakh study AC-driven Hamiltonian systems and show that directionality emerges due to symmetry breaking of a mixed phase space structure. Further, it is shown that the directed transport arises due to breaking of the symmetry of Levy flights near regular islands. This approach provides a new tool for manipulating and directing Hamiltonian systems.

The next set of papers deal with applications of long range dependent processes in economics and finance. When investigating economic time series, it is often difficult to decide whether it is stationary, has a deterministic or stochastic trend or whether it has long range dependence. Quite often, a combination of these could be present. Beran, Feng, Frankel, Hess and Ocker review in detail the semiparametric fractional autoregressive models which allow for simultaneous modeling of and distinction between the above components. The paper then applies the above model to economic time series, both real and simulated.

Teyssiere studies microeconomic models which capture the long range dependence observed in asset price volatilities. But these volatility processes differ from standard long range dependent processes in that they share only the properties of the second moment of the standard processes and not the properties of first moments. The models studied in this paper are able to replicate this type of long range dependence.

Estimating the presence of long range dependence from short samples is a problem fraught with pitfalls. Silverberg and Verspagen review different statistical estimators in the context of economic datasets. They argue that no single test successfully deals with all known problems. They also discuss in detail the relevance of long memory for the investigation of long-term economic growth.

The final set of papers deal with applications of long range dependence in biology especially neuroscience. Study of neuronal firing patterns is important to understand the behavior of the brain. Longtin, Laing and Chacron focus on the long range dependence of these patterns. They illustrate how such dependence can arise in systems that range from single cells to large assemblies of cells. In particular, attention is paid to the role of neural conduction delays and short term spatial memory.

Chen, Ding and Kelso study how long range dependence manifests itself in human sensorimotor coordination. They analyze time series that comprise observed timing errors when humans make finger movements in synchrony with periodic stimuli. They demonstrate that long range dependence is present in the error time series obtained from rhythmic coordination tasks and that the Hurst exponent varies with the type of task and coordination strategy.

Linkenkaer-Hansen investigates scaling and criticality in the human brain. The spontaneous large scale neuronal oscillations are viewed as an outcome of a self-organizing stochastic process. It is argued that the theory of self-organized criticality provides an unifying explanation for the large variability in amplitude, duration and recurrence of these oscillations.

Study of the statistical properties of heartbeat interval sequences has become an important field of research in recent years. Most of these studies concentrate on the long range dependence present in the second moments. Using a wavelet-based multifractal formalism, Ivanov shows that healthy human heartbeat dynamics exhibits a broad multifractal spectrum.

The final article deals with the presence of scaling and self-similarity in broadband network traffic. This phenomenon has been observed recently and has received considerable attention since then. Murali Krishna, Gadre and Desai review the multiplicative cascade first proposed by Kolmogorov to model turbulence and characterize its multifractal spectrum. Based on this, they propose a variable variance Gaussian multiplier process to model the inter arrival times in broadband LAN traffic and demonstrate its usefulness.

The editors would like to thank Ms. Angela Lahee at Springer for all her help in making this volume possible. GR would like to thank Mr. N. Hariharan for his help and the Homi Bahabha Fellowship for financial support. MD would like to thank ONR for financial support.

Bangalore and Boca Raton,
March 2003

Govindan Rangarajan
Mingzhou Ding

Table of Contents

Part I Theory

Part II Applications

List of Contributors

Patrice Abry
Laboratoire de Physique
UMR 5672 CNRS
Ecole Normale Supérieure de Lyon
46, allée d'Italie
F-69364, Lyon Cedex 7,
France
Patrice.Abry@ens-lyon.fr

Pierre-Olivier Amblard
Laboratoire des Images et
des Signaux
UMR 5083 CNRS
ENSIEG, BP 46
38402 Saint Martin d'Hères
Cedex
France
pierre-olivier.amblard@lis.inpg.fr

Jan Beran
Department of Mathematics and
Statistics
University of Konstanz
D-78457 Konstanz
Germany
beran@fmi.uni-konstanz.de

R.J. Bhansali
Department of Mathematical Sciences
University of Liverpool
Liverpool L69 7ZL
United Kingdom
sa17@liverpool.ac.uk

Pierre Borgnat
Laboratoire de Physique
UMR 5672 CNRS
Ecole Normale Supérieure de Lyon
46 allée d'Italie
69364 Lyon Cedex 7
France
pborgnat@ens-lyon.fr

Maurice J. Chacron
University of Ottawa
Physics Department
150 Louis Pasteur
Ottawa, Ontario
Canada K1N 6N5
mchacron@physics.uottawa.ca

Yanqing Chen
The Neurosciences Institute
San Diego
California 92121
USA
chen@nsi.edu

S. Denisov
School of Chemistry
Tel-Aviv University
Tel-Aviv 69978
Israel
denysov@post.tau.ac.il

Uday B. Desai
Department of Electrical Engineering
Indian Institute of Technology
Mumbai
India
ubdesai@ee.iitb.ac.in

Mingzhou Ding
Center for Complex Systems and Brain Sciences
Florida Atlantic University
Boca Raton
Florida 33431-0991
USA
ding@fau.edu

Yuanhua Feng
Department of Mathematics and Statistics
University of Konstanz
D-78457 Konstanz
Germany
yuanhua.feng@uni-konstanz.de

Patrick Flandrin
Laboratoire de Physique
(UMR 5672 CNRS)
ENS Lyon
46 allée d'Italie
69364 Lyon Cedex 07
France
flandrin@ens-lyon.fr

Günter Franke
Department of Economics
University of Konstanz
D-78457 Konstanz
Germany
Guenter.Franke@uni-konstanz.de

Vikram M. Gadre
Department of Electrical Engineering
Indian Institute of Technology
Mumbai
India
vmgadre@ee.iitb.ac.in

P. Gopikrishnan
Center for Polymer Studies
and Department of Physics
Boston University
Boston, MA 02215
USA
gopi@buphy.bu.edu

Rudolf Gorenflo
Department of Mathematics and Computer Science
Free University of Berlin
Arnimallee 3
D-14195 Berlin
Germany
gorenflo@math-fu.de

Plamen Ch. Ivanov
Center for Polymer Studies and Department of Physics
Boston University
Boston
USA
plamen@rainier.bu.edu

J.A. Scott Kelso
Center for Complex Systems and Brain Sciences
Florida Atlantic University
Boca Raton
Florida 33431-0991
USA
kelso@fau.edu

Joseph Klafter
School of Chemistry
Tel-Aviv University
Tel-Aviv 69978
Israel
klafter@post.tau.ac.il

Piotr S. Kokoszka
Mathematics and Statistics
Utah State University
3900 Old Main Hill
Logan, Utah 84322-3900
USA
piotr@math.usu.edu

P. Murali Krishna
Department of Electrical Engineering
Indian Institute of Technology
Mumbai
India
mkrishna@ee.iitb.ac.in

Carlo Laing
University of Ottawa
Physics Department
Ottawa, Ontario
Canada K1N 6N5
claing@science.uottawa.ca

Klaus Linkenkaer-Hansen
BioMag Laboratory
Engineering Centre
Helsinki University Central Hospital
P.O. Box 340
FIN-00029 HUS
Finland
klaus@biomag.hus.fi

André Longtin
University of Ottawa
Physics Department
Ottawa, Ontario
Canada K1N 6N5
alongtin@physics.uottawa.ca

Francesco Mainardi
Dipartimento di Fisica
Università di Bologna and INFN
Sezione di Bologna
Via Irnerio 46
I-40126 Bologna
Italy
mainardi@bo.infn.it

Benoit B. Mandelbrot
Sterling Professor of Mathematical Sciences
Yale University
New Haven
CT 06520-8283
USA
mandelbr@us.ibm.com

Dirk Ocker
Schweizer Verband der Raiffeisenbanken
CH-9001 St.Gallen
Switzerland

V. Plerou
Center for Polymer Studies
and Department of Physics
Boston University
Boston, MA 02215
USA
plerou@cgl.bu.edu

Hong Qian
Department of Applied Mathematics
University of Washington
Seattle, WA 98195
USA
qian@amath.washington.edu

Govindan Rangarajan
Department of Mathematics and
Centre for Theoretical Studies
Indian Institute of Science
Bangalore 560 012
India
rangaraj@math.iisc.ernet.in

M.A. Salinger
Department of Finance and Economics
School of Management
Boston University
Boston, MA 02215
USA
salinger@bu.edu

Pabitra N. Sen
Schlumberger-Doll Research
Ridgefield
CT 06877-4108
USA
psen@ridgefield.oilfield.slb.com

Michael F. Shlesinger
Physical Sciences Division
Office of Naval Research
800 N. Quincy St.
Arlington, VA 22217
USA
shlesim@onr.navy.mil

Gerald Silverberg
III and MERIT
University of Maastricht
The Netherlands
Gerald.Silverberg@MERIT.
unimaas.nl

H.E. Stanley
Center for Polymer Studies
and Department of Physics
Boston University
Boston, MA 02215
USA
hes@meta.bu.edu

Stilian Stoev
Department of Mathematics
Boston University
Boston, MA 02215
USA
sstoev@math.bu.edu

Murad. S. Taqqu
Department of Mathematics
Boston University
Boston, MA 02215
USA
murad@math.bu.edu

Gilles Teyssière
GREQAM & CORE
Centre de la Vieille Charité
F-13002 Marseille
France
gilles@ehess.cnrs-mrs.fr

M. Urbakh
School of Chemistry
Tel-Aviv University
Tel-Aviv 69978
Israel
urbakh@post.tau.ac.il

Bart Verspagen
ECIS
Eindhoven University of Technology
5600 MB Eindhoven
The Netherlands
b.verspagen@tm.tue.nl

Part I

Theory

Prediction of Long-Memory Time Series: A Tutorial Review

R.J. Bhansali[1] and P.S. Kokoszka[2]

[1] University of Liverpool, UK
[2] Utah State University, USA

Abstract. Two different approaches, called *Type-I* and *Type-II*, to linear least-squares prediction of a long-memory time series are distinguished. In the former, no new theory is required and a long-memory time series is treated *on par* with a standard short-memory time series and its multistep predictions are obtained by using the existing modelling approaches to prediction of such time series. The latter, by contrast, seeks to model the long-memory stochastic characteristics of the observed time series by a fractional process such that its dth fractional difference, $0 < d < 0.5$, follows a standard short-memory process. The various approaches to constructing long-memory stochastic models are reviewed, and the associated question of parameter estimation for these models is discussed. Having fitted a long-memory stochastic model to a time series, linear multi-step forecasts of its future values are constructed from the model itself. The question of how to evaluate the multistep prediction constants is considered and three different methods proposed for doing so are outlined; it is further noted that, under appropriate regularity conditions, these methods apply also to the class of linear long memory processes with infinite variance. In addition, a brief review of the class of non-linear chaotic maps implying long-memory is given.

1 Introduction

In this section we introduce the framework for the subsequent sections dealing with prediction and extensions of linear long memory processes.

If a time series is understood to be a set of data recorded chronologically, then 'memory' is not a characteristic of such a record; memory instead should be recognised as a property associated with a stochastic model used for describing the generating structure of this record. The class of discrete-time weakly stationary stochastic processes provides a basic but rich family of models towards this end. Thus, suppose that an observed time series of length T, $\{X_t\}$ $(t = 1, \ldots, T)$, $T > 1$, is a part realization of a weakly stationary process, $\{x_t\}$ $(t = \ldots, -1, 0, 1, \ldots)$ with mean $\mu_x = E(x_t)$, covariance function $R_x(u) = E[\{x_t - \mu_x\}\{x_{t+u} - \mu_x\}]$, $(t, u = \ldots, -1, 0, 1, \ldots)$, correlation function $r_x(u) = R_x(u)/R_x(0)$ and spectral density function

$$f_x(\lambda) = (2\pi)^{-1} \sum_{u=-\infty}^{\infty} R_x(u) \exp(-iu\lambda). \tag{1}$$

As in [61], we say $\{x_t\}$ is a short-memory process if the following conditions hold:

a) The covariance function of $\{x_t\}$ is absolutely summable:

$$\sum_{j=-\infty}^{\infty} |R_x(u)| < \infty. \tag{2}$$

b) The spectral density function is non-vanishing :

$$f_x(\lambda) \neq 0, \text{ all } \lambda \in [-\pi, \pi]. \tag{3}$$

Suppose in addition that $\mu_x = 0$. Under these conditions, the spectral density function, $f_x(\lambda)$, is uniformly continuous and bounded: $0 < m < f_x(\lambda) < M < \infty$, all $\lambda \in [0, \pi]$ and $\{x_t\}$ admits an infinite moving average representation and an infinite autoregressive representation as follows:

$$x_t = \sum_{j=0}^{\infty} b(j)\epsilon_{t-j}, \quad b(0) = 1, \tag{4}$$

$$\sum_{j=0}^{\infty} a(j)x_{t-j} = \epsilon_t, \quad a(0) = 1, \tag{5}$$

where $\{\epsilon_t\}$ is a sequence of uncorrelated random variables, each with mean 0 and variance σ^2 and the coefficients, $b(j)$ and $a(j)$, are absolutely summable:

$$\sum_{j=0}^{\infty} |b(j)| < \infty, \tag{6}$$

$$\sum_{j=0}^{\infty} |a(j)| < \infty. \tag{7}$$

For a proof of these results, see Theorem 3.8.4 of [10], who also shows that if $f_x(\lambda)$ is treated as being known exactly, the coefficients, $b(j)$ and $a(j)$, may be determined by the Wiener-Hopf spectral factorization procedure, see [11].

A popular class of short-memory processes is defined by the standard Autoregressive Moving Average model of order (p, q), ARMA(p, q) model, for short:

$$\sum_{j=0}^{p} \phi(j)x_{t-j} = \sum_{j=0}^{q} \theta(j)\epsilon_{t-j}, \tag{8}$$

where $\phi(0) = \theta(0) = 1$, $\{\epsilon_t\}$ is a sequence of independent Normal random variables, each with mean 0 and variance σ^2, and the real-valued coefficients, $\phi(j)$ and $\theta(j)$, are such that if

$$\Phi(z) = \sum_{j=0}^{p} \phi(j)z^j, \qquad \Theta(z) = \sum_{j=0}^{q} \theta(j)z^j, \tag{9}$$

repectively, denote their characteristic polynomials, then $\Phi(z) \neq 0$, $\Theta(z) \neq 0$, $|z| \leq 1$ and $\Phi(z)$ and $\Theta(z)$ do not have a common zero.

As is well-known, for this class of models, the spectral density function, $f_y(\lambda)$, of $\{y_t\}$ is specified to be a rational function:

$$f_y(\lambda) = \frac{\sigma^2}{2\pi} \frac{|\Theta(e^{-i\lambda})|^2}{|\Phi(e^{-i\lambda})|^2}. \tag{10}$$

Also, the correlation function, $r_x(u)$, tends to zero, as $u \to \infty$, at an exponential rate, that is, there exist constants, B and ρ such that, as $u \to \infty$,

$$|r_x(u)| \leq B\rho^{|u|}, \quad 0 < \rho < 1. \tag{11}$$

However, the condition (11) is not necessary for a process to possess short-memory and in this sense the definition given above is more general and enables the inclusion of the autoregressive model fitting approach, see [12], [13], within the group of statistical procedures based on the class of short-memory processes.

For a long-memory process, by contrast, the condition (2) does not hold, that is, the covariance function, $R_x(u)$, is not summable:

$$\sum_{u=-\infty}^{\infty} R_x(u) = \infty, \tag{12}$$

and $f_x(\lambda)$ has a singularity at the origin, and, for a constant, d, such that $0 < d < 0.5$, we may write

$$f_x(\lambda) \sim \lambda^{-2d} \Delta(1/|\lambda|), \quad \text{as} \quad \lambda \to 0, \tag{13}$$

where $\Delta(\lambda)$ is a slowly-varying function at infinity.

Following [14] and [15], and as in [61], we say that a discrete-time, weakly stationary process, $\{y_t\}$, has long-memory if the process, $\{x_t\}$, say, obtained by taking its *dth* fractional difference, $0 < d < 0.5$,

$$x_t = (1 - z)^d y_t, \tag{14}$$

where z denote the backshift operator, $z^j y_t = y_{t-j}$, has short-memory.

Now, since $\{x_t\}$ admits the infinite moving average and autoregressive representations given by (4) and (5), with coefficients, $b(j)$, and $a(j)$, satisfying (6) and (7), respectively, the long-memory process, $\{y_t\}$, also admits one-sided infinite moving average and autoregressive representations as follows:

$$y_t = \sum_{j=0}^{\infty} \psi(j)\epsilon_{t-j}, \; \psi(0) = 1, \tag{15}$$

$$\sum_{j=0}^{\infty} \pi(j) y_{t-j} = \epsilon_t, \; \pi(0) = 1, \tag{16}$$

where $\{\epsilon_t\}$ is as in (4) and (5), and if

$$\Psi(z) = \sum_{j=0}^{\infty} \psi(j) z^j, \tag{17}$$

$$\Pi(z) = \sum_{j=0}^{\infty} \pi(j) z^j, \tag{18}$$

denote the generating functions of the coefficients, $\psi(j)$ and $\pi(j)$, respectively, $\Psi(z)$ nad $\Pi(z)$ are given by:

$$\Psi(z) = (1-z)^{-d} \sum_{j=0}^{\infty} b(j) z^j, \tag{19}$$

$$\Pi(z) = (1-z)^{d} \sum_{j=0}^{\infty} a(j) z^j. \tag{20}$$

Hence, it follows that the $\psi(j)$ and $\pi(j)$ decrease to zero at a polynomial (geometric) rate: $\psi(j) \sim Bj^{d-1}$, $\pi(j) \sim Aj^{-d-1}$, as $j \to \infty$, and the $\psi(j)$ are square-summable but not absolutely summable:

$$\sum_{j=0}^{\infty} \psi(j)^2 < \infty, \tag{21}$$

but the $\pi(j)$ are absolutely-summable:

$$\sum_{j=0}^{\infty} |\pi(j)| < \infty. \tag{22}$$

Moreover both the covariance function, $R_y(u)$, and the correlation function, $r_y(u)$ of $\{y_t\}$ also decrease to zero as $u \to \infty$ at a polynomial rate, $r_y(u) \sim Bu^{2d-1}$. Many authors, , see [16], define a long-memory process in terms of this last property.

An advantage of defining the class of long-memory processes by (14) is that the resulting class of processes may conveniently be related to the popular class of non-stationary integrated processes, see [17], which is also defined as in (14) but with d allowed to take only integer values. On adopting the notation introduced by these authors, $\{y_t\}$ is said to follow an $ARFIMA(p,d,q)$ process, if it has long-memory but the fractionally-differenced process, $\{x_t\}$, in (14) follows an $ARMA(p,q)$ model defined by (8). For $d > 0.5$, however, the resulting process is non-stationary. If $-0.5 < d < 0$, then the process is stationary and it is often called an intermediate-memory or antipersistent process, see [8].

In Sect. 4 we discuss the question of how to define a long-memory process for the class of infinite variance stable processes for which the correlation function does not exist, see also [18]; moreover, in Sect. 5, we discuss non-linear chaotic maps with long-memory, in the sense of having correlations which decrease to zero at a polynomial rate. First, however, we discuss the question of linear least-squares prediction for the class of (linear) long-memory processes introduced above.

2 The Type-I and Type-II Approaches

It should be noted that from the point of view of linear least-squares prediction of a long-memory process, since (21) and (22) hold, the classical Wiener-Kolmogorov prediction theory continues to apply; see, for example, Theorem 4.2 of [19]. On the other hand, however, the long-memory process, $\{y_t\}$, is related to the short-memory process, $\{x_t\}$ by (14). We may accordingly distinguish between a *Type-I* and a *Type-II* approach to the linear least-squares prediction of long memory time series. In the former, the multistep forecasts of a long-memory time series are constructed without explicitly seeking to model its long-memory characteristics, that is, by approximating a long-memory time series by a suitable short-memory model, and, in essence, by employing the well-established modelling strategies for predicting the latter. The *Type-II* approach by contrast, explicitly seeks to model the long-memory stochastic characteristics by specifying that the dth difference, $0 < d < 0.5$, of an observed time series follows a standard short-memory model.

Below, we briefly describe how these two approaches may be implemented; for further details see [61].

2.1 Implementation of the Type-I Approach

For implementing the *Type-I* approach with an observed time series, there is a choice of modelling strategies, as further described below:

1). *Parameteric ARMA models.* A standard autoregressive-moving average model of order (p, q), defined by (8), is fitted to the observed time series, with the values of p and q determined by a model selection criterion such as AIC or BIC. [20] have applied this approach for multistep prediction of long-memory time series. This approach can be computationally demanding, however, since the various ARMA models of increasing orders would need to be fitted by a non-linear maximum likelihood procedure; as discussed by [21] in his review of order selection procedures for linear time series models, the use of a three-stage Hannan-Rissanen procedure, see [22], provides a way around this difficulty. A related but different possibility is investigated by [23], who use an $ARMA(1,1)$ model for multistep prediction of a long-memory time series, but estimate its parameters by an adaptive procedure, that is, see [62], by a multistep method such that the estimates are constructed anew for each step of prediction by a least-squares procedure involving minimisation of the sum of squares of h-step prediction errors; the classical maximum likelihood approach to estimating the parameters of an $ARMA$ model is, in this terminology, referred to as the one-step method. [24] and [25] are additional references on this approach; the former compares its effectiveness relative to the *Type-II* approach, while the latter examines the nature of approximation provided by an $ARMA(1,1)$ model for a long-mmeory process.

2). *The autoregressive model fitting approach.* For short-memory time series, this approach has been used successfully for spectral estimation, linear prediction and related applications; see [26], [12], [13], [27], among others. The observed time series is assumed to be a realization of a weakly stationary process admitting a one-sided infinite autoregressive representation, (5), whose coefficients, $a(j)$, are absolutely summable and satisfy (7); a model of order k is fitted by least-squares or by solving the Yule-Walker equations and the value of k is determined by AIC, or a related criterion. [28] has investigated the use of the autoregressive model fitting approach for multistep prediction of a long-memory time series. In her implementation, a 'plug-in' method is used for generating the multistep forecasts and this involves repeatedly iterating the model fitted for one-step prediction and then replacing the unknown future values by their own forecasts. An alternative Direct method may also be considered according to which a new autoregressive model is selected and fitted separately for each step of prediction; [29] has established an asymptotic optimality property for this method. It should be noted that this Direct method is related to yet slightly different from the adaptive method of [23] discussed above, in whose application a new model is not selected separately for each step of prediction, but rather the parameters of the model used for one-step prediction, namely an $ARMA(1,1)$ model, are re-estimated for each value of h; see [30] and [62] for a further discussion of the differences between these two methods and for an explanation of the motivation for considering the Direct method. Also, [61] compare the relative predictive performance of the *Type-I* plug-in and *Type-I* direct methods of fitting autoregressions on simulated long-memory time series.

3). *The spectral factorization procedure.* [11] has studied statistical properties of the Wiener-Hopf spectral factorization procedure when it is applied to a smooth periodogram estimator of the spectral density function of short-memory time series. Recently, [31] have considered its use for multistep prediction of long-memory time series.

2.2 Implementation of the Type-II Approach

As with the *Type-I* approach discussed above, there is a choice of different strategies concerning the nature of the stochastic model specified for $\{x_t\}$, the short-memory process defined by (14). Moreover, this choice is now wider because the use of the Exponential model, see [32], and related models specified in terms of the logarithm of the spectral density function of $\{x_t\}$ may be considered, and, in addition, if the principal aim is to simply estimate the long-memory parameter, d, rather than to predict the future values of $\{y_t\}$, then a stochastic model for $\{x_t\}$ need not be specified. Below, we describe the various modelling approaches together with qualitative features of the various models suggested for a long-memory process; in Sect. 2.2, we discuss the associated question of how to estimate the parameters of the specified models from an observed long-memory time series.

Stochastic Models for Long-Memory Processes. 1). *Parametric Models.* Here, the stochastic (covariance) structure of $\{x_t\}$ is fully specified, except for a finite number of parameters, but the values of these parameters are unknown and so is the actual number of parameters (the order of the model). Based on the success of the standard $ARMA(p,q)$ models for short-memory time series, the class of $ARFIMA(p,d,q)$ models is a popular choice of parametric models for long-memory processes. An alternative class of parametric models is obtained, see [33], by specifying that the spectral density function, $f_x(\lambda)$, of $\{x_t\}$ follows an exponential model of [32], that is,

$$f_y(\lambda) = \frac{\sigma^2}{2\pi}|\Pi(\lambda)|^{-2}, \tag{23}$$

where

$$\Pi(\lambda) = \left|1 - e^{-i\lambda}\right|^{-2d} \left[\exp\{\sum_{j=1}^{n} c(j)\cos(j\lambda)\}\right], \tag{24}$$

and $n \geq 1$ is the order of the model. and, thus, in this model, the logarithm of the spectral density function of $\{x_t\}$ is postulated to have a finite Fourier series expansion. An alternative specification involves postulating that $\log f_x(\lambda)$ is a polynomial function, which is one of the $non-ARFIMA$ models simulated by [20]; [33] also envisages the use of piecewise continuous functions in place of either Fourier cosine functions, or a polynomial function.

2). *Non-parametric Approach.* In this approach, only the behaviour of $f_y(\lambda)$ near the origin, as $\lambda \to 0$, is specified and little is explicitly assumed about the behaviour of $f_y(\lambda)$ away from the origin, that is, about the short-memory component of $\{y_t\}$.

Equation (13) gives the class of models being considered. A basic model of this class is obtained by specifying that $\Delta(1/|\lambda|)$ is a constant in a neighbourhood of the origin.

As discussed by [34], the specification (13) is equivalent to requiring that the covariance function, $R_y(s)$, of $\{y_t\}$ takes the following form:

$$R_y(s) \sim \Delta(s)s^{-(1-2d)}. \tag{25}$$

3). *Hierarchical Parametric Models.* According to this approach, which seeks to build on the success of the autoregressive model approach discussed in Sect. 2.1 for short-memory time series, $\{y_t\}$ is assumed to satisfy (14), but its dth fractional difference, $\{x_t\}$, is specified to follow a parameteric model with an infinite number of parameters. Hence, this approach is less general than the non-parametric approach but less restrictive than the parametric approach. Moreover, since this approach is based upon the existence of a one-sided infinite autoregressive and infinite moving average representations for $\{x_t\}$ discussed in Sect. 1, it is more plausible than a fully parametric approach, and at the same time, as discussed further in Sect. 3, it may still be applied for constructing linear least-squares predictors of the unknown future values of $\{y_t\}$. Two different families of hierarachical models may be considered:

a). *Fractional autoregressive, FAR, models.* In this approach, the short-memory process, $\{x_t\}$, of $\{y_t\}$, is assumed to admit representations (5) and (4), but with cefficients, $a(j)$ and $b(j)$ decreasing to zero at an exponential rate. Given a stretch of T observations from $\{y_t\}$, a fractional autoregressive model of order k, an $FAR(k,d)$ model, is fitted by a likelihood procedure discussed further in Sect. 2.2, where k is assumed to be a function of T and allowed to approach ∞ simultaneously but sufficiently slowly with it, see [35].

b). Fractional exponential, FEXP, models. This approach is basically equivalent to the FAR approach described above in a), but the short-memory process, $\{x_t\}$, of $\{y_t\}$ is assumed to follow an infinite-order Exponential model of [32]. Thus, now the spectral density function, $f_y(\lambda)$, of $\{y_t\}$, is given by (23) but with $n = \infty$ and the coefficients $c(j)$ assumed to approach zero sufficiently fast, either at an exponential rate or to satisfy a slightly weaker summability condition ensuring that, for all $\lambda \in [0, \pi]$, $\log f_x(\lambda)$ admits bounded derivatives of order β, $\beta \geq 1$. There has been much interest recently in this family of models, see, [36], Moulines and Soulier ([37], [38]) [39], among others, and it is motivated by the intuition that, under (14), $\log f_y(\lambda)$ admits the decomposition (23), and hence the behaviour of $\{y_t\}$ could be conveniently characterized by specifying that $\log f_x(\lambda)$, which controls the short-memory component, admits an infinite Fourier expansion. For an observed long-memory time series, this approach is implemented in a similar fashion to that described above for the FAR approach; the Fourier expansion for $\log f_x(\lambda)$ is truncated at a finite value, k, say, and the parameters of the resulting $FEXP$ model of order k, whose spectral density function is now given by (23) with $n = k$, are estimated, though a regression procedure discussed further in Sect. 2.2 is used for parameter estimation.

Parameter Estimation for Long-Memory Models. Denote the observed time series by $y_1, \ldots, y_T$, $T > 1$, which we suppose is a part-realizaton of a long-memory process, $\{y_t\}$, satisfying (14). Below, the question of how to estimate the long-memory parameter, d, together with the parameters of a stochastic model specified for the short-memory process, $\{x_t\}$, is addressed. The discussion of various methods is however necessarily brief since several different surveys reviewing the current state-of-the-art on this question are already available; see, [37] and [35], among others.

It is possible to group the underlying methods under two broad headings: A) Likelihood-based methods; B) Regression-based methods.

The periodogram function of the observed time series is given by

$$I_y(\lambda) = (2\pi T)^{-1} \left| \sum_{t=1}^{T} y_t \exp(-it\lambda) \right|^2 , \tag{26}$$

and both the likelihood-based and regression-based methods adopt the periodogram as a basic statistic.

The likelihood-based methods are used when either the parametric approach or the hierarchical model fitting approach involving the fitting of $FAR(k,d)$ models has been adopted. By contrast, the regression-based methods are used when either the non-parametric approach or the hierarchical model fitting approach involving the fitting of $FEXP$ models has been adopted. In addition, a semi-parametric approach is implemented by essentially adopting the non-parametric approach but using the Whittle likelihood for estimating the long-memory parameter, as further discussed below.

A convenient way of implementing the likelihood-based methods is to use the approximate Whittle likelihood function and to separately parametrize the variance, $\sigma^2 = var(\epsilon_t)$, of the innovation process, $\{\epsilon_t\}$, see (15), from the remaining parameters of the model. Under this assumption, the Whittle likelihood function of the data is given by:

$$L(y|\tau) = \int_{-\pi}^{\pi} \frac{I_y(\lambda)}{g(\lambda,\tau)} d\lambda, \tag{27}$$

where

$$g(\lambda,\tau) = \left| \sum_{j=0}^{\infty} \psi(j) \exp(-ij\lambda) \right|^2, \tag{28}$$

and τ denotes a finite-dimensional vector of 'spectral parameters' such that the infinite sequence, $\{\psi(j)\}$ $(j = 1, 2, \ldots)$, of the coefficients in the moving average representation of $\{y_t\}$ may be expressed as a function of these parameters. If the order, (p,q), of the $ARFIMA(p,d,q)$ model is treated as known, the $g(\lambda,\tau)$ occurring in (27) is given by

$$g(\lambda,\tau) = \left|1 - e^{-i\lambda}\right|^{-2d} \frac{|\Theta(e^{-i\lambda})|^2}{|\Phi(e^{-i\lambda})|^2}, \tag{29}$$

where the parameter vector, τ, is given by $\tau = [d, \phi(1), ..., \phi(p), \theta(1), ..., \theta(q)]'$.

Similarly, for the hierarchical fitting of $FAR(k,d)$ models, $\tau = [d, a(1), ...,$ $a(k)]'$, and

$$g(\lambda,\tau) = \left|1 - e^{-i\lambda}\right|^{-2d} |A_k(e^{-i\lambda})|^{-2}, \tag{30}$$

where

$$A_k(z) = 1 + a(1)z + ... + a(k)z^k. \tag{31}$$

Other approximations for the likelihood function of the data, including the use of the exact likelihood function, have also been suggested, and it is not necessary even to separately parametrize the innovation variance, σ^2; see [41], [40]. However, we do not pursue these question here beyond noting that the software for implementing the likelihood-based methods is available in standard statistical packages; for example, the *Splus* function `arima.fracdiff` can be used.

Next, consider the regression-based procedures. For the non-parametric approach, $f_y(\lambda)$, is specified by (13), with $\Delta(\lambda) = C$, a constant for all λ, and since this specification implies that $\log f_y(\lambda)$ is a linear function of $\log \lambda$ in a neighbourhood of the origin, $\lambda = 0$, an estimate of d is obtained by regressing $\log I_y(\lambda)$ on $\log \lambda$, or λ, in a neighbourhood of $\lambda = 0$. A correction for the 'bias' of $\log I_y(\lambda)$ in estimating $\log f_y(\lambda)$, see [42], is however necessary and it is made. In addition, because the behaviour of the periodogram function for frequencies near 0 is non-standard for a long-memory time series, see [43], periodogram ordinates which are either too close to or too far away from the 0 frequency are usually omitted by choosing two trimming numbers which determine the band of frequencies over which the actual regression is performed. This procedure is often called the *GPH* method, since it was originally suggested by [44]. A number of variations on this basic procedures have also been considered, especially the use of a 'smooth' periodogram in place of the 'raw' periodogram, and the inclusion of all frequencies near zero; we however do not discuss these variations here.

For the hierarchical model fitting approach involving the fitting of $FEXP$ models, an estimate of d is constructed by regressing $\log I_y(\lambda)$ on $\log|2\sin(\lambda/2)|$ and $\cos(j\lambda)$, $(j = 0, 1, ..., k)$. Moreover, an equivalent estimate of d may also be obtained, see [37], from the estimated regression coefficient in a simple linear regression of $\log I_y(\lambda)$ on ξ, where ξ is the difference between $\log|2sin(\lambda/2)|$ and its orthogonal projection on the linear space spanned by $\cos(j\lambda)$, $(j = 0, 1, ..., k)$.

As in the non-parametric approach, the specification (13) for $f_y(\lambda)$, with $\Delta(\lambda) = C$, a constant for all λ, is also adopted for the semi-parametric approach. However, see [45], [63], estimates of d and C are constructed from the Whittle likelihood, which is now computed only over a neighbourhood of the 0 frequency.

For a discussion of the asymptotic statistical properties of the estimates provided by these four approaches, see [35]. Below, we indicate some of the shortcomings of each approach:

If the order, (p, q), of an $ARFIMA(p, d, q)$ model is treated as known, the maximum likelihood estimator, $\hat{\tau}$, of τ is known to be asymptotically Normal and Fisher efficient. In practice, however, the order (p, q) is invariably unknown, and there is much evidence to suggest that when the actual order is misspecified, the estimates could be badly biased, see [46], among others. The question of model selection for the class of $ARFIMA(p, d, 0)$ models has been considered by [47]. An additional conceptual difficulty is, see [48], that an observed time series need not follow a 'true' $ARMA$ model.

A main difficulty with both non-parametric and semi-parametric approachesis that while each makes few prior assumptions concerning the stochastic structure of the observed time series, the resulting estimates have a slow rate of convergence, namely $T^{-\delta}$, where, no matter how smooth the short-memory spectral desity, $f_x(\lambda)$, is, $\delta < 0.4$. Moreover, an important practical difficulty is how to choose in an optimal fashion the band of frequencies over which the log-periodogrm regression is performed.

The $FEXP$ method yields an estimator with a faster rate of convergence than the non-parametric or semi-parametric methods, but without requiring that the short-memory process follows a parametric model. On the other hand, however, in the situation where the parametric approach is applicable, this and the FAR approach are not as efficient and have a larger variance.

[35] have studied the empirical behaviour of these approaches by a simulation study.

3 Multistep Prediction of the Long-Memory Time Series by the Type-II Approaches

The parametric and the hierarchical model fitting approaches specify the behaviour of the spectral density function, $f_y(\lambda)$, of the process, for all $\lambda \in [0, \pi]$, and are therefore suitable for constructing the multistep forecasts of a long-memory process by linear least-squares methods.

By their very nature, however, the non-parametric and semi-parametric approaches do not seek to model $f_y(\lambda)$ for all $\lambda \in [0, \pi]$ and hence in their original form they are not suitable for this purpose. [20] use the non-parametric approach for multistep prediction of a long-memory process by adopting a two-stage approach, in which at the first stage the long-memory component is estimated by a non-parametric method and then 'filtered' out and subsequently at the second stage a parametric $ARMA$ model is fitted to the resulting output series. This two-stage procedure is still parametric, however, and, even though its optimality is unclear, in the discussion below, it may be used in the same way as the parametric approach.

[64] consider the question of how to compute the prediction constants for multistep prediction of $\{y_t\}$ by linear least-squares methods. Three different methods are identified, namely the *Truncation Method*, *Type-II Plug-In Method* and *Type-II Direct Method.* Moreover, as the focus now is on the computation of the prediction constants from the estimated parameters, in the following exposition, we do not explicitly distinguish between the parametric or the hierarchical model fitting approaches.

In all three methods of computing the forecast coefficients, an estimate, $\hat{\Pi}(\lambda)$, say, of the transfer function, $\Pi(e^{-i\lambda})$, of the coefficients, $\pi(j)$, in the infinite autoregressive representation, (16), for $\{y_t\}$ is obtained first. Thus, for the likelihood-based estimation methods involving the fitting of either an $ARFIMA(p, d, q)$ model or an $FAR(k, d)$ model,

$$\hat{\Pi}(\lambda) = \hat{g}(\lambda, \tau)^{-1}, \tag{32}$$

where $\hat{g}(\lambda, \tau)$ is obtained from either (29) or (30) by replacing the unknown parameter vector, τ, by its estimate, $\hat{\tau}$, constructed by either of these two methods. Similarly, for the $FEXP$ method, $\hat{\Pi}(\lambda)$ is obtained from (24) by replacing the unknown parameters by their estimates constructed as described in Sect. 2.2.

For the Truncation method, which is similar to that suggested by [17], an estimate, $\hat{\pi}(j)$, say, of $\pi(j)$ is constructed by a discrete Fourier transform of the sequence $\Pi(2\pi t Q^{-1})$, $t = 0, 1, \ldots, Q-1$, as follows:

$$\hat{\pi}_y(j) = (2\pi Q)^{-1} \sum_{t=0}^{Q-1} \hat{\Pi}(2\pi t Q^{-1}) \exp(it2\pi t Q^{-1}) \quad (j = 0, 1, \ldots Q-1), \tag{33}$$

where Q is a large power of 2, e.g. $Q = 1024$. The Splus function `fft` can be used for this purpose. However, since only a finite past of $\{y_t\}$ has been observed, the rather long sequence of $\hat{\pi}(j)$ is truncated at a suitable finite value, K, say.

Next, consider the *Type-II Plug-In* method. This method is based on the innovations algorithm, see [16] and [20], and it is motivated by the observation that for each finite truncation point, K, the truncation method is not optimal in a maximum entropy sense, see [49], since it does not use all the information in the first K correlations of the fitted model about the unknown future values, and this method may be improved upon by using the appropriate finite order linear least-squares forecast coefficients based on the correlation function of the fitted model. For applying this method, and the *Type-II Direct* method, the estimated covariance and correlation functions, $\hat{R}_y(u)$, and $\hat{r}_y(u)$, say, implied by the fitted long-memory model is required. Let $\hat{f}_y(\lambda)$ denote the estimated spectral density corresponding to the fitted long-memory model, which may be computed from (23), but by replacing $\Pi(\lambda)$ by its estimate constructed as described above. Then, as in [64], the $\hat{R}_y(u)$ may be computed by a discrete Fourier transform of the sequence $\hat{f}(2\pi tQ^{-1})$, $t = 0, 1, \ldots, Q-1$, that is, by using the formula (33), but with $\hat{\Pi}(2\pi tQ^{-1})$ replaced by $\hat{f}(2\pi tQ^{-1})$. Also, $\hat{r}_y(u)$, may be computed as $\hat{r}_y(u) = \hat{R}_y(u)/\hat{R}_y(0)$.

Now let $S \geq 1$, denote a large integer, which is large enough for approximating the fitted model by an approprite finite order autoregression using the *AIC* criterion given below. As is well-known, the coefficients, $\hat{\pi}_{Ps}(j)$ $(j = 1, \ldots, s;\ s = 1, \ldots, S)$ and the residual error variance, $\hat{\sigma}^2(s)$, of an sth order autoregressive approximation to the fitted model are the solutions of the following equations, with $\hat{\pi}_{Ps}(0) = 1$:

$$\sum_{j=0}^{s} \hat{\pi}_{Ps}(j)\hat{r}_y(u-j) = 0, \quad (u = 1, \ldots, s), \tag{34}$$

$$\sum_{j=0}^{s} \hat{\pi}_{Ps}(j)\hat{R}_y(j) = \hat{\sigma}^2(s). \tag{35}$$

The value, k, say, of the order of an appropriate finite order autoregressive approximation to the fitted model is chosen by minimising the Akaike Information Criterion:

$$AIC(s) = T \ln \hat{\sigma}^2(s) + 2s \quad (s = 0, 1, \ldots, S), \tag{36}$$

and thus k is such that $AIC(s)$ takes its smallest value for $s = k$. For the chosen value of k, the *Type-II Plug-In* forecasts of the unknown future values are constructed by repeatedly iterating the fitted autoregressive approximation and replacing the unknown future values by their own forecasts; the details are omitted in the interest of space.

By contrast, the *Type-II Direct* forecasts are constructed, as discussed in Sect. 1, by developing a new autoregressive approximation, that is, the h-step prediction constants and the order of the model to be fitted, separately for each prediction lead time, h. A motivation for considering this approach in the present context is that if $h > 1$ and s is finite, the plug-in prediction constants obtained by solving (34) do not correspond to the sth order linear least-squares coefficients based on the $\hat{r}_y(u)$, and hence the corresponding h-step forecast may be improved upon by using the latter set of coefficients. As is readily verified, for an $h \geq 1$ and each finite s, the relevant h -step linear least-squares coefficients, $\hat{\pi}_{Dhs}(j)$ $(j = 1, \ldots, s;\ s = 1, \ldots, S)$ are the solutions of the following equations:

$$\sum_{j=1}^{s} \hat{\pi}_{Dhs}(j)\hat{r}_y(u-j) = -\hat{r}_y(h+u-1), \quad (u = 1, \ldots, s). \tag{37}$$

Also, the corresponding estimate of the h-step prediction error variance, $\hat{V}_{Dh}(s)$, is given by:

$$\sum_{j=1}^{s} \hat{\pi}_{Dhs}(j)\hat{R}_y(h+j-1) + \hat{R}_y(0) = \hat{V}_{Dh}(s). \tag{38}$$

For $h = 1$, (37) and (38) clearly reduce to (34) and (35); for $h > 1$, however, the two sets of prediction constants differ from each other.

The order of the autoregression to be fitted for each h may be determined by minimising the following h-step extension of the Akaike information criterion, see [29]:

$$AIC(h, s) = T \ln \hat{V}_{Dh}(s) + 2s \quad (s = 0, 1, \ldots, S). \tag{39}$$

Thus, an autoregression of order $k = \tilde{k}_h$ will be fitted, if the criterion (39) attains its smallest value for $s = k$.

The corresponding h-step predictor is then computed as follows:

$$\hat{y}_n(h) = -\sum_{j=1}^{k} \hat{\pi}_{Dhs}(j) y_{n+1-j}, \tag{40}$$

where $k = \tilde{k}_h$ is the autoregressive order selected by minimising the criterion (39).

[64] compare the relative predictive efficacy of the Truncation, *Type-II Plug-In* and *Type-II Direct* methods of evaluating the prediction constants for multistep prediction of several actual long-memory models and report that no one method necessrily outperforms the other two for all models and at all prediction lead times. The *Type II Direct* method may nevertheless be recommended because it frequently outperforms the other two methods and, even in the situations where it does not do so, the resulting loss in predictive efficiency is negligible.

4 Linear Long Memory Processes with Infinite Variance

The observed long-memory process, $\{y_t\}$, has so far been assumed to possess finite variance. However, many observed time series, notably those occurring in finance and hydrology, display both persistence and heavy tails, see [50], [51] and [52] for a detailed exposition of various empirical properties of such time series and a discussion of the many important theoretical challanges they pose.

A main problem is that since the covariances do not exist for such time series, the class of weakly stationary processes does not provide a suitable theoretical model and the class of strictly stationary processes should be considered instead. A difficulty now is how to define 'long memory' for the latter class of processes, since the definitions discussed in Sect. 1 do not apply. However, currently, there is no such universally accepted definition. One approach has been proposed by [18], according to whose definition, a strictly stationary process, $\{y_t\}$, is a long memory process if its normalized sums, $(\sum_{j=1}^{m} y_j^2)^{-1}(\sum_{j=1}^{m} y_j)^2$, diverge in probability to infinity as $m \to \infty$. A different approach has been proposed by [53] and [54] whose definition recognises the fact that there is a tendency for the realizations of long memory time series to stay above or below the mean for extended periods of time and hence focuses on "long strange segments". While definitions of this type are useful and allow broad classes of processes to be considered, it is not clear how they could be exploited for constructing mutltistep forecasts. The notion of long strange segments might be relevant for this purpose, but no theoretical results are currently availble in this direction.

A more fruitful approach, which does enable the development of effective prediction methods, is to focus instead on a subclass, $M_{d,\delta}$, say, of stictly stationary stable processes, $\{y_t\}$, satisfying (14) and with $\{x_t\}$ admitting a representation of the form (4), but in which $\{\epsilon_t\}$ is a sequence of independent identically distributed random variables with the common distribution belonging to the domain of attraction of a δ-stable law, $1 < \delta < 2$, and the $b(j)$ are absolutely summable coefficients satisfying (6). As discussed by [55], this subclass of processes does admit parameteric models like $ARFIMA, FAR$ and $FEXP$, discussed earlier in Sects. 1 and 2. A limitation of this approach, however, is that, see Chapters 6 and 7 of [51] within the class of all stictly stationary stable processes, the sub-class $M_{d,\delta}$ of processes forms only a subset which is disjoint from the sub-class of harmonizable processes, and moreover there are processes which do not belong to either of these two subclasses. Hence, it is not clear whether the $M_{d,\delta}$ class of processes is large enough to be suitable for all observed long-memory time series with infinite variance. This limitation notwithstanding, which does not apply to the class of Gaussian stationary processes because such processes admit both moving average and harmonizable representations, below we restrict attention to the $M_{d,\delta}$ class of processes. [56] show that a sufficient condition for a member of the $M_{d,\delta}$ class of processes to be strictly stationary is that $0 < d < (1 - \delta^{-1})$. The use of $Type - I$ approach for this class may be justified by the results of [57] who show that if instead of the criterion of minimum mean squared error of prediction, the optimality of the linear predictor is judged by the criterion of minimum dispersion, then the classical Wiener-Kolmogorov prediction theory still continues to apply. As regards the $Type - II$ approach, there has been much development on the question of parameter estimation and prediction for the family of $ARFIMA(p, d, q)$ models belonging to the $M_{d,\delta}$ class. Thus, [56] investigate certain stochastic properties of this family of models, while, [58] consider the approximate maximum likelihood estimation of the parameter of this family of models, but treat the order (p, q) as known. In addition, [59] has generalised the earlier results of [57] to this family of models and shown that the optimal linear predictor based on the infinite past coincides with the corresponding predictor in the Gaussian case. This result, and especially his Theorem 3.2 , provides a justification for using the *Truncation method* in the infinite variance case. An experimental justification for using the *Type II Direct and Plug-In* methods is provided by [64], who investigate the behaviour of these two methods and the truncation method by a simulation study. Also, [35] establish a consistency property for the parameters estimated by the FAR approach when $\{y_t\}$ is a member of the $M_{d,\delta}$ class of processes.

5 Long Memory Chaotic Time Series

By a chaotic time series we mean a deterministic sequence, $\{w_t\}$ $(t = 0, 1, \ldots)$, generated by iteratively applying a one-dimensional map of the following form:

$$w_{t+1} = \zeta(w_t) \ (t = 0, 1, \ldots), \tag{41}$$

where w_0 is real-valued and specifies the *initial condition* for the iterative scheme specified by (41), $\zeta : J \to J$ is a non-linear map (function) and J denotes a closed interval of the real line, $(-\infty, \infty)$. Such maps have attracted much attention in recent years and it is recognised that they can generate complex mathematical behaviour such as *strange attractors*, *period doubling*, *limit points* and *chaos*, and can be used to describe phenomena which appear random. We however only emphasise statistical time series aspects of such maps. Thus, we only consider measure-preserving and ergodic maps admitting a time-invariant density, $\chi(w)$, over the interval J. A main result in Ergodic theory states that if $\eta : J \to J$ is an integrable function, then, see [2], as $N \to \infty$,

$$\frac{1}{N} \sum_{t=0}^{N-1} \eta(\zeta^t(w)) \to \int_J \eta(w)\chi(w)dw, \tag{42}$$

where the convergence is almost surely with respect to the density $\chi(w)$. Thus, Ergodic theory enables a study of the long-term behaviour of a chaotic map. To relate the map (41) to a weakly stationary process, suppose that the initial condition, w_0, is generated from a distribution with probability density function, $\chi(w)$, the invariant density, and assume that the first two moments of this distribution exist. Then, $\{w_t, t = 0, 1, \ldots\}$, defines a weakly stationary process, with mean

$$\mu_w = \int_J w\chi(w)dw, \tag{43}$$

and covariance function

$$R_w(u) = \int_J \zeta^{u+2}(w)\chi(w)dw - (\mu_w)^2. \tag{44}$$

Closed-form analytic expressions for the invariant density and correlation function are now known for a range of different maps; see [3] for an expository discussion. Three well-known examples being the following:

1). *The Tent map.* Here $J = [0, 1]$ and

$$\zeta(w) = 1 - |2w - 1|. \tag{45}$$

The invariant density for this map coincides with that for the $Uniform[0, 1]$ distribution:

$$\chi(w) = 1, \quad \text{all } w \in J, \tag{46}$$

and its correlation function is given by:

$$r_w(u) = 0, u = 1, 2, \ldots. \tag{47}$$

2). *The Logistic map.* Here $J = [0, 1]$ and

$$\zeta(w) = 1 - |2w - 1|^2 = 4w(1 - w). \tag{48}$$

The invariant density for this map coincides with that for a $Beta(\frac{1}{2}, \frac{1}{2})$ distribution:

$$\chi(w) = \frac{1}{\sqrt{\pi^2 w(1 - w)}}, \quad \text{all } w \in J, \tag{49}$$

and its correlation function is given by (47).

3). *The Bernoulli map.* Here $J = [0, 1]$ and

$$\zeta(w) = 2w \bmod(1) \tag{50}$$

The invariant density for this map coincides with that for the *Uniform[0, 1]* distribution given by (46) and its correlation function is given by:

$$r_w(u) = 2^{-u}, u = 1, 2, \ldots. \tag{51}$$

Additional references include [5], [4], [9].

These results point to the limitations of using linear $ARMA$ models discussed in Sect. 1 for non-Gaussian and non-linear time series described by these maps and, for the Bernoulli map, for example, these methods lead to identifying an $AR(1)$ model but the mechanism generating the process is non-linear, with a similar remark applying to both the Tent and the Logistic map for which the traditional methods would lead to a purely random process as a stochastic model.

The chaotic maps mentioned above however still imply that the generating sequence has a short memory and has a correlation function, $r(u)$, which decays exponentially to 0, as $u \to \infty$. There has however lately been much development on devising chaotic *intermittency* maps exhibiting long or intermediate memory, that is, maps exhibiting correlation functions which decay at a polynomial rate, $r_y(u) \sim Bu^{-\alpha}$, $0 < \alpha < 2$, or even more slowly. A detailed expository discussion of such maps is given by [1]. Suffice to say here that early work on such maps is due to [7] whose motivation was to provide a dynamical model implying intermittency, that is, a phenomenon displaying long periods of laminar behaviour together with short bursts of erratic behaviour - see [60] for a further discussion of the latter concept. Following is a slightly generalised version of the *intermittency* map studied by these authors:

$$\zeta_\beta(w) = \begin{cases} w(1+2^\beta w^\beta) & (0 \le w \le \frac{1}{2}), \\ 2w-1 & (\frac{1}{2} < w \le 1). \end{cases} \tag{52}$$

where $\beta \in (0,1)$ is a free parameter of the map.

The invariant density for this map is of the following general form, with $J = [0,1]$:

$$\chi(w) = \frac{H(w)}{w^\beta}, \quad \text{all } w \in J, \tag{53}$$

where $H(w)$ is a uniformly bounded but unknown function of w. Moreover while a precise closed-form expression for the correlation function, $r_w(u)$, of $\{w_t\}$ is as yet unknown, it is known that if $\kappa_t = \Omega(w_t)$, where Ω denotes a Hölder continuous function, then there exists at least one such function, Ω, for which $r_\kappa(u) = O(u^{-\alpha})$, $\alpha = \frac{1}{\beta} - 1$; hence, $r_\kappa(u) = O(u^- a), a \in (0,\infty)$. It should be noted that for $\beta = 0$, the intermittency map, (52), reduces to the Bernoulli map, (50); for $\beta = 1$, however, the invariant density does not exist.

It is also possible to devise intermittency maps which imply a logarithmic rate of decay of correlations, see [1] for details. For an application of intermittency maps for modelling telecom network traffic, see [6].

Acknowledgement

Thanks are due to Mark Holland for helpful discussions relating to Sect. 5 and especially for introducing the authors to the intermittency maps discussed in that section.

References

1. R. J. Bhansali, M. Holland and P. Kokoszka: *Statistical properties of chaotic intermittency maps* In preparation (2002)
2. S. Chatterjee and M. R. Yilmaz: Statistical Science **7**, 49 (1992)
3. A. J. Lawrance: *Statistical Aspects of Chaos.* A Series of 7 Lectures for MSM4S2. School of Mathematics and Statistics, University of Birmingham, UK (2001)
4. A. J. Lawrance and N. Balakrishna: J. Royal Statist. Soc B **63**, 843 (2001)
5. A. J. Lawrance and N. M. Spencer: Scandinavian Journal of Statistics **25**, 371 (1998)
6. R. J. Mandragon, D. K. Arrowsmith and J. M. Pitts: Performance Evaluation **43**, 223 (2001)
7. P. Manneville and M. Pomeau: Comm. Math. Phys **74**, 189 (1980)
8. A. I. McLeod: J. Time Series Analysis **19**, 473 (1998)
9. H. Sakai and H. Tokumaru: IEEE Trans. Acoustics, Speech and Signal Processing **ASSP-28**, 588 (1980)
10. D. R. Brillinger: *Time Series: Data Analysis and Theory* (Holt, New York 1975)
11. R. J. Bhansali: Journal of the Royal Statistical Society, B **36**, 61 (1974)
12. K. N. Berk: The Annals of Statistics **2**, 489 (1974)

13. R. J. Bhansali: The Annals of Statistics **6**, 224 (1978)
14. J. R. M. Hosking: Biometrika **68**, 165 (1981)
15. C. W. J. Granger and R. Joyeux: J. Time Series Anal **1**, 15 (1980)
16. P. J. Brockwell and R. A. Davis: *Time Series: Theory and Methods* (Springer-Verlag, New York 1991)
17. G. E. P. Box and G. M. Jenkins: *Time Series Analysis: Forecasting and Control* (Holden Day, New York 1970)
18. C. C. Heyde and Y. Yang: J. Appl. Probab **34**, 939 (1997)
19. J. L. Doob: *Stochastic Processes*, (Wiley, New York 1953)
20. N. Crato and B. K. Ray: Journal of Forecasting **15**, 107 (1996)
21. R. J. Bhansali: *Developments in Time Series Analysis* (Chapman and Hall, London, 1993)
22. E. J. Hannan and J. Rissannen: Biometrika **69**, 81–94 (1982) correction note: **70**, 303 (1983)
23. G. C. Tiao and R. S. Tsay: J. Forecasting **13**, 109 (1994)
24. J. Brodsky and C. M. Hurvich: J. Forecasting **18**, 59 (1999)
25. G. K. Basak, N. H. Chan and W. Palma: Journal of Forecasting **20**, 367 (2001)
26. E. Parzen: *Multiple time series modelling* Multivariate Analysis II, (Academic Press, New York 1969)
27. R. Shibata: The Annals of Statistics **8**, 147 (1980)
28. B. K. Ray: Journal of Time Series Analysis **14**, 511 (1993)
29. R. J. Bhansali: Ann. Inst. Statist. Math. **48**, 577 (1996)
30. R. J. Bhansali: *Asymptotics, Nonparametrics and Time Series* (Marcel Dekker, New York 1999) pp. 201–225
31. J. Hidalgo and Y. Yajima: *Prediction of strong dependent processes in the frequency domain with application to signal processing*, preprint (1999)
32. P. Bloomfield: Biometrika **60**, 217 (1973)
33. J. Beran: Biometrica **80**, 817 (1993)
34. P. Hall: Fields Institute Communications **11**, 153 (1997)
35. R. J. Bhansali and P. S. Kokoszka: 'Estimation of the long-memory parameter: a review and an extension'. In: *Proceedings of the Symposium on Inference for Stochastic Process* (IMS Lecture Notes, IMS 2001) pp. 125–150
36. J. Beran: The Annals of Statistics **25**, 1852 (1997)
37. J. M. Bardet and E. Moulines and P. Soulier: 'Recent advances on the semi–parametric estimation of the long–range dependence coefficient'. In: *ESAIM Proceedings, Société de Mathématiques Appliquées et Industrielles* (1999) pp. 23–43,
38. E. Moulines and P. Soulier: Journal of Time Series Analysis **21**, 193 (2000)
39. C. M. Hurvich and J. Brodsky: Journal of Time Series Analysis **22**, 221 (2001)
40. R. Dahalhaus: Ann. Stat **17** , 4, 1749 (1989)
41. J. Beran: J. Royal Statist. Soc. B **57**, 659 (1995)
42. H. T. Davis and R. H. Jones: J. Amer. Statist. Assoc **63**, 141 (1968)
43. C. M. Huruvich and K. I. Beltrao: Journal of Time Series Analysis **14**, 455 (1993)
44. J. Geweke and S. Porter-Hudak: Journal of Time Series Analysis **4**, 221 (1983)
45. H. Künsch: Bernoulli **1**, 67 (1987)
46. M. S. Taqqu and V. Teverovsky: *A practical guide to heavy tails: Statistical techniques for analyzing heavy tailed distrubutions* (Birkhauser, Boston 1998) pp. 177–217
47. J. Beran, R. J. Bhansali and D. Ocker: Biometrika **85**, 921 (1998)
48. E. J. Hannan: Statistical Science **2**, 135 (1987)
49. M. B. Priestley: *Spectral Analysis and Time Series* vol. 1 (Academic Press, New York 1981)

50. S. Mittnik and S. Rachev: *Stable Paretian Models in Finance* (Wiley, New York 2000)
51. G. Samorodnitsky and M. S. Taqqu: *Stable Non-Gaussian Random Processes: Stochastic Models with Infinite Variance* (Chapman and Hall 1994)
52. R. Adler and R. Feldman and M. S. Taqqu: *A practical guide to heavy tails: Statistical techniques for analyzing heavy tailed distributions* (Birkhauser, Boston 1998)
53. S. T. Rachev and G. Samorondnitsky: Stochastic Processes and their Applications **93**, 119 (2001)
54. Mansfield, Rachev, Samorodnitsky: Ann. Math. Statist **23**, 193 (1952)
55. R. J. Bhansali: Athens Conference on Applied Probability and Time Series Springer **2**, 42 (1996)
56. P. S. Kokoszka and M. S. Taqqu: Stochastic Processes and their Applications **60**, 19 (1995)
57. D. B. H. Cline and P. J. Brockwell: Stochastic Processes and their Applications **19**, 281 (1985)
58. P. S. Kokoszka and M. s. Taqqu: The Annals of Statistics **24**, 1880 (1996)
59. P. S. Kokoszka: Probability and Mathematical Statistics **16/1**, 65 (1996)
60. E. Ott: *Chaos in Dynamical Systems*, (Cambridge University Press 1993)
61. R. J. Bhansali and P. S. Kokoxzka: *Prediction of long-memory time series: an overview*, Extadistica, (2002)
    note: forthcomming preprint available at
    http://www.liv.ac.uk/maths/SOR/HOME/RJBhansali.html
62. R. J. Bhansali: *Companion to Economic Forecasting* (Blackwell, New York 2002) pp. 206–221
63. P. M. Robinson: The Annals of Statistics **23**, 1630 (1995b)
64. R. J. Bhansali and P. S. Kokuszka: International Journal of Forecasting **18**, 181 (2002)

Fractional Brownian Motion and Fractional Gaussian Noise

Hong Qian

Department of Applied Mathematics, University of Washington
Seattle, WA 98195, USA

Abstract. Fractional Brownian motion is one of most cogent mathematical models for strongly correlated stochastic processes with self-similarity. In this article, we give a pedagogic introduction to this theory and investigate some of the statistical, geometric, and fractal properties of fractional Brownian motion and fractional Gaussian random fields. The connection between fractional Brownian motion and the renormalization group in statistical physics is emphasized.

1 Introduction

Random walk and Brownian motion are ubiquitous mathematical models for physical and biological processes [40,5]. The mathematical theory of Brownian motion developed in the early part of last century by Einstein, Kramers, Chandrasekhar, Uhlenbeck, and others has provided physicists and biologists with a powerful tool for analyzing a wide range of natural phenomena. The universal applicability of the model depends on the fact that the systems and processes we have studied in the past often have a large number of uncorrelated or weakly correlated components. However, complex natural structures and processes usually have long-range, strong spatial and temporal correlations. This gives the motivation for studying stochastic processes and random fields with long-range correlation.

The theory of fractional Brownian motion (fBm) [4,20] is a mathematical generalization of the classical theory of random walk and Brownian motion. The term "fractional" is related to fractional integration and differentiation [18]. In contrast to the classical Brownian motion which has independent increments, the fBm has a long-range, strong spatial and temporal correlation as its defining property. Although there is much scholarly work on the rigorous mathematics and statistics of fBm, relatively little has been developed on the applications of fBm as a mechanistic model for physical and biological problems.

The main objective of this article is to present an applied theory for fBm and its related problems. In particular, we introduce fBm according to an elegant approach given in [16,38] which emphasizes its connection with the theory of statistical physics [9,19]. We also present some results on the geometric shape and fractal dimension of fBm.

The concept of scaling invariance and the concept of renormalizing transformation [22,17] are two important ideas in modern statistical physics. They were developed into the theory of renormalization group by K.G. Wilson who was awarded the Nobel Prize in physics in 1982. The idea of scaling gives rise to the notion of critical exponent, or *fractal dimension*, which has become a fertile ground for interdisciplinary sciences [2,6].

2 Self-similarity, Fractional Gaussian Noise, and Fractional Brownian Motion

Fractional Gaussian noise (fGn) and fBm were origianlly introduced by Mandelbrot and van Ness [24] for modeling stochastic fractal processes [23]. The definition for fGn and fBm given by Gallavotti, Jona-Lasinio and independently by Sinai [16,38], however, is more general and insightful from the standpoint of statistical physics [22,17]. In the most general form, an fGn is a random field with E-dimensional random vectors defined in a d-dimensional space. It corresponds to the spin dimension E and space dimension d in the theory of critical phenomena in statistical physics.

It has been shown that the fGn is invariant under a semigroup (Kadanoff block) transformation with critical exponent H, known as the Hurst coefficient in the fields of engineering and stochastic fractal processes [24]. Furthermore, it was shown that other stationary processes approach asymptotically to fGn under the Kadanoff renormalization transformation with respective critical exponents. Hence, the space of all stationary processes is divided under the renormalization group transformation, with fGn as fixed points.

2.1 Self-similarity and Fractional Gaussian Noise

A self-similar [38] *fractional Gaussian noise* (fGn) is a series of identical Gaussian random variables $X_1, X_2, X_3, \ldots$, which has the following property [24]

$$\mathcal{A}_N[X] \triangleq \frac{X_1 + X_2 + X_3 + \ldots + X_N}{N^H} \sim X \tag{1}$$

where '$\sim$' denotes equality in the sense of probability distribution. An alternative but equivalent definition can be given in terms of relative dispersion [1]: $RD_N = N^{H-1} RD_1$ where

$$RD_N = \frac{\sqrt{Var[X_1 + X_2 + \ldots + X_N]}}{NE[X_1]}.$$

$E[\cdot]$ and $Var[\cdot]$ denote expectation value and variance of a random variable.

It is straightforward to show that when $H = \frac{1}{2}$, the X's are necessarily independent. In general, $0 < H < 1$ and the X's are correlated.

2.2 fBm, fGn and Their Correlation Functions

A *fractional Brownian motion* then is defined as the partial sum of the fGn:

$$B_k^H = X_1 + X_2 + X_3 + \ldots + X_k.$$

Hence, we have:

$$E[(B_h^H - B_k^H)^2] = E[(h-k)^{2H} X_1^2] = (h-k)^{2H}\sigma^2 \tag{2}$$

where we have assumed, without loss generality, $E[X_1] = 0$ and denoted $\sigma^2 = Var[X_1]$. We can also expand the lhs of (2) and obtain:

$$E[(B_h^H)^2] - 2E[B_h^H B_k^H] + E[(B_k^H)^2] = h^{2H}\sigma^2 - 2E[B_h^H B_k^H] + k^{2H}\sigma^2 = (h-k)^{2H}\sigma^2$$

Thus we have the correlation for fBm:

$$E[B_h^H B_k^H] = \frac{\sigma^2}{2}[h^{2H} - (h-k)^{2H} + k^{2H}]. \tag{3}$$

One can also determine the autocorrelation for the fGn:

$$\begin{aligned} E[X_n X_m] &= E[(B_n^H - B_{n-1}^H)(B_m^H - B_{m-1}^H)] \\ &= E[B_n^H B_m^H + B_{n-1}^H B_{m-1}^H - B_{n-1}^H B_m^H - B_n^H B_{m-1}^H] \\ &= \frac{\sigma^2}{2}[(n-m-1)^{2H} - 2(n-m)^{2H} + (n-m+1)^{2H}]. \end{aligned}$$

Using the stationarity of X we have

$$E[X_0 X_k] = \frac{\sigma^2}{2}[(k-1)^{2H} - 2k^{2H} + (k+1)^{2H}]. \tag{4}$$

This result also can be directly verified using the self-similarity [38]:

$$\frac{1}{N^{2H}} E\left[\left(\sum_{\ell=0}^{N-1} X_\ell\right)^2\right] = E[X_0^2] = \sigma^2.$$

One can furthermore verify that:

$$\frac{1}{N^{2H}} E\left[\sum_{s=kN}^{(k+1)N-1} X_s \sum_{t=0}^{N-1} X_t\right] = E[X_0 X_k].$$

That is for any integer $N \geq 1$, autocorrelation function $\rho(k) = E[X_0 X_k]$ uniquely satisfies:

$$\rho(k) = N^{-2H} \sum_{\ell=-(N-1)}^{N-1} (N - |\ell|)\rho(Nk + \ell). \tag{5}$$

For a Gaussian stochastic process, all the joint distributions for B^H's are multivariate Gaussian distributions. For any multivariate Gaussian random variables ξ_1, ξ_2, ..., ξ_N with $E[\xi_k] = 0$ ($k = 1, 2, ..., N$), the moments with order higher than 2 can all be expressed in terms of the second-order correlation known as the Isserlis theorem or the Wick's theorem in statistical physics [33]:

$$E[\xi_1 \xi_2 ... \xi_N] = \sum_P \prod_{j=1}^{N} E[\xi_j, \xi_{Pj}], \tag{6}$$

where the sum runs over *all* permutations P: $j \to Pj$ of the positive integers j. Specifically,

$$E[B_h^H B_i^H B_j^H B_k^H] = E[B_h^H B_i^H]E[B_j^H B_k^H] + E[B_h^H B_j^H]E[B_i^H B_k^H] + E[B_h^H B_k^H]E[B_i^H B_j^H]. \quad (7)$$

2.3 The Power Spectra of fGn and fBm

Equation (5) leads to one of the defining properties of fGn: power spectrum with a singularity

$$S(f) = \sum_{k=-\infty}^{\infty} \rho(k) e^{-2\pi i f k},$$

which has an inverse transformation:

$$\int_{-1/2}^{1/2} S(f) e^{2\pi i f k} df = \rho(k).$$

Noting the equality

$$\sum_{k=-\infty}^{\infty} e^{-2\pi i k f} = \sum_{k=-\infty}^{\infty} \delta(f-k)$$

which accounts for the alias phenomenon [39], we have for any positive integer N and $f \in [-1/2, 1/2]$:

$$\begin{aligned} N^{2H} S(f) &= \sum_{\ell=-(N-1)}^{N-1} (N - |\ell|) \sum_{k=[f-N/2]+1}^{[f+N/2]} \frac{1}{N} S\left(\frac{f-k}{N}\right) e^{2\pi i \ell (f-k)/N} \\ &= \sum_{k=[f-N/2]+1}^{[f+N/2]} \frac{1}{N} S\left(\frac{f-k}{N}\right) \sum_{\ell=-(N-1)}^{N-1} (N - |\ell|) e^{2\pi i \ell (f-k)/N} \\ &= \sum_{k=[f-N/2]+1}^{[f+N/2]} \frac{1}{N} S\left(\frac{f-k}{N}\right) \frac{1 - \cos(2\pi(f-k))}{1 - \cos(2\pi(f-k)/N)} \end{aligned}$$

which can be simplified into:

$$\sum_{k=[f-N/2]+1}^{[f+N/2]} \frac{S((f-k)/N)}{1 - \cos(2\pi(f-k)/N)} = \frac{N^{2H+1} S(f)}{1 - \cos(2\pi f)}. \quad (8)$$

Based on this iterative relation, it is easy to verify that [38]

$$S(f) = C(1 - \cos(2\pi f)) \sum_{m=-\infty}^{\infty} \frac{1}{|f+m|^{2H+1}}$$

satisfies (8) for any constant C.

3 The Continuous-Time Fractional Gaussian Noise and Fractional Brownian Motion

In the previous section we introduced the discrete-time fBm and its increment, fGn. We now introduce the continuous-time fBm and fGn (ctfGn). A ctfGn is a stationary stochastic process with the following defining property:

$$\frac{1}{T^H}\int_0^T X(t)dt \sim X(0) \qquad (T > 0),$$

where '$\sim$' again means equality in probability distribution; $E[X(t)] = 0$ and $E[X^2(t)] = \sigma^2$. Parallel to the derivation for (3) we have

$$E\left[B_t^H B_\tau^H\right] = \frac{\sigma^2}{2}\left[t^{2H} - 2(\tau - t)^{2H} + \tau^{2H}\right], \tag{9}$$

where

$$B_t^H = \int_0^t X(\xi)d\xi \tag{10}$$

is a continuous-time fractional Brownian motion (ctfBm). From (10) the ctfGn can be formally written as $X(t) = dB_t^H/dt$. Therefore, the covariance of $X(t)$ can be obtained from (9) [27]:

$$\begin{aligned}
E[X(t)X(\tau)] &= \frac{d^2}{dtd\tau}E\left[B_t^H B_\tau^H\right] = -\sigma^2\frac{d^2|t-\tau|^{2H}}{dtd\tau} \\
&= 2H(2H-1)\sigma^2|t-\tau|^{2H-2} - 2H\sigma^2|t-\tau|^{2H-1}\frac{d^2|t-\tau|}{dtd\tau} \\
&= 2H(2H-1)\sigma^2|t-\tau|^{2H-2} + 2H\sigma^2|t-\tau|^{2H-1}\delta(t-\tau),
\end{aligned}$$

and since $X(t)$ is stationary, we have autocorrelation function for ctfGn:

$$\rho(\tau) = E[X(0)X(\tau)] = 2H(2H-1)\sigma^2|\tau|^{2H-2} + 2H\sigma^2|\tau|^{2H-1}\delta(\tau). \tag{11}$$

For $H = 0.5$ (Wiener white noise), the first term in (11) is zero and the second term is Dirac's $\delta(\tau)$. For $H > 0.5$, the second term is zero. The integration of $\rho(t)$ has:

$$\int_{-\infty}^{\infty}\rho(t)dt = \begin{cases} \infty & 0.5 < H < 1 \\ \sigma^2 & H = 0.5 \\ 0 & 0 < H < 0.5\ . \end{cases} \tag{12}$$

Note that for $H < 0.5$, $\rho(t)$ is negative for all $t \neq 0$. The integrations of both terms in (11) do not converge in the traditional sense. One obtains the zero by a cancelation of two ∞'s.

We now calculate the spectral density function for the ctfGn:

$$S(f) = \int_{-\infty}^{\infty}\rho(t)e^{-2\pi ift}dt. \tag{13}$$

For $H > 0.5$, we have:

$$S(f) = 2H(2H-1)\sigma^2 C f^{1-2H}, \qquad (H > 0.5) \tag{14}$$

where the constant

$$C = \int_{-\infty}^{\infty} |\xi|^{2H-2} e^{-2\pi i\xi} d\xi.$$

Therefore ctfGn has a simple power-law spectral density function over all frequencies f. However, for discrete-time fGn this is not the case.

For anti-correlated fGn with $H < 0.5$, we note again that the Fourier transformtion of $\rho(t)$ does not exist in the traditional sense.

4 Estimations of Power Spectra and Their Statistical Accuracy

In applications, practical measurements only encounter discrete-time fractal time series. For a finite realization of the $\{B_k^H\}$, $y_0, y_1, y_2, ..., y_N$, the power spectrum of the motion is defined as

$$S_m(f) = \sum_{h=1}^{N}\sum_{k=1}^{N} y_h y_k e^{-i2\pi(k-h)f} = 2\sum_{k\geq h=1}^{N} y_h y_k \cos[2\pi(k-h)f] \tag{15}$$

where subscript 'm' stands for motion. If we treat (15) as a statistical estimation, then

$$\begin{aligned} E[S_m(f)] &= 2\sum_{k\geq h=1}^{N} E[B_h^H B_k^H]\cos[2\pi(k-h)f] \\ &= \sigma^2 \sum_{h\geq k=1}^{N} [h^{2H} - (k-h)^{2H} + k^{2H}]\cos[2\pi(k-h)f]. \end{aligned} \tag{16}$$

For large N, the summation in (16) can be evaluated asymptotically using the Euler-Maclaurin summation formula [3].

Similarly, for a finite realization of the fGn $\{X_k\}$, $x_0, x_1, x_2, ..., x_N$, the power spectrum of the noise is estimated by:

$$S_n(f) = 2\sum_{k>h=1}^{N} x_h x_k \cos[2\pi(k-h)f], \tag{17}$$

$$\begin{aligned} E[S_n(f)] &= \sigma^2 \sum_{k\geq h=1}^{N} [(k-h-1)^{2H} - 2(k-h)^{2H} + (k-h+1)^{2H}] \\ &\quad \times \cos[2\pi(k-h)f] \\ &= \sigma^2 \sum_{\ell=0}^{N} (N-\ell)[(\ell-1)^{2H} - 2\ell^{2H} + (\ell+1)^{2H}] \\ &\quad \times \cos(2\pi\ell f), \end{aligned}$$

and finally the variance in the estimated power spectrum $S_n(f)$:

$$
\begin{aligned}
E[S_n^2(f)] &= 4\sum_{k\geq h=1}^{N}\sum_{n\geq m=1}^{N} E[X_hX_kX_mX_n]\cos\left[2\pi(k-h)f\right]\cos\left[2\pi(n-m)f\right] \\
&= 4\sum_{\ell=0}^{N}\sum_{m=0}^{N} E[X_0X_\ell X_0X_m](N-\ell)(N-m)\cos\left(2\pi\ell f\right)\cos\left(2\pi m f\right), \\
Var[S_n(f)] &= E[S_n^2(f)] - E^2[S_n(f)] \\
&= 4\sum_{\ell,m=0}^{N}\{E[X_0X_\ell X_0X_m] - E[X_0X_\ell]E[X_0X_m]\} \\
&\quad \times (N-\ell)(N-m)\cos\left(2\pi\ell f\right)\cos\left(2\pi m f\right) \\
&= 4\sum_{\ell,m=0}^{N}\{E[X_0^2]E[X_\ell X_m] + E[X_0X_m]E[X_0X_\ell]\} \\
&\quad \times (N-\ell)(N-m)\cos\left(2\pi\ell f\right)\cos\left(2\pi m f\right).
\end{aligned}
$$

5 The Shape of Two-Dimensional fBm

fBm can also be used to model the geometric properties of polymers. The application of the theory of Brownian motion to polymers is one of the great successes of applied stochastic processes. Treating polymers as random walks, P.J. Flory developed a quantitative theory of polymer structures [15], for which he was awarded the Nobel Prize in chemistry in 1974. The random walk model remains the theoretical foundation for studying synthetic polymers and biological macromolecules today [28–30].

fBm can also be used to model the geometric properties of polymers like proteins. In a poor solvent, the size of such molecules grows asymptotic as N^ν where N is the number of units in the polymer, and $\nu < 1$. This is in contrast to a polymer in a good solvent, which has $\nu > 1$ [15]. We have recently obtained analytical results, asymptotically for large N, on the mean-square radius of gyration of a two-dimensional polymer modeled by fBm

$$\frac{\sigma^2 N^{2H}}{(2H+1)(2H+2)}, \tag{18}$$

with the asymmetricity in its shape

$$2 - \frac{1}{2(H+1)^2}\left[\frac{1}{2(1+H)^2} + \frac{2H+1}{4(4H+1)} - \frac{1}{4H+3} - \frac{\Gamma^2(2H+2)}{\Gamma(4H+4)}\right]^{-1}, \tag{19}$$

and in the spatial distributions of its kth unit [31]. The distribution is Gaussian, with variance:

$$2\sigma^2 N^{2H}\left[\frac{(1-k/N)^{2H+1} + (k/N)^{2H+1} - 1}{2H+1} + \frac{1}{2H+2}\right]. \tag{20}$$

Equation 20 gives the mean-square distance between the *kth* unit and the center of the molecule. These results generalize the classic work on random coil polymers [12,35].

6 The Fractal Geometry of fBm and fGn

There are many fractal geometric characteristics associated with an fBm or a fractional Gaussian random field. For example, the fractal dimension of fBm's graph in $(E + d)$-dimensions is $2 - H$ [43]. One can also obtain the fractal dimension of its sample paths in E-dimensional Euclidean space. Let us consider an N-step trajectory of a Gaussian process with zero mean and variance $\rho_n = \sigma^2 n^{2H}$ $(n = 1, 2, ..., N)$. This defines an fBm with Hurst coefficient H $(0 < H < 1)$. Hence, in an E-dimensional Euclidean space the probability of the fBm starting at the origin and at its *nth* step reaching an ϵ-ball in the neighborhood of $\mathbf{x}$ $(\mathbf{x} \in R^E)$ is:

$$\frac{\exp[-\mathbf{x}^2/(2\rho_n)]}{(2\pi\rho_n)^{E/2}} V_\epsilon, \tag{21}$$

where V_ϵ is the volume of the ϵ-ball. The N points have a radius of gyration $\sim N^H$, which statistically characterizes the average size of all the possible N-step trajectories. Conversely, for a given sphere of radius R, the average length of the trajectory within the sphere is $L \sim R^{1/H}$. Therefore, according to Mandelbrot's notion of a fractal set [23], the trajectory has a fractal dimension $d_f = \frac{d\ln(L)}{d\ln(R)} = 1/H$. The mathematically rigorous version of this result states that the Hausdorff dimension of an fBm is $\min(E, 1/H)$ [20,42].

Does an fBm eventually reach every part of the E-dimensional space? To answer this question, we note that if ϵ is sufficiently small, we can further assume that (21) is the probability of the fBm is reaching the ϵ-ball the first time since the probability of reaching it the second time is $\propto V_\epsilon^2$. We therefore have the probability of the fBm eventually, irrespective of n, reaching the neighborhood of $\mathbf{x}$

$$V_\epsilon \sum_{n=1}^{\infty} \frac{\exp[-\mathbf{x}^2/(2\rho_n)]}{(2\pi\rho_n)^{E/2}} \text{ which, asymptotically, } \sim V_\epsilon \sum_{n=1}^{\infty} \frac{1}{n^{HE}}. \tag{22}$$

For sufficiently small ϵ, this probability will be less than unity if $HE > 1$. Therefore, there is a finite probability that the fBm will not reach the neighborhood of $\mathbf{x}$. Note that $H = 1/d_f$; hence $d_f < E$ indicates that the fractal dimension of the trajectory is less than that of the Euclidean space. Conversely, if $d_f \geq E$, then with probability 1 the fBm will visit any small ϵ-ball centered at $\mathbf{x}$. In fact, it will visit it an infinite number of times.

This result has a strong intuitive physical interpretation: if the sample path of a correlated random walk has a fractal dimension d_f, it can fill the Euclid space with dimension $E \leq d_f$, but will not for $E > d_f$. For classic random walk with $H = 0.5$, the above result is the well-known Polya's theorem, which states that a random walker will visit every place in one and two dimensions, but not in three dimensions [14].

The mathematical investigation of fBm also provides insight into the problem of self-avoiding random walk (SAW) whose path cannot intersect itself in space. This is still a significant unsolved problem in the statistical physics of polymers. It is known that SAWs have asymptotic variance $\sim n^{6/(E+2)}$ where E is the dimension of the Euclidean space for the random walk [11]. Although fBm is not a faithful model for SAW, these two models agree asymptotically if $H = 3/(E+2)$. For $E = 2$ this corresponds to $H = \frac{3}{4}$. It will be interesting to find whether it is possible to use fBm as an approximated model for SAW. This entails defining a measure for the approximation and is intimately related to the mathematical concept of *intersectional local time*, in which the significance of $H = \frac{3}{4}$ has been specifically noted [34].

7 Nonlinear Block Transformation and Stability of Its Fixed Point

The theory of self-similar fBm developed by Gallavotti, Jona-Lasinio and Sinai [16,38] is both a theory of stochastic processes and a theory of nonlinear dynamical systems. The unifying theme of these two aspects is the distribution function and its transforming semigroup [21]. The nonlinear dynamical aspect of the renormalization group method has been actively investigated as singular perturbations of differential equations in recent years [10,13,26].

Consider an infinite-dimensional random vector $X_1, X_2, ..., X_k, ...$ The block transformation can be expressed either as a linear transformation for the joint distribution function $f_{\{X_k\}}(x_1, x_2, ..., x_k, ..)$, $\mathcal{P}_N$, called Frobenius-Perron operator [21],

$$\mathcal{P}_N[f_{\{X_k\}}](x_1, x_2, ..., x_k, ..) = \int \cdots \int \prod_{i=1}^{\infty}\prod_{j=1}^{N} dy_{ij} f_{\{X_k\}}(N^H x_1 - y_{11}, \\ y_{11} - y_{12}, ..., y_{1N-1} - y_{1N}, N^H x_2 - y_{21}, \\ ..., y_{2N-1} - y_{2N}, ..., N^H x_k - y_{k1}, ...) \qquad (23)$$

or a nonlinear transformation $\mathcal{A}_N^*$ for the singlet distribution $f_X(x)$ [19,38]:

$$\mathcal{A}_N^*[f_X](S) \triangleq f_X\left(\mathcal{A}_N^{-1} S\right), \qquad S \subset \mathbb{R}, \qquad (24)$$

where $\mathcal{A}_N$ is given in (1). $\mathcal{A}_N^*$ can not be given explicitly except when all the X's in (1) are independent. In that case,

$$\mathcal{A}_N^*[f_X](x) = N^H \int_{-\infty}^{\infty} \cdots \int_{-\infty}^{\infty} f_X(N^H x - y_1) f_X(y_1 - y_2) \cdots \\ f_X(y_{N-1} - y_N) dy_1 dy_2 \ldots dy_N. \qquad (25)$$

In general, instead of $\mathcal{A}^*$ let us consider the nonlinear transformation for the characteristic function $\varphi(s)$ of the density function $f_X(x)$:

$$\mathcal{R}_N[\varphi](s) = \varphi^N\left(\frac{s}{N^H}\right) e^{-N^{-2H}s^2\sum_{k=1}^{N}(N-k)\gamma_k}, \tag{26}$$

where

$$\varphi(s) = \int_{-\infty}^{\infty} f_X(x)e^{-isx}dx, \tag{27}$$

and γ_k characterizes the correlation between X's.

Solving the fixed point from $\mathcal{R}_N[\varphi^*] = \varphi^*$, we have $\varphi^*(s) = e^{-\alpha s^2}$ and $\forall N$

$$\alpha N^{1-2H} + N^{-2H}\sum_{k=1}^{N}\gamma_k = \alpha. \tag{28}$$

From (28) we have

$$\gamma_k = \alpha\left[(k-1)^{2H} - 2k^{2H} + (k+1)^{2H}\right], \tag{29}$$

which is precisely $E[X_0X_k]$ given in (4).

A linear stability analysis can be carried out for the fixed point φ^*. Let

$$\varphi(s) = \varphi^*(s)(1+\epsilon h(s)), \tag{30}$$

where ϵ is sufficiently small, $h(s)$ is an arbitrary function with $h(0) = h'(0) = 0$. Then

$$\begin{aligned}\mathcal{R}_N[\varphi](s) - \varphi^* &= \varphi^*(s)\left(1+\epsilon h\left(s/N^H\right)\right)^N - \varphi^*(s)\\ &= \epsilon N\phi^*(s)h\left(s/N^H\right) + O(\epsilon^2).\end{aligned}$$

Therefore the linear approximation near the fixed point

$$\mathcal{L}[h](s) \triangleq N\phi^*(s)h\left(s/N^H\right), \tag{31}$$

and

$$\lim_{m\to\infty}\mathcal{L}^m[h](s) = \lim_{m\to\infty}\left(N\varphi^*(s)\right)^m h\left(s/N^{mH}\right) \sim \frac{s^2}{2}h''(0)N^{(1-2H)m}. \tag{32}$$

Since $|\varphi^*(s)| \le |\varphi^*(0)| = 1$, the φ^* is stable for $H \ge 0.5$.

8 Discussion

fBm is a well-developed mathematical model of strongly correlated stochastic processes. It exhibits a wide-range of power-law behavior with critical exponents. It is closely related to the better-known Levy processes [24]. A larger literature for the latter exists. Levy processes also have found applications in numerious science and engineering problems [41] including glass relaxation [25] and diffusion in turbulent fluids [36,37]. Statistical and numerical aspects of fBm, as well as fitting the theoretical models, both discrete and continuous, to experimental time series have also been extensively investigated. The readers are refered to several recent works [7,8,44] and the references cited within.

Acknowledgements

I thank Gary Raymond and Mark Seligman for reading the manuscript and Professors James Bassingthwaighte and Donald Percival for discussions.

References

1. J.B.Bassingthwaighte, R.P.Beyer: Physica D **53**, 71 (1991)
2. J.B.Bassingthwaighte, L.S.Liebovitch, B.J.West: *Fractal Physiology* (Oxford University Press, New York 1994)
3. C.M.Bender, S.A.Orgzag: *Advanced Mathematical Methods for Scientists and Engineers* (McGraw-Hill, New York 1978)
4. J.Beran: *Statistics for Long-Memory Processes* (Chapman & Hall, New York 1994)
5. H.C.Berg: *Random Walks in Biology* (Princeton Univ. Press, New Jersey 1993)
6. A.Bunde, S.Havlin: *Fractals in Science* (Springer-Verlag, Berlin 1994)
7. D.C.Caccia, D.Percival, M.J.Cannon, G.M.Raymond, J.B.Bassingthwaighte: Physica A **246**, 609 (1997)
8. M.J.Cannon, D.Percival, D.C.Caccia, G.M.Raymond, J.B.Bassingthwaighte: Physica A **241**, 606 (1997)
9. M.Cassandro, G.Jona-Lasinio: Adv. Phys. **27**, 913 (1978)
10. L.-Y.Chen, N.Goldenfeld, Y.Oono: Phys. Rev. E. **54**, 376 (1996)
11. P.-G. de Gennes: *Scaling Concepts in Polymer Physics* (Cornell Univ. Press, Ithaca 1979)
12. P.Debye, F.Bueche: J. Chem. Phys. **20**, 1337 (1952)
13. S.-I. Ei, K.Fujii, T.Kunihiro: Ann. Phys. **280**, 236 (2000)
14. W.Feller: *Introduction to Probability Theory and Its Applications*, Vol. 1, 2nd Ed (John Wiley & Sons, New York 1957)
15. P.J.Flory: *Statistical Mechanics of Chain Molecules* (Hanser Publisher, Munich 1969)
16. G.Gallavotti, G.Jona-Lasinio: Comm. Math. Phys. **41**, 301 (1975)
17. N.Goldenfeld: *Lectures on Phase Transitions and the Renormalization Group* (Addision-Wesley, Reading, MA 1992)
18. R.Hilfer: *Applications of Fractional Calculus in Physics* (World Scientific, Singapore 2000)
19. G.Jona-Lasinio: Phys. Rep. **352**, 439 (2001)
20. J.-P.Kahane: *Some Random Series of Functions*, 2nd Ed. (Cambridge Univ. Press, London 1985)
21. A.Lasota, M.C.Mackey: *Chaos, Fractals, and Noise: Stochastic Aspects of Dynamics*, 2nd Ed. (Springer-Verlag, New York 1994)
22. S.K.Ma: *Modern Theory of Critical Phenomena* (Benjamin-Cummings Pub., Reading, MA 1976)
23. B.B.Mandelbrot: *The Fractal Geometry of Nature* (W.H. Freeman, San Francisco 1982)
24. B.B.Mandelbrot, J.W.van Ness SIAM Rev. **10**, 422 (1968)
25. E.W.Montroll, J.T.Bendler: J. Stat. Phys. **34**, 129 (1984)
26. B.Mudavanhu, R.E.O'Malley: Stud. Appl. Math. **107**, 63 (2001)
27. A.Papoulis: *Probability, Random Variables, and Stochastic Processes.* 3rd Ed. (McGraw-Hill, New York 1991)
28. H.Qian: J. Math. Biol. **41**, 331 (2000)

29. H.Qian: J. Phys. Chem. B **106**, 2065 (2002)
30. H.Qian, E.L.Elson: Biophys. J. **76**, 1598 (1999)
31. H.Qian, G.M.Raymond, J.B.Bassingthwaighte: J. Phys. A. Math. Gen. **31**, L527 (1998)
32. H.Qian, G.M.Raymond, J.B.Bassingthwaighte: Biophys. Chem. **80**, 1 (1999)
33. L.E.Reichl: *A Modern Course in Statistical Physics* (Univ. Texas Press, Austin 1980)
34. J.Rosen: J. Multivar. Anal. **23**, 37 (1987)
35. J.Rudnick, G.Gaspari: Science **237**, 384 (1987)
36. M.F.Shlesinger, J.Klafter, B.J.West: Physica A **140**, 212 (1986)
37. M.F.Shlesinger, B.J.West, J.Klafter: Phys. Rev. Lett. **58**, 1100 (1987)
38. Ya.G.Sinai: Theory Probab. Appl. **21**, 64 (1976)
39. G.Strang, T. Nguyen: *Wavelets and Filter Banks* (Wellesley-Cambridge, MA 1996)
40. N.Wax: *Selected Papers on Noise and Stochastic Processes* (Dover, New York 1954)
41. B.J.West, W.Deering: Phys. Rep. **246**, 1 (1994)
42. Y. Xiao: Osaka J. Math. **33**, 895 (1996)
43. Y. Xiao: Math. Proc. Camb. Phil. Soc. **122**, 565 (1997)
44. Z.M.Yin: J. Comput. Phys. **127**, 66 (1996)

Scaling and Wavelets: An Introductory Walk*

Patrice Abry

CNRS, UMR 5672, Physics Lab., Ecole Normale Supérieure de Lyon, 46, allée d'Italie, F-69364, Lyon cedex 7, France

Abstract. This chapter offers, first, an introductory walk through the notions related to scaling phenomena and intuitions behind are gathered to formulate a tentative definition. Second, it introduces the mathematical model of self-similar processes with stationary increments, understood as the canonical reference to describe scaling. Then, it shows how and why the wavelet transform constitutes a powerful and relevant tool for the analysis (detection, identification, estimation) of self-similarity. It is finally explained why self-similarity is too restrictive a model to account for the large variety of scaling encountered in empirical data and a review of the various models related to scaling – long range dependence, local Hölder regularity, fractal and multifractal processes, multiplicative or cascade processes – is proposed. Their interrelations and differences, as well as estimation issues, are discussed. A set of *Matlab* routines has been developed to implement the wavelet-based analysis for scaling described here. It is available at `www.ens-lyon.fr/~pabry`.

1 Introduction and Motivation

1.1 Scaling Phenomena

Power laws, scaling laws, scaling phenomena or, simply, scaling, recently became a very fashionable topic. Indeed, scaling behaviors were observed or studied or used as description paradigms in a large collection of research works covering a wide variety of different domains or applications. It is worth noting that those applications may be related either to natural phenomena or to data resulting from mankind's activities. For the first category, one can, for instance, mention hydrology [10] with the study of variabilities of water levels in rivers, hydrodynamic with the study of developed turbulence [24,20], statistical physics with the study of systems having long range interactions [44], microelectronics with $1/f$ noises in semi-conductors [9,23], geophysics and fault repartitions or geological layers [43,42], biology and physiology [52] with human rhythms variabilities, heart beat [45] or gait [26] for instance. For the second category, one can find human geography and population repartition in cities or continents [19], information flows on network and mainly computer network teletraffic [35], stock

* This article reviews works done in past years and still in progress, mainly in collaboration with D. Veitch, EMUlab, Univ. of Melbourne, Australia, and with Pierre Chainais, Patrick Flandrin, ENS Lyon, France. These works have been partially supported by the French CNRS grant *TL99035, Programme Télécommunications* and French MENRT grant *ACI jeune Chercheur 2329.*

market volatility or currency change rates fluctuations [29,51]. Very often, scaling in data is a crucial observation since it can be tied to key properties of the systems (e.g., high volatility in markets or large waiting times and congestions in computer network traffic or pathologies in body rythms...).

The notion of scaling, however, remains defined poorly or in a loose way and may be related to a variety of different properties of a system or a process. A possible common tentative definition for scaling can be formulated through a negative statement: there is no characteristic scale (of time or space...) in the studied system or process. In other words, this is no longer possible to identify any scale that plays a privileged role compared to others, or equivalently, all scales play identical roles and are of equal importance in the dynamics of the analyzed system or process. Scaling, therefore, correspond to situations where the *whole* can not be (statistically) distinguished from any of its *subpart*. This is commonly associated to the picture of geometric fractal object, obtained through the iteration of an identical construction procedure.

1.2 Random Walks and Self-similarity

From a data analysis (or signal processing) point of view, scaling in time series implies that the usual intuitive search techniques for characteristic scales are to abandoned and replaced instead by new ones aiming at evidencing relations, mechanisms between scales or involving a wide range of scales. This also means to abandon the use of models relying on the existence and definition of a characteristic scale (e.g, Markow chains, Poisson models, models with exponential autocorrelation functions,...). The canonical reference mathematical models for scaling are that of Random Walks and Self-Similar processes, and more particularly, the popular fractional Brownian motion. This will be introduced with details in Sect. 2.1.

1.3 Wavelets

The practical use and analysis of self-similar processes present however two major difficulties: they are non stationary and are characterized by long range dependence or long term correlations or long memory. Such statistical features turn the analysis of self-similar processes into an uneasy and non standard task. To overcome such difficulties, it has been shown recently in a collection of papers [32,33,46,11,31,2,4,47] that wavelet transforms constitute ideal tools for the analysis of scaling. Wavelet analysis will be introduced in Sect. 2.2. More precisely, the wavelet analysis can be considered as "matched" to self-similar processes in the sense that wavelet coefficients exactly reproduce, from scale to scale, the self-replicating statistical structure of such processes. This will be made explicit in Sect. 2.3. Section 2.4 explains how those statistical properties of the wavelet coefficients are to be used to design tools for the analysis (detection, identification and estimation) of scaling phenomena.

• BEYOND SELF-SIMILARITY

Self-similarity is a mathematically well-behaved model. Its definition however implies numerous constraints seen as limitations in the practical 0 of empirical data. Obviously, it cannot account for the large variety of scaling existing amongst actual empirical data. Section 3 will therefore allow a larger part to variations around self-similarity commonly used to describe scaling, such as long range dependence, fractal sample path, multifractal processes, multiplicative processes, infinitely divisible processes and cascades... and will underline their mutual interrelations, common denominators and differences.

• NOTE

This chapter mainly intends to be an introductory walk in the land of scaling phenomena and scaling laws. Its aim is to provide the reader with a synthetic and comprehensive overview of their related mathematical models and with a quick start to their wavelet-based analysis. Technical details as well as mathematical proofs can be found in references given along the text.

• MATLAB ROUTINES

All the analysis procedures (detection, identification, estimation) described here as well as synthesis ones presented elsewhere are implemented in MATLAB routines available at `www.ens-lyon.fr/~pabry` or `www.emulab.ee.mu.oz.au/~darryl`.

2 Self-similarity and Wavelets

2.1 Self-similarity

• RANDOM WALKS

The simplest model that can be thought of to model scaling phenomena is that of the standard random walk commonly involved is the pedagogical(!) description of a drunkard walk or more generally in that of diffusion phenomena. Let $X(t)$ denote some physical quantity of interest, a random walk consists in going from position X at time t to position $X + \delta X$ at time $t + \tau$ by making an elementary step (or increment) $\delta X(\tau, t)$:

$$X(t+\tau) = X(t) + \delta X(\tau, t)\,, \forall \tau \geq 0\,. \tag{1}$$

Without loss of generality, we assume in this whole text that $X(0) \equiv 0$ and $\mathbb{E}\delta X(\tau, t) \equiv 0, \forall\, \tau, \forall t$.

For normal (drunkard and) diffusion, one usually assumes for the increments the following statistical properties:

A1 : The $\{\delta X(\tau, t), t \in \mathbb{R}\}$ form random processes that are stationary with respect to the t variable. Their distributions are identical, do not depend on t but functionally depend on τ,

A2 : The steps $\{\delta X(\tau, t), t \in \mathbb{R}\}$ are mutually independent, i.e., for $t_1 \leq t_2 \leq t_3 \leq t_4$,

$$p_2[(X(t_4) - X(t_3)).(X(t_2) - X(t_1))] = p_1(X(t_4) - X(t_3))\, p_1(X(t_2) - X(t_1)).$$

where the $p_i(\cdot)$ denote the (joint) probability density functions. This means that the random variables $\delta X(\tau, t)$ and $\delta X(\tau', t')$ are statistically independent as soon as $t' > t + \tau$. In other words, once the step $\delta X(\tau, t)$ has been performed, one gains no extra information on the next following step.
A3 : The $\{\delta X(\tau, t), t \in \mathbb{R}\}$ form jointly Gaussian processes.

Though apparently simple and intuitive, those three properties together impose severe constraints on the walk X, and even define it in a unique manner as the ordinary Brownian motion. For instance, they imply a linear behavior of the variance of X (or of its increments) with respect to time:

$$\mathbb{E}|X(t) - \underbrace{X(0)}_{\equiv 0}|^2 = 2D_X|t|, \text{ or, equivalently } \mathbb{E}|\delta X(\tau, t)|^2 = 2D_X|\tau|. \quad (2)$$

Those behaviors, known as the celebrated Einstein's relations, constitute the signature of scale invariance or scaling phenomenon in the random walk: no characteristic scale exists or can be identified that would limit, or bound, or indicate a cut-off in the development of the walk nor plays any specific role.

However, empirical data very often exhibit significant departures from those linear behaviors. The so-called anomalous diffusion phenomena, for instance, are characterized by:

$$\mathbb{E}|X(t) - X(0)|^2 = 2D_X|t|^\gamma, \; 0 < \gamma < 2, \quad (3)$$

which can be seen as generalizations to (2) above, read as a power-law with exponent that takes the specific value 1. To account for the departure from a linear behavior, to bypass limitations resulting from **A1**, **A2** and **A3** and more generally to enlarge the framework of ordinary random walk and Brownian motion, one is naturally lead to that of self-similarity.

• SELF-SIMILAR PROCESSES

A process X is said to be statistically self-similar, with self-similarity parameter $H > 0$, if [41]:

$$\forall c > 0, \{c^H X(t/c), t \in \mathbb{R}\} \stackrel{fdd}{=} \{X(t), t \in \mathbb{R}\} \quad (4)$$

where $\stackrel{fdd}{=}$ means equality of all the finite dimensional distributions. This means that the sample paths (t, X) of the process $X(t)$ and those $(t/c, c^H X)$ of the process $c^H X(t/c)$ are statistically indistinguishable. In other words, the process X is statistically similar to any of its dilated templates. Therefore, no characteristic scale of time can be identified on these processes, self-similarity is hence a model for scaling behavior. This is illustrated on Fig. 1.

A major consequence of self-similarity also lies in the fact that the moments of the process, when they exist (we do nit discuss existence issues in this text), behave as power laws with respect to time,

$$\mathbb{E}|X(t)|^q = \mathbb{E}|X(1)|^q |t|^{qH}, \quad (5)$$

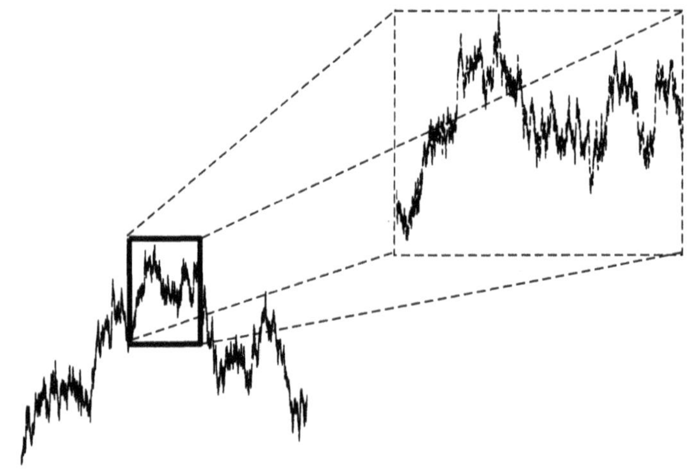

Fig. 1. Sample path of a self-similar process. Starting with the sample path of a self-similar process, if one performs a dilation of the time axis of factor $1/c$ and a dilation on the amplitude axis of factor c^H, one obtains a new sample path that is (statistically) indistinguishable from the original one.

whose exponents are all controlled by the self-similarity parameter H. Besides the connection between scaling and power law, those relations also show that self-similar processes are non stationary ones.

• SELF-SIMILAR PROCESSES WITH STATIONARY INCREMENTS

Because with actual empirical data, the use of non stationary models is a major difficulty, one often restricts the class of self-similar processes to that of self-similar processes with stationary increments (hereafter, H-sssi processes). A process is said to posses stationary increments if:

$$\{\delta X(\tau,t) \equiv X(t+\tau) - X(t), t \in \mathbb{R}\} \stackrel{fdd}{=} \{\delta X(\tau,0) \equiv X(0+\tau) - X(0) \equiv X(\tau)\}, \forall \tau. \tag{6}$$

Self-similarity and stationary increments together yield that the fundamental equation (5) above can be rewritten on the increments (for all finite moments):

$$\mathbb{E}|X(t+\tau) - X(t)|^q = \mathbb{E}|X(1)|^q |\tau|^{qH}. \tag{7}$$

In keeping with the general framework of random walks (cf. (1)), the class of H-sssi processes enlarges that of ordinary Brownian motion and ordinary random walk through the replacements of assumptions **A1, A2** and **A3** with:
B1: The $\{\delta X(\tau,t), t \in \mathbb{R}\}$ form random processes that are stationary with respect to the t variable.
B2: The steps $\{\delta X(\tau,t), t \in \mathbb{R}\}$ satisfy, $\forall H > 0$,

$$\forall c > 0, \{c^H \delta X(\tau/c, t/c), t \in \mathbb{R}\} \stackrel{fdd}{=} \{\delta X(\tau,t), t \in \mathbb{R},\}. \tag{8}$$

Obviously, **B1** is identical to **A1** while **B2** gathers **A2** and **A3** . This latest implies that the steps of the random walk no longer need to be independent (they can even exhibit long memory as detailed below) nor Gaussian (they can have any *stable* [41] marginal distribution, i.e., infinite variance distributions, see section 3). B. Mandelbrot, who significantly contribute to the promotion of the use of H-sssi processes (fractional Brownian motion [27], Lévy stable motions [29]) in applications (turbulence, finance, ...), named those two variations the *Joseph's effect* and the *Noah's effect*, respectively [29], after the celebrated biblical characters. One can, moreover, consider situations, problems or time series where the two difficulties occur jointly and simultaneously. Technically, H-sssi processes can be written as [41]:

$$X(t) = \int_{\mathbb{R}} f(t,u) M(du), \tag{9}$$

where $M(du)$ denotes an α stable stochastic measure (where α, $0 < \alpha \leq 2$, stands for the stability index, and $\alpha = 2$ corresponds to the Gaussian case). This definition means that the process X can be read as a weighted sum of independent α stable random variables. Technically, this implies that, for a fixed t, $X(t)$ is a α stable random variable, M therefore controls the marginals of the process, or in other words, its static properties. The weighting kernel function $f(t,u)$ controls the statistical dependences of X and hence its joint statistics, or in other words, its dynamical properties. For well chosen forms of the kernel, [18,34,41], X is a self-similar process with stationary increments. For instance, the celebrated fractional Brownian motion (fBm), which is, up to a normalization, the only Gaussian H-sssi process, corresponds to the choice of a Gaussian measure: $M(du) = dB(u)$ and of the specific *fractional integration* kernel $f(t,u) = (t-u)_+^d - (-u)_+^d$, where $(u)_+ = u$ if $u \geq 0$ and 0 else and where $d = H - 1/2$ controls the long memory of the process (hence its dependence structure) [27]. Lévy stable processes (that include the ordinary Brownian motion) are characterized by independent increments and correspond to $f(t,u) = 1$ if $0 \leq u \leq t$ and 0 elsewhere. Linear fractional stable motion is characterized by an infinite variance stable distribution $M(du)$ (i.e., α strictly smaller than 2) and by the same kernel as that of fBm with $d = H - 1/\alpha$; it therefore gathers simultaneously the two difficulties mentioned above.

• Self-similar processes with stationary increments and finite variance

Self-similarity, stationary increments and finite variance together impose that [41,27]:

$$0 < H < 1. \tag{10}$$

Hence, for H-sssi processes with finite variance, the choice $q = 2$ in the relation above (5) accounts for the *anomalous* behavior reported in (3) with $0 < \gamma = 2H < 2$.

Moreover, for H-sssi process X with finite variance (and with zero mean and $X(0) \equiv 0$), one can show that the covariance function necessarily takes the following form:

$$\mathbb{E}X(t)X(s) = \frac{\sigma^2}{2}\left(|t|^{2H} + |s|^{2H} - |t-s|^{2H}\right). \tag{11}$$

with $\sigma^2 = \mathbb{E}|X(1)|^2$ and that the correlation function of the increment process reads:

$$\mathbb{E}\delta X(\tau,t)\delta X(\tau,t+s) = \frac{\sigma^2}{2}\left(|\tau+s|^{2H} + |\tau-s|^{2H} - 2|s|^{2H}\right). \tag{12}$$

• LONG RANGE DEPENDENCE OR LONG TERM CORRELATIONS OR LONG MEMORY

From this relation, one can infer the asymptotic behavior of the covariance function on the increment process in the limit of large s (i.e., $s \to +\infty, s \gg \tau$):

$$\mathbb{E}\delta X(\tau,t)\delta X(\tau,t+s) \sim \frac{\sigma^2}{2}\,2H(2H-1)\,\tau^2 s^{2(H-1)}. \tag{13}$$

Such a power law decrease of the covariance function above refers to a notion known as long range dependence, or long term correlations [10,41]. A stochastic stationary process Y is said to be long range dependent if its spectrum behaves asymptotically as a power law in the limit of small frequencies or, equivalently, if its covariance function behaves asymptotically in the limit of large lag as a power law:

$$\left.\begin{array}{r} \Gamma_Y(\nu) \sim C|\nu|^{-\gamma},\ |\nu| \to 0, \quad 0 < \gamma < 1, \\ \mathbb{E}Y(t)Y(t+s) \sim C'|s|^{-\beta},\ |s| \to +\infty, 0 < \beta = 1-\gamma < 1. \end{array}\right\} \tag{14}$$

This asymptotic power law decrease of the autocorrelation function is to be compared to the exponential one encountered in more usual processes (like Markov processes): an exponential decrease, by definition, implies a characteristic time while a power law behavior does not, meeting again the intuition of the absence of characteristic scale of time. For long range dependent processes, the autocovariance function decreases so slowly that its sum diverges, i.e., for $A > 0$,

$$\int_A^\infty \mathbb{E}Y(t)Y(t+s)ds = \infty.$$

This implies that the correlation between any two samples of the process cannot be neglected without missing something crucial in the analysis of the process, no matter how far apart one from the other they are. Long range dependence among its increments constitute a major difficulty in the analysis of a self-similar process.

Among the statistical community, where its formal definition was first proposed, long memory is often studied through the so-called aggregation procedure. Let $Y^{(T)}(t)$ be the version of the process Y aggregated in a window of size T:

$$Y^{(T)}(t) = \frac{1}{T}\int_{tT}^{(t+1)T} Y(u)du. \tag{15}$$

The long memory of the process $Y(t)$ results in a asymptotical power law behavior of the variance of the aggregated process:

$$\mathbb{E}|Y^{(T)}(t)|^2 \sim T^{-\beta}, T \to \infty, \tag{16}$$

A large family of estimators has been based on this asymptotic property.

Long memory can be defined independently from self-similarity and in itself, is a model for scaling observed in the limit of large scales (see Sect. 3 for details). However, increments of H-sssi processes, when $H > 1/2$, exhibit long memory as proven by (13) above. Therefore, the two notions are subtly related.

• $1/f$-PROCESSES

So-called $1/f$-processes have also been widely used to model scaling phenomena. A stationary process is said to posses a $1/f$-spectrum if its spectrum behaves as a power law in a wide range of scales:

$$\Gamma_X(\nu) \sim C|\nu|^{-\gamma}, 0 \le \nu_m \le |\nu| \le \nu_M, \nu_M/\nu_m \gg 1, \gamma > 0. \tag{17}$$

In this setting, all frequencies are playing equivalent roles. Moreover, such spectrum satisfy $\forall \lambda > 0, \Gamma_X(\lambda\nu) \sim \lambda^{-\gamma}\Gamma_X(\nu)$, hence the connection with scale invariance.

Despite its being a non-stationary process, fBm is often *naturally* considered as the reference for $1/f$-Processes. However, connections are somchow intricate... The increment process of fBm can be regarded as the output of a linear time invariant filter whose input is fBm and whose impulse response reads $\psi_{\delta\tau}(t) = \delta(t+\tau) - \delta(t)$:

$$\delta X(\tau, t) = (\psi_{\delta\tau}(\cdot) * X(\cdot))(t),$$

where $*$ stands for the convolution. The standard Fourier relation of the linear time invariant filter yields:

$$\Gamma_{\delta X(\tau,\cdot)}(\nu) = |\Psi_{\delta\tau}(\nu)|^2 \Gamma_X(\nu),$$

where $\Psi_{\delta\tau}(\nu) = 1 - \exp(\imath 2\pi\tau\nu)$. In the limit $|\tau\nu| \ll 1$, $|\Psi_{\delta\tau}(\nu)| \sim |2\pi\tau\nu|$ and $\Gamma_{\delta X(\tau,\cdot)}(\nu) \sim |\nu|^{-(2H-1)}$, and hence, from the relation above one *heuristically* or *qualitatively* associate to fBm a spectrum of the form $\Gamma_X(\nu) \sim |\nu|^{-(2H+1)}$, hence a $1/f$-spectrum. This correspondence has been formulated in various frameworks including that of wavelets, see, e.g., [32].

• LOCAL REGULARITY OF THE PROCESS

Exploring the other limit, that of fine scales, i.e., $s \ll \tau, s \to 0$, one obtains that the autocovariance function of X behaves as:

$$\mathbb{E}\delta X(\tau, t)\delta X(\tau, t+s) \sim \sigma^2 |\tau|^{2H} (1 - |\tau|^{-2H}|s|^{2H}). \tag{18}$$

Such a power law behavior again traces back to the absence of characteristic scale (in the limit of small scales). Since $\delta X(\tau, t)$ is a stationary process, the equation above also straightforwardly yields:

$$\mathbb{E}|\delta X(\tau, t+s) - \delta X(\tau, t)|^2 \sim \sigma^2 |s|^{2H}, s \to 0, \tag{19}$$

which gives indications with respect to the local (ir)regularity of the sample paths of X.

Indeed, local regularity of sample paths of stochastic processes or of functions is usually measured in terms of Hölder exponent: this consists in comparing X at time t against a power law function. A process X is said to have Hölder regularity $h \geq 0$ at time t if their exists a local polynomial $P_t(s)$ of degree $n = \lfloor h \rfloor$ and a constant C such that:

$$|X(t+s) - P_t(s)| \leq C\,|s|^h. \tag{20}$$

For $0 \leq h < 1$, the regular part of X at time t reduces to $P_t(t) = X(t)$, yielding the following regularity characterization:

$$|X(t+s) - X(t)| \leq C\,|s|^h. \tag{21}$$

Heuristically, the Hölder h exponent describes the roughness of the sample path of X: h between 0 and 1 indicates that the sample path is everywhere continuous but nowhere differentiable, h close to 1 betrays a smooth and regular behavior and conversely, h close to 0 implies sharp roughness and large variability. For $1 < h < 2$, the same arguments apply to the first derivative of the sample path, and so on.

Self-similarity with stationary increments and finite variance, and more precisely the central relation (7) for the increments together with Kolmogorov's regularity criterion[1] shows that the local regularity of each sample path of the fractional Brownian motion (for which all moments exist for $q > -1$) is constant along time and controlled by the parameter H: $h = H$. From (19) above, one sees that the same holds for the increment process of fractional Brownian motion. Processes with sample paths characterized by a constant Hölder exponent, are often referred to as *monofractal* processes. Monofractal processes constitute therefore a model for scaling observed in the limit of small scales. For further details, see e.g., [37].

2.2 Wavelet Analysis

• CONTINUOUS WAVELET TRANSFORM

The wavelet coefficients of the so called continuous wavelet transform (CWT) [17,30] are defined as the results of comparisons, by means of inner products, between the process to be analyzed X and a family of functions, the wavelets $\psi_{a,t}$:

$$T_X(a,t) = \langle X, \psi_{a,t} \rangle, \quad (a,t) \in (\mathbb{R}^+, \mathbb{R}). \tag{22}$$

[1] Kolmogorov's criterion (see for example [38]): If $\{X(t) : t \in R\}$ is a stochastic process with values in a complete separable metric space (S,d), and if there exists positive constants β, C, ϵ such that for all $s,t \in R$

$$\mathbb{E}d(X_s, X_t)^\beta \leq C|s-t|^{1+\epsilon}$$

then there exists a continuous version of X. This version is Hölder continuous of order θ for each $\theta < \epsilon/\beta$.

The wavelets are dilated and translated templates of a reference pattern ψ called the mother wavelet:

$$\psi_{a,t}(u) = \frac{1}{|a|}\psi(\frac{u-t}{a}). \tag{23}$$

Figure 2 shows dilated templates of a single mother wavelet. Note that some definitions prefer a $1/\sqrt{|a|}$ instead of $1/|a|$ normalization term, mainly because it ensures energy preservation. For the analysis of scaling phenomena, however, the choice $1/|a|$ is more convenient. The function ψ is usually required to be bounded and to have time and frequency supports that are either bounded or decrease very fast, jointly in both domains, time and frequency. To ensure that

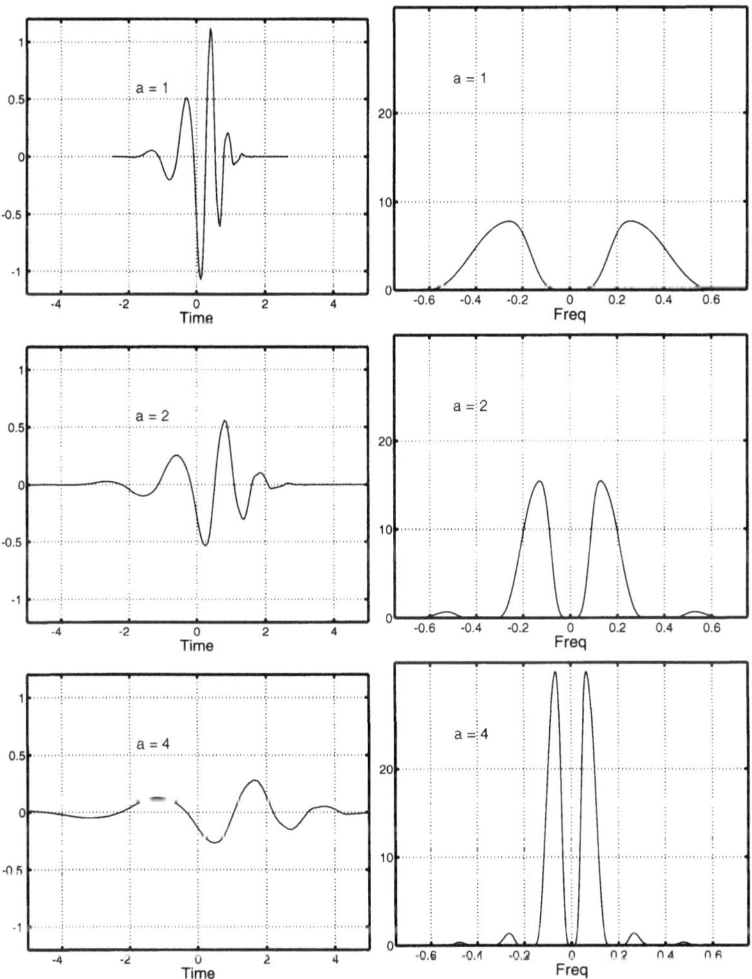

Fig. 2. Translated and dilated wavelets and their corresponding Fourier transforms. Left, dilated versions of the same mother-wavelet (Daubechies6) with dilation factors 1, 2 and 4, and, right, the corresponding Fourier transforms.

the wavelet transform is invertible, ψ moreover has to satisfy a so-called admissibility condition whose weak form reads:

$$\int_{\mathbb{R}} \psi(t)dt = 0. \tag{24}$$

Hence, ψ is a zero-mean function that has to oscillate and exist on a narrow time support. This is therefore a small wave: a *wavelet.*

This is because the mother wavelet ψ has a joint localization in time and frequency that the wavelet coefficients can be given the signification of frequency content of the data at a given time or of joint time-frequency content of the information in X.

The mother-wavelet is characterized by an integer N, called its number of vanishing (or zero) moments, defined as:

$$\forall m \in \{0, \dots, N-1\}, \quad \int_{\mathbb{R}} t^m \psi(t)dt = 0, \; \int_{\mathbb{R}} t^N \psi(t)dt \neq 0. \tag{25}$$

The admissibility condition above (cf. (24)) imposes $N \geq 1$. This means that, for a mother-wavelet with N vanishing moments, the wavelet coefficients of a polynomial of degree $P < N$ are strictly zero. More generally, it means that the wavelet coefficients $T_X(a,t)$ of a process X at time t are only sensitive to the part of the local behavior of X which is more irregular than that of a polynomial of degree N. In other words, the higher N, the less the wavelet coefficients are sensitive to regular parts of the time series. The number of vanishing moments of the mother wavelets also controls the behavior of its Fourier transform at the origin:

$$|\Psi(\nu)| \sim |\nu|^N, \; |\nu| \to 0. \tag{26}$$

Figure 3 shows examples of wavelets with different vanishing moments.

- Discrete wavelet transform

One also defines the coefficients of the discrete wavelet transform (DWT) as a discrete subset of the $T_x(a,t)$:

$$d_X(j,k) = T_X(a = 2^j, t = 2^j k) = \langle X, \psi_{j,k} \rangle, \; (j,k) \in (\mathbb{Z}^+, \mathbb{Z}), \tag{27}$$

where $\psi_{j,k}(u) = 2^{-j}\psi(2^{-j}u - k)$. This discrete subset of points is usually called the dyadic grid of the discrete wavelet transform and this definition also usually implies that the mother wavelet is constructed through a multiresolution analysis [17,30]. The major interest in the use of the discrete wavelet transform lies in the facts that the $\{\psi_{j,k}, (j,k) \in (\mathbb{Z}^+, \mathbb{Z})\}$ form (possibly orthonormal) basis of $L^2(\mathbb{R})$ (so that the DWT is a non redundant representation of X) and that the $d_X(j,k)$ can be computed with fast pyramidal recursive algorithm whose computational costs is of the order of that of a fast Fourier transform. In all the methods and algorithms proposed here, the DWT is always used. For further details on wavelet transforms, the reader is referred to [17,30].

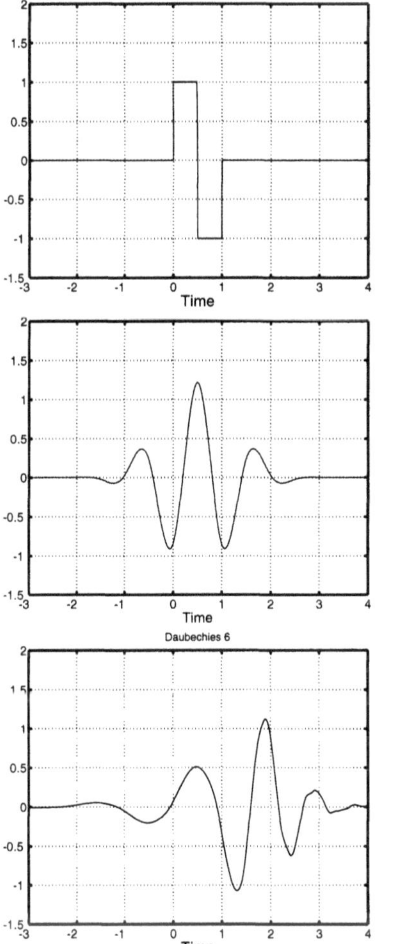

Fig. 3. Vanishing moments. Examples of mother wavelet with respectively, 1 (Haar, or poor man's, wavelet), 3 (B-spline wavelet) and 6 (Daubechies6) vanishing moments.

• MULTIRESOLUTION ANALYSIS AND SCALING

Self-similar processes are usually studied through the analysis of their increments, and mainly (8) and so is local regularity (cf. (19)). Long memory in turn is analyzed through the aggregation procedure (cf. relation (16)). In spirit both techniques already are multiresolution analysis since they consist in analyzing the evolution of a resolution or scale dependent quantity (increments or aggregation) with that of the scale.

The wavelet based analysis of scaling proposes to replace increments or aggregation with wavelet coefficients and intuition behind this substitution can be formulated as follows. First, consider the wavelet coefficients of a process X obtained with a particular choice for the wavelet-like function $\psi(t) = \delta(t + \tau_0) - \delta(t)$

and note that they are identical to its increments:

$$X(t+\tau) - X(t) \equiv T_X(a,t), \text{ with } \tau = a\tau_0.$$

Increments can therefore be thought of as a specific example of wavelet coefficients obtained from a particular mother wavelet, which has a poor spectral localization and only one vanishing moment (i.e., $N = 1$). Moreover, let us note that selecting a mother-wavelet with N vanishing moment amounts to compute increments of order N (i.e., increments of increments of ...).

Consider now the wavelet coefficients of a long range dependent process Y, obtained with a particular choice for the mother wavelet $\psi(t) = 1$ for $-T_0/2 < t < T_0/2$ and 0 elsewhere[2] and note that they are identical to the aggregated process:

$$Y^T(t) \equiv T_Y(a,t), \text{ with } \tau = aT_0.$$

Aggregation can therefore be thought of as a specific case of wavelet decomposition which has a poor spectral localization and zero vanishing moment (i.e., $N = 0$).

While the increments and the aggregation procedure compute differences and averages respectively, a wavelet, being a band-pass function, naturally performs a difference of averages and therefore gathers the two techniques in a single one. And not only, the wavelet transform unifies these two techniques in a single framework but it also brings generalization, versatility and robustness through the choice of the mother-wavelet. One can, indeed, naturally think of using mother-wavelets with better joint time and frequency localizations and higher numbers of vanishing moment, resulting in better statistical properties, this will be further discussed in Sect. 2.4.

SUMMARY. All what the reader unfamiliar with wavelets needs to have in mind to follow the remainder of this text is that the relevance of the wavelet transform for the analysis of self-similar processes relies on two ingredients:

I1) the wavelet basis is designed from a dilation operator, $\psi_{a,0}(u) = \frac{1}{|a|}\psi(\frac{u}{a})$;

I2) the mother wavelet is characterized by a strictly positive integer N, its number of vanishing moments, cf. (25).

2.3 Self-similarity and Wavelets: Theory

Let X be a H-sssi process. Its DWT coefficients $d_X(j,k)$ and its CWT coefficients $T_X(a,t)$ have the following statistical properties (for proofs, see, e.g., [8,32,11,46,31]).

P1 Self-Similarity: The $d_X(j,k)$ (and the $T_X(a,t)$) reproduce, in an exact manner, the self-similarity of the process:

$$\{2^{-jH} d_X(j,k), k \in \mathbb{Z}\} \stackrel{d}{=} \{d_X(0,k), k \in \mathbb{Z}\}. \tag{28}$$

[2] Strictly speaking, this *box* or *indicator* function is not a wavelet, since this is not a band pass function.

$$\forall c > 0, \{c^H T_X(a/c, t/c), t \in \mathbb{R}, a \in \mathbb{R}^+\} \stackrel{fdd}{=} \{T_X(a,t), t \in \mathbb{R}, a \in \mathbb{R}^+\}. \quad (29)$$

These two relations result, fundamentally, from the fact that wavelets are designed using a dilation operator (ingredient I1 above). It is, moreover, interesting to note that this last relation has strong and obvious analogies to that satisfied by increments (cf. (8)).

P2 Non Stationarity: Though self similar processes are non stationary, their $\{d_X(j,k), k \in \mathbb{Z}\}$ form stationary sequences at each octave j. Identically, their $\{T_X(a,t), t \in \mathbb{R}\}$ form stationary processes at each scale a. This is again analogous to the stationarity of the increments and this is deeply connected to the fact that $N \geq 1$ (cf. ingredient I2 above).

P3 Long Range Dependence: It can be shown that the covariance function of any two wavelet coefficients on the dyadic grid can be asymptotically bounded as, $|2^j k - 2^{j'} k'| \to +\infty$,

$$|\mathrm{Cov}\ d_X(j,k), d_X(j',k')| \leq C|2^j k - 2^{j'} k'|^{-2(N-H)}. \quad (30)$$

This shows the key role played by the number of vanishing moments N. Increasing N allows to increase the rate of decrease of the covariance function and therefore to reduce as much as desired the range of correlation amongst the wavelet coefficients. More precisely, it can be shown that, in the Gaussian case, if $N > H + 1/2$, long range dependence that exists amongst increments of X when $H > 1/2$ is turned into short range dependence. Note that obtaining this last property requires the simultaneous use of both ingredient I1 and I2 above.

P3ID Idealization: The "decorrelation effect", i.e., the reduction of the range of dependence of the wavelet coefficients under the increase of N, is idealized as follows:

any two wavelet coefficients of X, on the dyadic grid $\{d_X(j,k), k \in \mathbb{Z}, j \in \mathbb{Z}^+\}$, can be regarded as independent one from the other.

This idealization is used to provide approximated but analytical studies of the performance of the estimators proposed below.

Summary: Together, properties **P1** and **P2** imply that, for all finite moments:

$$\mathbb{E}|d_X(j,k)|^q \equiv C_q 2^{qjH}, \forall j, \quad (31)$$

where $C_q = \mathbb{E}|d_X(0,0)|^q$. Those relations are reminiscent of the fundamental equations (5) and (7) and yield the same constraints on the process X: power laws are to hold for all finite moments (e.g., for all $q > -1$ in the Gaussian case or for all $\alpha > q > -1$ in the α-stable case) and for all scales 2^j, moreover, all the exponents of the power laws are controlled by the single parameter H.

2.4 Self-similarity and Wavelets: Application

• Intuition

Self similar processes with stationary increments and finite variance are traditionally studied through their increments, mainly through (7), with $q = 2$.

However, the practical use of such an equation requires that the mathematical expectation be estimated, usually from a single observation of finite length. The existence of long term correlations amongst increments, however, substantially increase difficulties in the practical issue of estimation. For instance, the use of the standard sample variance estimator (that replaces statistical averages with time averages) presents remarkably poor statistical estimation performances [10].

The wavelet rewriting of (7), see (31) above with $q = 2$ can be used as a new starting point:

$$\mathbb{E}|d_X(j,k)|^2 = C2^{2jH}. \tag{32}$$

From **P1**, wavelet coefficients exactly reproduce self-similarity. From properties **P2** and **P3**, they form, at each scale 2^j, stationary sequences with short range and weak statistical dependence. On condition that N is high enough, they do not suffer any more from long range dependence. They are therefore statistically better behaved than increments and offer a versatile and convenient tool for the analysis of self-similarity. For instance, the standard sample variance estimator of the wavelet coefficients is a very satisfactory estimator for the ensemble average.

- Log-scale Diagram

To study scaling and more specifically self-similarity with wavelets, we define the following quantities:

$$\begin{aligned} Y_j &= \log_2\left(\frac{1}{n_j}\sum_{k=1}^{n_j}|d_X(j,k)|^2\right), \\ \sigma_j^2 &= \operatorname{Var} Y_j, \end{aligned} \tag{33}$$

where the n_js denote the numbers of wavelet coefficients available at octaves js. Then, we form the plots of the $Y_j = \log_2\left(1/n_j\sum_{k=1}^{n_j}|d_X(j,k)|^2\right)$, together with their error bars (σ_j^2), versus $\log_2 2^j = j$. In those plots, that we proposed to call *Logscale Diagrams*, straight lines evidence the existence of self-similarity and the measurement of their slopes allows for an estimation of the parameter H. Figure 4 proposes examples of logscale diagrams for the sample paths of fractional Brownian motion and of a Long Range Dependent process.

- Estimation Issues

Precisely, the estimator $\hat{H}$ for H is defined through a weighted linear fit:

$$\hat{H} = \sum_j w_j Y_j/2, \tag{34}$$

where $\sum_j$ runs over the range $\{j_1, \ldots, j_2\}$ of octaves where the linear fit is to be performed, this range is to be chosen a priori. The w_j satisfy the usual relations,

$$\left.\begin{aligned} \textstyle\sum_j jw_j &= 1 \\ \textstyle\sum_j w_j &= 0 \\ w_j &= (1/\lambda_j)(S_0 j - S_1)/(S_0S_2 - S_1^2) \\ S_m &= \textstyle\sum_{j=j_1}^{j_2}\lambda_j^{-1}j^m (m = 0,1,2), \end{aligned}\right\} \tag{35}$$

the λ_js being arbitrary numbers.

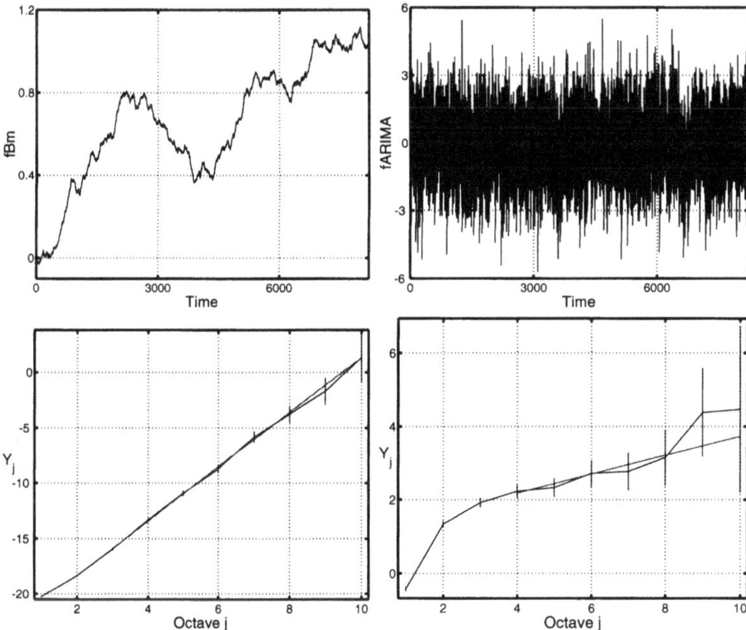

Fig. 4. Examples of Logscale Diagrams. Top left, sample path of a fractional Brownian motion; top right, sample path of a Long Range Dependent process (a fARIMA process). The fractional Brownian motion is an exact self-similar process, this can be seen on its Logscale diagram through the fact that the linear behavior of the log of the variance of the wavelet coefficients against the log of the scale holds for all scales (bottom left). The measurement of the slope enables moreover to precisely estimate the self-similarity parameter. The logscale diagram of the LRD process (bottom right) shows scaling behavior that holds only in the limit of large scales.

The statistical performance of this estimator have been studied in detail in [47,4]. For Gaussian processes, analytical calculations relying on the idealization of exact independence of the wavelet coefficients **P3ID**, show that a residual bias can be determined and therefore subtracted to produce an unbiased estimator. Let n denote the number of samples of the analyzed process X, neglecting the practical border effects resulting from the computation of the wavelet coefficients, the n_js behave as $n_j = 2^{-j}n$. For Gaussian processes and under **P3ID**, we showed that [47]:

$$\left.\begin{aligned} \sigma_j^2 \equiv \text{Var}\, Y_j \simeq = 2(\log_2 e)^2/n_j \\ \text{Var}\, \hat{H} \simeq 2(\log_2 e)^2 \textstyle\sum_j w_j^2 2^j/n. \end{aligned}\right\} \tag{36}$$

The variance of the estimator asymptotically decreases as the inverse of the analyzed number of samples. Numerical simulations showed [47] that the actual statistical performance are very close to the idealized ones, even for non Gaussian processes. This approximate, however realistic, estimation of the variance of $\hat{H}$ enables us to have confidence intervals on the estimation of the parameter H.

The choice of the weights w_j remains to be specified through the choice of the λ_js. It is well-known that the variance of the linear fit is minimal on condition that the λ_js match the covariance structure of the Y_js. Assuming the idealization **P3ID**, one obtains that the Y_js are independent. The choice $\lambda_j = \text{cste } n_j^{-1}$, therefore, ensures that the quantities $(\sum_j w_j^2/n_j)$ and Var $\hat{H}$ takes close to minimal values. The estimator $\hat{H}$ is systematically implemented with that particular choice.

• Additional properties

Thanks to its number of vanishing moments, the wavelet-based analysis of self-similarity, moreover, benefits from robustness against non-stationarities. For instance, if deterministic smooth trends (like a linear trend or an oscillating trend) are superimposed to self-similar processes, this may significantly complicate the detection and analysis of self-similarity. Because wavelet coefficients are blind to polynomials (of degree smaller than N) and *only feels* the most irregular parts of a process, increasing N will *cancel out* the possibly superimposed trends and therefore enable a relevant analysis of scaling in data without requiring any a priori processing. The possibility of performing various discrete wavelet transforms with different N and of comparing the resulting analysis and estimations is hence a key feature of the wavelet-based analysis of scaling against non stationarities. Comparison of wavelet-based analysis performed using mother-wavelet with different number of vanishing moments will allow to detect those trends and perform relevant analysis of self-similarity (see [2,47,4] for details).

From another perspective, non-stationarities and scaling may have practical effects and consequences that are practically very close and similar so that it may be difficult to distinguish one phenomena from the other. We have also shown that the wavelet framework offers a convenient way to design a statistical test allowing to discriminate actual scaling from some non stationary effects [49].

Finally, other interesting features of this wavelet based analysis lies in the facts that it is simple both conceptually and practically (DWT plus linear fits) and that it has a low computational cost thanks to the recursive pyramidal algorithm underlying the DWT. This is of importance when dealing with large sample of data, as is often the case when dealing with scaling and allowed us to propose real time on line algorithm for the analysis of scaling [39].

3 Beyond Self-similarity

3.1 Practical Limitations

Self-similar processes with stationary increments and finite variance, and more specifically their Gaussian version, the fractional Brownian motion, are very attractive models to describe scaling in empirical data and they are used and quoted in numerous and various applications. This is mainly because they are mathematically well-defined and well-documented and they moreover fulfill the intuition of scaling phenomena in a very satisfactory manner. Their major practical quality is their *simplicity*: each and every of their (scaling) properties, (self-similarity, long-range dependence, fractal sample paths) is controlled by the single H. This parameter is therefore used in applications to describe, sometimes confusingly, either global or local scaling properties, i.e., either long memory or fractality... The major practical drawback of H-sssi processes lies in ... their *simplicity*: it is very unlikely that the numerous and various types of scaling encountered in the many different applications where they occur can all be described by a unique model depending on a single parameter. More precisely, exact self-similarity implies a number of specific properties (as summarized in (5) or (31)) and significant departures from those properties can be observed in the analysis of actual empirical data: *i*) moments of different orders may have scaling exponents that are not controlled by a single parameter, or more simply some moments may not present scaling or, even more simply, may not exist at all; *ii*) when scaling are observed, they may not exist over the whole range of scales as in the self-similar case, but only in a given range of scales, or only asymptotically in the limit of large scales or in the limit of small scales; *iii*) power-law behavior of the moments may not exist despite scaling behavior. In this section, we will explore those variations and describe some related mathematical models.

Scaling phenomena may also occur or exist in point processes and might be fruitfully studied through wavelets as well. This has been discussed in [1] and will not be addressed here.

3.2 Beyond Finite Variance

In previous sections, we assumed that the variance of the process X, as well as all higher moments, existed. However, one may encounter situations where scaling and self similarity are valid but where the variance of the process, for instance, and therefore all higher moments, are infinite. For those situations, the model of Gaussian self-similar processes as well as the analysis presented above and based on the variance of the wavelet coefficients are no longer relevant. Such situations can be modeled using α-stable self-similar processes, see Sect. 2.1. A wavelet-based analysis of α-stable self-similar processes can be conducted but the log of the variance of the wavelet coefficients, $\log_2 \mathbb{E}|d_X(j,k)|^2$, has to be replaced by the quantity $\mathbb{E} \log_2 |d_X(j,k)|$, a random quantity with finite variance. This has been discussed at length in [3,5].

3.3 Beyond Scaling over All Scales: Long Range Dependence, 1/f-Processes and Local Regularity

As said, a major consequence of self-similarity lies in the fact that the scaling behavior holds for all the scales (see (31) or (32)). Practically however, scaling may exist for the second order statistics (namely the variance) of the process, but may be observed only in a given, large but finite, range of scales, or in the asymptotic limits of small or large scales, rather than in the whole range of scales.

For instance, one may empirically observe the following asymptotic behavior:

$$\mathbb{E}|d_X(j,k)|^2 \simeq C2^{j(\gamma-1)}, 2^{j_m} \leq j \leq 2^{j_M}, 2^{j_M}/2^{j_m} \gg 1.$$

This is to be read as a weakened version of (32) and this is the signature of a scaling behavior in a wide, but finite, range of scales therefore corresponding to $1/f$-processes, with power law exponent γ (cf. (17)).

One may also empirically observe the following asymptotic behavior (see, for example, Fig. 4(right)):

$$\mathbb{E}|d_X(j,k)|^2 \simeq C2^{j2H}, 2^j \to +\infty.$$

This is to be seen as a weakened version of (32) and this is the signature of scaling that exist only for the largest scales of the process. This is reminiscent of (16) and tells us that the data are not self-similar but rather present some long term correlations properties and can therefore be modeled as a stationary long range dependent process [47,4].

Another possibility is to empirically observe an asymptotic scaling behavior in the limit of fine scales:

$$\mathbb{E}|d_X(j,k)|^2 \simeq C2^{j2H}, 2^j \to 0.$$

This is again to be seen has a weakened version of (32) and is reminiscent of Hölder regularity behavior (cf. (19)). This means that the data are not self similar but rather that their sample paths are characterized by a local regularity h controlled by H and that remains constant along time. This therefore betrays a fine scale scaling property.

For those situations, $1/f$-processes, long range dependence or local regularity, the analysis and estimation of the exponent can be performed with the logscale diagram, as in the self-similar case, except that linear fits are to be performed over a finite chosen range of scales, respectively. The question of automatically choosing or *detecting* the relevant range of scales is subtle and has been addressed e.g., in [50] for the LRD case.

3.4 Beyond Second-Order Statistics – Multiplicative and Multifractal Processes

• MULTISCALING

Another practical major limitation of self-similar processes lies in the fact that the exponents of the power-laws for all the moments are controlled by the single parameter H, see (31). It is, however, quite common on empirical data to observe, in a given, finite, range of scales, a behavior of the type:

$$\mathbb{E}|d_X(j,k)|^q \simeq C2^{jH(q)}, 2^{j_m} < 2^j < 2^{j_M}, \tag{37}$$

where the exponents $H(q)$ may significantly depart from the linear qH behavior. We proposed to refer to this observation as *multiscaling*.

• MULTIPLICATIVE PROCESSES

Modeling multiscaling implies a major change of paradigm: the additive structure underlying a random walk (cf. (1)) has to be abandoned and replaced by a multiplicative scheme. In other words, being at position X at time t results from a collection of elementary steps that are no longer added up together but instead multiplied one to the other. The canonical reference for multiplicative processes are the celebrated Mandelbrot's c-adic cascade processes [28]. Their construction is based on the iteration of a sequence of operations. At iteration j, one has c^j segments, to which are associated numbers $X_{j,k}$, $k = 0, \ldots, c^j - 1$. At stage $j+1$, one divides segment (j,k) into c subsegments to which are associated new numbers $X_{j+1,l} = W_{j+1,l} X_{j,k}$ where $l = (k-1) * c + p$, with $p = 0, \ldots, c-1$. The multipliers $W_{j,k}$ are i.i.d. positive random variables. Usually, $\mu(t)$ denote the measure obtained in the limit of an infinite number of iterations and $X(t) = \int_0^t \mu(du)$ the corresponding process. There exists numerous variations around this scheme that all share the spirit of multiplicative *cascade*.

• MULTIFRACTAL PROCESSES

An important consequence of the Mandelbrot's multiplicative cascade procedure lies in the fact that the resulting motions $X(t) = \int_0^t \mu(du)$ are multifractal processes. In other words, they present sample paths with Hölder exponents $h(t)$ that vary very widely, irregularly and erratically from point to point and with each realizations. Those fluctuations of the local regularity are often described through the so-called multifractal spectrum $D(h)$, (which consists of the Hausdorff dimension of the set of points where the local regularity take the value h).

An important practical consequence of multifractality is that quantities called partition functions,

$$(1/n_j) \sum_{k=1}^{n_j} |d_X(j,k)|^q$$

present in the limit of small scales power law behaviors,

$$(1/n_j) \sum_{k=1}^{n_j} |d_X(j,k)|^q \sim C_q 2^{jH(q)}, 2^j \to 0.$$

Reading the partition functions $(1/n) \sum_{k=1}^{n} |d_X(j,k)|^q$ as estimators of the moments $\mathbb{E}|d_X(j,k)|^q$, the scaling relation above is very close to the equation defining multiscaling in the limit of small scales. Therefore multifractal can be seen as the very example for multiscaling.

Moreover, for the c-adic cascades, the multifractal spectrum $D(h)$ can be obtained from the function $H(q)$ through a Legendre transform. In that case, $H(q)$, and therefore $D(h)$, are controlled by the probability density function of the multipliers W. Further details on multifractal are beyond the scope of this chapter and the interested reader is referred to e.g., [30,37].

• Estimation issues

To test multiscaling or multifractal in empirical time series and estimate the corresponding $H(q)$ exponents, one forms the quantities $Y_j^{(q)}$ that can be read both as generalization of the Y_j (cf. (33)) to statistics of order different than 2 and as sample time average estimators for the ensemble averages:

$$Y_j^{(q)} = (1/n_j) \sum_{k=1}^{n_j} |d_X(j,k)|^q. \tag{38}$$

An extension of the estimation procedure described in Sect. 2 has been proposed to estimate the $H(q)$ exponents: it mainly consists in measuring slopes in $\log_2 2^j = j$ *vs* $\log_2 Y_j^{(q)}$ through **non weighted** linear regressions. For details, see [4,48]. The estimation of the $H(q)$ exponents for a multifractal process, synthesized according to the definitions proposed in [16], is illustrated in Fig. 5.

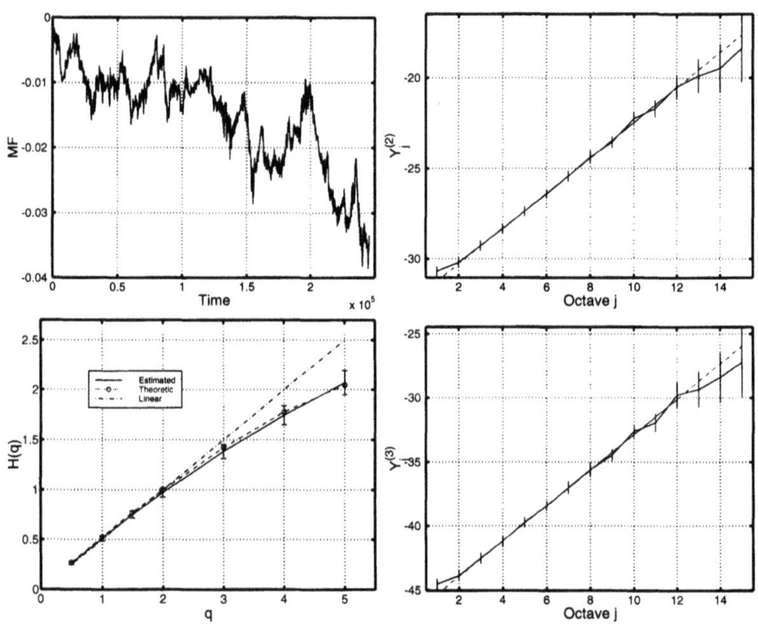

Fig. 5. Multiscaling Analysis of a Multifractal Process. Top left, sample path of a multifractal process synthetized according to the technique proposed in [16]; right column, Structure Functions ($Y_j^{(q)}$ vs j) of orders 2 and 3; bottom left, estimated exponents $H(q)$.

A key practical issue is to define a statistical test that would allow to test whether $H(q)$ is a linear function of q or not. It would enable to decide whether the analyzed data are to be modeled with an additive or multiplicative scheme. Up to our knowledge, this has hardly ever been addressed theoretically.

3.5 Beyond Power Laws – Infinitely Divisible Cascade

• INTUITION

Multiscaling offers an extension to self-similarity insofar as a collection of exponents rather than a single one is needed to describe data. Yet, it maintains a major feature of self-similarity: moments behaves as power laws of the scales. However, when analyzing actual data, it may very well be observed that this is not the case, see e.g., [12,20,48,52]. To account for those situations, the infinitely divisible cascade model, on which we concentrate now, proposes to gain an extra degree of freedom by giving up the requirement that moments behaves a priori as power laws of the scales. The equations below summarize the connections between self-similarity, multiscaling and infinitely divisible cascade:

$$\begin{array}{lll} \text{Self-Sim.} & \mathbb{E}|d_X(j,k)|^q = C_q(2^j)^{qH} & = C_q \exp(qH \ln(2^j)) \\ \text{MultiScaling} & \mathbb{E}|d_X(j,k)|^q = C_q(2^j)^{H(q)} & = C_q \exp(H(q) \ln(2^j)) \\ \text{Inf. Div. Casc.} & \mathbb{E}|d_X(j,k)|^q = - - - - - & = C_q \exp(H(q) n(2^j)), \end{array} \qquad (39)$$

where the functions $n(2^j)$ and $H(q)$ need not a priori be the function $\ln 2^j$ and the linear function qH, respectively.

• DEFINITION

The concept of infinitely divisible cascade was first introduced by B. Castaing in [12,13] and rephrased in the wavelet framework in [7,48]. We briefly recall its intuition, definition and relations to other models. Starting again from the self-similar case, one can write the probability density function (pdf) of the wavelet coefficients at scale $a = 2^j$, as a dilated version of the pdf of the wavelet coefficients at a larger scale a': $p_a(d) = (1/\alpha_0)\, p_{a'}(d/\alpha_0)$ where the dilation factor is unique: $\alpha_0 = (a/a')^H$. In the cascade model, the key change is that there is no longer a unique factor but a collection of dilation factors α; consequently p_a will result from a weighted sum of dilated incarnations of $p_{a'}$:

$$p_a(d) = \int G_{a,a'}(\ln \alpha)\, \frac{1}{\alpha}\, p_{a'}\left(\frac{d}{\alpha}\right) d\ln \alpha.$$

The weighting function $G_{a,a'}$ is called the kernel or the *propagator* of the cascade. Obviously, if $G_{a,a'}$ is a Dirac function, $G_{a,a'}(\ln \alpha) = \delta(\ln \alpha - H \ln(a/a'))$, infinitely divisible cascade reduces to self-similarity, therefore understood as a special case. The definition of the cascade above shows that the pdf's of $\underline{p}_a$ and $\underline{p}_{a'}$ of the log wavelet coefficients $\ln |d|$ are related by a convolution with the propagator:

$$\begin{aligned} \underline{p}_a(\ln \alpha) &= \int G_{a,a'}(\ln \alpha)\, \underline{p}_{a'}(\ln |d| - \ln \alpha)\, d\ln \alpha \\ &= (G_{a,a'} * \underline{p}_{a'})(\ln \alpha). \end{aligned} \qquad (40)$$

If cascades exist between scales a and a'' and between scales a'' and a', then a cascade between scales a and a' exists, and the corresponding propagator results from the convolutions of the two propagators: $G_{a,a'} = G_{a,a''} * G_{a'',a'}$. Infinite divisibility (also called continuous self similarity) means that no scale between a and a' plays any characteristic role (i.e., a'' in the above statement can be any scale between a and a'). Infinite divisibility therefore implies that the propagator consists of an elementary function G_0 convolved with itself a number of times, where that number depends on a and a':

$$G_{a,a'}(\ln \alpha) = [G_0(\ln \alpha)]^{*(n(a)-n(a'))} .$$

Using the Laplace transform $\tilde{G}_{a,a'}(q)$ of $G_{a,a'}$, this can be rewritten as $\tilde{G}_{a,a'}(q) = \exp\{H(q)(n(a)-n(a'))\}$, with $H(q) = \ln \tilde{G}_0(q)$ and $a := 2^j$; this implies that $\mathbb{E}|d_X(j,k)|^q = C_q \exp\{H(q)n(2^j)\}$, thus validating (39). The main consequences of infinitely divisible cascade read therefore:

$$\ln \mathbb{E}|d_X(j,k)|^q = H(q)n(2^j) + K_q \tag{41}$$

$$\ln \mathbb{E}|d_X(j,k)|^q = \frac{H(q)}{H(p)} \ln \mathbb{E}|d_X(j,k)|^p + \kappa_{q,p}. \tag{42}$$

This last equation implies that for any p and q, the moment of order q behave as power-law of the moment of order p. This is sometimes referred to as "extended self-similarity", in turbulence mainly. Note moreover, that, in the relation (41) above, there is some arbitrariness, indeed:

$$\begin{aligned} H(q)n(a) + K_q &= \left(\tfrac{H(q)}{\beta}\right)(\beta n(a) + \gamma) + (K_q - \tfrac{H(q)\gamma}{\beta}) \\ &= H'(q)n'(a) + K'_q \end{aligned} \tag{43}$$

where $\beta \neq 0$ and γ are arbitrary constants. It clearly indicates that the function $H(q)$ is defined up to a multiplicative constant while n is defined up to multiplicative and additive constants.

If it is moreover required that the function $n(a) \equiv \ln a$, the infinitely divisible cascade is called *scale invariant* and this implies that:

$$\tilde{G}_{a,a'}(q) = (a/a')^{\ln \tilde{G}_0(q)} \text{ and } \mathbb{E}|d_X(j,k)|^q = (2^j)^{\ln \tilde{G}_0(q)},$$

proving that scale invariant infinitely divisible cascade reduces to multiscaling with exponents being controlled by the propagator: $H(q) = \ln \tilde{G}_0(q)$. In this framework, multiscaling, or multifractal, is therefore understood as a special case of infinitely divisible cascade. In a scale invariant infinitely divisible cascade, one can also inquire on whether $H(q)$ is a linear function of q or not, in which case the cascade reduces to the even more special case of self-similarity. It is, therefore, natural to consider the function $H(q)/q$ and to test its constancy.

• A FUNDAMENTAL FEATURE

The Infinitely Divisible Cascade model is hence a natural extension to the multiscaling and self-similarity ones, it is important to note however that it maintains a fundamental feature that already existed in the two previous models (cf. the set of equations (39): the dependence of the moments in the variables q

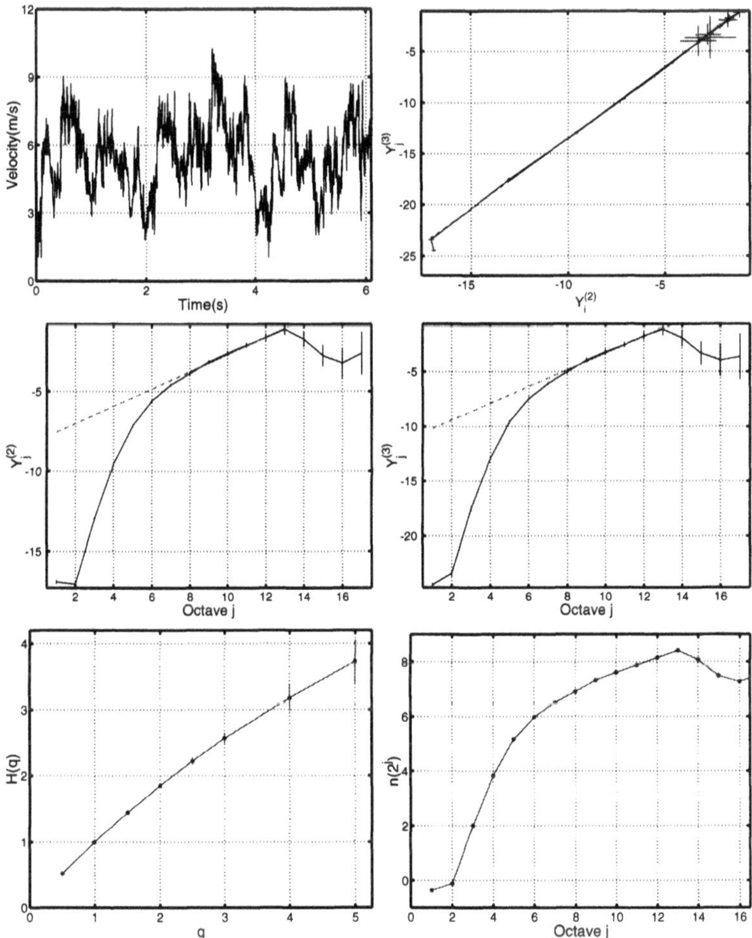

Fig. 6. Infinitely Divisible Scaling Analysis for Turbulent Velocity Signals. Top left, Velocity Time Series; Second Raw, the Structure Functions of orders 2 and 3 show clear departures from strict power laws; top right, Extended Self-Similarity ($Y_j^{(3)}$ vs $Y_j^{(2)}$), the straight line show that an Infinitely Divisible Cascade Model gives a satisfactory description of the Velocity Time Series; bottom left and right respectively, estimated functions $H(q)$ and $n(2^j)$.

(order of the moment) and 2^j (scale) is separable. This key feature can actually be considered as a *practical or operational definition* of scaling in time series.

• ESTIMATION ISSUES

To analyze Infinite Divisibility in empirical time series, we propose to form the diagrams $Y_j^{(q)}$ versus $Y_j^{(p)}$, that constitute natural extensions to logscale diagrams. Again, those diagrams come with confidence intervals for the $Y_j^{(q)}$s. Testing the validity of the model empirical data amounts to test for the existence of an affine relation amongst the $Y_j^{(q)}$ versus $Y_j^{(p)}$, i.e., straight lines in the diagrams. The estimation of the corresponding $H(q)$ and $n(a)$ parameter functions is performed through a non weighted linear fit between the random variables $Y_j^{(q)}$ versus $Y_j^{(p)}$ (see (41) and (42)):

$$\begin{aligned} \hat{H}(q)/H(p) &= \text{slope}_{q,p}, \ \forall q \\ \hat{K}_q &= \text{intercept}_{q,p}, \ \forall q \\ \hat{n}(a = 2^j) &= \frac{1}{H(p)} \left\langle \frac{1}{\text{slope}_{q,p}} \left(Y_j^{(q)} - \hat{K}_q \right) \right\rangle_q + K_p, \end{aligned} \tag{44}$$

where $\langle . \rangle_q$ stands for simple means on q-values. Details on the difficulties (best choice of p, arbitrariness from (43), ...) of the procedure as well as on its statistical performances are given in [4,15]. Fig. 6 illustrates this estimation procedure on turbulence velocity data[3] recorded in jet turbulence at a R_λ Reynolds number of the order of 600.

4 Conclusion

In this article, we showed that scaling phenomena in empirical data may be described through a large variety of mathematical models. We gave an introductory and comprehensive overview of those models, that can be read as variations on the self-similarity reference. We pointed out however that multiscaling, multifractal and cascades imply the replacement of the additive random walk framework with multiplicative constructions.

Then, we have shown how and why the wavelet transform offers a versatile, powerful and efficient tool to perform the analysis of scaling: with only little a priori on the nature and properties of the empirical data, it allows to *detect* the existence of scaling in data, to *identify* the nature of the detected scaling, to *estimate* the corresponding parameter. Wavelets may also be used for the numerical synthesis of stochastic processes with a priori prescribed (scaling) properties. This has not been detailed here and the interested reader is referred to [4].

This article is expected to propose to the reader a *quick start* on what scaling in time series may mean and on what to do with wavelets to analyze them. A set of MATLAB routines enables a practical use of all the techniques presented here. Those techniques have been fruitfully used for the analysis of hydrodynamic turbulence [14,15] and computer network traffic [4,48].

References

1. P. Abry, P. Flandrin: 'Point processes, long-range dependence and wavelets'. In: *Wavelets in Medicine and Biology*, ed. by A. Aldroubi and M. Unser (CRC Press, Boca Raton (FL) 1996) pp.413–438

[3] Data collected by C. Baudet, in Physic Lab. at Ecole Normale Supérieure de Lyon, France. The calculation of R_λ is based on the Taylor microscale.

2. P. Abry, D. Veitch: IEEE Trans. on Info. Theory, **44**, 2 (1998)
3. P. Abry, L. Delbeke and P. Flandrin: IEEE-ICASSP-99, Phoenix (AZ) (1999)
4. P. Abry, P. Flandrin, M.S. Taqqu and D. Veitch: *Wavelets for the analysis, estimation and synthesis of scaling data* A chapter in [35]
5. P. Abry, B. Pesquet-Popescu, M. S. Taqqu: 'Wavelet Based Estimators for Self Similar α-Stable Processes'. In: *Int. Conf. on Signal Proc., 16th World Computer Congress, Beijing, China, 2000*
6. P. Abry, P. Gonçalvès and J. Lévy Véhel: Traité Information, Commandes, Contrôle, Hermès Editions, Paris, France, April (2002)
7. A. Arnéodo, J.F. Muzy, S.G. Roux: J. Phys. II France, **7**, 363 (1997)
8. R. Averkamp and C. Houdré: IEEE Trans. on Info. Theory, Vol.**44**, 1111 (1998)
9. V. Bareikis and R. Katilius: *Noise in Physical Systems and* $1/f$ *Fluctuations* (World Scientific 1995)
10. J. Beran: *Statistics for Long-Memory Processes* (Chapman and Hall, New York 1994)
11. S. Cambanis and C. Houdré: IEEE Trans. on Info. Theory, **41**, 628 (1995)
12. B. Castaing, Y. Gagne, E. Hopfinger: Physica D, **46**, 177 (1990)
13. B. Castaing: J. Phys. II France, **6**, 105 (1996)
14. P. Chainais, P. Abry, and J. F. Pinton: Phys. Fluids, **11**, 3524 (1999)
15. P. Chainais: Cascades log-infiniment divisibles et analyse multirésolution. Application à l'étude des intermittences en turbulence. PhD Thesis, Ecole Normale Supérieure de Lyon (2001)
16. P. Chainais, R. Riedi and P. Abry: 'Infinitely Divisible Processes'. In: *Stochastic processes and Their Applications* (May 2002)
17. I. Daubechies: *Ten Lectures on Wavelets* (SIAM, Philadelphia 1992)
18. L. Delbeke and P. Abry: Stochastic Processes and their Applications, **86**, 177 (2000)
19. P. Frankhauser: Population, **4**, 1005 (1997)
20. U. Frisch: *Turbulence. The legacy of A. Kolmogorov* (Cambridge University Press, UK 1995)
21. A. C. Gilbert, W. Willinger and A. Feldmann: *IEEE Trans. on Info. Theory* **45**, 971 (1999)
22. P. Gonçalvès and R. H. Riedi: Proc. 17ème Colloque GRETSI, Vannes, France (1999)
23. M. Keshner: proc. of the IEEE, **70**, 212 (1982)
24. A. N. Kolmogorov: 'a) Dissipation of energy in the locally isotropic turbulence, b) The local structure of turbulence in incompressible viscous fluid for very large Reynolds number, c) On degeneration of isotropic turbulence in an incompressible viscous liquid'. In: *Turbulence, Classic papers on statistical theory*, ed.by S.K. Friedlander and L. Topper (Interscience publishers 1941) pp.151-161
25. W. E. Leland, M. S. Taqqu, W. Willinger, and D. V. Wilson: IEEE/ACM Trans. on Networking, **2**, 1 (1994)
26. L. Liebovitch and A. Todorov: The American Physiological Society, 1446 (1996)
27. B. B. Mandelbrot and J. W. Van Ness: SIAM Rev, **10**, 422 (1968)
28. B. B. Mandelbrot: J. of Fluid Mech., **62**, 331 (1974)
29. B. B. Mandelbrot: *Fractales, Hasard et Finance* (Flammarion 1997)
30. S. Mallat: *A Wavelet Tour of Signal Processing* (Academic Press, Boston 1997)
31. E. Masry: IEEE Trans. on Info. Theory, **39**, 260 (1993)
32. P. Flandrin: IEEE Trans. on Info. Theory, **38**, 910 (1992)

33. P. Flandrin: 'Fractional Brownian motion and wavelets'. In: *Wavelets, Fractals, and Fourier Transforms* ed. by M. Farge, J.C.R. Hunt and J.C. Vassilicos (Clarendon Press, Oxford 1993) pp.109–122
34. B. Pesquet-Popescu: Signal Processing **75** (1999)
35. K. Park and W. Willinger: *Self-Similar Network Traffic and Performance Evaluation* (Wiley ,Interscience Division 2000)
36. R. Riedi, M.S. Crouse, V.J. Ribeiro and R.G. Baraniuk: IEEE Trans. on Info. Theory, **45**, 992 April (1999)
37. R. H. Riedi: 'Multifractal processes'. In: *Long range dependence: theory and applications* ed. by Doukhan, Oppenheim and Taqqu (2002)
38. L. C. G. Rogers and D. Williams: *Diffusions, Markov processes and martingales*, 2nd edition, vol 1 (Foundations, Wiley 1994)
39. M. Roughan, D. Veitch, and P. Abry: IEEE Trans. on Networking, **8**, 467 (2000)
40. M. Roughan, D. Veitch, J. Yates, M. Ashberg, H. Elgelid, M. Castro, M. Dwyer and P. Abry: 'Real-Time Measurement of Long-Range Dependence in ATM Networks'. In: *Passive and Active Measurement Workshop, Hamilton, New-Zealand 2000*
41. G. Samorodnitsky and M. S. Taqqu: *Stable Non-Gaussian Processes: Stochastic Models with Infinite Variance* (Chapman and Hall, New York, London 1994)
42. A. Saucier: 'Méthodes multifractales pour l'analyse d'images et de signaux'. In: [6]
43. D. Sornette: *Critical Phenomena in Natural Sciences* (Springer 2000)
44. A. Torcini and M. Antoni: Physical Review E, **59**, 2746 (1999)
45. M. Teich, S. Lowen, B. Jost and K. Vibe-Rheymer: *Heart rate Variability: Measures end Models* (preprint 2000)
46. A. H. Tewfik and M. Kim: IEEE Trans. Info. Theory, **38**, 904 (1992)
47. D. Veitch and P. Abry: IEEE Trans. on Info. Theory, **45**, 878, April (1999)
48. D. Veitch, P. Abry, P. Flandrin and P. Chainais: 'Infinitely Divisible Cascade Analysis of Network Traffic Data'. In: *Proceedings of ICASSP, Istanbul, June 2000*
49. D. Veitch and P. Abry: 'A statistical test for the constancy of scaling exponents'. In: itshape IEEE Trans. on Sig. proc. 2001
50. D. Veitch, P. Abry and M. S. Taqqu: *On the automatic selection of the onset of scaling* (Preprint 2002)
51. C. Walter: *Lois d'échelle en finance* in [6]
52. B. West and A. Goldberger: American Scientist, **75**, 354 (1987)

Wavelet Estimation for the Hurst Parameter in Stable Processes*

Stilian Stoev and Murad. S. Taqqu

Boston University, Boston, USA

Abstract. In this paper we review some recent results on the statistical properties of wavelet-based estimators of the Hurst parameter when the time series are heavy tailed with α-stable infinite variance distributions. We focus on two models: *linear fractional stable motion* (LFSM, in short) and *fractional autoregressive moving average* (FARIMA, in short) time series with SαS innovations. Using the wavelet coefficients of the LFSM process or these of the FARIMA time series, we define two types of estimators for the Hurst parameter, for each of the two models. These estimators are obtained by using a weighted linear regression, involving wavelet coefficients of the process over a set of scales.

For the case of LFSM when the scales are fixed, the two types of wavelet estimators are *strongly consistent*, as the sample size tends to infinity. These estimators are also *asymptotically normal* under certain conditions on the mother wavelet and the index of stability α.

In the case of FARIMA time series, the resulting estimators will be consistent in probability, provided that the scales *as well as* the sample size tend to infinity. Under conditions on the mother wavelet, the index α and the rate of increase of the estimation scales as a function of the sample path, these wavelet estimators are also *asymptotically normal.*

We illustrate the basic properties of these estimators by using simulated paths of LFSM processes and FARIMA time series with SαS innovations.

1 Introduction

Recently the self-similarity has proved to be a useful notion in the context of computer networks (see for example Leland, Taqqu, Willinger and Wilson [16] and the book of Park and Willinger [17]). *Self-similar processes* are random processes whose probability distribution is invariant under a judicious time and space scale. More precisely, the random process $X(t)$ is self-similar with self-similarity parameter H if for all $c > 0$, $X(ct)$ and $c^H X(t)$ have the same finite-dimensional distributions. The level of self-similarity is characterized by the exponent H, often called the Hurst exponent. The finite-dimensional distributions of $X(t)$ can be Gaussian. But they can also be *heavy-tailed*, and thus in particular, $\mathbb{P}\{|X(t)| > x\} \sim Cx^{-\alpha}$ for large x, where C is a constant and

* This research was supported in part by the NSF Grants DMS-0102410 and ANI-9805623 at Boston University. Stilian Stoev was partially supported by the Institute of Pure and Applied Mathematics at UCLA during the program "Large Scale Communication Networks" 2002.

$\alpha < 2$. In this case, $X(t)$ take high values with large probability. Whereas in the Gaussian case the probability tails of the process decrease exponentially fast, in the heavy-tailed case they decrease like a power function $x^{-\alpha}$ with $\alpha < 2$. This implies, in particular, that $X(t)$ does not have finite variance and hence that its dependence structure cannot be described by covariances. Heavy-tailed processes are treated, for example, in Samorodnitsky and Taqqu [21] and Adler, Feldman and Taqqu [3]. We will suppose here that the heavy-tailed process $X(t)$ is SαS (symmetric α–stable) with $0 < \alpha < 2$. This means that all linear combinations of its finite-dimensional distributions are SαS random variables ξ, that is, random variables, whose characteristic function is $\mathbb{E}e^{i\theta\xi} = e^{-\sigma^\alpha|\theta|^\alpha}$, where $\sigma > 0$ is a scale parameter. When $\alpha = 2$, ξ and hence $X(t)$ is Gaussian.

The estimation of the Hurst parameter H of the process X, given a set of observations $X(1), X(2), \ldots, X(N)$, is an important statistical problem. Many techniques have been proposed in the finite variance case for self-similar processes X as well as for *asymptotically* self-similar or long-range dependent ones (see, for example, Fox and Taqqu [11], Dahlhaus [8], Robinson [19,20]). Since most of these methods use the covariance structure of the process X or involve the power spectrum of X, they do not work for heavy-tailed infinite variance processes.

The wavelet transform has recently proved to be a very useful tool in studying the scaling structure of a stochastic process X and in estimating its Hurst parameter H. Abry and Veitch [1] have proposed an estimator of the Hurst parameter of a second-order time series X, based on the scaling behavior of the wavelet coefficients of X. Wavelet methods are very useful in practice, because they are computationally efficient and *semi-parametric* in nature, that is, they do not require an explicit parametric knowledge of the distributions of the process.

Wavelet estimators were studied for the case of infinite variance processes as well see, for example, Abry, Delbeke and Flandrin [2]. So far, however, there have been very few rigorous results on the statistical properties of these estimators of the Hurst parameter H for the cases of infinite variance processes X (see e.g. Pipiras, Taqqu and Abry [18]). In this paper we review some recent results on wavelet-based estimation of the Hurst exponent H for symmetric α-stable (SαS, in short) processes. We focus on two classes of SαS models: the *linear fractional stable motion* (LFSM) and the *fractional autoregressive moving average* (FARIMA) with SαS innovations.

The paper is organized as follows. In Sect. 2 we define the two SαS models considered in this paper namely, LFSM and FARIMA time series, and recall some of their properties. The goal is to estimate their Hurst coefficient. In Sects. 3 and 4 we introduce two types of wavelet-based estimators and present some recent results on their asymptotic properties. In Sect. 5 we discuss the practical implementation of these estimators and in Sect. 6 we illustrate their behavior using simulated paths of LFSM processes and FARIMA time series.

2 The Time Series

In this section, we briefly recall the definitions of the LFSM process and the FARIMA time series with SαS innovations and then indicate some connections between these two models.

2.1 Linear Fractional Stable Motion

Consider the stochastic integral

$$X_{H,\alpha}(t) = \int_{\mathbb{R}} \{(t-s)_+^{H-1/\alpha} - (-s)_+^{H-1/\alpha}\} M_\alpha(ds), \tag{1}$$

where $H \in (0,1)$ and $M_\alpha(ds)$ is a SαS random measure with control measure ds. In practice, the above integral is approximated by a discrete sum:

$$\begin{aligned} X_{H,\alpha}(t) &\approx \int_{-M}^{t} (t-s)^{H-1/\alpha} M_\alpha(ds) + \int_{-M}^{0} (-s)^{H-1/\alpha} M_\alpha(ds) \\ &\stackrel{d}{\approx} \sum_{k=-mM}^{[tm]} \left(t - \frac{k}{m}\right)^{H-1/\alpha} \left(\frac{1}{m}\right)^{1/\alpha} Z_k + \\ &\quad \sum_{k=-mM}^{-1} \left(-\frac{k}{m}\right)^{H-1/\alpha} \left(\frac{1}{m}\right)^{1/\alpha} Z_k, \end{aligned} \tag{2}$$

where $m, M \in \mathbb{N}$, $[t]$ denotes the integer part of $t \in \mathbb{R}$ and the random variables $(1/m)^{1/\alpha} Z_k$, $k \in \mathbb{Z}$ are independent SαS with scale parameter $(1/m)^{1/\alpha}$. The above expression converges in finite-dimensional distributions to the LFSM process $X_{H,\alpha}(t)$, $t > 0$, as the discretization parameters m and M go to infinity. Since $(1/m)^{1/\alpha} Z_k$ replaces $M_\alpha(ds)$ in the discrete sum approximation, one can view the terms $M_\alpha(ds)$ in (1) as a SαS white noise, that is, independent SαS infinitesimal random elements with scale parameters $\{ds\}^{1/\alpha}$ assigned at each $s \in \mathbb{R}$. In particular, for a (Borel) set A, $M_\alpha(A)$ is a SαS random variable with characteristic function $\mathbb{E}e^{i\theta M_\alpha(A)} = \exp\{-|\theta|^\alpha \int_{\mathbb{R}} 1_A(s) ds\} = \exp\{-|\theta|^\alpha |A|\}$, where $|A|$ denotes the Lebesgue measure of the set A. The approximation in (2) will be used in the sequel to generate paths of LFSM.

The kernel function $f_{H,\alpha}(t,s) := (t-s)_+^{H-1/\alpha} - (-s)_+^{H-1/\alpha}$ in (1) governs the temporal dependence structure of the process $\{X_{H,\alpha}(t),\ t \in \mathbb{R}\}$, because it controls the magnitude of the weights assigned to the random noise $\{M_\alpha(ds),\ s \in \mathbb{R}\}$. In particular, the parameter $d = H - 1/\alpha$ controls the regimes of long-term dependence of the process $X_{H,\alpha} = \{X_{H,\alpha}(t),\ t \in \mathbb{R}\}$, since $f_{H,\alpha}(t,s) \sim \mathrm{const}\, t^{d-1}$, as $t \to \infty$.

The process $X_{H,\alpha}$ will be called a linear fractional stable motion. It has stationary increments and is self similar with Hurst parameter H. The stochastic integral in (1) is well-defined (in the sense of convergence in probability) for all $H \in (0,1)$ and $t \in \mathbb{R}$, because $\int_{\mathbb{R}} |f_{H,\alpha}(t,s)|^\alpha ds < \infty$. For more details on stable stochastic integrals, stable self-similar processes and, in particular, the class of LFSM processes we refer to the monograph Samorodnitsky and Taqqu [21].

When $0 < H < 1/\alpha$, the exponent $H - 1/\alpha$ in (1) is *negative* and the LFSM process $X_{H,\alpha}$ is said to be *negatively dependent.* Whereas the case $1/\alpha < H < 1,\ 1 < \alpha < 2$, is referred to as the *long-range dependence* case of LFSM. This terminology is adapted from the Gaussian case of fractional Brownian motion (see e.g. Ch. 7 in Samorodnitsky and Taqqu [21]). We now present some intuition for the distinction between the path behavior of the LFSM process in these two cases (see Fig. 1).

In the case of *negative dependence* of LFSM, the kernel function $f_{H,\alpha}(t,s)$ has singularities at $s = 0$ and $s = t$. These singularities magnify the large fluctuations in the noise $M_\alpha(ds)$ at $s \approx 0$ and $s \approx t$ and cause extremely large jumps in the paths of the process $X_{H,\alpha}(t)$. These jumps, however, are quickly balanced by other extremely large jumps with the opposite sign as the time t increases, which are due to *new* large fluctuations in the noise $M_\alpha(ds)$. In the limit, this effect leads to a process with everywhere discontinuous paths.

In the case of *long-range dependence*, the kernel function $f_{H,\alpha}(t,s)$ is bounded and it decays very slowly, as $s \to -\infty$. Thus large fluctuations in the noise terms $M_\alpha(ds)$ have *long-term effect* on the values of the process $X_{H,\alpha}(t)$. And because these fluctuations in the noise are not magnified (in contrast with the case of *negative dependence*), the paths of the LFSM process in this case can be made to be *continuous* with probability one.

2.2 FARIMA(p, d, q) with SαS Innovations

Fractional autoregressive moving average (FARIMA, in short) time series with SαS, $\alpha \in (1,2)$ innovations were studied in Kokoszka and Taqqu [14] as an infinite variance counterpart to the FARIMA model introduced by Granger and Joyeux [12]. These models, denoted FARIMA(p,d,q), $p,q \in \mathbb{N}$, extend the usual ARIMA(p,d,q) models, replacing the integer differencing exponent d with an arbitrary fractional real number. We suppose that

$$0 < d < 1 - 1/\alpha, \quad \text{and} \quad 1 < \alpha < 2. \tag{3}$$

A fractional ARIMA time series $Y = \{Y_k,\ k \in \mathbb{Z}\}$ with SαS innovations is defined as the stationary solution to the back-shift operator equation

$$\Phi_p(B)Y_k = \Theta_q(B)(I - B)^{-d} Z_k,\ k \in \mathbb{Z}, \tag{4}$$

where the innovations Z_ks are iid standard SαS random variables, I is the identity operator ($IY_k = Y_k$), B is the Backward operator ($BY_k := Y_{k-1}$), $(I-B)^{-d}$ is defined through the formal Taylor series expansion $(I-B)^{-d} = \sum_{k=0}^{\infty} \frac{\Gamma(d+k)}{k!\Gamma(d)} B^k$ and $\Phi_p(z)$, $\Theta_q(z)$ are real polynomials of degrees p, q, respectively, with roots outside the unit disk $\{z \in: |z| \le 1\}$. Relations (4) can be viewed as a set of equations in $Y = \{Y_k,\ k \in \mathbb{Z}\}$. The above assumptions and the condition (3)

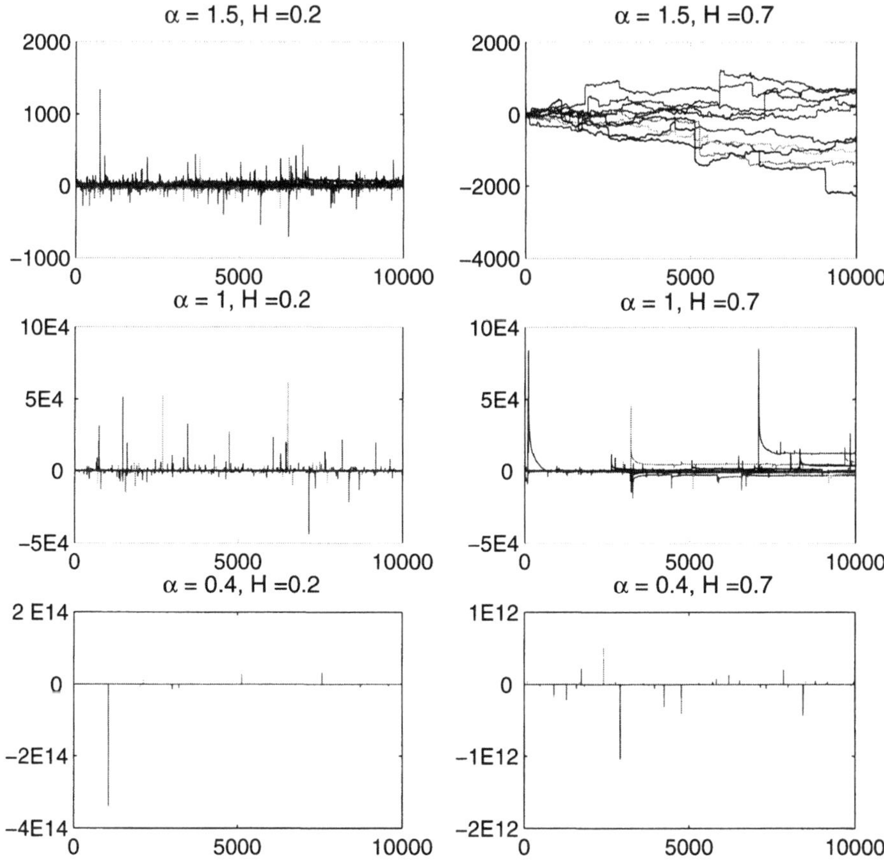

Fig. 1. Each graph displays ten independent simulated paths of LFSM of length 10 000, where the symbol $a\mathrm{E}b$ denotes the number $a \times 10^b$. To generate these paths we used the discrete sum approximation in (2) with parameters $M = 600$ and $m = 256$ (see the Remarks below). The vertical scale is different from one graph to the other. Observe the difference in path behavior between "high" and "low" values of α, that is, "heavy" and "very heavy" tails, respectively. (There are some extremely large fluctuations when the values of α are very low, e.g. $\alpha = 0.4$.) Also note the difference between *negative dependence* ($H < 1/\alpha$) and *long-range dependence* ($H > 1/\alpha$) cases. Because of the nature of the kernel of LFSM, in the cases of negative dependence, the jumps are quickly balanced by jumps with opposite signs, which results in visual spikes on the graphs. In the case of long-range dependence ($\alpha = 1.5$, $H = 0.7$), the paths are more dependent on the past and no significant spikes are observed.

guarantee the existence and uniqueness of the solution Y to (4). This solution, by Corollary 3.1 in Kokoszka and Taqqu [14]

$$Y_k = \sum_{j=0}^{\infty} c(j) Z_{k-j} = \sum_{j=-\infty}^{k} c(k-j) Z_j, \quad k \in \mathbb{Z}, \tag{5}$$

where the coefficients $c(j)$ satisfy the relation

$$\left|\frac{c_j}{j^{d-1}} - \frac{\Theta_q(1)}{\Phi_p(1)\Gamma(d)}\right| = O(j^{-1}), \quad \text{as } j \to \infty. \tag{6}$$

The polynomials Φ_p and Θ_q can be used to model a variety of short term dependence structures of the heavy-tailed time series $Y_k,\ k \in \mathbb{Z}$. For example, if $p = q = 1$, $\Phi_1(B) = I - \phi_1 B$, $\Theta_1(\theta) = I - \theta_1 B$, then (4) becomes

$$Y_k - \phi_1 Y_{k-1} = (I - B)^{-d}(Z_k - \theta_1 Z_{k-1})$$

(see Fig. 2). On the other hand, because of Relation (6), the parameter d governs the long-term behavior of the FARIMA model. In fact, since $0 < d < 1 - 1/\alpha$ and $c(j) \sim j^{d-1}$, as $j \to \infty$, the moving average coefficients $c(j)$ decay very slowly, so that $\sum_j |c_j| = \infty$. This indicates that the process Y is strongly dependent in time and the parameter $d > 0$ can be interpreted as the long-range dependence exponent of Y. In view of (7) and (8) below, the FARIMA model can be used as an alternative to the *long-range dependent* LFSM ($H - 1/\alpha > 0,\ 1 < \alpha < 2$) model when the time-series data is *not* exactly self-similar and has non-trivial *short-term dependence* behavior. In the rest of this section we discuss the relationship between these two models. (For more details and applications of FARIMA with heavy-tailed innovations see Kokoszka and Taqqu [15].)

2.3 Asymptotic Self-similarity. Connection Between LFSM and FARIMA

Consider a LFSM process $X_{H,\alpha}$ and a FARIMA(p,d,q) time series with SαS innovations Y defined by (1) and (5), respectively, where

$$H = d + 1/\alpha,\ 0 < d < 1 - 1/\alpha, \ \text{ and } \ 1 < \alpha < 2. \tag{7}$$

The parameter $H = d + 1/\alpha$ will be called the *Hurst parameter* of the FARIMA time series. This is because, as established by Astrauskas [4],

$$\left\{\frac{1}{C\lambda^H}\sum_{k=1}^{[\lambda t]} Y_k,\ t \ge 0\right\} \stackrel{f.d.d.}{\longrightarrow} \left\{X_{H,\alpha}(t),\ t \ge 0\right\}, \quad \text{as } \lambda \to \infty, \tag{8}$$

where $C > 0$ is a constant and $\stackrel{f.d.d.}{\longrightarrow}$ denotes convergence in finite dimensional distributions. In fact, the results of Astrauskas [4] include the case of heavy-tailed innovations, instead of merely stable ones (see e.g. Avram and Taqqu [5]).

In the SαS case, a stronger result than (8) can be established. Namely, for all $t \in \mathbb{R}$ let

$$\widetilde{X}(t) := \sum_{m\in\mathbb{Z}} K(t-m) \sum_{j\in(0,m]} Y_j,$$

where $\sum_{j\in(0,m]} Y_j$ means $-\sum_{j\in(m,0]} Y_j$ when $m < 0$ and $K(t),\ t \in \mathbb{R}$ is an integrable function with compact support which satisfies Relation (20) below.

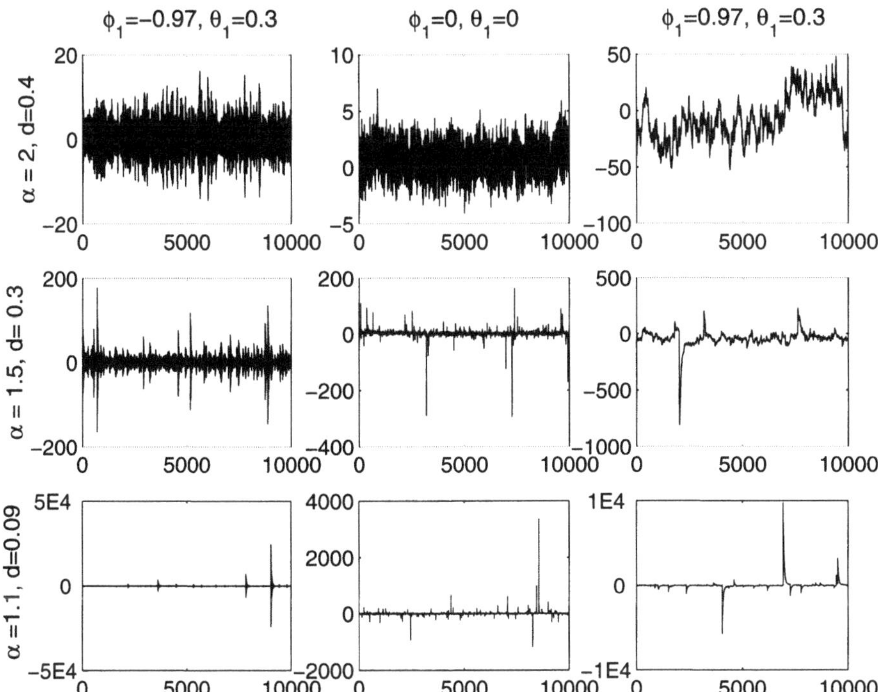

Fig. 2. Each graph displays one simulated path of FARIMA$(1, d, 1)$ with SαS innovations of length $N = 10\,000$, where $\Phi_1(z) = 1 - \phi_1 z$ and $\Theta_1(z) = 1 - \theta_1 z$. To generate these paths we used truncated moving averages: the first sum in (5) is taken from 0 to M with $M = 2^{20} - N$ (for more details, see Stoev and Taqqu [24]). The vertical scale is different from one graph to the other and the symbol aEb denotes the number $a \times 10^b$. The paths in the first column correspond to the so-called *weak short-range dependence* case and those in the last column to the *strong short-range dependence* case (see Sect. 6.6).

Then the processes Y and $X_{H,\alpha}$ can be defined on the same probability space, so that

$$\left\| \widetilde{X}(\lambda t)/C\lambda^H - X_{H,\alpha}(\lambda t)/\lambda^H \right\|_\alpha \leq \text{const}\,(1 \vee |t|)^H \lambda^{-d/(d+1)}, \quad \text{as } \lambda \to \infty, \quad (9)$$

(see Proposition 5.1 in Stoev and Taqqu [22]). The "norm" $\|\zeta\|_\alpha$ used in (9) stands for the *scale parameter* σ of the SαS random variable ξ, that is, instead of writing $\mathbb{E}e^{i\theta\xi} = e^{-\sigma^\alpha|\theta|^\alpha}$, one often writes $\mathbb{E}e^{i\theta\xi} = e^{-\|\xi\|_\alpha^\alpha|\theta|^\alpha}$ (see e.g. Relation (2.8.1) in Samorodnitsky and Taqqu [21]).

We note that (9) implies the convergence in (8). Indeed, let $K(t) = 1_{[0,1)}(t)$, then we have that $\widetilde{X}(t) = \sum_{k=1}^{[t]} Y_k$, $t \geq 0$. Now since $d > 0$, by (9), for all $t \geq 0$, we get $\widetilde{X}(\lambda t)/C\lambda^H - X_{H,\alpha}(\lambda t)/\lambda^H \xrightarrow{P} 0$, as $\lambda \to 0$, where $\xrightarrow{P}$ denotes convergence in probability.

Relation (9) shows that the continuous-time interpolation $\{\widetilde{X}(t),\ t \geq 0\}$ of the integrated FARIMA time series *and* the linear fractional stable motion process $\{X_{H,\alpha}(t),\ t \geq 0\}$ are *asymptotically equivalent*, as $t \to \infty$. This fact suggests interpreting the long-range dependent FARIMA(p,d,q) time series as an *asymptotically self-similar* process with asymptotic self-similarity parameter H. The normalization exponent $H = d + 1/\alpha$ is in fact the Hurst parameter of the time series $Y_k,\ k \in \mathbb{Z}$. We will indicate how one can use Relation (9) in Sect. 4 to obtain *consistent* and *asymptotically normal* estimators of the Hurst parameter H.

3 Wavelet Estimators for the Self-similarity Parameter *H* of LFSM

In this section we briefly review existing wavelet techniques for estimation of the self-similarity exponent H of the LFSM process $X_{H,\alpha}$ defined in (1) (for more details, see e.g. Pipiras, Taqqu and Abry [24] and the references therein).

Let $\psi \in L^2(\mathbb{R}),\ \|\psi\|_{L^2} = 1$ be a function with *compact support* and $Q \geq 1$ zero moments, that is,

$$\int_{\mathbb{R}} t^q \psi(t)dt = 0, \text{ for } q = 0, 1, \ldots, Q-1, \quad \int_{\mathbb{R}} t^Q \psi(t)dt \neq 0. \tag{10}$$

The function ψ will be called a wavelet. The random variables

$$D_{X_{H,\alpha}}(n,k) := \int_{\mathbb{R}} n^{-1}\psi(n^{-1}t - k)X_{H,\alpha}(t)dt, \quad n \in \mathbb{N},\ k \in \mathbb{Z} \tag{11}$$

are the wavelet coefficients of the LFSM process $X_{H,\alpha}$ at a scale n and location k. If $n = 2^j,\ j \in \mathbb{N}$ then the $2^{j/2}D_{X_{H,\alpha}}(2^j,k),\ j \in \mathbb{N},\ k \in \mathbb{Z}$, coincide with the so-called discrete wavelet transform coefficients of the signal $X_{H,\alpha}$ (see Daubechies [9]). Delbeke and Abry [10] have shown that for the case of LFSM, the integrals in (11) are finite with probability one, provided that the function ψ is *bounded* and

$$H - 1/\alpha + 1 > 0, \quad \alpha \in (0,2]. \tag{12}$$

They have also established that the wavelet coefficients $D_{X_{H,\alpha}}(n,k)$ have the following *scaling* and *stationarity* properties.

- *(stationarity)* For all $n_1, \ldots, n_m \in \mathbb{N}$, the process

$$\{(D_{X_{H,\alpha}}(n_1,k), \ldots, D_{X_{H,\alpha}}(n_m,k)),\ k \in \mathbb{Z}\}$$

is strictly stationary.

- *(scaling)* For all $c > 0$ and $n_1, \ldots, n_m \in \mathbb{N}$,

$$\begin{aligned}\{(D_{X_{H,\alpha}}(cn_1,k), \ldots, D_{X_{H,\alpha}}(cn_m,k))\}_{k\in\mathbb{Z}} &=_d \\ \{c^H(D_{X_{H,\alpha}}(n_1,k), \ldots, D_{X_{H,\alpha}}(n_m,k))\}_{k\in\mathbb{Z}}&\end{aligned} \tag{13}$$

For more details on the proofs of these properties, see Delbeke and Abry [10] or Corollaries 3.1 and 3.2 in Stoev and Taqqu [22]. In the sequel we shall assume that the wavelet ψ is bounded and that the condition (12) is satisfied, so that the wavelet coefficients $D(n,k)$ are well-defined. Observe that (12) always holds when $1 < \alpha \leq 2$. (For all $n \in \mathbb{N},\ k \in \mathbb{Z}$, the random variable $D(n,k)$ is *non-trivial* and SαS (see e.g. Lemma 4.1 in Stoev and Taqqu [22]).)

We now introduce two types of estimators of the self-similarity exponent H.

3.1 The "power" and "log" Estimators

Let $\{D(n_j,k),\ k = 1,\dots,N_j\},\ N_j \in \mathbb{N},\ j = 1,\dots,m$, be a collection of wavelet coefficients of the LFSM process $X_{H,\alpha}$, where $n_1 < n_2 < \dots < n_m$, is a set of *fixed* integer scales. Consider the statistics

$$Y_\beta(N_j) = \frac{1}{\beta}\log\Big(\frac{1}{N_j}\sum_{k=1}^{N_j}|D(n_j,k)|^\beta\Big), \qquad Y_{\log}(N_j) = \frac{1}{N_j}\sum_{k=1}^{N_j}\log|D(n_j,k)|, \tag{14}$$

where $0 < \beta < \alpha$. The stationary sequences of wavelet coefficients $\{D(n,k),\ k \in \mathbb{Z}\},\ n \in \mathbb{N}$ are ergodic (see e.g. Sect. 4 in Stoev and Taqqu [22]). Hence, as $N_j \to \infty$,

$$Y_\beta(N_j) \xrightarrow{a.s.} \beta^{-1}\log(\mathbb{E}|D(n_j,0)|^\beta) \text{ and } Y_{\log}(N_j) \xrightarrow{a.s.} \mathbb{E}\log|D(n_j,0)|,$$

where the expectations $\mathbb{E}|D(n,k)|^\beta$ and $\mathbb{E}\log|D(n,k)|$ are finite, because the random variables $D(n,k)$ are SαS and $0 < \beta < \alpha$. Now by the scaling properties of the $D(n,k)$, we have $D(n_j,0) =_d n_j^H D(1,0)$ and, as $N_j \to \infty$, we obtain

$$\begin{aligned} Y_\beta(N_j) &\xrightarrow{a.s.} H\log(n_j) + \beta^{-1}\log(\mathbb{E}|D(1,0)|^\beta), \\ Y_{\log}(N_j) &\xrightarrow{a.s.} H\log(n_j) + \mathbb{E}(\log|D(1,0)|), \end{aligned} \tag{15}$$

for all $j = 1,\dots,m$.

The Relations (15) suggest estimating H by using a weighted linear regression of $Y_\beta(N_j)$ (or $Y_{\log}(N_j)$), $j = 1,\dots,m$ against the vector $\log(n_j),\ j = 1,\dots,m$. Equivalently, we define the following two estimators of the parameter H

$$\begin{aligned} H_\beta(\{n_j\},\{N_j\}) &:= \sum_{j=1}^m w_j Y_\beta(N_j) \\ &\text{and} \\ H_{\log}(\{n_j\},\{N_j\}) &:= \sum_{j=1}^m w_j Y_{\log}(N_j), \end{aligned} \tag{16}$$

where the vector $(w_1,\dots,w_m) \in \mathbb{R}^m$ is such that

$$\sum_{j=1}^m w_j = 0 \quad \text{and} \quad \sum_{j=1}^m w_j\log(n_j) = 1. \tag{17}$$

We will call the estimators H_β and $H_{\log}$, the "power" and "log" estimators, respectively.

3.2 Asymptotic Properties of the "power" and "log" Estimators

The estimators H_β and $H_{\log}$ are *strongly consistent*, as $N \to \infty$, more precisely we have the following.

Proposition 1. *Assume that* $0 < \alpha \le 2$, $0 < \beta < \alpha$ *and that the condition* (12) *holds. Then, for any fixed collection of scales* $n_1 < n_2 < \cdots < n_m \in \mathbb{N}$, *we have that*

$$H_\beta \xrightarrow{a.s.} H \quad and \quad H_{\log} \xrightarrow{a.s.} H, \tag{18}$$

as $N_j \to \infty$, $j = 1, \ldots, m$, *where* H_β *and* $H_{\log}$ *are defined in* (16).

The proof of this result uses the ergodicity of the wavelet coefficients $\{D(n,k), k \in \mathbb{Z}\}$ of LFSM and follows directly from Relations (15), (16) and the fact that $\sum_{j=1}^m w_j(H\log(n_j) + \text{const}) = H$.

One also has *asymptotic normality* (see Theorem 3.1 in Pipiras, Taqqu and Abry [18]).

Theorem 1. *Let* $1 < \alpha < 2$, $0 < \beta < \alpha/2$ *and* $Q - H > 1/\alpha(1-\alpha)$. *Assume that* $0 < n_1 < n_2 < \cdots < n_m$ *is a collection of fixed integer scales and that the parameters* N_j, $j = 1, \ldots, m$ *are such that* $N_j \sim N/n_j$, *as* $N \to \infty$. *Then, as* $N \to \infty$,

$$\sqrt{N}(H_\beta - H) \xrightarrow{d} \mathcal{N}(0, \sigma_\beta^2) \quad and \quad \sqrt{N}(H_{\log} - H) \xrightarrow{d} \mathcal{N}(0, \sigma_{\log}^2).$$

4 Wavelet Estimators for the Hurst Exponent of a FARIMA Time Series

In this section we discuss two estimators of the Hurst parameter H of a FARIMA time series with SαS innovations. These estimators do not require knowledge of the stability exponent α and are based on the "power" and "log" wavelet estimators introduced in Sect. 3. In addition, when the parameter $\alpha \in (1,2]$ is given, we obtain estimators of the fractional differencing exponent d of the FARIMA time series, by using the fact that $d = H - 1/\alpha$. We start by defining a wavelet transformation of a FARIMA time series.

To analyze a discrete time series $X = \{X_m, k \in \mathbb{Z}\}$ by using wavelets, one should consider a continuous-time interpolation $\widetilde{X}(t)$ of the process X. In our context it is convenient to use the following construction,

$$\widetilde{X}(t) := \sum_{m \in \mathbb{Z}} X_m K(t-m), \tag{19}$$

where the function K is such that

$$\sum_{k \in \mathbb{Z}} K(t-m) = 1, \ a.e. \tag{20}$$

The interpolated process $\widetilde{X} = \{\widetilde{X}(t), t \in \mathbb{R}\}$ is well-defined provided that the function K has a compact support, for example. In practice, ψ is typically chosen to be a wavelet generating multiresolution analysis of $L^2(\mathbb{R})$. In such a case, one usually chooses K to be equal to a *scaling function* φ, corresponding to the wavelet ψ (see Sect. 5 below or Sect. 3 in Stoev and Taqqu [22]). For concreteness, in the rest of this section we assume that $K \equiv \varphi$ and ψ are Daubechies *scaling function* and *wavelet*, respectively. The following results, however, are valid in a greater generality (see Stoev and Taqqu [22]).

Consider the integrated time series $X_m = \sum_{k\in(0,m]} Y_k,\ m \in \mathbb{Z}$, of a FARIMA process $Y = \{Y_k,\ k \in \mathbb{Z}\}$ (Recall that for $m < 0$, the notation $\sum_{k\in(0,m]} Y_k$ means $-\sum_{k\in(m,0]} Y_k$.) Let $\widetilde{D}(n,k) := D_{\widetilde{X}}(n,k),\ n \in \mathbb{N},\ k \in \mathbb{Z}$ be the wavelet coefficients of the interpolated process $\widetilde{X}$, defined as in (11). The random variables $\widetilde{D}(n,k)$ will be called the *wavelet coefficients of the integrated FARIMA time series* $X_m,\ m \in \mathbb{Z}$. These wavelet coefficients are well-defined because the functions $\varphi, \psi \in L^2(\mathbb{R})$ have compact supports. We shall use these wavelet coefficients to estimate the Hurst exponent of the time series Y.

4.1 The Estimators $\widetilde{H}_\beta$ and $\widetilde{H}_{\log}$

The asymptotic equivalence in (9), implies that for all $a, n \in \mathbb{N},\ k \in \mathbb{Z}$,

$$\left\| \frac{\widetilde{D}(an,k)}{Ca^H} - \frac{D(an,k)}{a^H} \right\|_\alpha \le C(n) a^{-d/(d+1)}, \tag{21}$$

where $D(n,k)$ are the wavelet coefficients of a LFSM process $X_{H,\alpha}$, which is suitably defined on the same probability space (see Proposition 5.2 in Stoev and Taqqu [22]). The Relation (21) shows that the wavelet coefficients $\widetilde{D}(n,k)$ of the integrated FARIMA time series are *asymptotically equivalent* to the wavelet coefficients $D(n,k)$ of an lfsm process as the scale n goes to infinity. This fact and Relations (14), (15) and (16) suggest defining the following two estimators of the Hurst parameter $H = d + 1/\alpha$ of the FARIMA time series,

$$\widetilde{H}_\beta := \sum_{j=1}^m w_j \widetilde{Y}_\beta(a, N_j) \quad \text{and} \quad \widetilde{H}_{\log} := \sum_{j=1}^m w_j \widetilde{Y}_{\log}(a, N_j), \tag{22}$$

where $a \in \mathbb{N}$,

$$\begin{aligned} \widetilde{Y}_\beta(a, N_j) &:= \beta^{-1} \log\Big(N_j^{-1} \sum_{k=1}^{N_j} |\widetilde{D}(an_j,k)|^\beta\Big) \text{ and} \\ \widetilde{Y}_{\log}(a, N_j) &:= N_j^{-1} \sum_{k=1}^{N_j} \log|\widetilde{D}(an_j,k)|. \end{aligned}$$

In these expressions the *scales* $n_1 < n_2 < \cdots < n_m$ are fixed, $w_j,\ j = 1, \ldots, m$ satisfy Relation (17) and N_j denotes the number of wavelet coefficients available at the scale $an_j,\ j = 1, \ldots, m$, so that up to asymptotically negligible edge effects, we have

$$N_j \sim N/an_j, \quad \text{as } N/a \to \infty,$$

where N denotes the length of the available time series.

4.2 Asymptotic Properties of the Estimators $\widetilde{H}_\beta$ and $\widetilde{H}_{\log}$

The following result shows that the estimators $\widetilde{H}_\beta$ and $\widetilde{H}_{\log}$ are *weakly consistent.*

Proposition 2. *Assume that* $0 < d < 1 - 1/\alpha$, $1 < \alpha \le 2$, $0 < \beta < \alpha$ *and that the scales* $n_1 < n_2 < \cdots < n_m$ *are fixed. Then, as* $a \to \infty$ *and* $N_j \to \infty$, $j = 1, \ldots, m$, *we have*

$$\widetilde{H}_\beta = \widetilde{H}_\beta(a, \{N_j\}) \xrightarrow{P} H \quad \text{and} \quad \widetilde{H}_{\log} = \widetilde{H}_{\log}(a, \{N_j\}) \xrightarrow{P} H. \tag{23}$$

The proof of this result is given in Sect. 6.6.

Using the Relation (21) one can also establish the *asymptotic normality* of the estimators $\widetilde{H}_\beta$, $0 < \beta < \alpha/2$ and $\widetilde{H}_{\log}$ in (22) by using the asymptotic normality of the estimators H_β, $0 < \beta < \alpha/2$ and $H_{\log}$ for the parameter H of LFSM (see (16)). More precisely, we have the following result (see Theorem 5.1 in Stoev and Taqqu [22]).

Theorem 2. *Let* $1 < \alpha < 2$, $0 < \beta < \alpha/2$ *and* $Q - H > 1/\alpha(1-\alpha)$. *Assume that the integer scales* $n_1 < n_2 < \cdots < n_m$ *are fixed and suppose that* $a = a(N) \in \mathbb{N}$ *is such that, as* $N \to \infty$,

$$\frac{N}{a(N)} \to \infty \quad \text{and} \quad \frac{N^\delta}{a(N)} \to 0,$$

where $\delta \in (0,1)$. *If*

$$\delta \ge \frac{d+1}{d+1+2\beta d} \quad \left(\text{or } \delta > \frac{d+1}{2d+1}, \quad \text{resp.}\right)$$

then, as $N \to \infty$,

$$\sqrt{\frac{N}{a(N)}}(\widetilde{H}_\beta - H) \xrightarrow{d} \mathcal{N}(0, \sigma_\beta^2) \quad \left(\text{or} \quad \sqrt{\frac{N}{a(N)}}(\widetilde{H}_{\log} - H) \xrightarrow{d} \mathcal{N}(0, \sigma_{\log}^2), \text{ resp.}\right).$$

When the index of stability $\alpha \in (1,2)$ is given, then one can also define the estimators $\widetilde{d}_\beta := \widetilde{H}_\beta - 1/\alpha$ and $\widetilde{d}_{\log} := \widetilde{H}_{\log} - 1/\alpha$ of the fractional differencing exponent d. The consistency and asymptotic normality results for the estimators $\widetilde{H}_\beta$ and $\widetilde{H}_{\log}$ imply directly similar results for the estimators $\widetilde{d}_\beta$ and $\widetilde{d}_{\log}$, respectively (see e.g. Corollary 5.1 in Stoev and Taqqu [22]).

5 On Computing the Estimators in Practice

In this section we will discuss only the case of LFSM. The case of FARIMA is analogous. In practice, the wavelet coefficients $D(n,k) := D_{X_{H,\alpha}}(n,k)$ are usually not available. One typically has a discrete-time path $X(1), \ldots, X(N)$, $N \in \mathbb{N}$, of the process $X(t) := X_{H,\alpha}(t)$, $t \in \mathbb{R}$ and then the goal is to estimate the parameter H by using this dataset. The wavelet estimators defined in (16) cannot be computed from these data, because they involve the wavelet coefficients of the continuous-time process. However, in practice, the wavelet coefficients $D(n,k)$ can be approximated efficiently by using the Mallat's algorithm, also known as the fast discrete wavelet transform (FDWT, in short) (see e.g. Ch. 5 in Daubechies [9]). Moreover, the computational complexity of the FDWT algorithm is of the order $\mathcal{O}(N)$ and thus it can be used in applications involving *very large* datasets.

Assume that φ and ψ is a pair of a scaling function and a wavelet, which generate a multiresolution analysis of $L^2(\mathbb{R})$. Suppose also that the functions φ and ψ have compact supports, the wavelet ψ has at least one zero moment and that $K := \varphi$ satisfies (20). Examples of such functions φ and ψ are the Daubechies wavelets (see e.g. Stoev and Taqqu [22]). When the Mallat's FDWT algorithm is *initialized* with the data $X(1), \ldots, X(N)$, it computes coefficients

$$\widetilde{w}(j,k) = \int_{\mathbb{R}} 2^{-j/2}\psi(2^{-j}t-k)\widetilde{X}_{H,\alpha}(t)dt, \quad k=1,\ldots,N_j, \ j=1,\ldots,J, \tag{24}$$

where $\widetilde{X}_{H,\alpha}(t) := \sum_{m=1}^{N} X(m)\varphi(t-m)$ and $N_j \sim N/2^j$. Hence the coefficients $\widetilde{w}(j,k)$ equal exactly (up to asymptotically negligible *edge effects*) the discrete wavelet transform coefficients of the interpolated process $\widetilde{X}_{H,\alpha}$, defined in (19) with $K = \varphi$. Given our choice of normalization (compare (11) with (24)), we use instead of $\widetilde{w}(j,k)$, $\widetilde{D}(2^j,k) = 2^{-j/2}\widetilde{w}(j,k)$.

We will initialize the Mallat's FDWT algorithm with the data $X(1), \ldots, X(N)$ to obtain the coefficients $\widetilde{D}(2^j,k)$, $j \in \mathbb{N}$, $k \in \mathbb{Z}$, as indicated above, and use the $\widetilde{D}(2^j,k)$ as approximations to the wavelet coefficients $D(2^j,k)$ of the LFSM process $X_{H,\alpha}$ (see (11)). Using the *scaling property* of the coefficients $D(2^j,k)$ (13) and the *self-similarity* of the process $X_{H,\alpha}$, it can be shown that the relative approximation error vanishes, as $j \to \infty$, that is,

$$\left\| \frac{\widetilde{D}(2^j,k)}{2^{jH}} - \frac{D(2^j,k)}{2^{jH}} \right\|_{\alpha} \longrightarrow 0, \tag{25}$$

(see Proposition 2.4 in Stoev, Pipiras and Taqqu [22]).

We then compute the estimators $\widetilde{H}_{\beta}$ and $\widetilde{H}_{\log}$ of the Hurst parameter H by using Relation (16), where the coefficients $D(n_j,k)$ are replaced by $\widetilde{D}(2^{p_j},k)$, $k = 1,\ldots,N_j$ and the scales $n_j = 2^{p_j}$, $p_j \in \mathbb{N}$ are now *dyadic*. By using (25) one can prove that the estimators $\widetilde{H}_{\beta}$ and $\widetilde{H}_{\log}$ are *consistent* and *asymptotically normal* as the scales n_j, $j = 1,\ldots,m$ go to infinity together with the sample size N (for more details, see Stoev, Pipiras and Taqqu [24]).

In practice, however, we encounter two major problems: *(a)* how to choose the scales $n_1,\ldots,n_m$, *and (b)* how to choose the weights w_j, $j = 1,\ldots,m$ in (17), so that the estimators $\widetilde{H}_{\beta}$ and $\widetilde{H}_{\log}$ have the lowest *biases* and *variances*. The problem *(a)* involves the resolution of a bias-variance trade-off due the fact that the number of wavelet coefficients at the scale n decreases as n grows. We investigate this problem in Sect. 6 by using simulated data (see Figs. 6 and 9), however, we do not have a satisfactory general solution. On the other hand, problem *(b)* admits a simple practical solution, which we will now describe.

The estimators H_β and $H_{\log}$ in (16) can be equivalently obtained by using a generalized least squares regression of the vector $\mathbf{Y}_\beta = (Y_\beta(N_1) \ \cdots \ Y_\beta(N_m))'$ ($\mathbf{Y}_{\log}$, resp.) against the matrix $\mathbf{X} = (\mathbf{L}\ \mathbf{E})$, $\mathbf{L} = (\log n_1 \ \cdots \log n_m)'$, $\mathbf{E} = (1 \ \cdots \ 1)'$. This generalized least squares method involves the use of a $(m \times m)$ *error covariance matrix* $\mathbf{G}$. If the matrix $\mathbf{G}$ coincides with the covariance matrix $\mathbf{\Sigma}_\beta$ ($\mathbf{\Sigma}_{\log}$, resp.) of the vector $\mathbf{Y}_\beta$ ($\mathbf{Y}_{\log}$, resp.), then the corresponding estimator H_β ($H_{\log}$, resp.) will have the smallest variance among the class of estimators given by (16). Thus, in practice, problem *(b)* is reduced to the problem of choosing a matrix $\mathbf{G}$, which approximates well the covariance matrix $\mathbf{\Sigma}_\beta$ ($\mathbf{\Sigma}_{\log}$, resp.) (see Bardet [6], for the Gaussian case). If many independent replications of the LFSM process are available then $\mathbf{\Sigma}_\beta$ (or $\mathbf{\Sigma}_{\log}$, resp.) can be approximated, for example, by using the sample covariance matrix of the vectors $\mathbf{Y}_\beta$ (or $\mathbf{Y}_{\log}$, resp.) (see Sect. 6.1).

When an estimate of $\mathbf{\Sigma}_\beta$ (or $\mathbf{\Sigma}_{\log}$, resp.) is not available, one can still reduce the sample variance of the estimator by using the matrix $\mathbf{G} := \mathrm{diag}\{n_1, \ldots, n_m\}$. The intuition behind this choice is based on the fact that the wavelet coefficients $D(n_j, k)$, $k = 1, \ldots, N_j$ are *weakly dependent* and hence $Var(Y_\beta(N_j))/N \sim$ const n_j (or $Var(Y_{\log}(N_j))/N \sim$ const n_j, resp.) $j = 1, \ldots, m$, as $N \to \infty$. This choice of the matrix $\mathbf{G} = \mathrm{diag}\{n_1, \ldots, n_m\}$ is equivalent to the variance reduction method proposed by Abry and Veitch [1].

Remarks

1. In Sect. 6.1 we illustrate the importance of problem *(b)* by studying the variance of the estimators $\widetilde{H}_\beta$ and $\widetilde{H}_{\log}$ for three choices of the matrix $\mathbf{G}$ (see Figs. 3 and 4).
2. Observe that the discrete wavelet transform coefficients $\widetilde{D}(2^j, k)$ of the integrated FARIMA time series $X(k)$, $k \in \mathbb{Z}$, defined in Sect. 4 can be also obtained by using Mallat's algorithm. In this case however, the problem *(a)* of choosing the scales n_j is even more delicate (see Sect. 7 in Stoev and Taqqu [22] and below).

6 Computer Simulations

In this section we illustrate the behavior of the wavelet estimators defined in Sects. 3 and 4 over simulated paths of LFSM and FARIMA processes.

6.1 The Case of LFSM

We generate *independently* $n = 100$ paths $X(1), \ldots, X(N)$ of length $N = 2^{14} - 600 = 15\,784$ of an LFSM process $X_{H,\alpha}$ for several values of the self-similarity parameter $H \in (0,1)$ and the stability exponent $\alpha \in (0,2]$. We used the Riemann sum approximation of the integral in (1), proposed on p. 371 in Samorodnitsky and Taqqu [21] with discretization parameters $m = 128$ and $M = 600$ (see also the Remarks below).

In the sequel we study the *biases* and the *standard deviations* of the estimators $\widetilde{H}_\beta$ and $\widetilde{H}_{\log}$ as functions of the parameters: $\mathbf{G}$ (regression weight matrix); Q (number of zero moments of the wavelet ψ); $n_1, \dots, n_m$ (estimation scales) and β (the exponent of the estimator $\widetilde{H}_\beta$). (A preliminary analysis, which also includes the case of a finite-impulse response-based estimators, can be found in Stoev, Pipiras and Taqqu [24].) In all cases the biases and the standard deviations are computed from samples of $n = 100$ independent replication of the estimators. We now present four types of analyses. We shall refer to $\mathbf{Y}_\beta$ and $\mathbf{Y}_{\log}$ as $\mathbf{Y}_\kappa$, where $\kappa \in \{\beta, \log\}$.

6.2 The Choice of the Regression Weight Matrix G

We consider three cases for the matrix $\mathbf{G}$ (see Sect. 5), namely: *(i) the OLS case* (ordinary least squares regression) $\mathbf{G} = \mathbf{I}$, where $\mathbf{I}$ denotes the identity matrix; *(ii) the WLS case* (weighted least squares regression) $\mathbf{G} = \mathrm{diag}\{n_1, \dots, n_m\}$; *(iii) the GLS case* (generalized least squares regression) $\mathbf{G} = \widehat{\mathbf{\Sigma}}_\kappa$, where $\widehat{\mathbf{\Sigma}}_\kappa$ denotes the sample covariance matrix of the vectors $\mathbf{Y}_\kappa = (Y_\kappa(N_j))_{j=1,\dots,m}$, $\kappa \in \{\beta, \log\}$, computed from samples of $n = 100$ independent replications. The rest of the parameters are chosen as follows:

- *scales*: $(n_1, n_2, \dots, n_m) = (2^3, 2^4 \dots, 2^{11})$
- *type of the wavelet* ψ: Daubechies wavelet with $Q = 3$ zero moments.
- *the exponent* β: $\beta = 0.9 \times \alpha/2$.

As seen on Figs. 3 and 4, the behaviors of the biases of the estimators $\widetilde{H}_\beta$ and $\widetilde{H}_{\log}$, as functions of the parameter α, are qualitatively comparable in the three cases: OLS, WLS and GLS. The biases of the estimators in all cases (OLS, WLS and GLS) sharply increase, when the heavy tail index α decreases below 1. In fact, the estimators $\widetilde{H}_\beta$ and $\widetilde{H}_{\log}$ become essentially unusable for very low values of α.

When $1 < \alpha \leq 2$, however, the biases of the estimators remain low and comparable for all choices of the matrix $\mathbf{G}$ (OLS, WLS and GLS) and for all values of H that we consider. We believe that the precise value of the threshold corresponding to the sharp increase of the bias is $\alpha = 1/(H+1)$, because the wavelet coefficients of LFSM are not well-defined when $0 < \alpha \leq 1/(H+1)$ (see Delbeke and Abry [10]).

Figs. 3 and 4 show that the standard deviations of the estimators $\widetilde{H}_\beta$ and $\widetilde{H}_{\log}$ in the OLS case are always greater than the corresponding standard deviations in the WLS and GLS cases. This difference is significant and, in practice, one should use, for example, WLS or GLS instead of OLS regression weights. Furthermore, the standard deviations of the estimators $\widetilde{H}_\beta$ and $\widetilde{H}_{\log}$ in the GLS case are always smaller than these in the WLS case. However, the difference between these standard deviations is relatively small and it decreases as α increases. Thus, in practice, the WLS regression weights can be used successfully when a good estimate of the matrix $\widehat{\mathbf{\Sigma}}_\kappa$ is not available. Observe that in the

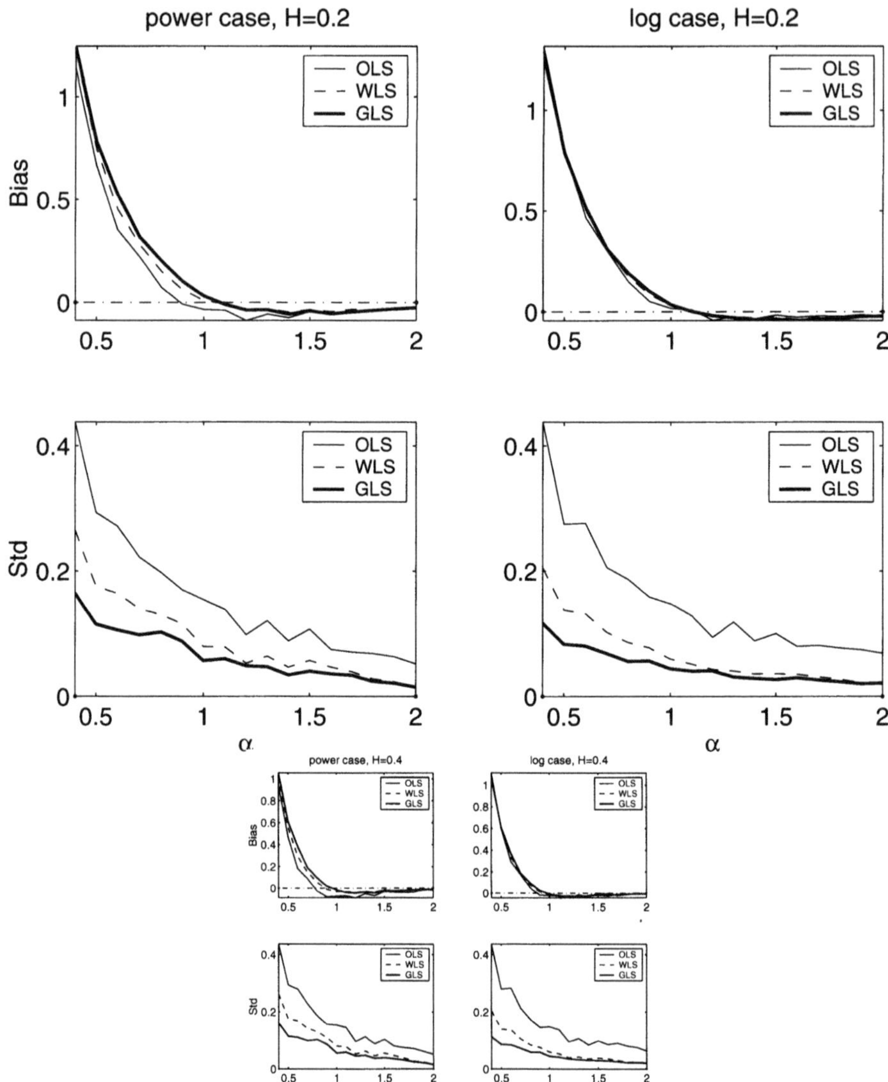

Fig. 3. Bias and standard deviations for samples of 100 *independent* estimates $\widetilde{H}_\beta$ (*power case*) and $\widetilde{H}_{\log}$ (*log case*) of the parameter H for LFSM process, where $H = 0.2,\ 0.4$ and $\alpha = 0.4, 0.5, \ldots, 1.9, 2$.

case of GLS, the estimate $\widehat{\boldsymbol{\Sigma}}_\kappa$ of the regression weight matrix involves multiple paths of the LFSM process, whereas in the WLS case, the estimators can be computed by using one path of the LFSM process.

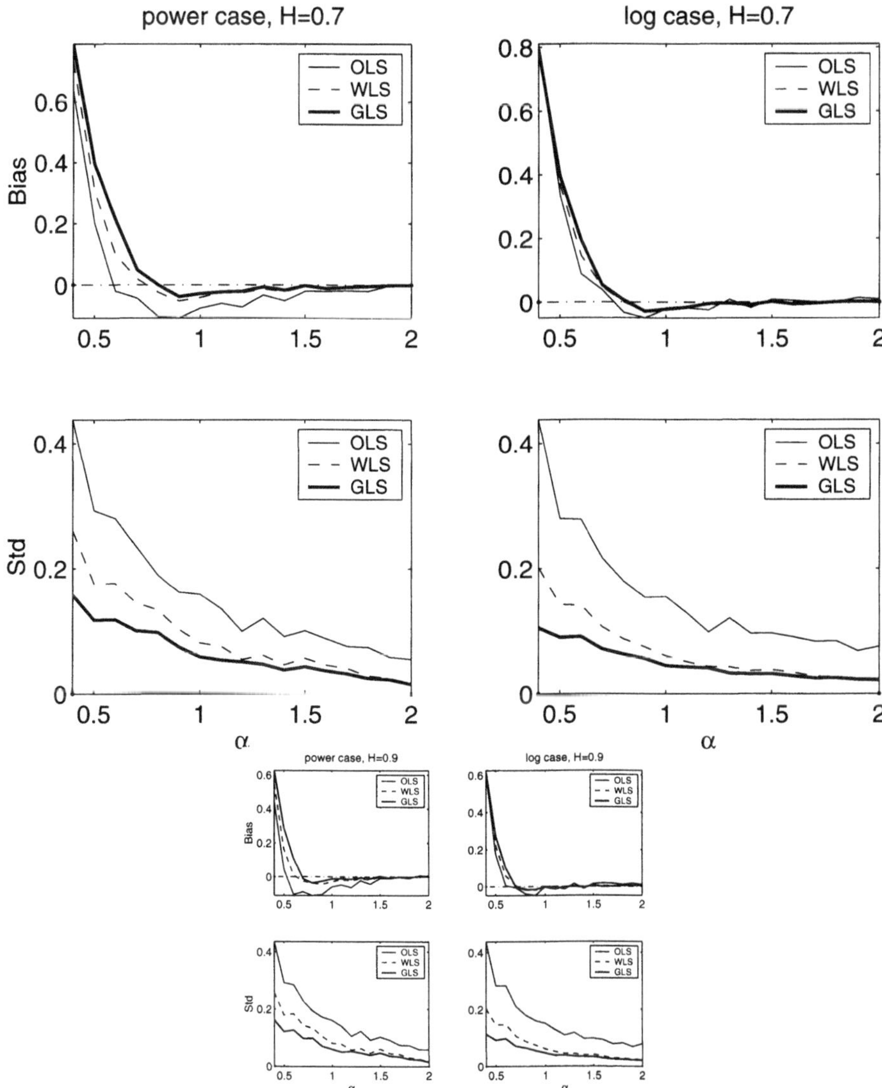

Fig. 4. Bias and standard deviations for samples of 100 *independent* estimates $\widetilde{H}_{\beta}$ (*power case*) and $\widetilde{H}_{\log}$ (*log case*) of the parameter H for LFSM process, where $H = 0.7,\ 0.9$ and $\alpha = 0.4, 0.5, \ldots, 1.9, 2$.

6.3 Number of Zero Moments of the Wavelet ψ

We focus only on Daubechies wavelets ψ, which have compact support, and study the behavior of our estimators for wavelets with different number Q of zero moments. As seen on Figs. 5, the biases of the estimators $\widetilde{H}_{\beta}$ and $\widetilde{H}_{\log}$ are essentially *independent* of the number of zero moments of the wavelet ψ. Furthermore, the standard deviations of the estimators for large values of Q are

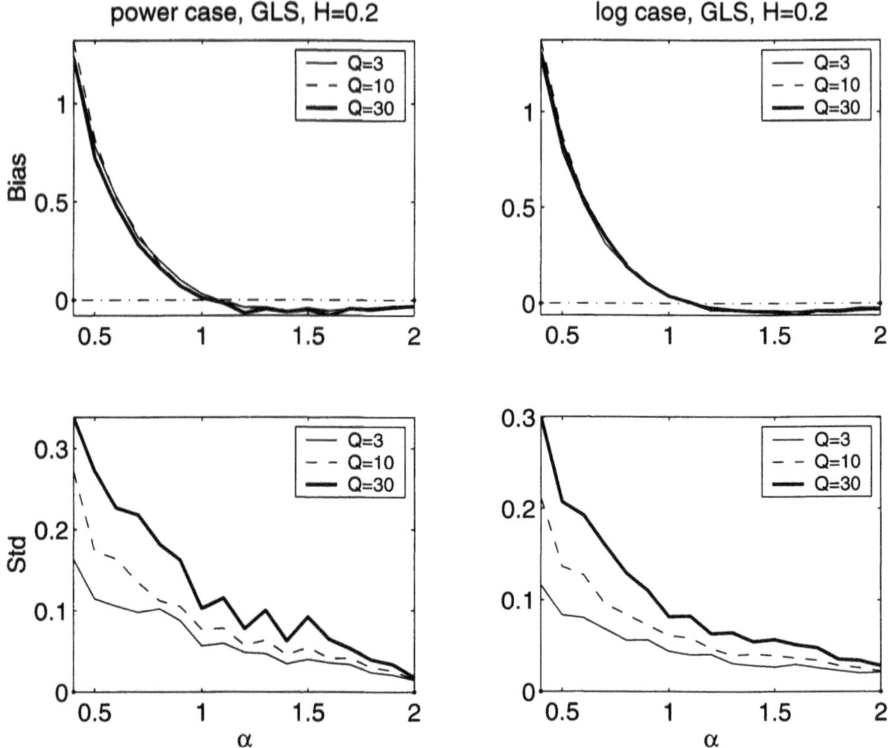

Fig. 5. Bias and standard deviations for samples of 100 *independent* estimates $\widetilde{H}_\beta$ (*power case*) and $\widetilde{H}_{\log}$ (*log case*) for LFSM with self-similarity parameter $H = 0.2$ and stability exponent $\alpha = 0.4,\ 0.5, \ldots, 1.9, 2$. All estimators are obtained by using *GLS* regression.

always greater than the corresponding standard deviations for small values of Q. This can be explained by the fact that for larger Q the support of the wavelet ψ is wider and hence there are fewer wavelet coefficients available on each estimation scale. Consequently, the statistics $\widetilde{Y}_\beta(\widetilde{N}_j)$ and $\widetilde{Y}_{\log}(\widetilde{N}_j)$, $j = 1, \ldots, m$ will have greater variance and hence the estimators $\widetilde{H}_\beta$ and $\widetilde{H}_{\log}$ will have greater standard deviations.

Remarks

1. In the simulation of the paths of LFSM, we chose the parameters m, M and $N = 15\,794$ so that $m(M + N)$ is a power of two, where N is the path-length. This choice allows us to use the Fast Fourier Transform algorithm to compute *efficiently* the involved Rieman sums (for more details, see Stoev and Taqqu [22]). Our choice was $m(M + N) = 2^{21}$. (On Fig. 1 we display only the first 10 000 observations of the path.)
2. The length of the simulated paths of LFSM ($N = 15\,784$) (used in this section) is not sufficient to be able to make reliable conclusions about the

asymptotic variances of the estimators $\widetilde{H}_\beta$ and $\widetilde{H}_{\log}$ as functions of Q. In practice the larger the number of zero moments Q, the larger the sample variance of the estimators.

3. Recall that the asymptotic normality results for the estimators $\widetilde{H}_\beta$ and $\widetilde{H}_{\log}$ required the condition $Q - H > 1/\alpha(\alpha - 1)$, where $\alpha > 1$. Based on our simulation results, we believe that this condition is merely *technical* and not essential in practice.

6.4 The Coice of Scales $n_1, \ldots, n_m$

Figure 6 contains boxplots of 100 replications of the estimators $\widetilde{H}_\beta$ and $\widetilde{H}_{\log}$ for $H = 0.7$, $\alpha = 0.5, 1, 1.5, 2$ and for several choices of the estimation scales $n_1, \ldots, n_m$. We chose dyadic scales $(n_1, n_2, \ldots, n_m) = (2^{j_1}, 2^{j_1+1}, \ldots, 2^{j_2})$, where $2^{j_2} = 2^{11}$ is held fixed and j_1 varies in the range $1, 2, \ldots, 8$.

As seen on Fig. 6, in the cases when $\alpha = 1, 1.5, 2$, the choices $j_1 = 1$ and $j_1 = 2$ lead to systematic biases of the estimators. This effect is related to the relatively large approximation error present at the smallest scales, due to the initialization of the Mallat's fast discrete wavelet transform algorithm. In principle, as the scale j_1 increases, the relative approximation error decreases and the estimators become less *biased* (see also (21), above, or Proposition 2.4 in Stoev, Pipiras and Taqqu [24]). On the other hand, when j_1 grows, fewer scales and wavelet coefficients are involved in the estimation, which leads to an increase of the *standard deviations* of the estimators. This bias-variance trade-off is illustrated on Figs. 6 (see also Fig. 9). One notices that the effect of the initialization error becomes negligible when the scale j_1 is greater than 2. Thus, in practice, for the case of LFSM, we use all available dyadic scales, with the exception of the smallest ones. As seen on Fig. 6, in the cases $\alpha = 0.5$, the bias-variance trade-off, due to the initialization error of the FDWT, is also present at small scales. In these cases, however, the estimators $\widetilde{H}_\beta$ and $\widetilde{H}_{\log}$ have also a large positive bias, independently of the choice of the scales (see also Figs. 3 and 4).

6.5 The Choice of the Exponent β

On Fig. 7 we display the bias of the *power case* estimators $\widetilde{H}_\beta$ when the exponent β varies in the interval $(0, \alpha)$. We also display the bias of the estimator $\widetilde{H}_{\log}$ (*log case*) and 95% confidence intervals for the bias of the estimators $\widetilde{H}_\beta$, which are computed by using normal approximation from a sample of 100 independent replications. We present only the case of WLS estimators, since the results in the case GLS are similar. Observe also that the corresponding confidence intervals in the WLS case are only slightly *wider* than these in the GLS case (see Figs. 3 and 4).

As seen on Fig. 7 in the infinite variance cases for the LFSM process ($\alpha = 0.5$, 1 and 1.5), the biases of the estimators $\widetilde{H}_\beta$ and $\widetilde{H}_{\log}$ are *comparable*, provided that $0 < \beta < \alpha_0$ for some threshold $\alpha_0 < \alpha$. These biases are relatively *low* in the

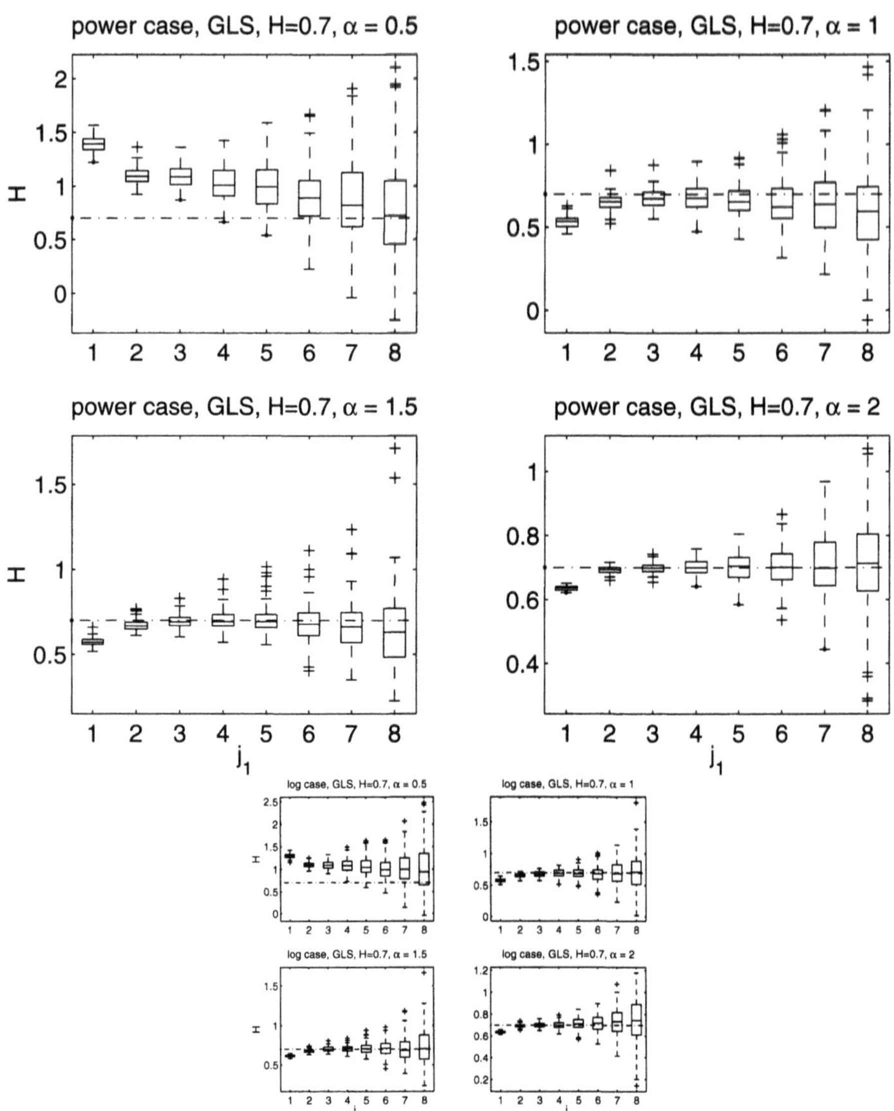

Fig. 6. Boxplots for samples of 100 *independent* estimates $\widetilde{H}_\beta$ (*power case*) and $\widetilde{H}_{\log}$ (*log case*) for the parameter H of LFSM with stability exponent α. All estimators are computed by using *GLS* regression and dyadic estimation scales $(n_1, n_2, \ldots, n_m) = (2^{j_1}, 2^{j_1+1}, \ldots, 2^{11})$, where $j_1 = 1, 2, \ldots, 8$.

cases $\alpha = 1$ and 1.5. On the other hand, the bias of the estimator $\widetilde{H}_\beta$ *sharply* deviates from the bias of the estimator $\widetilde{H}_{\log}$ when the exponent β becomes larger, in all infinite variance scenarios. We believe that the exact value of the threshold α_0 is $\alpha/2$, because the statistics $Y_\beta(N_j)$ have finite variances only when $0 < \beta < \alpha/2$ (see (14) and (16)).

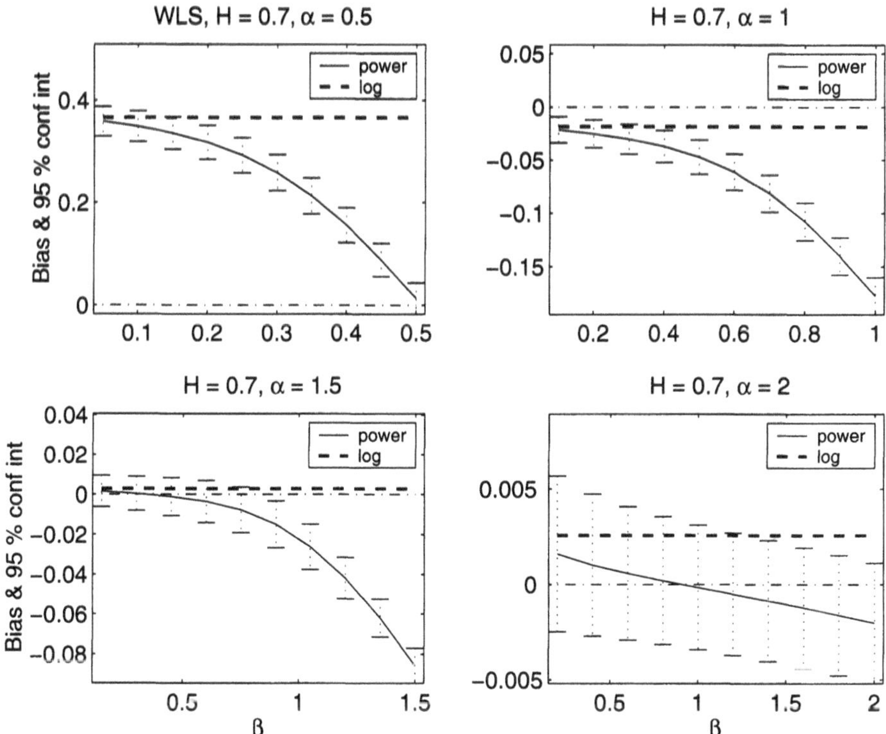

Fig. 7. Bias and 95% confidence intervals for the bias of the estimator $\widetilde{H}_\beta$ for the parameter H of LFSM with stability exponent α, as a function of the exponent β, $0 < \beta < \alpha$. The confidence intervals are based on normal approximation from samples of 100 *independent* estimates $\widetilde{H}_\beta$. The broken line represents the value of the estimator $\widetilde{H}_{\log}$. Observe that the position of the 0 and the units in the vertical scales differ from graph to graph.

In the Gaussian case ($\alpha = 2$) the biases and the standard deviations of both estimators $\widetilde{H}_\beta$ and $\widetilde{H}_{\log}$ are considerably lower than in the infinite variance cases of LFSM. Observe also that according to the corresponding confidence intervals the biases of $\widetilde{H}_\beta$ and $\widetilde{H}_{\log}$ are essentially equal, for most of the values of the exponent β.

6.6 The Case of FARIMA Time Series

In this section we illustrate the behavior of the wavelet estimators presented in Sect. 4 for the case of a long-range dependent FARIMA$(1, d, 1)$ time series with SαS innovations. We do this by generating $n = 100$ paths of $Y(1), \ldots, Y(N)$ (the study in Stoev and Taqqu [22] involved $n = 50$ paths). For simplicity we shall assume that the heavy-tail index α is *known* and we will focus on the estimators $\widetilde{d}_\beta = \widetilde{H}_\beta - 1/\alpha$ and $\widetilde{d}_{\log} = \widetilde{H}_{\log} - 1/\alpha$ of the fractional differencing exponent $d = H - 1/\alpha$ of the FARIMA time series. We study the behavior of these estimators when the values of the short-range dependence (srd, in short) parameters ϕ and θ, $\phi, \theta \in (-1,1)$ vary, where $\Phi_1(z) = 1-\phi z$ and $\Theta_1(z) = 1-\theta z$ (see (4)). We perform two types of analyses.

We first obtained the estimators $\widetilde{d}_\beta$ and $\widetilde{d}_{\log}$ using a fixed set of scales, for a range of values of the parameters ϕ and θ, in the cases of *intermediate* ($N = 10\,000$) and *long* ($N = 100\,000$) time series. Then, in a second analysis, we obtained $\widetilde{d}_\beta$ and $\widetilde{d}_{\log}$, with $(\phi, \theta) = (0.9, 0.1)$ for a set of increasing scales. As explained below, the value $(\phi, \theta) = (0.9, 0.1)$ corresponds to a "bad" situation which involves the presence of "strong short-range dependence" because $\phi = 0.9$ is very close to 1.

Experimenting with various choices of the regression weight matrix $\mathbf{G}$, the values of the exponent $\beta \in (0, \alpha/2)$ and Q (the number of zero moments of the wavelet ψ), we obtain results which are similar to those in the LFSM case (see Sect. 6.1 above). More precisely:

- The variance of the estimators $\widetilde{d}_\beta$ and $\widetilde{d}_{\log}$ is *significantly reduced* when WLS or GLS regression weights are used instead of OLS regression weights.
- The estimators $\widetilde{d}_\beta$ and $\widetilde{d}_{\log}$ are qualitatively comparable in terms of their biases and sample variances when $0 < \beta < \alpha/2$.
- The use of higher number of zero moments Q for the wavelet ψ does not improve significantly the biases of the estimators $\widetilde{d}_\beta$ and $\widetilde{d}_{\log}$ and in fact leads to an increase in the sample variances of the estimators.

Thus in the sequel we discuss only the estimators $\widetilde{d}_{\log}$, which are computed by using WLS regression weight matrix $\mathbf{G} = \mathrm{diag}\{n_1, \ldots, n_m\}$ and Daubechies wavelet ψ with $Q = 3$ zero moments. The estimation scales $n_1, n_2, \ldots, n_m$ are described in the captions of the figures.

As seen on Fig. 8, the estimator $\widetilde{d}_{\log}$ is essentially unbiased, for a wide range of values of ϕ and θ. Also, as N grows, the sample variance of the estimators improves rapidly (see the plot of the standard deviation on Fig. 8). However, in the cases when ϕ or θ are close to 1, the bias of the estimators is considerable and does not decrease as N increases. Moreover, the bias grows sharply as ϕ or θ approach 1. Thus, empirically, on Fig. 8 we can distinguish two regions for the parameters ϕ and θ, that is, when ϕ and θ are *away* from 1 and when ϕ or θ are *close* to 1. We refer to them as *weak short-range dependence* and *strong short-range dependence* regions.

Note that if θ and ϕ are not close to 1, but θ and/or ϕ are close to -1, the estimators are still good, even though the moduli of the roots of the polynomials $\Phi_1(z) = 1-\phi z$ and/or $\Theta_1(z) = 1-\theta z$ are close to 1. The same effect was encountered for a periodogram-type estimator in the case of a long-range dependent Gaussian FARIMA time series in Taqqu and Teverovsky [25] (see also Hurvich, Deo and Brodsky [13]).

Figure 9 illustrates the estimators in a *strong short-range dependence* case when the lowest scale $n_1 = 2^{j_1}$ increases. We note that the bias of the estimator $\widetilde{d}_{\log}$ decreases rapidly as j_1 grows. However, when $n_1 = 2^{j_1}$ is very large, fewer wavelet coefficients are available and consequently the estimators have greater

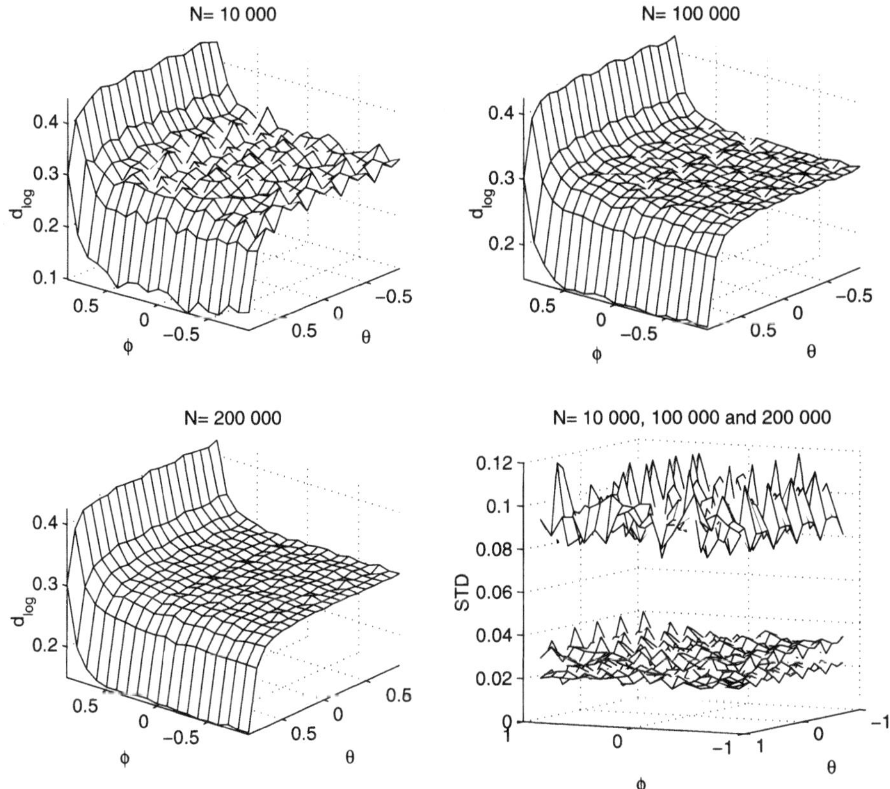

Fig. 8. Means and standard deviations for samples of 100 *independent* estimates $\widetilde{d}_{\log}$ for a FARIMA$(1, d, 1)$ series of length N, with $d = 0.3$, $\alpha = 1.5$, $\phi = -0.9, -0.8, \ldots, 0.8, 0.9$, $\theta = -0.9, -0.8, \ldots, 0.8, 0.9$, scales $n_j = 2^{4+j}$, $j = 1, \ldots, 10$ and ψ is a Daubechies wavelet with $Q = 3$ zero moments.

variances. On Fig. 9 we see that this bias-variance trade-off is rather sharp and that it is qualitatively the *same* for various values of the stability exponent α and the parameter d. This trade-off is well pronounced even in the Gaussian case ($\alpha = 2$). Moreover, the values j_1, which give the best estimators in terms of mean square error seem not to depend significantly on the parameters α and d, but mostly on the short-range dependence parameters ϕ and θ.

Note that in contrast to the case of LFSM the problem of the choice of *optimal* estimation scales does not depend only on the error due to the *initialization* of the Mallat's algorithm. It depends also on the amount of "short-range dependence" present, which affects the high frequencies (small j). In the case of a *strong short-range dependence* FARIMA$(1, d, 1)$, for example, one must discard a greater number of highest frequency scales (see Figs. 6 and 9).

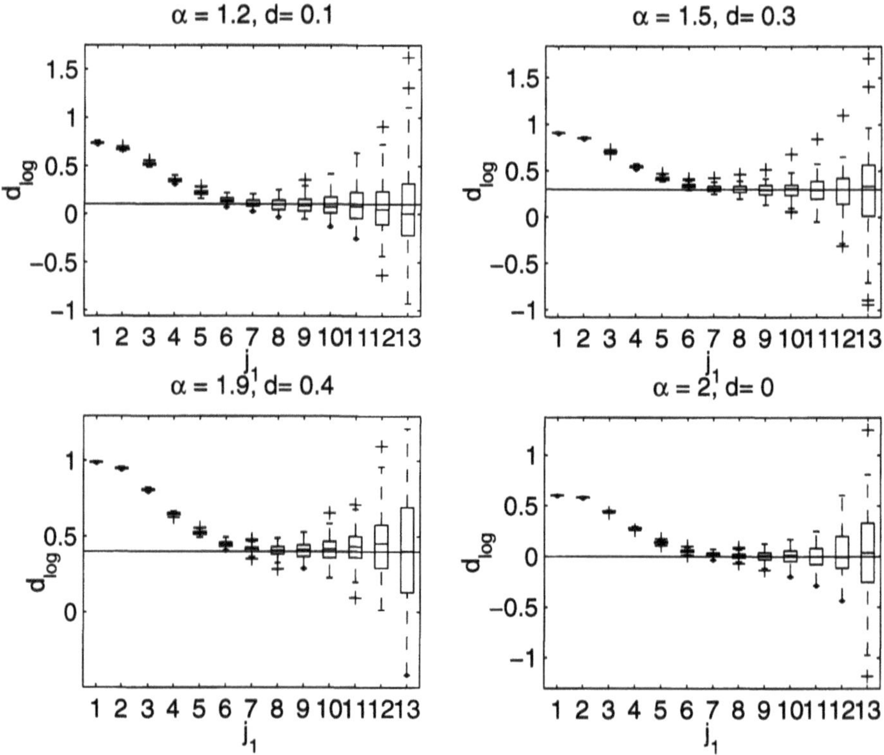

Fig. 9. Boxplots of samples of 100 *independent* estimates $\widetilde{d}_{\log}$ for a FARIMA$(1, d, 1)$ series of length $N = 2^{18}$, with $\phi = 0.9$ and $\theta = 0.1$ (*strong short-range dependence* scenario), scales $n_k = 2^{j_1}2^{k-1}$, $j_1 = 1, \dots, 13$, $k = 1, \dots, 18 - j_1$, (all available dyadic scales from 2^{j_1} to 2^{18} are used), and ψ is a Daubechies wavelet with $Q = 3$ zero moments.

Appendix

Proof of Proposition 2: We first focus on the "power" estimator $\widetilde{H}_\beta$ in the case $0 < \beta \leq 1$. By (17), $\widetilde{H}_\beta$ can be expressed as

$$\widetilde{H}_\beta(a, \{N_j\}) = \sum_{j=1}^{m} w_j \widetilde{Y}_\beta^0(j), \tag{26}$$

where

$$\widetilde{Y}_\beta^0(j) := \frac{1}{\beta} \log\Big(\frac{1}{N_j}\sum_{k=1}^{N_j}\Big|\frac{\widetilde{D}(an_j, k)}{Ca^H}\Big|^\beta\Big).$$

Let now

$$\xi(N_j, a) := \frac{1}{N_j}\sum_{k=1}^{N_j}\Big|\frac{\widetilde{D}(an_j, k)}{Ca^H}\Big|^\beta,$$

so that $\widetilde{Y}^0_\beta(j) = \beta^{-1}\log(\xi(N_j,a))$, and

$$\eta(N_j,a) := \frac{1}{N_j}\sum_{k=1}^{N_j}\Big|\frac{D(an_j,k)}{a^H}\Big|^\beta,$$

where the $D(n,k)$s denote the wavelet coefficients of an LFSM process, involved in (21). By the scaling property (13), the stationarity and the ergodicity of the wavelet coefficients $D(n,k),\ k\in\mathbb{Z}$, we have that for all $j=1,\dots,m$,

$$\eta(N_j,a) =_d \frac{1}{N_j}\sum_{k=1}^{N_j}|D(n_j,k)|^\beta \xrightarrow{a.s.} \mathbb{E}|D(n_j,0)|^\beta, \tag{27}$$

as $N_j\to\infty$ (see also (15)).

Now by (21) and (30), using the triangle inequality, we obtain that

$$\begin{aligned}\mathbb{E}|\xi(N_j,a)-\eta(N_j,a)| &\le \frac{1}{N_j}\sum_{k=1}^{N_j}\mathbb{E}\left|\Big|\frac{\widetilde{D}(an_j,k)}{ca^H}\Big|^\beta - \Big|\frac{D(an_j,k)}{a^H}\Big|^\beta\right| \\ &\le \frac{C_{\alpha,\beta}}{N_j}\sum_{k=1}^{N_j}\Big\|\frac{\widetilde{D}(an_j,k)}{ca^H} - \frac{D(an_j,k)}{a^H}\Big\|_\alpha^\beta \qquad (28)\\ &\le \text{const}\, a^{-d/(d+1)}, \qquad (29)\end{aligned}$$

where $C_{\alpha,\beta}$ is a constant (see e.g. Relation (2.8.1) in Samorodnitsky and Taqqu [21]). Thus, as $a\to\infty$, $\xi(N_j,a)-\eta(N_j,a)\xrightarrow{P}0$. By (27), we also have that $\eta(N_j,a)\xrightarrow{P}\mathbb{E}|D(n_j,0)|^\beta$, where $\mathbb{E}|D(n_j,0)|^\beta$ is a positive constant (see Lemma 4.1 in Stoev and Taqqu [22]). The last two convergences imply that $\xi(N_j,a)\xrightarrow{P}\mathbb{E}|D(n_j,0)|^\beta>0$, as $a\to\infty$, $N_j\to\infty$. Hence by the continuity of the log finction, using Slutsky's Theorem (see Corollary 5.2 in Billingsley [7]), we obtain that

$$\widetilde{Y}^0_\beta(j) = \frac{1}{\beta}\log(\xi(N_j,a)) \xrightarrow{P} \frac{1}{\beta}\log(\mathbb{E}|D(n_j,0)|^\beta) = H\log(n_j)+\text{const}\,,$$

as $a\to\infty$, $N_j\to\infty$. In view of Relation (26), the last convergence implies (23).

In the case of the $\widetilde{H}_\beta$ estimator when $1<\beta<\alpha$ *and* in the case of $\widetilde{H}_{\log}$ estimator we use the inequalities (31) and (32), respectively, given in Lemma 1, to obtain bounds on the term $\mathbb{E}|\xi(N_j,a)-\eta(N_j,a)|$, as in (28). The rest of the proof in these two cases remains the same. □

The following lemma was used in the proof of the preceding proposition.

Lemma 1. *Let $0<\beta<\alpha$, $\alpha\in(0,2]$. Let also ξ and η be non-trivial jointly $S\alpha S$ random variables. Then the following inequalities hold.*

(a) If $0<\beta\le1$, then

$$\mathbb{E}||\xi|^\beta-|\eta|^\beta| \le C_{\alpha,\beta}\|\xi-\eta\|_\alpha^\beta. \tag{30}$$

(b) If $1 < \beta < \alpha$, *then*

$$\mathbb{E}\big||\xi|^\beta - |\eta|^\beta\big| \le C_{\alpha,\beta}\Big(\|\xi\|_\alpha^{\beta-1}\|\xi-\eta\|_\alpha + \|\eta\|_\alpha\|\xi-\eta\|_\alpha^{\beta-1}\Big). \tag{31}$$

(c) For all ϵ, $0 < \epsilon < (1 \wedge \alpha)/2$,

$$\mathbb{E}\big|\log|\xi| - \log|\eta|\big| \le C_\alpha(\epsilon)\|\xi-\eta\|_\alpha^\epsilon/(\|\xi\|_\alpha \wedge \|\eta\|_\alpha)^\epsilon, \tag{32}$$

where $x \wedge y := \min\{x,y\}$. *The constants* $C_{\alpha,\beta}$ *and* $C_\alpha(\epsilon)$ *are positive and depend only on* (α,β) *and* (α,ϵ), *respectively.*

Proof. Part *(a)* follows directly from the inequality $\big||x|^\beta - |y|^\beta\big| \le |x-y|^\beta$, valid for all $x, y \in \mathbb{R}$, $0 < \beta \le 1$ by using the fact that $\mathbb{E}|\xi|^\beta = C_{\alpha,\beta}\|\xi\|_\alpha^\beta$, for any S$\alpha$S r.v. ξ and $\beta \in (0,\alpha)$.

For part *(b)*, by adding and subtracting the term $|\xi|^{\beta-1}|\eta|$ and using that $0 < \beta - 1 < 1$, we obtain

$$\begin{aligned}\mathbb{E}\big||\xi|^\beta - |\eta|^\beta\big| &\le \mathbb{E}|\xi|^{\beta-1}\Big||\xi| - |\eta|\Big| + \mathbb{E}|\eta|\Big||\xi|^{\beta-1} - |\eta|^{\beta-1}\Big| \\ &\le \mathbb{E}|\xi|^{\beta-1}|\xi-\eta| + \mathbb{E}|\eta||\xi-\eta|^{\beta-1}.\end{aligned}$$

Now using the Hölder inequality we bound above the last expression by

$$(\mathbb{E}|\xi|^\beta)^{(\beta-1)/\beta}(\mathbb{E}|\xi-\eta|^\beta)^{1/\beta} + (\mathbb{E}|\eta|^\beta)^{1/\beta}(\mathbb{E}|\xi-\eta|^\beta)^{(\beta-1)/\beta},$$

which equals the right-hand side of (31).

For the proof of part *(c)*, we refer to Lemma 6.6 in Stoev, Pipiras and Taqqu [24]. □

References

1. P. Abry and D. Veitch: IEEE Transactions on Information Theory **44**, 2 (1998)
2. P. Abry, L. Delbeke and P. Flandrin: IEEE International Conference on Acoustics, Speech and Signal Processing **III**, 1581 (1999)
3. R. Adler, R. Feldman, and M. S. Taqqu: *A Practical Guide to Heavy Tails: Statistical Techniques and Applications* (Birkhäuser, Boston 1998)
4. A. Astrauskas: Lithuanian Mathematical Journal **23**, 127 (1983)
5. F. Avram and M. S. Taqqu: *Weak convergence of moving averages with infinite variance* Dependence in Probability and Statistics, 399 (1986) E. Eberlein and M. S. Taqqu(eds): Boston, Birkhäuser
6. J. M. Bardet: Journal of Time Series Analysis **21**, 497 (2000)
7. P. Billingsley: *Convergence of Probability Measures* (Wiley, New York 1968)
8. R. Dahlhaus: The Annals of Statistics **17**, 1749 (1989)
9. I. Daubechies: *Ten Lectures on Wavelets* SIAM Philadelphia. CBMS-NSF series, Vol **61**, (1992)
10. L. Delbeke and P. Abry: Stochastic Processes and their Applications **86**, 177 (2000)
11. R. Fox and M. S. Taqqu: The Annals of Statistics **14**, 517 (1986)
12. C. W. J. Granger and R. Joyeux: Journal of Time Series Analysis **1**, 15 (1980)

13. C. Hurvich, R. Deo and J. Brodsky: Journal of Time Series Analysis **19**, 19 (1998)
14. P. S. Kokoszka and M. S. Taqqu: Stochastic Processes and their Applications **60**, 19 (1995)
15. P. S. Kokoszka and M. S. Taqqu: Journal of Econometrics **73**, 79 (1996)
16. W. E. Leland, M. S. Taqqu, W. Willinger and D. V. Wilson: *Ethernet traffic is self-similar: stochastic modelling of packet traffic data* (Preprint 1993)
17. K. Park and W. Willinger: *Self-Similar Network Traffic and Performance Evaluation* (J. Wiley and Sons, Inc., New York, eds 2000)
18. V. Pipiras, M. S. Taqqu and P. Abry: *Asymptotic normality for wavelet-based estimators of fractional stable motion* (Preprint 2001)
19. P. M. Robinson: The Annals of Statistics **22**, 515 (1994)
20. P. M. Robinson: The Annals of Statistics **23**, 1048 (1995)
21. G. Samorodnitsky and M. S. Taqqu: *Stable Non-Gaussian Processes: Stochastic Models with Infinite Variance* (Chapman and Hall, New York, London 1994)
22. S. Stoev and M. S. Taqqu: *Asymptotic self-similarity and wavelet estimation for long-range dependent FARIMA time series with stable innovations* (Preprint 2002*a*)
23. S. Stoev and M. S. Taqqu: *Simulation methods for linear fractional stable motion and FARIMA using the Fast Fourier Transform* (Preprint 2002*b*)
24. S. Stoev, V. Pipiras, and M. S. Taqqu: *Estimation of the self-similarity parameter in linear fractional stable motion* Signal Processing (2002) To appear.
25. M. S. Taqqu and V. Teverovsky: 'Semi-parametric graphical estimation techniques for long-memory data'. In: *Athens Conference on Applied Probability and Time Series Analysis*, ed. by P. M. Robinson and M. Rosenblatt, Volume II: Time Series Analysis in Memory of E. J. Hannan', Vol. 115 of *Lecture Notes in Statistics*, (Springer-Verlag, New York 1996) pp.420–432

From Stationarity to Self-similarity, and Back: Variations on the Lamperti Transformation

Patrick Flandrin[1], Pierre Borgnat[1], and Pierre-Olivier Amblard[2]

[1] Laboratoire de Physique (UMR 5672 CNRS), ENS Lyon, 46 allée d'Italie, 69364 Lyon Cedex 07, France
[2] Laboratoire des Images et des Signaux (UMR 5083 CNRS), ENSIEG, BP 46, 38402 Saint Martin d'Hères Cedex, France

Abstract. The Lamperti transformation defines a one-to-one correspondence between stationary processes on the real line and self-similar processes on the real half-line. Although dating back to 1962, this fundamental result has further received little attention until a recent past, and it is the purpose of this chapter to survey the Lamperti transformation and its (effective and/or potential) applications, with emphasis on variations which can be made on the initial formulation. After having recalled basics of the transform itself, some results from the literature will be reviewed, which can be broadly classified in two types. In a first category, classical concepts from stationary processes and linear filtering theory, such as linear time-invariant systems or ARMA modeling, can be given self-similar counterparts by a proper "lampertization" whereas, in a second category, problems such as spectral analysis or prediction of self-similar processes can be addressed with classical tools after stationarization by a converse "delampertization". Variations and new results will then be discussed by investigating consequences of the Lamperti transformation when applied to weakened forms of stationarity, and hence of self-similarity. Different forms of locally stationary processes will be considered this way, as well as cyclostationary processes for which "lampertization" will be shown to offer a suitable framework for defining a stochastic extension to the notion of *discrete scale invariance* which has recently been put forward as a central concept in many critical systems. Issues concerning the practical analysis (and synthesis) of such processes will be examined, with a possible use of Mellin-based tools operating directly in the space of scaling data.

1 Introduction

In a seminal paper published in 1962 [19], J.W. Lamperti introduced key concepts related to what is now referred to as *self-similar processes*. Among other important results, he first pointed out the one-to-one connection which exists between self-similar processes and stationary processes, *via* a transform which essentially consists in a proper warping of the time axis. This result has often been quoted in the literature (e.g., in [3], [30] or [34]), but rarely used and even discussed *per se*. Notable exceptions are the contributions of Burnecki et al. [8] who proved unicity, and of Nuzman and Poor [25,26] who explicitly (and extensively) took profit of it for linear estimation issues concerning fractional Brownian motion (fBm).

The transform that Lamperti initially pushed forward in 1962 has, since then, been rediscovered from time to time, under different forms. For instance, Gray and Zhang re-established in [17] a weakened form of Lamperti's theorem, upon which they based a discussion on specific classes of self-similar processes, referred to as *multiplicative stationary processes*. From a very close (yet independent) perspective, Yazici and Kashyap advocated in [37] the use of a transform – which indeed identifies to Lamperti's – for constructing related classes of self-similar processes referred to as *scale stationary processes*, a concept which had also been briefly discussed and commented in [12]. More recently, Vidács and Virtamo proposed in [35,36] an original ML estimation scheme for fBm parameters, which basically relies on a geometrical sampling of the data, i.e., on a pre-processing guaranteeing a stationarization in the spirit of the Lamperti approach.

Recognizing both the importance of the Lamperti transform and the sparsity of its coverage in the literature, the purpose of this text is to offer a guided tour of existing material in a unified form, and also to discuss new extensions. More precisely, the text is organized as follows. In Sect. 2, basics of stationarity and self-similarity are first recalled, and the Lamperti transform is introduced. The ability of this transform to put self-similar and stationary processes in a one-to-one correspondence is then proved, and a number of consequences are detailed, with respect to covariances, spectra, long-range dependence and scale-covariant generating systems for self-similar processes. Some examples and applications are dealt with in Sect. 3, including either stationary processes (random phase tones, Ornstein-Uhlenbeck, ARMA) and their self-similar counterparts, or self-similar processes (fractional Brownian motion, Euler-Cauchy) and their stationary counterparts. Sect. 4 is then devoted to variations on the original approach, obtained by applying the Lamperti transform to weakened forms of stationarity or self-similarity. Following a brief introduction of relevant concepts such as multiplicative harmonizability or scale-invariant Wigner spectra, special emphasis is put on the newly introduced notion of *stochastic discrete scale invariance* which is shown to be the Lamperti image of cyclostationarity.

2 The Lamperti Transformation

2.1 Stationarity and Self-similarity

The notion of stationarity is basic in the study of many stochastic processes. Heuristically, the idea of stationarity is equivalent to that of statistical invariance under time shifts, and this concept has proven most useful in many steady-state applications. From a different perspective, scale invariance (or self-similarity) is also ubiquitous in many natural and man-made phenomena (landscape texture, turbulence, network traffic, ...). The underlying idea is in this case that a function is scale invariant if it is identical to any of its rescaled versions, up to some suitable renormalization in amplitude.

To make these ideas more precise, let us first introduce two basic operations.

Definition 1. *Given some number $\tau \in \mathbb{R}$, the shift operator $\mathcal{S}_\tau$ operates on processes $\{Y(t), t \in \mathbb{R}\}$ according to:*

$$(\mathcal{S}_\tau Y)(t) := Y(t+\tau). \tag{1}$$

Definition 2. *Given some numbers $H > 0$ and $\lambda > 0$, the renormalized dilation operator $\mathcal{D}_{H,\lambda}$ operates on processes $\{X(t), t > 0\}$ according to:*

$$(\mathcal{D}_{H,\lambda}X)(t) := \lambda^{-H}\, X(\lambda t). \tag{2}$$

Using these operators in the context of stochastic processes, and introducing the notation "$\stackrel{d}{=}$" for equality of all finite-dimensional distributions, the definitions of stationarity and self-similarity follow as:

Definition 3. *A process $\{Y(t), t \in \mathbb{R}\}$ is said to be stationary if*

$$\{(\mathcal{S}_\tau Y)(t), t \in \mathbb{R}\} \stackrel{d}{=} \{Y(t), t \in \mathbb{R}\} \tag{3}$$

for any $\tau \in \mathbb{R}$.

Definition 4. *A process $\{X(t), t > 0\}$ is said to be self-similar of index H (or "H-ss") if*

$$\{(\mathcal{D}_{H,\lambda}X)(t), t > 0\} \stackrel{d}{=} \{X(t), t > 0\} \tag{4}$$

for any $\lambda > 0$.

Such an equality holds in the usual sense for homogeneous functions proportional to $t^H, t > 0$, and it is useful to remark that, whenever $\{X(t), t > 0\}$ is H-ss, then the modulated process $\{X_{H'}(t), t > 0\}$ such that

$$X_{H'}(t) := t^{H'}\, X(t) \tag{5}$$

is $(H + H')$-ss.

2.2 The Transform

Definition 5. *Given some number $H > 0$, the Lamperti transform $\mathcal{L}_H$ operates on processes $\{Y(t), t \in \mathbb{R}\}$ according to:*

$$(\mathcal{L}_H Y)(t) := t^H\, Y(\log t), t > 0, \tag{6}$$

and the corresponding inverse Lamperti transform $\mathcal{L}_H^{-1}$ operates on processes $\{X(t), t > 0\}$ according to:

$$(\mathcal{L}_H^{-1} X)(t) := e^{-Ht}\, X(e^t), t \in \mathbb{R}. \tag{7}$$

The Lamperti transform is invertible, which guarantees that $(\mathcal{L}_H^{-1}\mathcal{L}_H Y)(t) = Y(t)$ for any process $\{Y(t), t \in \mathbb{R}\}$, and $(\mathcal{L}_H \mathcal{L}_H^{-1} X)(t) = X(t)$ for any process $\{X(t), t > 0\}$. We can however remark that, given two different parameters H_1 and H_2, we only have

$$(\mathcal{L}_{H_2}^{-1}\mathcal{L}_{H_1} Y)(t) = e^{-(H_2-H_1)t}\, Y(t), \tag{8}$$

and, in a similar way, it is immediate to establish that

$$(\mathcal{L}_{H_2}\mathcal{L}_{H_1}^{-1} X)(t) = t^{H_2-H_1}\, X(t). \tag{9}$$

2.3 From Stationarity to Self-similarity, and Back

Lemma 1. *The Lamperti transform (6)-(7) guarantees an equivalence between the shift operator (1) and the renormalized dilation operator (2) in the sense that, for any* $\lambda > 0$*:*

$$\mathcal{L}_H^{-1}\mathcal{D}_{H,\lambda}\mathcal{L}_H = \mathcal{S}_{\log\lambda}. \tag{10}$$

Proof – Assuming that $\{Y(t), t \in \mathbb{R}\}$ is stationary and using the Definitions 1, 2 and 5, we may write

$$\begin{aligned}(\mathcal{L}_H^{-1}\mathcal{D}_{H,\lambda}\mathcal{L}_H Y)(t) &= (\mathcal{L}_H^{-1}\mathcal{D}_{H,\lambda})(t^H\, Y(\log t)) \\ &= \mathcal{L}_H^{-1}(\lambda^{-H}(\lambda t)^H Y(\log \lambda t)) \\ &= e^{-Ht}(s^H Y(\log \lambda s))_{s=e^t} \\ &= Y(t + \log\lambda) \\ &= (\mathcal{S}_{\log\lambda} Y)(t).\end{aligned}$$

QED

This observation is the key ingredient for establishing a one-to-one connection between self-similarity and stationarity. This fact – which, while first stated in 1962 [19], has been rediscovered many times, see e.g., [37] – is referred to as Lamperti's theorem and reads as follows:

Theorem 1. *If* $\{Y(t), t \in \mathbb{R}\}$ *is stationary, its Lamperti transform* $\{(\mathcal{L}_H Y)(t), t > 0\}$ *is H-ss. Conversely, if* $\{X(t), t > 0\}$ *is H-ss, its inverse Lamperti transform* $\{(\mathcal{L}_H^{-1} X)(t), t \in \mathbb{R}\}$ *is stationary.*

Proof – Let $\{Y(t), t \in \mathbb{R}\}$ be a stationary process. Using Definition 3 and Lemma 1, we have for any $\lambda > 0$,

$$\{Y(t), t \in \mathbb{R}\} \overset{d}{=} \{(\mathcal{S}_{\log\lambda} Y)(t) = (\mathcal{L}_H^{-1}\mathcal{D}_{H,\lambda}\mathcal{L}_H Y)(t), t \in \mathbb{R}\} \tag{11}$$

and it follows from Definition 4 that the Lamperti transform $X(t) := (\mathcal{L}_H Y)(t)$ is H-ss since

$$\{X(t), t > 0\} \overset{d}{=} \{(\mathcal{D}_{H,\lambda} X)(t), t > 0\} \tag{12}$$

for any $\lambda > 0$.

Conversely, let $\{X(t), t > 0\}$ be a H-ss process. Using Definition 4 and Lemma 1, we have for any $\lambda > 0$,

$$\{X(t), t > 0\} \overset{d}{=} \{(\mathcal{D}_{H,\lambda} X)(t) = (\mathcal{L}_H \mathcal{S}_{\log\lambda}\mathcal{L}_H^{-1} X)(t), t > 0\} \tag{13}$$

and it follows from Definition 3 that the inverse Lamperti transform $Y(t) := (\mathcal{L}_H^{-1} X)(t)$ is stationary since

$$\{Y(t), t \in \mathbb{R}\} \overset{d}{=} \{(\mathcal{S}_{\log\lambda} Y)(t), t \in \mathbb{R}\} \tag{14}$$

for any $\lambda > 0$. QED

The Lamperti transform establishes therefore a one-to-one connection between stationary and self-similar processes, and it is worth noting that it is in fact the unique transform to permit such a connection [8]. A graphical illustration of this one-to-one correspondence is given in Fig. 1.

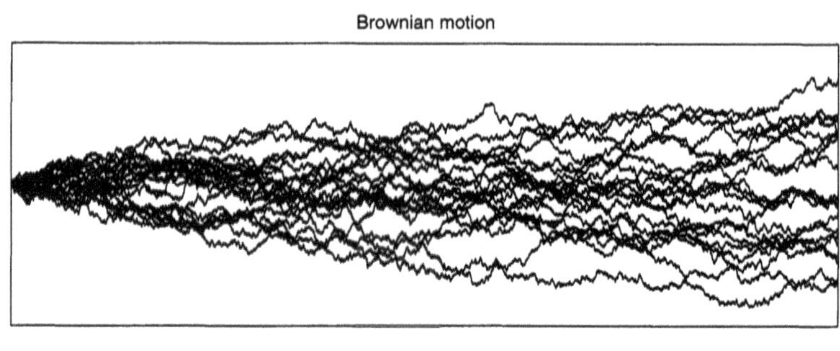

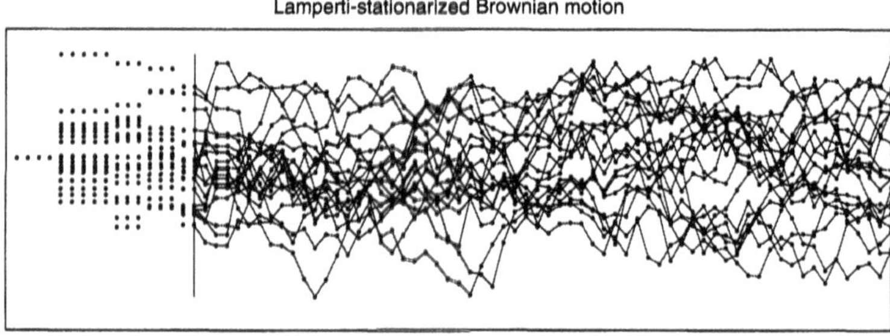

Fig. 1. A graphical illustration of Lamperti's theorem – Whereas sample paths of (nonstationary and self-similar) Brownian motion (top) reveal a time-dependence of variance as a square-root function of time, their "lampertized" versions (bottom) essentially lie within a band of constant width, in accordance with the stationarity properties induced by the inverse Lamperti transform.

Using (9), one can remark that, if $\{Y(t), t \in \mathbb{R}\}$ is stationary, the transformed process $\{(\mathcal{L}_{H_2}^{-1}\mathcal{L}_{H_1}Y)(t), t \in \mathbb{R}\}$ cannot be stationary, unless $H_1 = H_2$. In a similar way, making use of the remark on processes as in (5), the composition rule given in (8) shows that, if $\{X(t), t > 0\}$ is H-ss, the transformed process $(\mathcal{L}_{H_2}\mathcal{L}_{H_1}^{-1}X)(t)$ is $(H + H_2 - H_1)$-ss.

2.4 Consequences

Covariances and Spectra. As a direct consequence of Theorem 1, statistical properties of self-similar processes can be inferred from those of their Lamperti counterparts, and vice-versa. In particular, if we restrict to zero-mean second-order processes and if we introduce the notation $\mathbf{R}_X(t,s) := \mathbb{E}X(t)X(s)$, it is straightforward to establish that, for any process $\{X(t), t > 0\}$, the covariance function of its inverse Lamperti transform is given by:

$$\mathbf{R}_{\mathcal{L}_H^{-1}X}(t,s) = e^{-H(t+s)}\,\mathbf{R}_X(e^t, e^s) \tag{15}$$

for any $t, s \in \mathbb{R}$.

Conversely, for any process $\{Y(t), t \in \mathbb{R}\}$, the covariance function of its Lamperti transform reads

$$\mathbf{R}_{\mathcal{L}_H Y}(t,s) = (ts)^H\,\mathbf{R}_Y(\log t, \log s) \tag{16}$$

and, if $Y(t)$ happens to be stationary, we then have $\mathbf{R}_Y(t,s) = \gamma_Y(t-s)$ (with $\gamma_Y(.)$ a non-negative definite function), leading to:

$$\mathbf{R}_{\mathcal{L}_H Y}(t,s) = (ts)^H\,\gamma_Y(\log(t/s)). \tag{17}$$

Two corollaries to Theorem 1 are therefore as follows:

Corollary 1. *Any second-order H-ss process $\{X(t), t > 0\}$ has necessarily a covariance function of the form*

$$\mathbf{R}_X(t,s) = (ts)^H\,c_H(t/s) \tag{18}$$

for any $t, s > 0$, with $c_H(\exp(.))$ a non-negative definite function.

In the specific case where $H = 0$, we recover this way the class of "multiplicative stationary processes" introduced in [17], whereas the more general factorization given by (18) has been pointed out, e.g., in [12] and [37].

Corollary 2. *Given a second-order H-ss process $\{X(t), t > 0\}$, the power spectrum density of its stationary counterpart $(\mathcal{L}_H^{-1}X)(t)$ is the Mellin transform of the scale-covariant function c_H given in Eq. (18).*

Proof – Starting from (15) and using (18), it is immediate to establish that

$$\mathbf{R}_{\mathcal{L}_H^{-1}X}(t+\tau/2, t-\tau/2) = c_H(e^\tau),$$

from which it follows that the power spectrum density $\mathbf{\Gamma}_{\mathcal{L}_H^{-1}X}(f)$ of the inverse Lamperti transform of $X(t)$ is such that

$$\begin{aligned}
\mathbf{\Gamma}_{\mathcal{L}_H^{-1}X}(f) &:= \int_{-\infty}^{+\infty} \mathbf{R}_{\mathcal{L}_H^{-1}X}(t+\tau/2, t-\tau/2)\,e^{-i2\pi f\tau}\,d\tau \\
&= \int_{-\infty}^{+\infty} c_H(e^\tau)\,e^{-i2\pi f\tau}\,d\tau \\
&= \int_{0}^{+\infty} c_H(\theta)\,\theta^{-i2\pi f-1}\,d\theta \\
&= (\mathcal{M}c_H)(i2\pi f),
\end{aligned}$$

with

$$(\mathcal{M}X)(s) := \int_0^{+\infty} X(t)\,t^{-s-1}\,dt \tag{19}$$

the Mellin transform [5].

QED

Long-Range Dependence. In the case of stationary processes, long-range dependence (LRD), or long-memory, is usually associated with a slow power-law decay of the correlation function [3] but, more generally, it may also be defined as follows:

Definition 6. *A second-order stationary process $\{Y(t), t \in \mathbb{R}\}$ is said to be long-range dependent if its normalized correlation function*

$$\tilde{\gamma}_Y(\tau) := \gamma_Y(\tau)/\gamma_Y(0) \tag{20}$$

is not absolutely summable:

$$\int_0^{+\infty} |\tilde{\gamma}_Y(\tau)| \, d\tau = \infty. \tag{21}$$

In the case of nonstationary processes, a generalization of this definition can be given as follows [1,23]:

Definition 7. *A second-order nonstationary process $\{X(t), t > 0\}$ is said to be LRD if its normalized covariance function*

$$\tilde{\mathbf{R}}_X(t,s) := \frac{\mathbf{R}_X(t,s)}{(\mathbf{R}_X(t,t)\,\mathbf{R}_X(s,s))^{1/2}} \tag{22}$$

is such that

$$\int_0^{+\infty} |\tilde{\mathbf{R}}_X(t, t+\tau)| \, d\tau = \infty \tag{23}$$

for any fixed t.

Starting from (17), we get that

$$\tilde{\mathbf{R}}_{\mathcal{L}_H Y}(t,s) = \tilde{\gamma}_Y(\log(t/s)), \tag{24}$$

and it follows from a direct calculation that a (nonstationary) H-ss process $\{X(t), t > 0\}$ will be LRD in the sense of Definition 7 if and only if its (stationary) Lamperti counterpart is such that

$$\int_1^{\infty} |\tilde{\gamma}_{\mathcal{L}_H^{-1} X}(\log \tau)| \, d\tau = \infty. \tag{25}$$

Conversely, a stationary process $\{Y(t), t \in \mathbb{R}\}$ will be LRD in the sense of Definition 6 if and only if its nonstationary (H-ss) Lamperti counterpart is such that

$$\int_1^{\infty} |\tilde{\mathbf{R}}_{\mathcal{L}_H Y}(t, \lambda t)| \, d\lambda/\lambda = \infty, \tag{26}$$

or, equivalently (since, from (24), we have $\tilde{\mathbf{R}}_{\mathcal{L}_H Y}(t, \lambda t) = \tilde{\mathbf{R}}_{\mathcal{L}_H Y}(t, t/\lambda)$ for any $\lambda > 0$),

$$\int_0^{1} |\tilde{\mathbf{R}}_{\mathcal{L}_H Y}(\lambda t, t)| \, d\lambda/\lambda = \infty. \tag{27}$$

Scale-Covariant Systems. In classical linear system theory, it is well-known that linear *filters* are those linear operators $\mathcal{H}$ which are shift-covariant, i.e., such that

$$\mathcal{H}\mathcal{S}_\tau = \mathcal{S}_\tau \mathcal{H} \tag{28}$$

for any $\tau \in \mathbb{R}$. By analogy, it is natural to introduce systems which preserve self-similarity, according to the following definition:

Definition 8. *A linear operator $\mathcal{G}$, acting on processes $\{X(t), t > 0\}$, is said to be scale-covariant if it commutes with any renormalized dilation, i.e., if*

$$\mathcal{G}\mathcal{D}_{H,\lambda} = \mathcal{D}_{H,\lambda}\mathcal{G} \tag{29}$$

for any $H > 0$ and any $\lambda > 0$.

Proposition 1. *If an operator $\mathcal{G}$ is scale-covariant, then it necessarily acts on processes $\{X(t), t > 0\}$ as a multiplicative convolution, according to*

$$(\mathcal{G}X)(t) = \int_0^{+\infty} g(t/s)\, X(s)\, ds/s. \tag{30}$$

Proof – Let $k(t, s)$ be the kernel of some operator $\mathcal{G}$ acting on processes $\{X(t), t > 0\}$. We then have, for any $t > 0$,

$$\begin{aligned}(\mathcal{G}\mathcal{D}_{H,\lambda}X)(t) &= \int_0^{+\infty} k(t,s)\, \lambda^H\, X(\lambda s)\, ds \\ &= \lambda^{H-1} \int_0^{+\infty} k(t/\lambda, s)\, X(s)\, ds\end{aligned}$$

and

$$(\mathcal{D}_{H,\lambda}\mathcal{G}X)(t) = \lambda^H \int_0^{+\infty} k(\lambda t, s)\, X(s)\, ds.$$

It follows that imposing the scale-covariance of $\mathcal{G}$ for any process $X(t)$ (in the sense of Definition 8) amounts to equating the two above expressions, and thus to require that

$$k(t,s) = k(t/\lambda, s/\lambda)/\lambda \tag{31}$$

for any $t, s > 0$ and any $\lambda > 0$. In particular, the specific choice $\lambda = s$ leads to

$$k(t,s) = k(t/s, 1)/s =: g(t/s)/s, \tag{32}$$

which concludes the proof.

QED

Corollary 3. *Scale-covariant operators preserve self-similarity.*

Proof – Let $(\mathcal{G}X)(t)$ be the ouput of a scale-covariant system whose input $\{X(t), t > 0\}$ is H-ss. We then have from (4) and (29):

$$\{(\mathcal{D}_{H,\lambda}\mathcal{G}X)(t) = (\mathcal{G}\mathcal{D}_{H,\lambda}X)(t), t > 0\} \stackrel{d}{=} \{(\mathcal{G}X)(t), t > 0\}, \tag{33}$$

thus guaranteeing that $\{(\mathcal{G}X)(t), t > 0\}$ is H-ss.

QED

Corollary 4. *The Lamperti transform maps linear filters onto scale-covariant systems.*

Proof – The output $\{Z(t), t \in \mathbb{R}\}$ of a linear filter $\mathcal{H}$ of impulse response $h(.)$ is given by the convolution

$$Z(t) := (\mathcal{H}Y)(t) = \int_{-\infty}^{+\infty} h(t-s)\, Y(s)\, ds \tag{34}$$

for any input process $\{Y(t), t \in \mathbb{R}\}$. Using (5), we may write

$$\begin{aligned}(\mathcal{L}_H Z)(t) &= t^H\, Z(\log t)\\ &= t^H \int_{-\infty}^{+\infty} h(\log t - s)\, Y(s)\, ds\\ &= t^H \int_0^{+\infty} h(\log(t/v))\, Y(\log v)\, dv/v\\ &= \int_0^{+\infty} (t/v)^H\, h(\log(t/v))\, (\mathcal{L}_H Y)(v)\, dv/v\\ &= \int_0^{+\infty} (\mathcal{L}_H h)(t/v)\, (\mathcal{L}_H Y)(v)\, dv/v\end{aligned}$$

and it thus follows that, when "lampertized," the input-ouput relationship (34) is transformed into

$$(\mathcal{L}_H Z)(t) = \int_0^{+\infty} (\mathcal{L}_H h)(t/s)\, (\mathcal{L}_H Y)(s)\, ds/s, \tag{35}$$

taking on the form of a scale-covariant system, according to (30).

QED

Fourier transforming (34) leads to a product form for the input-output relationship of linear filters in the frequency domain:

$$(\mathcal{F}Z)(f) = (\mathcal{F}h)(f)\, (\mathcal{F}X)(f), \tag{36}$$

with $\mathcal{F}$ the Fourier transform operator, defined by

$$(\mathcal{F}X)(f) := \int_{-\infty}^{+\infty} X(t)\, e^{-i2\pi f t}\, dt. \tag{37}$$

In a very similar way, Mellin transforming (35) leads to a product form too, as expressed by:

$$(\mathcal{M}\mathcal{L}_H Z)(s) = (\mathcal{M}\mathcal{L}_H g)(s)\,(\mathcal{M}\mathcal{L}_H Y)(s). \tag{38}$$

Continuing along this analogy, H-ss processes can be represented as the output of scale-covariant systems, as stationary processes are outputs of linear filters. More precisely, stationary processes $\{Y(t), t \in \mathbb{R}\}$ are known to admit the Cramér representation [28]

$$Y(t) = \int_{-\infty}^{+\infty} e^{i2\pi f t}\, d\xi(f), \tag{39}$$

with spectral increments $d\xi(f)$ such that

$$\mathbb{E} d\xi(f)\overline{d\xi(\nu)} = \delta(f-\nu)\, d\mathbf{S}_Y(f)\, d\nu, \tag{40}$$

and $d\mathbf{S}_Y(f) = \mathbf{\Gamma}_Y(f)\, df$ in case of absolute continuity with respect to the Lebesgue measure. Stationarity being preserved by linear filtering, stationary processes admit an equivalent representation as in (34):

$$Y(t) = \int_{-\infty}^{+\infty} h(t-s)\, dB(s), \tag{41}$$

with $\mathbb{E} dB(t) dB(s) = \sigma^2\, \delta(t-s)\, dt\, ds$, and therefore:

$$d\mathbf{S}_Y(f) = \sigma^2\, |(\mathcal{F}h)(f)|^2\, df. \tag{42}$$

Applying the Lamperti transformation to (41) ends up with the relation

$$(\mathcal{L}_H Y)(t) = \int_0^{+\infty} (\mathcal{L}_H h)(t/s)\, (\mathcal{L}_H dB)(s)/s. \tag{43}$$

Comparing with (35), this corresponds to the output of a linear scale-covariant system whose input is such that

$$\begin{aligned}
\mathbb{E}(\mathcal{L}_H dB)(t)(\mathcal{L}_H dB)(s) &= \mathbb{E} t^H dB(\log t)\, s^H dB(\log s)\\
&= \sigma^2\, (ts)^H\, \delta(\log(t/s))\, dt\, ds\\
&= \sigma^2\, t^{2H+1}\, \delta(t-s)\, dt\, ds,
\end{aligned}$$

and it follows that

Proposition 2. *Any H-ss process $\{X(t), t > 0\}$ can be represented as the output of a linear scale covariant system of impulse response $g(.)$:*

$$X(t) = \int_0^{+\infty} g(t/s)\, dV(s)/s, \tag{44}$$

with

$$\mathbb{E} dV(t) dV(s) = \sigma^2\, t^{2H+1}\, \delta(t-s)\, dt\, ds. \tag{45}$$

Corollary 5. *Given the representation (44), the covariance function of a H-ss process $\{X(t), t > 0\}$ can be expressed as in Eq. (18), with:*

$$c_H(\lambda) = \sigma^2 \lambda^{-H} \int_0^{+\infty} g(\theta)\, g(\theta/\lambda)\, d\theta/\theta^{2H+1}. \tag{46}$$

Corollary 6. *Given the representation (44), the power spectrum density of the stationary counterpart $\{(\mathcal{L}_H^{-1}X)(t), t \in \mathbb{R}\}$ of a H-ss process $\{X(t), t > 0\}$ is given by*

$$\mathbf{\Gamma}_{\mathcal{L}_H^{-1}X}(f) = \sigma^2 \, |(\mathcal{M}g)(H + i2\pi f)|^2. \tag{47}$$

3 Examples and Applications

Examples and applications of the Lamperti transformation can be broadly classified in two types. One can for instance be interested in "lampertizing" (according to (6)) some specific stationary processes $\{Y(t), t \in \mathbb{R}\}$ and constructing this way classes of specific self-similar processes. From a reversed perspective, one can use the inverse transform (7) for "delampertizing" self-similar processes and making them amenable to the large body of machineries aimed at stationary processes. In this case, some desired operation $\mathcal{T}$ on H-ss processes $\{X(t), t > 0\}$ can rather be handled via the commutative diagram

$$\begin{array}{ccc} X(t) & \stackrel{?}{\longrightarrow} & (\mathcal{T}X)(t) \\ | & & \uparrow \\ \text{inverse Lamperti} & & \text{Lamperti} \\ \downarrow & & | \\ (\mathcal{L}_H^{-1}X)(t) & \longrightarrow & (\tilde{\mathcal{T}}\mathcal{L}_H^{-1}X)(t) \end{array} \tag{48}$$

according to which the overall operation is decomposed as

$$\mathcal{T} = \mathcal{L}_H \tilde{\mathcal{T}} \mathcal{L}_H^{-1}, \tag{49}$$

where the companion operation $\tilde{\mathcal{T}}$ acts on stationary processes.

3.1 Tones and Chirps

Besides white noise, maybe the simplest example of a stationary process is

$$Y_0(t) := a \cos(2\pi f_0 t + \varphi), \tag{50}$$

with $a, f_0 > 0$ and $\varphi \in \mathcal{U}(0, 2\pi)$. "Lampertizing" such a random phase "tone," i.e., applying (6) to (50), leads to

$$X_0(t) := (\mathcal{L}_H Y_0)(t) = a\, t^H \cos(2\pi f_0 \log t + \varphi). \tag{51}$$

The transformed process takes therefore on the form of a (random phase) "chirp," in the sense of, e.g., [9,22]. One can remark that $X_0(t) = \mathrm{Re}\{a\, e^{i\varphi}\, m_s(t)\}$ with $s = H + i2\pi f_0$ and $m_s(t) := t^s$ the basic building block of the Mellin transform (see Fig. 2).

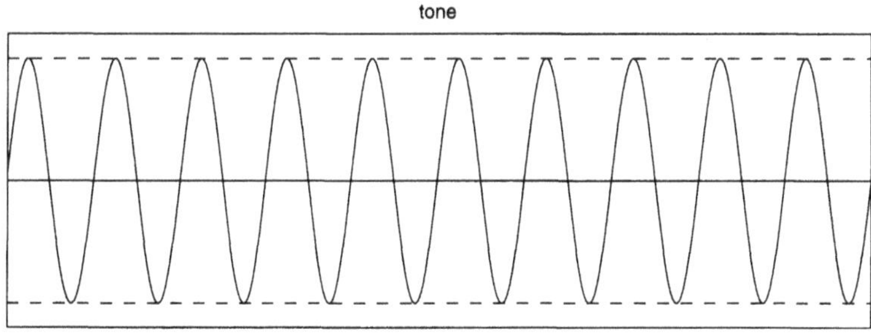

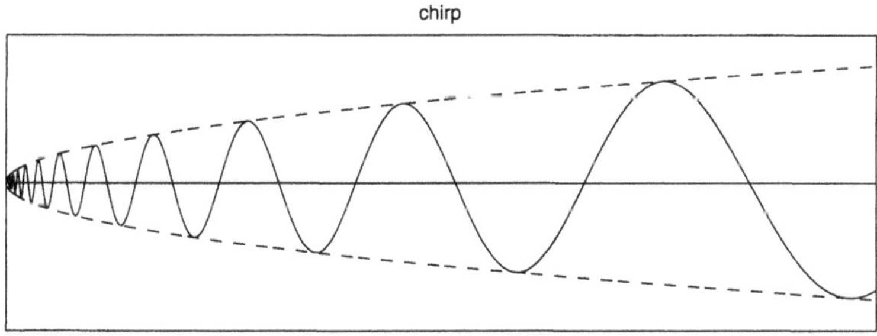

Fig. 2. Tones and chirps – The Lamperti transform of a pure tone (top) is a "chirp" (bottom) with a power-law amplitude modulation and a logarithmic frequency modulation. Said in other words, the Lamperti transform maps the Fourier basis onto a Mellin basis.

3.2 Fractional Brownian Motion

If we consider second-order processes $\{X(t), t > 0\}$ which are not only H-ss but also have stationary increments (or, "H-sssi" processes), it is well-known that their covariance function is necessarily of the form

$$\mathbf{R}_X(t,s) = \frac{\sigma^2}{2}\left(t^{2H} + s^{2H} - |t-s|^{2H}\right), \tag{52}$$

with $\sigma^2 := \mathbb{E}X^2(1)$.

If we further assume Gaussianity and if we restrict to $0 < H < 1$, we end up with the only family of *fractional Brownian motions* (fBm) $B_H(t)$ [20]. This offers an extension of ordinary Brownian motion $B(t) \equiv B_{1/2}(t)$, known to have uncorrelated increments, to situations where increments may be correlated (negatively if $0 < H < 1/2$ and positively if $1/2 < H < 1$).

Since fBm is H-ss, its covariance function (52) can be factorized according to (18), with

$$c_H(\lambda) = \frac{\sigma^2}{2}\left[\lambda^H + \lambda^{-H}\left(1 - |1-\lambda|^{2H}\right)\right]. \tag{53}$$

By application of (15) to (52), the covariance function of the inverse Lamperti transform $\{Y_H(t) := (\mathcal{L}_H^{-1} B_H)(t), t \in \mathbb{R}\}$ expresses as

$$\mathbf{R}_{Y_H}(t,s) = e^{-H(t+s)}\,\frac{\sigma^2}{2}\left(e^{2Ht} + e^{2Hs} - |e^t - e^s|^{2H}\right), \tag{54}$$

and it is immediate to reorganize terms so that $\gamma_{Y_H}(\tau) := \mathbf{R}_{Y_H}(t, t+\tau)$ reads:

$$\gamma_{Y_H}(\tau) = \sigma^2\left(\cosh(H|\tau|) - [\sinh(|\tau|/2)]^{2H}/2\right). \tag{55}$$

This stationary covariance function is plotted in Fig. 3, as a function of the Hurst parameter H. If we let $H = 1/2$ in (55), we readily get

$$\gamma_{Y_{1/2}}(\tau) = \sigma^2\, e^{-|\tau|/2}, \tag{56}$$

in accordance with the known-fact that the (Ornstein-Uhlenbeck) process whose stationary covariance function is given by (56) is the Lamperti image of the ordinary Brownian motion [30]. The stationary counterpart of fBm appears therefore as a form of *generalized Ornstein-Uhlenbeck* (gOU) process.

As a remark, it is worth noting that resorting to fBm increments rather than to the self-similar process itself guarantees stationarity (and, hence, eases further processing), but at the expense of facing *long-range dependence* (LRD) when $1/2 < H < 1$. In contrast, it follows from (55) that

$$\gamma_{Y_H}(\tau) \sim \frac{\sigma^2}{2}\left(e^{-H\tau} + 2He^{-(1-H)\tau}\right) \propto e^{-\min(H,1-H)\tau} \tag{57}$$

when $\tau \to \infty$, which means that the stationary counterpart of fBm is indeed *short-range dependent* for any $H \in (0,1)$, since its correlation function decreases exponentially fast at infinity. This result is nevertheless consistent with the fact that, according to (25), fBm itself is LRD in the sense of Definition 7 for any $H \in (0,1)$ since, for any τ_*,

$$\int_{\tau_*}^{\infty} |\tilde{\gamma}_{Y_H}(\tau)|\, e^{\tau}\, d\tau \sim \int_{\tau_*}^{\infty} e^{[1-\min(H,1-H)]\tau}\, d\tau = \infty. \tag{58}$$

As shown in [25,26], using the Lamperti transformation in the context of linear estimation of self-similar processes makes possible a number of manipulations (such as whitening or prediction) which otherwise prove much more difficult to handle. Indeed, it first follows from (57) that the power spectrum density of the (stationary) Lamperti counterpart of fBm reads

$$\boldsymbol{\Gamma}_{Y_H}(f) = \frac{\sigma^2}{H^2 + 4\pi^2 f^2}\left|\frac{\Gamma((1/2) + i2\pi f)}{\Gamma(H + i2\pi f)}\right|^2. \tag{59}$$

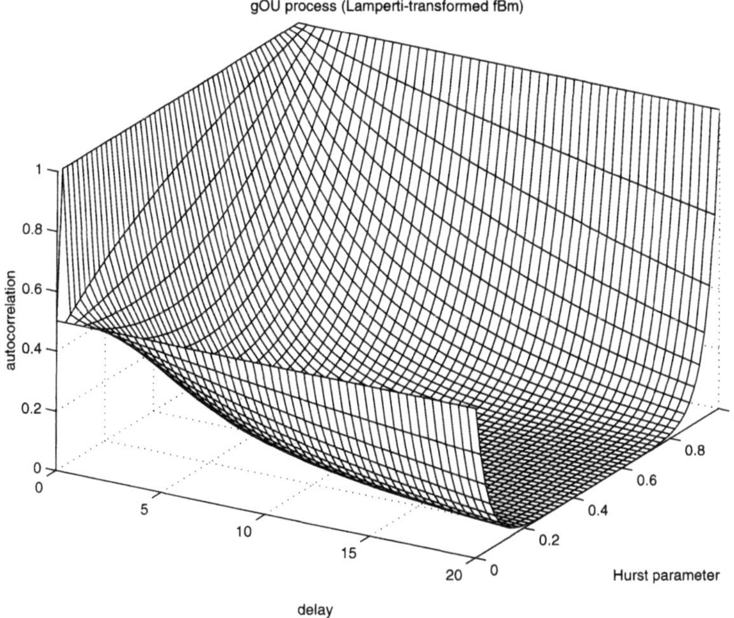

Fig. 3. Stationary covariance function of generalized Ornstein-Uhlenbeck processes – "Delampertizing" fractional Brownian motion (fBm) ends up with a stationary process, referred to as a *generalized Ornstein-Uhlenbeck* (gOU) process, whose covariance function is plotted here as a function of the Hurst parameter H. It is worth noting that this covariance decays exponentially fast for any $H \in (0,1)$, indicating that gOU is always *short-range dependent*.

Given this quantity, it becomes possible to get its spectral factorization and to write $\mathbf{\Gamma}_{Y_H}(f) = |\Phi_+(f)|^2$, with $\Phi_+(f)$ the transfer function of a causal filter. (One can remark that, instead of the exact fBm, we could have considered its (Barnes-Allan [2]) version $\{\tilde{B}_H(t), t > 0\}$, with

$$\tilde{B}_H(t) := \frac{1}{\Gamma(H+1/2)} \int_0^t (t-s)^{H-1/2}\, dB(s). \tag{60}$$

This corresponds to an H-ss process (with nonstationary increments) that admits a representation as in (44), with

$$g(\theta) = \frac{1}{\Gamma(H+1/2)} (\theta - 1)^{H-1/2} u(\theta - 1), \tag{61}$$

(where $u(.)$ stands for the unit step function) and, as first established in [12], it follows from Corollary 6 that the Lamperti counterpart $\{\tilde{Y}_H(t), t \in \mathbb{R}\}$ of $\{\tilde{B}_H(t), t > 0\}$ has a power spectrum density which turns out to exactly identify with (59).)

Considering the above mentioned factorization of (59) and using representations of H-ss processes as given by Proposition 2, it then becomes possible [25,26] to re-derive representations formulæ for fBm on a finite interval using a finite interval of ordinary Bm (and vice-versa), as well as to get explicit prediction formulæ for fBm (including a new one for the case $H < 1/2$).

3.3 Ornstein–Uhlenbeck Processes

The Ornstein-Uhlenbeck process $\{Y_{1/2}(t), t \in \mathbb{R}\}$ is solution of the Langevin equation:

$$dY(t) + \alpha\, Y(t)\, dt = dB(t), \tag{62}$$

with $\alpha = 1/2$.

Lamperti transforming the general Langevin equation (62), and using appropriate differentiation rules (as justified in [26]), we get

$$\begin{aligned}(\mathcal{L}_H dY)(t) &= t^H\, dY(\log t)\\ &= t^{H+1}\, d(Y(\log t))\\ &= t^{H+1}\, d(t^{-H}\, X(t))\\ &= t^{H+1}\, [t^{-H}\, dX(t) - H\, t^{-H-1}\, X(t)\, dt]\\ &= t\, dX(t) - H\, X(t)\, dt,\end{aligned}$$

with $X(t) := (\mathcal{L}_H Y)(t)$. It thus follows that the H-ss process $\{X(t), t > 0\}$ is solution of

$$t\, dX(t) + (\alpha - H)\, X(t)\, dt = dV(t), \tag{63}$$

where $dV(t) := (\mathcal{L}_H dB)(t)$ is such that $\mathbb{E} dV(t) dV(s) = \sigma^2\, t^{2H+1}\, \delta(t-s)\, dt\, ds$, and is thus covariance-equivalent to $d\tilde{V}(t) := t^{H+1/2}\, dB(t)$.

Indeed, for a given $\alpha > 0$, Ornstein-Uhlenbeck processes admit the integral representation

$$Y_\alpha(t) = \int_{-\infty}^{t} e^{-\alpha(t-s)}\, dB(s), \tag{64}$$

whose Lamperti transform reads

$$X_{\alpha,H}(t) := (\mathcal{L}_H Y_\alpha)(t) = t^{H-\alpha} \int_0^t s^\alpha\, dB(\log s). \tag{65}$$

Noting that $dB(\log t)$ is covariance-equivalent to $t^{-1/2} dB(t)$, we end up with

$$X_{\alpha,H}(t) = t^{H-\alpha} \int_0^t s^{\alpha-1/2}\, dB(s), \tag{66}$$

an expression which can be equivalently rewritten as

$$X_{\alpha,H}(t) = \int_0^{+\infty} [(t/s)^{H-\alpha}\, u(t/s-1)]\, [s^{H+1/2}\, dB(s)]/s. \tag{67}$$

We recognize in (66) the form resulting from the approach described in [24], whereas (67) enters the framework of the general representation (44), with the explicit identification $g(\theta) := \theta^{H-\alpha}\, u(\theta - 1)$ and $dV(t) := s^{H+1/2}\, dB(t)$.

Given $\alpha > 0$, the (Ornstein-Uhlenbeck) solution $Y_\alpha(t)$ of the general Langevin equation (62) is known to have a (stationary) covariance function $\gamma_{Y_\alpha}(\tau)$ which reads

$$\gamma_{Y_\alpha}(\tau) = \sigma^2 \, e^{-\alpha|\tau|}, \tag{68}$$

thus generalizing (56). It readily follows from (17) that the Lamperti transform (65)-(67), solution of (63), admits the (nonstationary) covariance function

$$\begin{aligned} \mathbf{R}_{X_{\alpha,H}}(t,s) &= \sigma^2 \, (ts)^H \, e^{-\alpha|\log(t/s)|} \\ &= \sigma^2 \, (\min(t,s))^{H+\alpha} \, (\max(t,s))^{H-\alpha}. \end{aligned}$$

Letting $\alpha = H$ in the above expression, we get $\mathbf{R}_{X_{H,H}}(t,s) = (\min(t,s))^{2H}$, in trivial generalization of the ordinary Brownian situation, corresponding to $H = 1/2$. In the special case where $\alpha = 1/2$, it follows from the composition rule (9) that the solution $X_{1/2,H}(t)$ of (63) is given by

$$X_{1/2,H}(t) := (\mathcal{L}_H Y_{1/2})(t) = (\mathcal{L}_H \mathcal{L}_{1/2}^{-1} B)(t) = t^{H-1/2} \, B(t), \tag{69}$$

which, as expected, identifies to $B(t)$ too if $H = 1/2$.

In the general case of arbitrary α and H, $\{X_{\alpha,H}(t), t > 0\}$ has been put forward [24] as a versatile two-parameter model, in which H controls self-similarity whereas α may be related to long-range dependence. Indeed, we know from (68) that $\tilde{\gamma}_{Y_\alpha}(\tau) = e^{-\alpha|\tau|}$, and it follows from Definition 7 and (25) that $\{X_{\alpha,H}(t), t > 0\}$ will be LRD if $\alpha < 1$.

3.4 Euler–Cauchy Processes

Whereas it is known that Brownian motion is not differentiable in the classical sense, the Langevin equation is usually written as the stochastic (first-order) differential equation

$$\frac{dY}{dt}(t) + \alpha \, Y(t) = W(t), \tag{70}$$

where the "white noise" $W(t)$ (such that $\mathbb{E} W(t) \, W(s) = \sigma^2 \, \delta(t-s) \, dt \, ds$) plays formally the role of a "derivative" for Brownian motion.

The interpretation of (70) is that $Y(t)$ is the output of a first-order linear system whose input is white noise. As such, it may constitute a building block for more complicated systems (with elementary sub-systems in cascade and/or in parallel [18,23]), and it can also be generalized to higher orders, as in ARMA(p,q) processes of the form

$$\sum_{n=0}^{p} \alpha_n \, Y^{(n)}(t) = \sum_{n=0}^{q} \beta_n \, W^{(n)}(t), \tag{71}$$

with the notation $Y^{(n)}(t) := (d^n X/dt^n)(t)$.

Such (stationary) processes have (self-similar) Lamperti counterparts that are solutions of a generalization of (63) [37].

Lemma 2. *Let $\{Y(t), t \in \mathbb{R}\}$ be a stationary process, with $\{X(t) := (\mathcal{L}_H Y)(t), t > 0\}$ its Lamperti transform. Given a set of coefficients $\{\alpha_n, n = 0, \dots N\}$, one can find another set of coefficients $\{\alpha'_n, n = 0, \dots N\}$ such that the Lamperti transform of the linear process*

$$Z(t) = \sum_{n=0}^{N} \alpha_n \, Y^{(n)}(t) \tag{72}$$

takes on the form

$$(\mathcal{L}_H Z)(t) = \sum_{n=0}^{N} \alpha'_n \, t^n \, X^{(n)}(t). \tag{73}$$

Proof – From the definition and the linearity of the Lamperti transform, we may write:

$$\begin{aligned}(\mathcal{L}_H Z)(t) &= \sum_{n=0}^{N} \alpha_n \, (\mathcal{L}_H Y^{(n)})(t) \\ &= \sum_{n=0}^{N} \alpha_n \, t^H \, Y^{(n)}(\log t).\end{aligned}$$

Iterating the differentiation rule

$$Y^{(1)}(t) = t \, \frac{d}{dt}(Y(\log t)),$$

there exist coefficients $\gamma_j(n)$, functionnally dependent on the α_n's, such that the quantity $Y^{(n)}(\log t)$ admits an expansion of the form

$$Y^{(n)}(\log t) = \sum_{j=0}^{N} \gamma_j(n) \, t^j \, \frac{d^j}{dt^j}(Y(\log t)).$$

After a suitable re-organization of terms, we have therefore

$$(\mathcal{L}_H Z)(t) = \sum_{n=0}^{N} \delta_n \, t^{H+n} \, \frac{d^n}{dt^n}(t^{-H} X(t)), \tag{74}$$

with $X(t) = (\mathcal{L}_H Y)(t)$ and

$$\delta_n := \alpha_n \sum_{j=n}^{N} \gamma_n(j).$$

The above expression (74) can be simplified further by remarking that

$$\frac{d}{dt}(t^{-H} X(t)) = -H \, t^{-H-1} \, X(t) + t^{-H} \, X^{(1)}(t),$$

thanks to which there exist coefficients $\mu_k(n)$, functionnally dependent on the δ_n's, such that

$$\frac{d^n}{dt^n}(t^{-H}X(t)) = \sum_{k=0}^{n} \mu_k(n)\, t^{-H+k-n}\, X^{(k)}(t),$$

thus leading to the claimed result (73), with

$$\alpha'_n := \delta_n \sum_{k=0}^{n} \mu_k(n).$$

QED

It follows from this Lemma that [37]

Proposition 3. *The stationary ARMA process (71) has an H-ss Lamperti counterpart, referred to as an Euler-Cauchy process, which is solution of an equation of the form*

$$\sum_{n=0}^{p} \alpha'_n\, t^n\, X^{(n)}(t) = \sum_{n=0}^{q} \beta'_n\, t^n\, \tilde{W}^{(n)}(t), \tag{75}$$

with $\tilde{W}(t) = t^{H+1/2}\, W(t)$ *and* $t > 0$.

Proof – The proof, which follows directly from the application of Lemma 2 to both sides of (71), is completed by noting that $(\mathcal{L}_H W)(t) = t^H\, W(\log t)$ has for covariance function

$$\begin{aligned}\mathbb{E}(\mathcal{L}_H W)(t)\,(\mathcal{L}_H W)(s) &= \sigma^2\,(ts)^H\, \delta(\log(t/s)) \\ &= \sigma^2\, t^{2H}\, \delta(t/s-1) \\ &= \sigma^2\, t^{2H+1}\, \delta(t-s) \\ &= \mathbb{E}\tilde{W}(t)\, \tilde{W}(s),\end{aligned}$$

with $\tilde{W}(t) = t^{H+1/2}\, W(t)$.

QED

4 Variations

Given Lamperti's theorem, it is easy to develop variations on the same theme by relaxing in some way the strict notion of scale invariance, or of stationarity.

4.1 Nonstationary Tools

Multiplicative Harmonizability. In the case of nonstationary processes $\{Y(t), t \in \mathbb{R}\}$, a Cramér representation of the type (39) stills holds, but with non orthogonal increments:

$$\mathbb{E}d\xi(f)\overline{d\xi(\nu)} = d^2\mathbf{\Phi}_Y(f,\nu), \tag{76}$$

i.e., with spectral masses which are not located along the only diagonal of the frequency-frequency plane. Provided that the Loève's condition

$$\iint_{-\infty}^{+\infty} |d^2 \mathbf{\Phi}_Y(f, \nu)| < \infty \tag{77}$$

is satisfied, the corresponding nonstationary processes are referred to as *harmonizable*, and such that

$$\mathbf{R}_Y(t, s) = \iint_{-\infty}^{+\infty} e^{i2\pi(ft - s\nu)}\, d^2 \mathbf{\Phi}_Y(f, \nu). \tag{78}$$

A companion concept of *multiplicative harmonizability* can be introduced in the case of processes $\{X(t), t > 0\}$ deviating from exact self-similarity [6]. This readily follows from the "lampertization" of (39) which, together with (77), leads

$$(\mathcal{L}_H Y)(t) = \int_{-\infty}^{+\infty} t^{H + i2\pi\sigma}\, d\xi(\sigma), \tag{79}$$

whereas the restriction of this general expression to the special case of independent spectral increments leads to the representation considered, e.g., in [12,17,37]. Provided that (77) holds, multiplicatively harmonizable processes $\{Y(t), t > 0\}$ have a (nonstationary) covariance function such that

$$\mathbf{R}_Y(t, s) = \iint_{-\infty}^{+\infty} t^{H + i2\pi f}\, s^{H - i2\pi\nu}\, d^2 \mathbf{\Phi}_Y(f, \nu). \tag{80}$$

Time-Dependent Spectra. In the general nonstationary case, (multiplicatively) harmonizable processes have a second-order structure which is described by a two-dimensional function, either in the time-time plane (covariance function) or in the frequency-frequency plane (spectral distribution function). These two equivalent descriptions can be supplemented by *mixed* time-frequency representations interpreted as time-dependent spectra. Starting from (78) and assuming further that me may write $d^2 \mathbf{\Phi}_Y(f, \nu) = \tilde{\mathbf{\Phi}}_Y(f, \nu)\, df\, d\nu$, a proper symmetrization of the covariance function, followed by a partial Fourier transform, leads to:

$$\int_{-\infty}^{+\infty} \mathbf{R}_Y(t + \tau/2, t - \tau/2)\, e^{-i2\pi f\tau}\, d\tau = \int_{-\infty}^{+\infty} \tilde{\mathbf{\Phi}}_Y(f + \nu/2, f - \nu/2)\, e^{i2\pi t\nu}\, d\nu. \tag{81}$$

Both sides of the above equation equivalently define the so-called *Wigner-Ville spectrum* (WVS) [14], thereafter labelled $\mathbf{W}_Y(t, f)$.

By construction, the WVS is a nonstationary extension of the classical power spectrum density, and it reduces to the latter in the stationary case: if we have $\mathbf{R}_Y(t, s) = \gamma_Y(t - s)$ or, equivalently, $\tilde{\mathbf{\Phi}}_Y(f, \nu) = \delta(f - \nu)\, \mathbf{\Gamma}_Y(f)$, we simply get

$$\mathbf{W}_Y(t, f) = \int_{-\infty}^{+\infty} \gamma_Y(\tau)\, e^{-i2\pi f\tau}\, d\tau = \mathbf{\Gamma}_Y(f) \tag{82}$$

for all t's. Among the many other properties of the WVS [14], one can cite those related to marginalizations, according to which:

$$\int_{-\infty}^{+\infty} \mathbf{W}_Y(t,f)\,df = \mathbf{R}_Y(t,t) \tag{83}$$

and

$$\int_{-\infty}^{+\infty} \mathbf{W}_Y(t,f)\,dt = \tilde{\mathbf{\Phi}}_Y(f,f). \tag{84}$$

Conventional mixed representations of nonstationary processes are based on Fourier transforms, but alternative forms based on Mellin transforms can also be considered, which prove especially useful in the case of self-similar processes. A motivation for their introduction can be found in the interconnection which exists between the Fourier, Mellin and Lamperti transforms, and which is expressed in the following lemma:

Lemma 3. *The Fourier transform of a process $\{Y(t), t \in \mathbb{R}\}$ can be equivalently expressed as the Mellin transform of its Lamperti transform, according to*

$$(\mathcal{F}Y)(f) = (\mathcal{M}\mathcal{L}_H Y)(H + i2\pi f). \tag{85}$$

A consequence of this equivalence is that, in the case of self-similar processes, the Mellin transform plays, with respect to scaling, a role similar to that played by the Fourier transform with respect to shifting. More precisely, we have:

Proposition 4. *The Mellin spectrum of H-ss processes $\{X(t), t > 0\}$ is invariant under renormalized dilations $\mathcal{D}_{H,k}$, for any $k > 0$:*

$$|(\mathcal{M}X)(H + i2\pi f)|^2 = |(\mathcal{M}\mathcal{D}_{H,k}X)(H + i2\pi f)|^2. \tag{86}$$

Proof – Assuming that $\{X(t), t > 0\}$ is an H-ss process, its inverse Lamperti transform $\{(\mathcal{L}_H^{-1}X)(t), t \in \mathbb{R}\}$ is stationary, and thus such that

$$|(\mathcal{F}\mathcal{L}_H^{-1}X)(f)|^2 = |(\mathcal{F}\mathcal{S}_{\log k}\mathcal{L}_H^{-1}X)(f)|^2 \tag{87}$$

for any $k > 0$. We can then deduce from Lemma 1 and 3 that:

$$\begin{aligned}
|(\mathcal{M}X)(H + i2\pi f)|^2 &- |(\mathcal{F}\mathcal{L}_H^{-1}X)(f)|^2 \\
&= |(\mathcal{F}\mathcal{S}_{\log k}\mathcal{L}_H^{-1}X)(f)|^2 \\
&= |(\mathcal{F}\mathcal{L}_H^{-1}\mathcal{D}_{H,k}X)(f)|^2 \\
&= |(\mathcal{M}\mathcal{D}_{H,k}X)(H + i2\pi f)|^2,
\end{aligned}$$

whence the claimed result.

QED

Using the notation $\mathbf{R}_{Y,t}(\tau) := \mathbb{E}Y(t+\tau/2)Y(t-\tau/2)$, we have

$$\begin{aligned}\mathbf{W}_Y(t,f) &= (\mathcal{F}\mathbf{R}_{Y,t})(f)\\ &= (\mathcal{M}\mathcal{L}_H\mathbf{R}_{Y,t})(H+i2\pi f)\\ &= \int_0^{+\infty} \tau^H\,\mathbf{R}_Y(t+\log\tau^{+1/2}, t-\log\tau^{-1/2})\,\tau^{-H-i2\pi f-1}\,d\tau,\end{aligned}$$

whence

$$\begin{aligned}\mathbf{W}_Y(\log t,f) &= \int_0^{+\infty} \mathbf{R}_Y(t+\log\tau^{+1/2}, t-\log\tau^{-1/2})\,\tau^{-i2\pi f-1}\,d\tau\\ &= \int_0^{+\infty} \mathbb{E}Y(\log(t\tau^{+1/2}))Y(\log(t\tau^{-1/2}))\,\tau^{-i2\pi f-1}\,d\tau\\ &= t^{-2H}\,\underline{\mathbf{W}}_{\mathcal{L}_H Y}(t,f),\end{aligned} \tag{88}$$

with

$$\underline{\mathbf{W}}_X(t,f) := \int_0^{+\infty} \mathbf{R}_X(t\tau^{+1/2}, t\tau^{-1/2})\,\tau^{-i2\pi f-1}\,d\tau. \tag{89}$$

The above quantity $\underline{\mathbf{W}}_X(t,f)$ is referred to as a *scale-invariant Wigner spectrum* [12], since we have, for any H-ss process $\{X(t), t>0\}$ and any $k>0$:

$$\begin{aligned}\underline{\mathbf{W}}_{\mathcal{D}_{H,k}X}(t,f) &= t^{2H}\,\mathbf{W}_{\mathcal{L}_H^{-1}\mathcal{D}_{H,k}X}(\log t,f)\\ &= t^{2H}\,\mathbf{W}_{\mathcal{S}_{\log k}\mathcal{L}_H^{-1}X}(\log t,f)\\ &= t^{2H}\,\mathbf{W}_{\mathcal{L}_H^{-1}X}(\log(kt),f)\\ &= k^{-2H}\,\underline{\mathbf{W}}_X(kt,f).\end{aligned}$$

Proposition 5. *In the case of H-ss processes $\{X(t), t>0\}$, the scale-invariant Wigner spectrum is a separable function of its two variables which can be factorized as:*

$$\underline{\mathbf{W}}_X(t,f) = t^{2H}\,\mathbf{\Gamma}_{\mathcal{L}_H^{-1}X}(f). \tag{90}$$

Proof – We know from (88) that

$$\underline{\mathbf{W}}_X(t,f) = t^{2H}\,\mathbf{W}_{\mathcal{L}_H^{-1}X}(\log t,f). \tag{91}$$

If $\{X(t), t>0\}$ is H-ss, its inverse Lamperti transform $\{(\mathcal{L}_H^{-1}X)(t), t\in\mathbb{R}\}$ is stationary by construction, and Eq. (82) guarantees therefore that the WVS of the latter reduces to its power spectrum density for any t.

QED

4.2 From Global to Local

Locally Stationary Processes. Rather than resorting to processes that are exactly (second-order) stationary, one can make use of the weakened model

$$\mathbf{R}_Y(t,s) = m_Y\left(\frac{t+s}{2}\right)\gamma_Y(t-s), \tag{92}$$

with $m_Y(t) \geq 0$ and $\gamma_Y(.)$ a non-negative definite function. This corresponds to a class of nonstationary processes referred to as *locally stationary* [31], since their symmetrized covariance function is given by

$$\mathbf{R}_Y(t+\tau/2, t-\tau/2) = m_Y(t)\,\gamma_Y(\tau), \tag{93}$$

i.e., as an ordinary stationary covariance function that is allowed to fluctuate as a function of the local time t. From an equivalent perspective, the WVS of a locally stationary process expresses simply as a modulation in time of an ordinary power spectrum, since it factorizes according to:

$$\mathbf{W}_Y(t,f) = m_Y(t)\,\mathbf{\Gamma}_Y(f).$$

When properly "lampertized," locally stationary processes are therefore such that:

$$\mathbf{R}_{\mathcal{L}_H Y}(t,s) = m_Y(\log\sqrt{ts})\,(ts)^H\,\gamma_Y(\log(t/s))$$

and

$$\underline{\mathbf{W}}_{\mathcal{L}_H Y}(t,f) = m_Y(t)\,t^{2H}\,\mathbf{\Gamma}_Y(f).$$

thus generalizing the forms given in (17) and (90), respectively.

Locally Self-similar Processes. Another possible variation is to accommodate for deviations from strict self-similarity, as it may be the case with *locally self-similar processes*, i.e., those processes whose scaling properties are governed by a time-dependent function $H(t)$ in place of a unique constant Hurst exponent H. When dealing with second-order Gaussian processes, a useful framework for such a situation has been developed [1,27], referred to as *multifractional Brownian motion* (mBm). Such processes admit the harmonizable representation

$$B_{H(t)}(t) = \int_{-\infty}^{+\infty} \frac{e^{i2\pi\xi t}-1}{|\xi|^{H(t)+1/2}}\,d\xi(f),$$

with $H : [0,\infty) \to [a,b] \subset (0,1)$ any Hölder function of exponent $\beta > 0$, and it has been shown [1] that their covariance function generalizes that of fBm according to

$$\mathbf{R}_{B_{H(t)}}(t,s) = \frac{\sigma^2}{2}\left(t^{h(t,s)} + s^{h(t,s)} - |t-s|^{h(t,s)}\right), \tag{94}$$

with

$$h(t,s) := H(t) + H(s).$$

Using (16) and proceeding as in Sect. 3.2, it is easy to show that

$$\mathbf{R}_{Y_H}\left(t+\frac{\tau}{2}, t-\frac{\tau}{2}\right) = \sigma^2 e^{(\eta(t,\tau)-2H)t}\left(\cosh[\eta(t,\tau)|\tau|/2] - [2\sinh(|\tau|/2)]^{\eta(t,\tau)}/2\right), \quad (95)$$

with

$$\eta(t,\tau) := h(\exp(t+\tau/2), \exp(t-\tau/2)),$$

and where $\{Y_H(t), t \in \mathbb{R}\}$ stands for the inverse Lamperti transform of mBm, computed with some fixed exponent $H \in (0,1)$:

$$Y_H(t) := (\mathcal{L}_H^{-1} B_{H(t)})(t).$$

If we formally consider the case where $H(t) := H + \alpha \log t$, we have $\eta(t,\tau) = 2(H+\alpha\tau)$ and the process

$$\tilde{Y}(t) := e^{-(H+\alpha t)t} B_{H(t)}(t)$$

turns out to be such that

$$\mathbf{R}_{\tilde{Y}}\left(t+\frac{\tau}{2}, t-\frac{\tau}{2}\right) = \sigma^2\left(\cosh[(H+\alpha t)|\tau|] - [2\sinh(|\tau|/2)]^{2(H+\alpha t)}/2\right). \quad (96)$$

Comparing with (55), it appears that the above covariance is identical to that of a gOU process, with H replaced, *mutatis mutandis*, by $H + \alpha t$. The interpretation of this result is that, when $H(t)$ admits locally the logarithmic approximation $H(t) := H + \alpha \log t$, "lampertizing" a mBm with the time-dependent exponent $H + \alpha t$ ends up with a process which can be approximated by a (tangential) locally stationary process of gOU type, locally controlled by the same exponent.

4.3 Discrete Scale Invariance

According to Definition 4, self-similarity usually refers to an invariance with respect to *any* dilation factor λ. In some situations however, this may be a too strong requirement (or assumption) for capturing scaling properties which are only observed for some preferred dilation factors (think of the triadic Cantor set [11], for which exact replication can only be achieved for scale factors $\{\lambda = 3^k, k \in \mathbb{Z}\}$, or of the Mellin "chirps" of the form (51) for which scale invariance applies for $\{\lambda = (\exp 1/f_0)^k, k \in \mathbb{Z}\}$ only).

Such a situation, which is referred to as *discrete scale invariance* (DSI), has in fact been recently put forward as a central concept in the study of many critical systems [32], and it has received much attention in a deterministic context. The purpose of this Section is to show that the Lamperti transform may be instrumental in the definition and the analysis of processes which are DSI in a stochastic sense [6,7].

Definitions

Definition 9. *A process* $\{Y(t), t \in \mathbb{R}\}$ *is said to be periodically correlated of period* T_0 *(or "*T_0*-cyclostationary") if*

$$\{(\mathcal{S}_{T_0} Y)(t), t \in \mathbb{R}\} \overset{d}{=} \{Y(t), t \in \mathbb{R}\}. \tag{97}$$

Definition 10. *A process* $\{X(t), t > 0\}$ *is said to possess a discrete scale invariance of index* H *and of scaling factor* $\lambda_0 > 0$ *(or to be "*(H, λ_0)*-DSI") if*

$$\{(\mathcal{D}_{H,\lambda_0} X)(t), t > 0\} \overset{d}{=} \{X(t), t > 0\}. \tag{98}$$

It naturally follows from these two definitions that T_0-cyclostationary processes are also T-cyclostationary for any $T = kT_0, k \in \mathbb{Z}$, and that (H, λ_0)-DSI processes are also (H, λ)-DSI for any $\lambda = \lambda_0^k, k \in \mathbb{Z}$.

Given Definition 9, second-order T_0-cyclostationary processes $\{Y(t), t \in \mathbb{R}\}$ have a covariance function $\mathbf{R}_Y(t, t+\tau)$ which is periodic in t of period T_0, and which can thus be decomposed in a Fourier series according to

$$\mathbf{R}_Y(t, t+\tau) = \sum_{n=-\infty}^{\infty} \mathbf{C}_n(\tau)\, e^{i2\pi nt/T_0}. \tag{99}$$

One deduces from this representation that the spectral distribution function of T_0-cyclostationary processes takes on the form:

$$\begin{aligned}
\tilde{\mathbf{\Phi}}_Y(f, \nu) &= \iint_{-\infty}^{+\infty} \mathbf{R}_Y(t, s)\, e^{-i2\pi(ft - \nu s)}\, dt\, ds \\
&= \iint_{-\infty}^{+\infty} \mathbf{R}_Y(t, t+\tau)\, e^{-i2\pi((f-\nu)t - \nu\tau)}\, dt\, d\tau \\
&= \sum_{n=-\infty}^{\infty} \mathbf{c}_n(\nu)\, \delta(\nu - (f - n/T_0)),
\end{aligned}$$

with

$$\mathbf{c}_n(\nu) := \int_{-\infty}^{+\infty} \mathbf{C}_n(\tau)\, e^{-i2\pi\nu\tau}\, d\tau. \tag{100}$$

In contrast with the stationary case for which $\tilde{\mathbf{\Phi}}_Y(f, \nu)$ is entirely concentrated along the main diagonal $\nu = f$ of the frequency-frequency plane, the spectral distribution function of cyclostationary processes is also non-zero along all the equally spaced parallel lines defined by $\nu = f - n/T_0, n \in \mathbb{Z}$.

More on the theory of cyclostationary processes can be found, e.g., in [16].

Characterization and Analysis. It has been stated in Theorem 1 that the Lamperti transformation establishes a one-to-one correspondence between stationary and self-similar processes. An extension of this result is given by the following Theorem:

Theorem 2. *If* $\{Y(t), t \in \mathbb{R}\}$ *is* T_0*-cyclostationary, its Lamperti transform* $\{(\mathcal{L}_H Y)(t),\, t > 0\}$ *is* (H, e^{T_0})*-DSI. Conversely, if* $\{X(t), t > 0\}$ *is* (H, e^{T_0})*-DSI, its inverse Lamperti transform* $\{(\mathcal{L}_H^{-1} X)(t), t \in \mathbb{R}\}$ *is* T_0*-cyclostationary.*

Proof – Let $\{Y(t), t \in \mathbb{R}\}$ be a T_0-cyclostationary process. From Definition 9, we have $\{Y(t), t \in \mathbb{R}\} \overset{d}{=} \{Y(t + T_0), t \in \mathbb{R}\}$ and, using (6), we may write

$$\begin{aligned}(\mathcal{L}_H Y)(e^{T_0} t) &= (e^{T_0} t)^H \, Y(\log t + T_0) \\ &\overset{d}{=} e^{H T_0} \, t^H \, Y(\log t) \\ &= (e^{T_0})^H \, (\mathcal{L}_H Y)(t),\end{aligned}$$

thus proving that $\{(\mathcal{L}_H Y)(t), t > 0\}$ is (H, e^{T_0})-DSI.

Conversely, let $\{X(t), t > 0\}$ be a (H, e^{T_0})-DSI process. From Definition 10, we have $\{X(e^{T_0} t), t > 0\} \overset{d}{=} \{e^{H T_0} \, X(t), t > 0\}$ and, using (7), we may write

$$\begin{aligned}(\mathcal{L}_H^{-1} X)(t + T_0) &= e^{-Ht} \, e^{-H T_0} \, X(e^{T_0} \, e^t) \\ &\overset{d}{=} e^{-Ht} \, X(e^t) \\ &= (\mathcal{L}_H^{-1} X)(t),\end{aligned}$$

thus proving that $\{(\mathcal{L}_H^{-1} X)(t), t \in \mathbb{R}\}$ is T_0-cyclostationary.

QED

Since DSI processes result from a "lampertization" of cyclostationary processes, the form of their covariance function can readily be deduced from the general correspondence (16) when applied to the specific form (99). We get this way that (H, λ_0)-DSI processes $\{X(t), t > 0\}$ have a covariance function such that:

$$\mathbf{R}_X(t, kt) = (kt)^H \sum_{n=-\infty}^{\infty} \mathbf{C}_n(\log k) \, t^{H + i 2\pi n / T_0}, \tag{101}$$

with $T_0 = \log \lambda_0$. Plugging this expression into (89), we also get an expansion for the corresponding scale-invariant Wigner spectrum:

$$\underline{\mathbf{W}}_X(t, f) = \sum_{n=-\infty}^{\infty} \mathbf{c}_n(f - n/2T_0) \, t^{2H + i 2\pi n / T_0}.$$

While such representations might suggest to make use of Mellin-based tools for analyzing DSI processes by working directly in the observation space, Theorem 2 offers another possibility of action by first "delampertizing" the observed scaling data so as to make them amenable to more conventional cyclostationary techniques (see, e.g., [15,29]). This is in fact the procedure followed in [6,7], where the existence of stochastic DSI is unveiled by marginalizing an estimated cyclic spectrum computed on the "delampertized" data.

Examples. *Weierstrass-Mandelbrot* – Let us consider the process

$$X(t) = \sum_{n=-\infty}^{\infty} \lambda^{-Hn} G(\lambda^n t) \, e^{i\varphi_n}, \tag{102}$$

with $\lambda > 1$, $0 < H < 1$, $\varphi_n \in \mathcal{U}(0, 2\pi)$ i.i.d. random phases and $G(.)$ a 2π-periodic function. We get this way a generalization of the (randomized) Weierstrass-Mandelbrot function [4], the latter corresponding to the specific choice:

$$G(t) = 1 - e^{it}. \tag{103}$$

It is immediate to check that

$$\begin{aligned}
(\mathcal{D}_{H,\lambda} X)(t) &= \lambda^{-H} \sum_{n=-\infty}^{\infty} \lambda^{-Hn} G(\lambda^n \lambda t) \, e^{i\varphi_n} \\
&= \sum_{n=-\infty}^{\infty} \lambda^{-Hn} G(\lambda^n t) \, e^{i\varphi_{n-1}} \\
&\stackrel{d}{=} X(t),
\end{aligned}$$

thus guaranteeing that $\{X(t), t > 0\}$ is (H, λ)-DSI. In a similar way, Lamperti transforming (102) leads to

$$\begin{aligned}
(\mathcal{L}_H^{-1} X)(t) &= e^{-Ht} \sum_{n=-\infty}^{\infty} \lambda^{-Hn} G(\lambda^n e^t) \, e^{i\varphi_n} \\
&= \sum_{n=-\infty}^{\infty} e^{-H(t+n\log\lambda)} \, G(e^{t+n\log\lambda}) \, e^{i\varphi_n} \\
&= \sum_{n=-\infty}^{\infty} (\mathcal{L}_H^{-1} G)(t + n\log\lambda) \, e^{i\varphi_n},
\end{aligned}$$

from which we deduce that

$$\begin{aligned}
(\mathcal{S}_{\log\lambda} \mathcal{L}_H^{-1} X)(t) &= \sum_{n=-\infty}^{\infty} (\mathcal{L}_H^{-1} G)(t + \log\lambda + n\log\lambda) \, e^{i\varphi_n} \\
&= \sum_{n=-\infty}^{\infty} (\mathcal{L}_H^{-1} G)(t + n\log\lambda) \, e^{i\varphi_{n-1}} \\
&\stackrel{d}{=} (\mathcal{L}_H^{-1} X)(t),
\end{aligned}$$

evidencing therefore that the "delampertized" process $\{(\mathcal{L}_H^{-1} X)(t), t \in \mathbb{R}\}$ becomes $\log\lambda$-cyclostationary, as expected from Theorem 2. In the case where the phases would not be randomly chosen, but all set to the same given value (say, 0), the "delampertized" version of (102) would simply takes the form of a periodic function [33] (see Fig. 4).

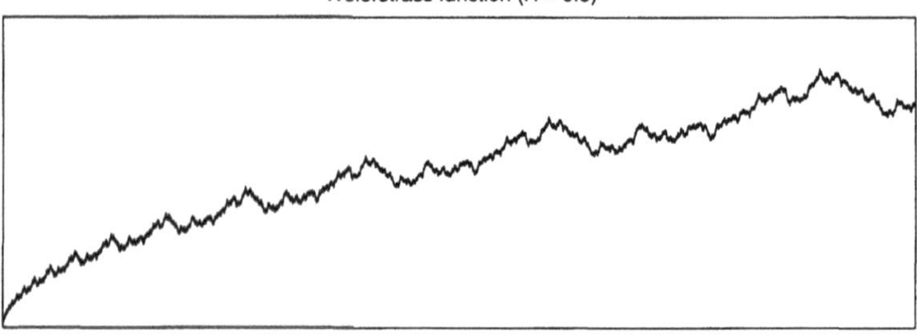

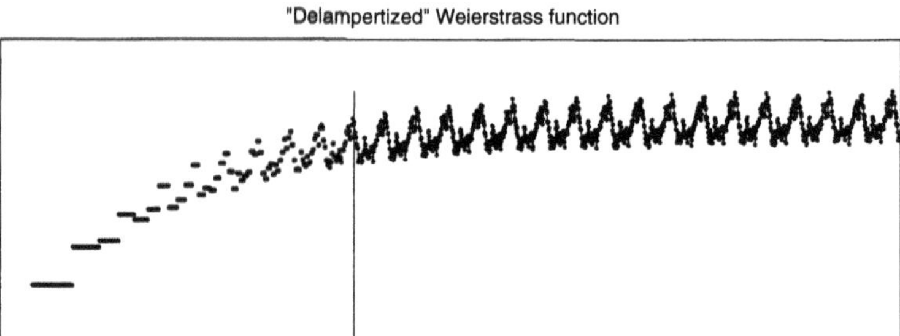

Fig. 4. Discrete scale invariance and cyclostationarity – The Weierstrass function (top) is not scale-invariant for all dilation factors, as evidenced by log-periodic fluctuations. "Delampertizing" this function ends up in a periodic function (bottom).

As a DSI process, $\{X(t), t > 0\}$ is necessarily nonstationary. However, introducing the notation

$$\Delta Z(t, \tau) := Z(t + \tau) - Z(t)$$

for the increment process of a given process $Z(t)$, we readily get from the definition (102) that

$$\Delta X(t, \tau) = \sum_{n=-\infty}^{\infty} \lambda^{-Hn} \, \Delta G(\lambda^n t, \lambda^n \tau) \, e^{i\varphi_n},$$

and it follows that the variance of this increment process expresses as

$$\begin{aligned} \mathbb{E}|\Delta X(t, \tau)|^2 &= \sum_{n,m=-\infty}^{\infty} \lambda^{-H(n+m)} \, \Delta G(\lambda^n t, \lambda^n \tau) \, \overline{\Delta G(\lambda^m t, \lambda^m \tau)} \\ &\qquad \times \iint_0^{2\pi} e^{i(\varphi_n - \varphi_m)} \, \frac{d\varphi_n}{2\pi} \, \frac{d\varphi_m}{2\pi} \\ &= \sum_{n=-\infty}^{\infty} \lambda^{-2Hn} \, |\Delta G(\lambda^n t, \lambda^n \tau)|^2 . \end{aligned}$$

In the specific case of the usual Weierstrass-Mandelbrot process defined through (103), it is interesting to note that

$$|\Delta G(\lambda^n t, \lambda^n \tau)|^2 = 2\ (1 - \cos \lambda^n \tau)\,,$$

evidencing the fact that the increment process $\Delta X(t, \tau)$ has a variance which does not depend on time t [4].

Parametric models – It has been shown in Sect. 3.4 that continuous-time H-ss processes can be obtained from Euler-Cauchy systems driven by some appropriately modulated white noise. Since such systems result from the lampertization of classical ARMA systems, it follows that varying their coefficients in a log-periodic way in time will generate DSI processes, in exactly the same way as cyclostationary processes can be obtained as the output of a (nonstationary) ARMA system with periodic time-varying coefficients.

The problem of getting corresponding models in discrete-time would need a specific discussion, and it will not be addressed here. Referring to [6,7] for some further details and illustrations, we will only mention that two preliminary approaches have been considered so far, both based on the idea of introducing a log-periodicity in the coefficients of a discrete-time model. In the first approach, the discretization is obtained by integrating the evolution of a continuous-time Euler-Cauchy system between two time instants, leading to an approximate form of DSI. In the second approach, a fractional difference operator (discretized, e.g., as in [38]) is introduced, and it is cascaded with a nonstationary AR filter whose coefficients are log-periodic.

5 Conclusion

The Lamperti transform is a simple way of connecting the two key concepts of stationarity and self-similarity, which are both known to be ubiquitously encountered in many applications. As such, it has been shown to offer simpler alternative viewpoints on some known problems, while providing new insights in their understanding. From a more innovative perspective, it has also been advocated as a new starting point for the analysis, modelling and processing of situations which depart from strict stationarity and/or self-similarity.

The purpose of this text was to collect and develop a number of general results related to the Lamperti transform and to support its revival but, of course, much work is still to be done in different directions. One can think of a number of further natural (e.g., multidimensional) extensions, as well as of the need for efficient algorithmic tools, based in particular on genuinely discrete-time formulations of the transform.

References

1. A. Ayache, S. Cohen, and J. Lévy-Véhel: *The covariance structure of multifractional Brownian motion, with application to long-range dependence* IEEE Int. Conf. on Acoust., Speech and Signal Proc. ICASSP'00, Istanbul (TR) (2000)
2. J. A. Barnes and D. W. Allan: Proc. IEEE **54**, 176 (1966)
3. J. Beran:*Statistics for Long-Memory Processes* (Chapman and Hall 1994)
4. M. V. Berry and Z. V. Lewis: *Proc. R. Soc. Lon. A* **370**, 459 (1980)
5. P. Bertrand, J. Bertrand and J. P. Ovarlez: 'The Mellin transform'. In: *The Transforms and Applications Handbook*, ed. by A. D. Poularikas, (CRC Press 1996)
6. P. Borgnat, P. Flandrin and P. O. Amblard: *Stochastic discrete scale invariance* IEEE Sig. Proc. Lett, submitted (2001)
7. P. Borgnat, P. Flandrin and P. O. Amblard: IEEE Workshop on Stat. Signal. Proc **SSP01**, 66 (2001)
8. K. Burnecki, M. Maejima and A. Weron: Yokohama Math. J. **44**, 25 (1997)
9. E. Chassande-Mottin and P. Flandrin: Appl. Comp. Harm. Anal. **6**, 252 (1999)
10. L. Cohen: IEEE Signal Proc. Lett. **5**, 292 (1998)
11. K. Falconer: *Fractal Geometry* (Wiley, New York 1990)
12. P. Flandrin: 'Scale-invariant Wigner spectra and self-similarity'. In: *Signal Processing V: Theories and Applications* ed. by L. Torres et al (Elsevier 1990) 149
13. P. Flandrin: IEEE Trans. on Info. Th. **38**, 910 (1992)
14. P. Flandrin: *Time-Frequency/Time-Scale Analysis* (Academic Press, San Diego 1999)
15. W. A. Gardner: IEEE Signal Proc. Mag. **8**, 14 (1991)
16. E. Gladyshev: Th. Prob. Appl. **8**, 173 (1963)
17. H. L. Gray and N. F. Zhang: J. Time Series Anal **9**, 133 (1988)
18. J. Koyama and H. Hara: Chaos, Solitons and Fractals **3**, 467 (1993)
19. J. Lamperti: Trans. Amer. Math. Soc. **104**, 62 (1962)
20. B. B. Mandelbrot and J. W. Van Ness: SIAM Rev. **10**, 422 (1968)
21. G. J. Maskarinec and B. Onaral: IEEE Trans. on Circuits and Systems **41**, 75 (1994)
22. Y. Meyer and H. Xu: Appl. Comp. Harm. Anal. **4**, 366 (1997)
23. E. Noret: Contribution à l'étude de signaux auto-similaires et à mémoire longue au moyen de systèmes linéaires non stationnaires. PhD Thesis, Université de Nantes (1999)
24. E. Noret and M. Guglielmi: 'Modélisation et synthèse d'une classe de signaux auto-similaires et à mémoire longue'. In: *Proc. of Fractals in Engineering* (Delft NL 1999) pp. 301–315
25. C. Nuzman and H. V. Poor: 'Transformed spectral analysis of self-similar processes'. *33rd Annual Conf. on Info. Sci. Syst.* (CISS'99, Baltimore MD 1999)
26. C. Nuzman and H. V. Poor: J. Appl. Prob. **37**, 429 (2000)
27. R. Peltier and J. Lévy-Véhel: *Multifractional Brownian motion: definition and preliminary results* INRIA Research Report No. 2645 (1995)
28. M. B. Priestley: *Spectral Analysis and Time Series* (Academic Press 1981)
29. R. Roberts, W. Brown and H. Loomis: IEEE Sig. Proc. Mag., 38 (1991)
30. G. Samorodnitsky and M. S. Taqqu: *Stable Non-Gaussian Processes : Stochastic Models with Infinite Variance* (Chapman and Hall 1994)
31. R. Silverman: IEEE Trans. on Info. Th. *IT-3*, 182 (1957)
32. D. Sornette: Physics Reports **297**, 239 (1998)
33. C. Tricot: *Curves and Fractal Dimension* (Springer Verlag 1995)

34. W. Vervaat: Bull. Int. Stat. Inst. **52**, 199 (1987)
35. A. Vidács and J. T. Virtamo: 'ML estimation of the parameters of fBm traffic with geometrical sampling'. 5th Int. Conf. on Broadband Comm (BC'99, Hong-Kong 1999)
36. A. Vidács and J. T. Virtamo: IEEE Infocom 2000, Tel-Aviv (2000) pp. 1791–1796
37. B. Yazici and R. L. Kashyap: IEEE Trans. on Signal Proc. **45**, 396 (1997)
38. W. Zhao and R. Rao: 'On modeling self-similar random processes in discrete-time'. In: *IEEE Int. Symp. on Time-Freq. and Time-Scale Anal.* (TFTS'98, Pittsburgh 1998) pp. 333–336

Fractal Sums of Pulses and a Practical Challenge to the Distinction Between Local and Global Dependence

Benoit B. Mandelbrot

Sterling Professor of Mathematical Sciences, Yale University, New Haven, CT 06520-8283 USA

Abstract. The partly random fractal sums of pulses (PFSP) are a family of random functions that depend on a kernel function K and at least two positive parameters C and δ. Given K, the construction of $F(t; C, \delta)$ consists in adding N affine versions of a pulse as follows. The pulse height Λ and its width W are random variables related by w/λ^{δ} = a constant. The width is distributed according to the Poisson measure $Cw^{-1}dwdt$ in the "address plane" of coordinates w and t. For finite C, the increments of $F(t; C, \delta)$ fail to be strongly mixing therefore they exhibit global dependence. Indeed some PFSP resemble the icon of global dependence, which is fractional Brownian motion (FBM) with $H \neq 1/2$. When the presence of strong mixing must be tested empirically, many tests rely on the comparison of two exponents of diffusion: that of a r.f. $X(t)$ and of a "shuffled" r.f. $\tilde{X}(t)$ whose increments for $\Delta t = 1$ (say) are independent and follow the same distribution as $X(t)$. In FBM, the diffusion exponent is H for the process itself and $\frac{1}{2}$ for its shuffled variant. Therefore, $H \neq 1/2$ is a token of global dependence. For the Lévy stable motion (LSM) to the contrary, the diffusion exponent is the same as for the independent process with the same marginal distribution. The PFSP are not so clear-cut. The dependence is always global. But consider those tests of globality versus locality that, like R/S testing, are founded on the exponent of diffusion. Those tests will classify the dependence in many PFSP as local. Therefore, the PFSP are challenging borderline cases, while the conceptual fact is important, more important is the concrete fact that their rich properties and the absence of arbitrary grids make them excellent candidates for modeling phenomena that combine global dependence with long distribution tails. Furthermore, related structures that are discussed elsewhere, namely, the multifractal measures obtained as products of pulses, are grid-free and provide a great improvement over the now-classical multifractal measures generated by multiplicative cascades in a grid.

1 Introduction

1.1 "Normal" and "Anomalous" Fluctuations, the Noah and Joseph Effects and the Distinction Between the Local and the Global Form of Statistical Dependence

The "normal" model of natural fluctuations is the Wiener Brownian motion process (WBM). By this standard, however, many natural fluctuations exhibit clear-cut "anomalies". Some are due to large discontinuities ("Noah Effect"), others to non-negligible global statistical dependence ("Joseph Effect"). I have long argued - especially in [7] and [9] - that the geometric features of surprisingly many of these anomalous aspects of nature are fractal. For example, for many large "Noah" discontinuities, the tail probability distribution follows a power-law. That is, for large u, a discontinuity U that exceeds the value u has a probability of the form $Pr(U > u) \sim u^{-\alpha}$, with α a positive constant. Second example: for large lags s, many globally correlated "Joseph" fluctuations have a correlation function of the power-law form $C(s) \sim 2H(2H-1)s^{2H-2}$, with $1/2 < H < 1$. [7] shows that one can model various instances of the Noah effect by the classical process of Lévy stable motion (LSM), and various instances of the Joseph effect by the process of fractional Brownian motion (FBM).

In the classical cases of WBM, FBM and LSM, the distinction between local and global dependence is straightforward. The increment of WBM and LSM are independent, hence locally (short-run) dependent. But if $H \neq \frac{1}{2}$, the increments of FBM are globally (long-run) dependent. This contrast makes the task of laying down a boundary between local and global that can be implemented by a statistical test.

LSM and FBM, however, are far from exhausting the anomalies found in nature. In particular, many phenomena exhibit *both* the Noah and the Joseph effects and fail to be represented by either LSM or FBM. Hence, fractal modeling of nature demands "bridges," namely, r.f.'s that combine the infinite variance feature that is characteristic of LSM and the global dependence feature that is characteristic of FBM. Such is the case of a new family of random functions described in this paper and called "partly random sums of pulses" or PFSP. A very different and more classical bridge, is *fractional Lévy motion.* It is interesting mathematically, but not versatile and has found no concrete use.

Two general issues come up.

Firstly, when the tail exponent α satisfies $\alpha < 2$, the exponents of PFSP satisfy $\alpha = 1/H$. This identity strongly links the tail and the correlation. As a result, while the PFSP are globally dependent, their dependence seems from certain viewpoints to be only local.

Furthermore, the mathematical theories of LSM and FBM are known to exhibit striking formal parallels as well as often-noted discrepancies. One major discrepancy is in the allowable range of the exponent H which is defined by the self-affinity condition that the distribution of $T^{-H}F(T\rho)$ is independent of T. For LSM, $1/2 < H < \infty$, while for FBM, $0 < H < 1$. This mismatch poses a conceptual challenge. Being unexpected, the formal parallels between LSM and FBM are sometimes described as miraculous. But in fact they have deep roots worth exploring. Some such roots are brought forward by the PFSP.

1.2 Inspiration for the Partly Random Fractal Sums of Pulses (PFSP)

The PFSP focus on a property of *Lévy flight* or *Lévy stable motion*, modify it, achieve a major generalization. LSM is well-known to be the sum of an infinity of step-functions whose sizes follow a power-law distribution. This paper preserves this classical distribution of sizes but replaces the step-functions by suitable affine reductions or dilations of more general "templates", obtained as graphs of a *kernel function* K. The result can be described as being an *affine convolution.* A template whose kernel function only varies on a (bounded) interval will be called *pulse.*

1.3 Sketch of the Construction of the PFSP

A PFSP built with only positive pulses, involves a single parameter C. It will be denoted as $F(t; C, \delta)$ and constructed as follows. A pulse's height and location are ruled by the same distribution as in a Lévy flight: the probability of the point $\{\lambda, t\}$ being found in an elementary rectangle of the (λ, t) address plane is $C\lambda^{-\delta-1}d\lambda dt$. This probability is carried over from the original paper on cutouts or tremas on the line, [6].

Recall that in a Lévy flight, the exponent δ is constrained to satisfy $0 < \delta < 2$. Many applications view this limitation as unduly restrictive. A major immediate novelty is that in an PFSP, the constraints are far weaker: either $\delta > 0$ or $\delta > 1$, depending on the kernel function K. The pulse's width W (the length of the smallest interval in which the pulse varies) satisfies $W = \sigma|\Lambda^{\delta}|$, where $\sigma > 0$ is a scale constant, implying that $Pr\{W > w\} \propto w^{-1}$. The reason why the resulting $F(t, C, \delta)$ is called *semi-random* is because Λ and W are functionally related. In fully random PFSP, as discussed in [4], Λ and W are statistically independent.

When the pulses can be either positive or negative, there are two parameters C' and C'' and the PFSP will be denoted as either $F(t : C', C'', \delta)$ or as $F(t; C, \delta)$ where C represents the combination of two posoitive real numbers C' and C''.

1.4 The Roles of δ as Tail Exponent and $H = 1/\delta$ as Self-affinity Exponent

The key property of the PFSP is that for many kernels the same exponent δ plays a fundamental role for both longtailedness and global dependence.

A first role of δ is that, over any time increment Δt, the increment of F follows a power-low probability distribution of exponent δ.

A second role of δ is that, by design, all PFSP are self-affine, namely, invariant under an affine rescaling of the form $T^{-H}X(T\rho)$ with the exponent $H = 1/\delta$. (In physicists' language, each PFSP defines its own "class of universality" with respect to a suitable affinity).

As a result, there is no counterpart to the standard study of attraction of a random function $X(t)$ to a limit such as Wiener or fractional Brownian motion or Lévy motion.

1.5 When $\delta < 2$, the Link Between the Tail and Self-affinity Exponents Complicates the Testing for Global Dependence

Assume $\delta < 2$, write $X = F(t+1, c, \delta) - F(t, c, \delta)$, and "shuffle" the X to obtain a sequence of independent r.v. $\tilde{X}$ with the same distribution as X. That sum is attracted by the LSM. Therefore, both the original X and the shuffled $\tilde{X}$ have the same self-affinity exponent $H = 1/\delta$. Tests like R/S which are based on the comparison of those exponents will declare the X to be (R/S) independent. But we know that they are globally dependent.

1.6 Lateral Rescaling and Lateral Attractors

It is important to introduce an alternative "lateral" rescaling, and show that each semi-random PFSP has an interesting "lateral attractor." This attractor is sometimes a Lévy flight (LSM), a conclusion that might have been vaguely expected, because the construction starts with the Lévy measure. But there exists another possibility, the lateral attractor may be a fractional Brownian motion (FBM), which is a surprise. This is readily implemented when $\delta > 2$. The possibilities of merging FBM and LSM into a broader common universe is enlightening. It establishes the existence of deep links - already mentioned and described as desirable - between two independent theories that are known to be very different yet strikingly parallel. Furthermore, the attractor may also be neither LSM nor FBM.

For finite C, the increments of the $F(t; C, \delta)$ are globally dependent, but when the lateral limit is LSM, they become independent. Thus, the PFSP bring altogether new conceptual light on the probabilistic notion of global dependence and the related notion that a process is attracted by another process. Indeed, depending on the K, global dependence is expressed in either of two ways: a) by a special exponent H familiar in such known contexts as FBM, or b) by a prefactor rather than a special exponent.

1.7 Summary

This paper is far from exhaustion but shows the great variety of distinct behaviors that can be found in an PFSP, as we vary δ and three properties of the kernels:

a) Discontinuous versus very smoothly continuous;

b) Canceling (i.e., vanishing outside the interval in which they vary) versus non-canceling.

c) Atoms versus bursts. This is a useful but elusive distinction. When the pulse is made of a rise and fall followed later by another rise and fall, it can be decomposed into a burst of two indecomposable or atomic pulses.

The resulting templates are exemplified in Table 1: cylinders (one discontinuous *rise* followed after the time W by one discontinuous *fall*), multiple steps, cones (uniform rate rise followed by uniform rate fall), and uniform rate rises. Other templates are discussed in forthcoming papers.

1.8 Related Papers

They investigate: A) additional classes of PFSP with kernels for which convergence demands second differences;

Table 1. Effect of the pulse shape over the "lateral" limit. As C tends to infinity, the limit can be either LSM, a stable Lévy motion, or FBM, a fractional Brownian motion. The Table is explained in the text. The empty box corresponds to PFSP for which the first increments diverge. A further renormalization leads to converging second increments that will be described in a forthcoming publication.

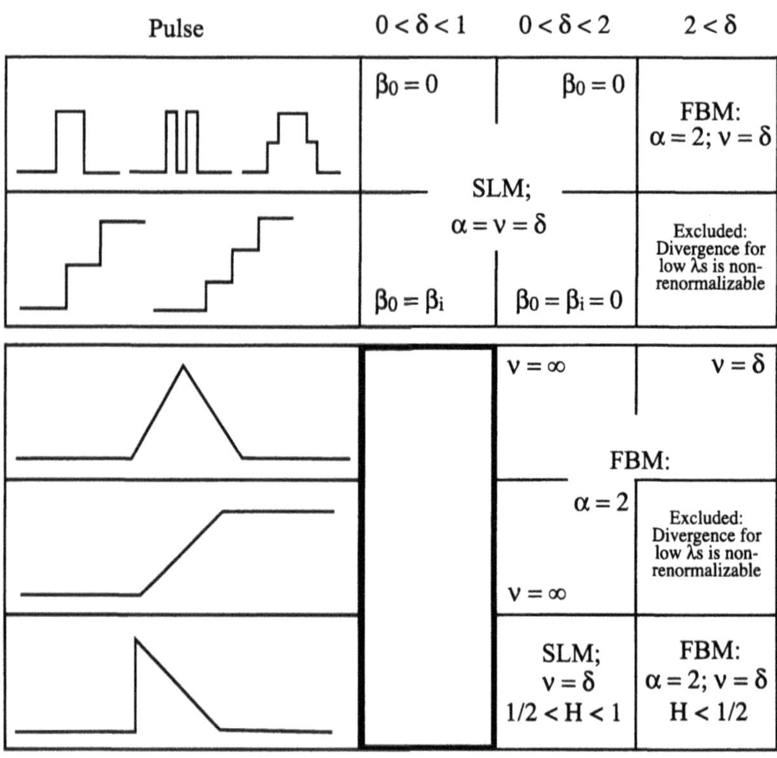

B) semi-random PFSP with $W = \sigma \Lambda^{\theta}$, where $\theta \neq \delta$; they are not self-affine.

C) PFSP that are fully random in the sense that W is statistically independent of Λ with the measure $\propto w^{-\theta-1} dw$, where $0 < \theta < 1$; they are shown in [4] to be self-affine with $H = (1 - \delta)/\theta$;

D) fractal sums of *micropulses* (FSM) which generate (FBM); see [2,3];

E) multifractal products of pulses [1]; and

F) concrete applications [5].

2 Definitions

2.1 Stationarity and Self-affinity

The function $F(t)$ is said to have *stationary increments* if the translated function

$$F(t_0 + t) - F(t_0)$$

has the same distributions for all values of t_0. A function $F(t; C, \delta)$ is said to be *self-affine of exponent* $H > 0$ when the rescaled function

$$\rho^{-H}[F(t_0 + \rho t; C, \delta) - F(t_0; C, \delta)]$$

has the same distributions for all values of t_0 and $\rho > 0$. Some authors denote self-affinity by the term, self-similarity, which I proposed long ago but abandoned because it is inappropriate.

2.2 Pulse Templates, Pulses, Affine Convolutions, and Fractal Sums of Pulses

The graph of $K(t)$, a one-dimension function of a one-dimensional variable, will be called *generator*, or *pulse template*, if $K(t)$ is constant outside an interval; we shall set the shortest such interval to be of length 1.

A *pulse* is a translated affine transform $K\left(\frac{t-t_n}{w_n}\right)$ of $K(t)$, where λ_n, t_n, and w_n are called the pulse's height, position, and width.

A *sum of pulses* is a function of the form

$$F(t) = \sum \lambda_n K\left(\frac{t - t_n}{w_n}\right).$$

Figure 1 is an example of a sum of these pulses. When $F(t)$ is a self-affine function, it will be called a *fractal sum of pulses*, FSP.

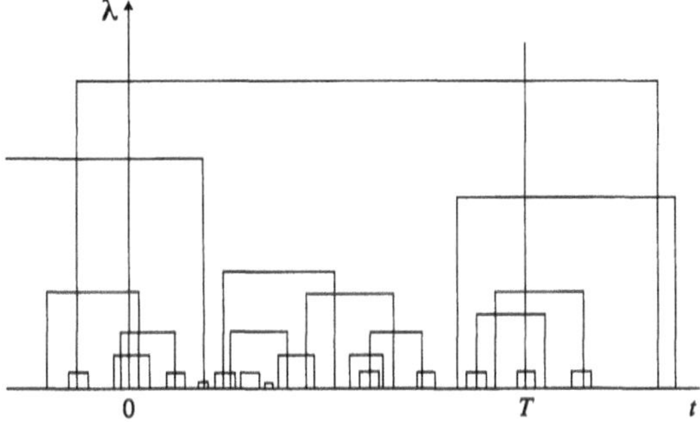

Fig. 1. Addition of cylindrical pulses. This diagram is extremely schematic and its only purpose is to classify the pulses from the viewpoint of the increment of a PFSP from t=0 to t=T. Therefore there was no need to either implement the very special relation between the pulses' heights and widths, or to attempt to draw the PFSP as the sum of infinite numbers of both infinitely small and infinitely large pulses. From the viewpoint of the increment over a finite time interval, three possibilities arise. Some pulses are constant in the interval, therefore contribute nothing. Other pulses' up and down increments cancel out. The only pulses that matter are those for which the interval from t=0 to t=T contains only the initial or the final discontinuity but not both.

An *affine convolution* of the sequence $\{\lambda_n, t_n\}$ by the kernel $K(\cdot)$ is obtained when the pulse heights and widths are linked by a relation $w_n = \sigma\lambda_n^\delta$, where $\sigma > 0$ and $\delta > 0$ are prescribed. Thus, the PFSP are affine convolutions of sequences $\{\lambda_n, t_n, w_n\}$.

2.3 The Lévy Measure for the Probability Distribution of Pulse Height and Position

The simplest template is a step function for all n. The distribution of $\{\lambda_n, t_n\}$ that insures self-affinity in that case was discovered by Paul Lévy and is classical. The same distribution continues to be required in all FSP. Let the plane of coordinates t and λ be called *address plane.* Take a rectangle $[\lambda, \lambda+d\lambda]\times[t, t+dt]$ in the address plane, such that $[\lambda, \lambda + d\lambda]$ does not contain $\lambda = 0$. Given an exponent $\delta > 0$ and two scale factors C' and C'', the Lévy measure is

$$C'\lambda^{-\delta-1}d\lambda dt, \text{ if } \lambda > 0, \text{ and } \mathrm{C''|\lambda|^{-\delta-1}d\lambda dt}, \text{ if } \lambda < 0.$$

Giving C' and C'' is, of course, equivalent to giving the overall scale $C' + C''$ and the skewness factor $C'/(C' + C'')$.

To define an FSP, the probability of finding an address point (λ, t) in the elementary rectangle is set equal to the Lévy measure. The number of address points in a domain $\mathcal{D}$ in the address space is taken to be a Poisson random variable whose expectation is the integral of the Lévy measure over $\mathcal{D}$. The total number of pulses is countably infinite.

2.4 Semi-random Pulses and the Probability Distribution of Pulse Widths and Position

The simplest pulses are the step functions used by Paul Lévy to generate the LSM. The pulses examined in this paper and illustrated in Table 1 and generalize the step functions. The height Λ is random with the Lévy measure, but $|\Lambda|$ fully determines the width W. The resulting processes are called *semi-random* and denoted as PFSP. To insure that the PFSP is self-affine, one must take

$$W = \sigma|\Lambda|^\delta, \text{ where } \sigma > 0.$$

The resulting joint measures of width and position are

$$C'w^{-1}d\lambda dw \text{ and } \mathrm{C''w^{-1}d\lambda dw}.$$

They are independent of δ. So is the probability distribution of W, namely,

$$Pr\{W > w\} \propto w^{-1}.$$

The units in which Λ and W are measured are arbitrary and unrelated. If those units are identical, either $C' + C''$ or σ can be normalized to 1 by changing the unit. However, up to scale, the distribution of an PFSP is determined by

$(C' + C'')\sigma$ and the skewness factor $C'/(C' + C'')$. In the sequel, an important role is played by lateral limit theorems that are expressed most conveniently by fixing $C' + C''$ and allowing $\sigma \to \infty$. The consequences of $\sigma \to \infty$ are obvious when the pulses are "cylindrical:" made of a *rise* followed, after a span of $\Delta t = w$, by a *fall* of equal absolute value and opposite sign. Clearly, the contribution to $F(t + T; C, \delta) - F(t; C, \delta)$ from a pulse such that $w > T$ is not a pulse but a solitary rise or fall. Therefore, as $\sigma \to \infty$, each cylindrical pulse reduces to a rise or a fall, and the fact that a pulse has a bounded support becomes less and less significant.

3 Self-affinity and the Exponent $H = 1/\delta$; Existence of Global Dependence

3.1 The Self-affinity Property of all PFSP

An *admissible* combination of a pulse shape and a value of δ is a combination such that the semi-random PFSP construction yields a well-defined random process $F(t; C, \delta)$. For other values of δ, the construction diverges and δ is called *excluded.* When the construction converges, it is easy to see that the resulting PFSP are self-affine with $H = 1/\delta$. This formula is familiar from the Lévy case when the pulses are step functions but with a striking novelty: $\delta > 2$ provokes an irreducible divergence for LSM. But for PFSP, there is no divergence, that is, δ is unrestricted.

In one case to be described elsewhere, the construction of $F(t + T; C, \delta) - F(t; C, \delta)$ diverges, but that of the second difference $[F(t+T; C, \delta) - F(t; C, \delta)] - [F(t; C, \delta) - F(t - T; C, \delta)]$ converges and is self-affine.

3.2 First Corollary of Self-affinity: Each PFSP Defines a Special Domain of Attraction, Hence the Standard Limit Problem Concerning Random Processes Becomes Degenerate

In the study of random functions, the standard next issue is whether or not there exists an exponent H such that, setting $F(0; C, \delta) = 0$,

$$\text{weak} \lim_{\mathrm{T} \to \infty} \mathrm{T}^{-\mathrm{H}} \mathrm{F}(\rho \mathrm{T}; \mathrm{C}, \delta)$$

is a non-degenerate function of ρ, called the "attractor" of F. The most familiar attractions are WBM and LSM in the case of independence and FBM in the case of dependence.

When $F(t)$ is a PFSP, the standard limit problem does not arise, since in distribution $T^{-H}F(\rho T; C, \delta)$ is independent of T. That is, each PFSP defines its own domain of attraction of exponent $H = 1/\delta$.

Standard domains of attraction to an PFSP. Given a self-affine attractor $X(t)$, the next challenge is to describe its domain of attraction. This is the collection of r.f.'s $G(t)$ for which the rescaling (or renormalizing) function $A(T)$ can be selected so as to insure that

$$\text{weak} \lim_{T \to \infty} A^{-1}(T) G(\rho T) = X(t).$$

Thus far, my study of the domain of attraction of PFSP has limited itself to r.f.'s that are themselves sums of pulses, but involve a density other than Lévy's or a relation other than $w = \sigma\lambda^{\delta}$.

Clearly, weak $\lim_{T\to\infty} T^{-H} F(\rho T; C, \delta)$ is unchanged if the distribution of pulse heights is changed over a bounded interval, for example if pulse height is restricted to $\lambda > \varepsilon > 0$.

For the next obvious change, the example of LSM suggests replacing the constants C' and C'' in the Lévy density by functions $C'(\lambda)$ and $C''(\lambda)$ that vary slowly for $\lambda \to \infty$. And FBM suggests replacing the relation $w = \sigma\lambda^{\delta}$ by $w = \sigma(\lambda)\lambda^{\delta}$, where $\sigma(\lambda)$ is also slowly varying. These changes lead to a nonself-affine generalized PFSP. The questions is whether or not, as $T \to \infty$, there exists a rescaling $A(T)$ that makes $A^{-1}(T)G(T)$ converge weakly to a self-affine PFSP. I have been content with verifying (see Sect. 7.6) that a sufficient condition is that $C'(\lambda)\sigma(\lambda) \to 1$; in that case, $T^{1/\delta}A(T)$ is the inverse function of $T = \sigma(\lambda)\lambda^{\delta}$, implying that $A(T)$ is slowly varying. The condition $C'(\lambda)\sigma(\lambda) \to 1$ is demanding, resulting in a narrow domain of attraction, a notion discussed in Sect. 3.4.

3.3 Second Corollary of Self-affinity: All PFSP Are Globally Dependent

A corollary of Sect. 3.2 is that if F is a PFSP, then $T^{-H}F(\rho T)$ fails to converge to a standard attractor relative to asymptotically independent increments, namely, either WBM or LSM. This implies that all semi-random PFSP must fail to satisfy the usual criteria that express that dependence between increments is local.

In fact, they are uniformly globally dependent. For example, define for each t the following two functions

- the rescaled finite past $T^{-1/\delta}[F(t) - F(t - Tp; C, \delta)]$
- and the rescaled finite future $T^{-1/\delta}[F(t + Tp; C, \delta) - F(t; C, \delta)]$.

Some pulses contribute to both past and future, hence these random functions of p are not statistically independent. Because of self-affinity, their joint distributions are independent of T. This means that strong mixing is contradicted uniformly for all T.

3.4 Thoughts on the Role of Limit Theorems, Given That, in the Case of PFSP, the Standard Limit Problem Is Degenerate

In the light of Sect. 3.3, let us compare the "attractands" with their attractor. The process of going to the limit to destroy idiosyncratic features of the attractand while preserving "universal" features is expected. This is the case for the most important attractors namely, the nonrandom attractor for the laws of large numbers, the Gaussian r.v. for the central limit theorems, and WBM for the functional central limit theorems. These attractors' domains of attraction are very broad, being largely characterized by the absence of global independence and of significant probabilities for large values. By contrast, when a basin of attraction is narrow, the attractor yields specific information about the attractands. Thus, there is a sharp contrast between broad universality with little information and narrow universality with extensive information.

Down to specifics. For LSM, assuming that the attractands are broadly dependent, their tails must be long and strictly constrained. For FBM, assuming that the tails are short, the dependence must be global and strictly constrained. Similarly, the sufficient condition $C'(\lambda)\sigma(\lambda) \to 1$ in Sect. 3.2 defines for each PFSP a (partial) domain of attraction that is tightly constrained. The novelty is that with *the tails and the global dependence must involve the same exponent.*

The resulting variety of forms creates a use for additional limit problems that would put order by destroying some of the PFSP's overabundant specifications. That is, the finding in Sect. 3.3 must spur the search for alternative limit problems, for which the domains of attraction are broader, therefore reveal more "universal" properties of the PFSP. Sect. 4 will advance one possibility.

4 The Concept of "Lateral Limit Problem" and the Exponent α; for PFSP, the Lateral Attractor Can Be Either Uniscaling ($\alpha = \delta = 1/H$) or Pluriscaling ($\alpha = \min[2, \delta] \neq 1/H$)

4.1 Background of the New "Lateral" Limit Problem

Neither the random walk nor the Poisson process of finite density is self-affine, but both are attracted in the usual way to the Wiener-Brownian motion $B(t)$, which is self-affine with $H = 1/2$. That is, writing $F(0) = 0$, it is true in both cases that $T^{-1/2}F(\rho T)$ depends on T for $T < \infty$, but not in the limit $T \to \infty$. Since the replacement of T by ρT transforms discrete time and F into real variables, rescaling time before taking a limit is necessary in the case of a random walk. But in the Poisson case the standard passage to the limit can be rephrased to avoid rescaling time. One imagines N independent and identically distributed Poisson processes $F_n(t)$, then one forms

$$\tilde{F}_N(t) = \sum_{n=1}^{N} F_n(t),$$

and one finds that,

$$\mathrm{weaklimit}_{N\to\infty} \mathrm{N}^{-1/\alpha}\tilde{\mathrm{F}}_{\mathrm{N}}(\mathrm{t}) = \mathrm{B}(\mathrm{t}), \text{ with } 1/\mathrm{H} = \alpha = 2.$$

This rephrasing of the passage to the limit is not important in the Poisson case, but it has the virtue that it continues to make sense in the case of PFSP. The theory of the addition of independent identically distributed random variables tells us that, for fixed t, one must have $0 < \alpha \leq 2$ and the limit is a stable rv: Gaussian for $\alpha = 2$, and Lévy stable for $0 < \alpha < 2$. Table 1 lists the values of α for a selection of pulse shapes, and Sect. 7 gives an example of derivation.

4.2 The Lateral Limit Problem as Applied to PFSP; Reason for the Term "Lateral"

Let us now observe that forming $\tilde{F}_N(t)$ for a semi-random PFSP, and then letting $N \to \infty$ is equivalent to viewing $F(t)$ as a function of both t and $C' + C''$. Then we keep C'/C'' fixed and let $C' + C'' \to \infty$ and $C'' \to \infty$. One can think of the axis of $C' + C''$ as orthogonal to the axis of t, hence the term *lateral.*

For example, replacing C' by $C'_1 + C'_2$ and C'' by $C''_1 + C''_2$ can be interpreted as follows. The pulses corresponding to C'_1 and C''_1 can be called *red* and said to add up to $F_R(t)$, and the pulses corresponding to C'_2 and C''_2 can be called *blue* and said to add up to $F_B(t)$; all pulses together yield $\tilde{F}(2, t) = F_R(t) + F_B(t)$. In order to represent $F(N, t)$ graphically on a page, one must rescale it by the factor $N^{-1/\alpha}$.

4.3 An Important Corollary of the Results in Table 1: Global Dependence Can Be "Pluriscaling" ($H \neq 1/\alpha$), Like in the Limit Case FBM; but It Can Also Be "Uniscaling" ($H = 1/\alpha$), Like in the Limit Case LSM

Table 1 uses α to denote the usual exponent of stability. A first glance shows familiar r.f.'s among the lateral limits. The LSM has independent increments and satisfies $0 < \alpha < 2$ and $H = 1/\alpha$. The FBM, except for $H = 1/2$, has globally dependent Gaussian increments, so that $\alpha = 2$, but satisfies $H \neq 1/\alpha$.

In the nineteen sixties and for some time later, I carried out studies of global dependence that are reproduced in [7]. They concentrated on these two examples (broadened by fractional Lévy motion) and on the R/S statistic. Those examples made me think that global dependence can be tested and measured by a single exponent H. Given α, I thought that global dependence could be defined by $H \neq 1/\alpha$ and measured by $H - 1/\alpha$. It is a pleasure to note that I was prudent enough *not* to write of independence and dependence but rather of (R/S)-independence and (R/S)-dependence. But I confess having expected that any example beyond this latter distinction would belong to "mathematical pathology."

The PFSP show that I was thoroughly mistaken.

5 Address Diagrams and the Mechanism of Non-linear Global Dependence in PFSP

5.1 Address Function and Diagram; Characteristic Function of F

Once again, each semi-random pulse is represented by a point in the address space $\{t, \lambda\}$. Therefore, when a pulse template includes one or more time-intervals, one can define an address function and an address diagram, as seen momentarily. Examples are provided by this paper's illustrations.

When there is one time interval it will be denoted by $[0, T]$. The *address* function of $[0, T]$, denoted by $\varphi(t, \lambda \mid 0, T)$, will be the contribution of the pulse represented by $\{t, \lambda\}$ to $F(T; C, \delta)$, where we set $F(0; C, \delta) = 0$, and – as may be the case - to other increments of F as well.

Since $\{t, \lambda\}$ has a Poisson distribution, and the contributions of the address points to F are additive, the logarithmic characteristic function of F is simply the integral

$$\psi(\xi) = C(C', C'') \int (e^{i\xi\varphi(t,\lambda|0,T)} - 1)|\lambda|^{-\delta-1} d\lambda$$

carried over the address space. Here,

$$C(C', C'') = C' \text{ for } \lambda > 0 \text{ and } \mathrm{C}(\mathrm{C}', \mathrm{C}'') = \mathrm{C}'' \text{ for } \lambda < 0.$$

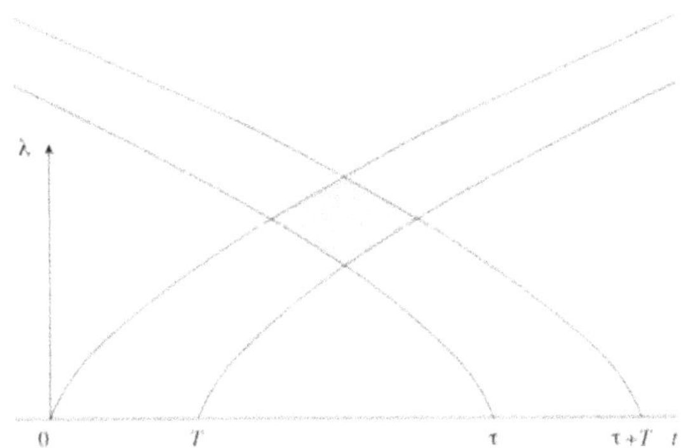

Fig. 2. This is an illustration of the address plane in which a cylindrical pulse of height k and center t is represented by a point (t, k). In the interval from t = 0 and $t = T$, the increment of the corresponding PFSP is affected by the pulses with an address located in "domain 1". All the other pulses do not matter: those with an address in the hatched domain because they are constant, and those with an address in "domain 2" because they go up and down but cancel out between 0 and T. The presence of this domain 2 is critically important and the PFSP was specifically "invented" to include this feature. It allows the tail exponent to exceed the bound relative to the Lévy stable processes, namely $\alpha < 2$. When the lateral limit for $C\delta\theta$ is Lévy stable, the domain 2 disappears in the limit.

When pulses are constant outside an interval, it follows that $\varphi = 0$ over a large *excluded* part of the address space. The analytic form of φ may depend on (λ, t), making it convenient to integrate $\psi(\xi)$ separately over a number of distinct domains. The boundaries of these domains will be said to form an *address diagram*. The case when the pulses are cylinders is illustrated in Fig. 2. The address points in the dotted arrows have no effect.

5.2 Joint Address Diagrams and Pictorial Illustration of the Reason for Interdependence Between the ΔF Corresponding to Two Non-overlapping Intervals; Its Strongly Non-linear Character

The two intervals being $[0, T]$ and $[t, t+T']$, with $t \geq T$, is denoted by cross BUG. It it is important to fully understand the simplest case when the pulse reduces to two steps, either cancelling (up then down or down then up) or noncancelling (up and up or down and down). The case $0 < \delta < 1$ raises no convergence problem.

For both of these pulses $\varphi(t, \lambda \mid 0, T)$ and $\varphi(t, \lambda \mid t, t+T')$ are either $= \lambda$ or $= 0$. Therefore, dotted areas of Fig. 3 have no effect and the domain of integration of φ splits into three parts: The first is denoted by horizontal hatching. It affects only $[0, T]$ and defines a r.v. Δ_{LL}. The second is denoted by cross-hatching. It affects both $[0, T]$ and $[t, t+T']$ and defines a r.v. Δ_O. The third is denoted by vertical hatching. It affects only $[t, t+T]$ and defines a r.v. Δ_{RR}.

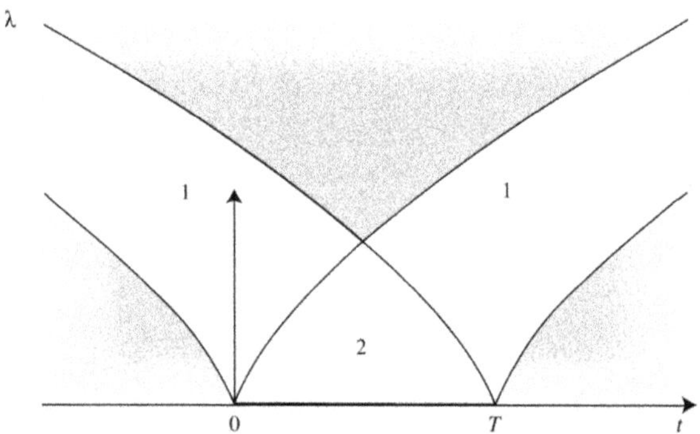

Fig. 3. Reason why the increments of a PSFP over two disjoint time intervals manifest a significant level of statistical dependence. The reason is that those increments include a common additive component contributed by pulses with an address in the hatched domain. As a function of the distance between the two intervals, the resulting dependence decreases following a very specific power-law. One finds that the exponents of self-affinity and of dependence are linked. This is why, from the viewpoint of the concept of global (long-range) dependence, the PFSP processes described in this paper present a challenge.

That is,

$$\begin{aligned} \Delta_L &= F(T; C, \delta) - F(0; C, \delta) = \Delta_{LL} + \Delta_O \\ \Delta_R &= F(t+T; C, \delta) - F(t; C, \delta) = \pm \Delta_O + \Delta_{RR}. \end{aligned}$$

Once again, Δ_{LL}, Δ_O and Δ_{RR} are independent and the symbol $\pm$ means $+$ when the pulse is noncancelling and $-$ when the pulse is cancelling.

The origin and the nature of the resulting dependence between Δ_{LL} and Δ_{RR} are clearest when t is several times larger than T and easiest to follow if $C' = C''$ and their common value C is made to increase from 0.

The theory of Lévy stable variables correctly suggests the following: except for small C, the order of magnitude $\lambda(C)$ of the largest contribution to Δ_{LL} and Δ_{RR}, hence to Δ_L and Δ_R, satisfies $C[\lambda(C)]^\delta = 1$, i.e., $\lambda(C) \sim C^{1/\delta}$.

For small C, below the cross-over at $C \sim T$, both Δ_L and Δ_R continue to be dominated by $\lambda \sim \pm T^{1/\delta}$. To the contrary, the addend Δ_O comes exclusively from tail values of λ, and is small. The dependence is therefore small, except when $t = T$, which corresponds to neighboring intervals.

When C is very large, $\Delta_{LL} \propto C^{1/\delta}$ and $\Delta_{RR} \propto C^{1/\delta}$, while $\Delta_O \propto C^{1/2}$. Again, the dependence of Δ_L and Δ_R is small.

We are left with the midrange where C is of the order of t. There, both Δ_L and Δ_R are of the order of magnitude of the largest contributing λ. Once again, the dependence between Δ_L and Δ_R of the order of 1 if the same largest λ contributes to Δ_0, hence *both* Δ_L and Δ_R. Otherwise the dependence between Δ_L and Δ_R is small. Strong dependence has a probability of the order of 1, say 1/3.

As $C \to \infty$, the midrange where t is of the order of C also $\to \infty$, which shows that Δ_L and Δ_R become independent and do so in non-uniform fashion.

6 Discussion of Table 1: Effects of Pulse Shape on the Admissible δ, and on the Lateral Attractor

Changing the pulse shape greatly affects an PFSP. First, it affects the domain of admissible δ's. Next, it affects the attractor of $F(t)$ and in particular the dependence of α on δ. Several examples are summarized in Table 1.

The left column of Table 1 illustrates a selection of pulse templates. Other templates are examined in follow-up papers. A striking immediate observation is that a discontinuous PFSP can have continuous lateral attractors, and continuous PFSP can have discontinuous lateral attractors.

The right column states that $\delta > 2$ is excluded when the pulse is non-cancelling but is admissible when the pulse is cancelling. In the latter case, the lateral attractor is FBM. In any event, $H < 1/2$ is incompatible with LSM.

The column second from the right shows that $1 < \delta < 2$ is admissible for all pulse templates and all values of δ, and that it yields varied and interesting results. Combining proven facts with inferences based on compelling heuristics, one is tempted to infer the following.

- For continuous pulses having left and right derivatives bounded away from 0 and ∞, the lateral attractor is FBM.
- For discontinuous or mixed pulses, the lateral attractor is LSM.

The column shows that the dependence of the limit on the template is far more complicated for $0 < \delta < 1$ than for $1 < \delta < 2$. When the pulse template is discontinuous, a formal extrapolation to $0 < \delta < 1$ of the results relative to LSM with $1 < \delta < 2$ is both meaningful and correct. When the pulse is continuous, the extrapolation to $0 < \delta < 1$ of the results relative to FBM with $1 < \delta < 2$ is meaningless, in fact, the construction diverges. As shown in a follow-up paper, one finds meaningful results by considering the second difference

$$\Delta\Delta F = [F(t+T;C,\delta) - F(t;C,\delta)] - [F(t;C,\delta) - F(t-T;C,\delta)].$$

The case $1 < \delta < 2$; *sensitive dependence of the lateral attractor on the pulse templates; a bridge between FBM and LSM.* Table 1 indicates that the attractor is FBM in the case of a smooth conical pulse, but it is LSM if the pulse has a discontinuity, however small, which occurs, for example, when the pulse is a stepped cone, namely a staircase made of many steps up followed by many steps down. "Intuition" suggests that, for finite C, these two pulse forms could not make much of a difference. This is confirmed by a comparative examination of the address diagrams. The diagram corresponding to the stepped cones approximates that of the smooth cone, except for large λ's. As the number of steps increase, so does the quality of the approximation, and it also spreads up to increasingly large λ's.

7 Proofs of the Claims in Table 1 for the Cylindrical Pulses

The parameter σ is set to 1 in this Section. The address points with $\lambda > 0$ and with $\lambda < 0$ contribute to two independent parts of $F(T;C,\delta)$. The calculations leading to their characteristic functions are the same except for different values C' and C''. We restrict ourselves to the case $\lambda > 0$ and $C' = 1$.

7.1 The Logarithms of the Characteristic Function (l.c.f.) of $F(T;C,\delta)$.

From Fig. 1 it is clear that the address points that contribute to $F(T;C,\delta)$ fall into two domains described as left and right and denoted by $\mathcal{D}_L$ and $\mathcal{D}_R$. The strip $(\lambda, \lambda + d\lambda)$, where $\lambda > 0$, makes the following contribution to the l.c.f.

$$T[(e^{i\xi\lambda} - 1) + (e^{-i\xi\lambda} - 1)]\lambda^{-\delta-1} d\lambda \text{ if } \lambda > \mathrm{T}^{1/\delta},$$
$$\lambda^{\delta}[(e^{i\xi\lambda} - 1) + (e^{-i\xi\lambda} - 1)]\lambda^{-\delta-1} d\lambda \quad \text{ if } \quad \lambda < \mathrm{T}^{1/\delta}.$$

Integrating over λ and transforming to the rescaled variables $x = \lambda T^{-1/\delta}$ and $y = \xi T^{1/\delta}$ yields

$$\int_1^\infty [(e^{iyx} - 1) + (e^{-iyx} - 1)]x^{-\delta-1} dx + \int_0^1 [(e^{iyx} - 1) + (e^{-iyx} - 1)]x^{-1} dx.$$

This expression converges for all $\delta > 0$ and is the l.c.f. of a rescaled r.v. independent of T. Therefore, the rescaled increment $T^{-1/\delta} F(T;C,\delta)$ has a distribution independent of T. This is a property of self-affinity with $H = 1/\delta$. We know the mechanisms of self-affinity: in FBM, it is caused by global dependence without long-tailedness, in LSM it is caused by long-tailedness without global dependence, and in PFSP it is caused by both long-tailedness and global dependence, acting together with the same value of H. The next issue is to separate the long-tailedness and dependence aspects.

7.2 The Attractor in the Case $\delta < 2$. Lateral Attraction to Symmetric Lévy Stable Increments with $\alpha = \delta$

Given N independent r.v.'s $F_n(T; C, \delta)$ with the above distribution, the behavior of $\Delta\tilde{F}_N = \sum_{n=1}^{N} F_n(T; C, \delta)$ depends sharply on the value of δ. When $\delta < 2$, the l.c.f. of $N^{-1/\delta}\Delta\tilde{F}_N$ can be written in the form

$$T \int_{(T/N)^{1/\delta}}^{\infty} (e^{i\xi u} + e^{-i\xi u} - 2) u^{-\delta-1} du$$
$$+N \int_0^{T^{1/\delta}} (e^{i\xi\lambda N^{-1/\delta}} + e^{-i\xi\lambda N^{-1/\delta}} - 2)\lambda^{-1} d\lambda.$$

When $N \to \infty$, the first term converges to the well-known l.c.f. of a symmetric Lévy stable r.v. with the stability parameter $\alpha = \delta$ and the scale parameter proportional to $T^{1/\delta}$. The second term is of order $N^{1-2/\delta}$, therefore converges to zero. Hence, for fixed T, $F(T)$ belongs to the symmetric δ-stable domain of attraction.

7.3 The Dependence Structure of $F(T; C, \delta)$ in the Case $\delta < 2$. Lateral Attraction to a Lévy Stable r.f.'s with $\alpha = \delta$ and Independent Increments

We proceed to the multidimensional structure of the PFSP. We show that the multidimensional distributions of the PFSP are attracted to that of a symmetric LSM. To prove it we need to find the limit in distribution of a linear combination

$$\sum_{i=1}^{k} \xi_i N^{-1/\delta} \sum_{n=1}^{N} F_n(T_i; C, \delta),$$

where $F_n(T_i; C, \delta), i = 1, 2, \ldots, k$, are nonoverlapping increments of an PFSP copy F_n, over (possibly different) time spans T_i. The limit should be the corresponding linear combination of independent δ-stable variables with scale parameters proportional to the respective $T_i^{1/\delta}$. Here, we consider increments over two disjoint time spans T_1 and T_2, i.e., $k = 2$. The general case is not mathematically more involved but requires an overload of notation. Our assertion follows from the same reasoning as in the one-dimensional case, if we show that the expression

$$N \int_{\mathcal{D}} [(e^{i(\xi_1-\xi_2)\lambda N^{-1/\delta}} - 1) - (e^{i\xi_1\lambda N^{-1/\delta}} + e^{-i\xi_2\lambda N^{-1/\delta}} - 2)]\lambda^{-\delta-1} dt d\lambda$$

converges to zero, where $\mathcal{D}$ is the dotted region depicted in Fig. 3. But this expression is again of order $N^{1-2/\delta}$, which establishes the result.

7.4 The Attractor in the Case $\delta > 2$. Lateral Attraction to Gaussian Increments

This case is very different, since the variance $EF^2(T; C, \delta)$ is finite, therefore the rescaled sum $N^{1/2}[\tilde{F}_N - E\tilde{F}_N]$ is asymptotically Gaussian, i.e. stable with $\alpha = 2$.

7.5 The Dependence Structure of $F(T; C, \delta)$ in the Case $\delta > 2$. Lateral Attraction to a Fractional Brownian r.f.'s with $H = 1/\delta < 1/2$

Examination of the characteristic function of a linear combination of nonoverlapping increments of the PFSP shows that it has the second derivative at 0, i.e. every linear combination has the second moment, and multidimensional distributions of the PFSP are attracted to multidimensional Gaussian distributions. Note that this Gaussian process in the limit must have stationary increments and be self-affine with the constant $H = 1/\delta$ since these properties are preserved under convolution and convergence in distribution. It is well-known that FBM with $H = 1/\delta$ is the unique Gaussian process satisfying these requirements.

7.6 Some Semi-random PFSP Belonging to the Domain of Standard Attraction of a Semi-random Self-affine PFSP

We replace the constants C' and σ with the slowly varying (at ∞) functions $\gamma(\lambda)$ and $\sigma(\lambda)$ such that the function $w = \sigma(\lambda)\lambda^{\delta}$ is monotonically increasing. Writing the inverse function of $w(\lambda)$ as $\lambda = w^{1/\delta} L(w)$ yields the identity

$$L(w) = \sigma^{-1/\delta}(w^{1/\delta} L(w)),$$

which is an implicit equation for $L(w)$ and will be used momentarily without having to be solved.

When $\lambda > 0$, the strip $(\lambda, \lambda + d\lambda)$ makes the following contribution to the l.c.f.

$$\begin{aligned} &T[(e^{i\xi\lambda} - 1) + (e^{-i\xi\lambda} - 1)]\gamma(\lambda)\lambda^{-\delta-1} d\lambda \qquad \text{if} \quad \lambda > \mathrm{T}^{1/\delta}\mathrm{L}(\mathrm{T}), \\ &\sigma(\lambda)\lambda^{\delta}[(e^{i\xi\lambda} - 1) + (e^{-i\xi\lambda} - 1)]\gamma(\lambda)\lambda^{-\delta-1} d\lambda \qquad \text{if} \quad \lambda < \mathrm{T}^{1/\delta}\mathrm{L}(\mathrm{T}). \end{aligned}$$

Integrating over λ and transforming to the rescaled variables $x = \lambda T^{-1/\delta} L^{-1}(T)$ and $y = \xi T^{1/\delta} L(T)$ yields

$$\begin{aligned} &\int_1^{\infty} [(e^{iyx} - 1) + (e^{-iyx} - 1)] x^{-\delta-1} \{\gamma[x T^{1/\delta} L(T)] L^{-\delta}(T)\} dx \\ &+ \int_0^1 [(e^{iyx} - 1) + (e^{-iyx} - 1)] x^{-1} \{\gamma[x T^{1/\delta} L(T)] \sigma[x T^{1/\delta} L(T)]\} dx. \end{aligned}$$

The integral over $(1, \infty)$ converges for all $\delta > 0$. Assuming that the functions γ and σ are such that also the second integral is finite (e.g. γ and σ are both bounded in the neighborhood of zero), the above expression gives the l.c.f. of a rescaled r.v. $T^{-1/\delta} L^{-1}(T) F$.

This l.c.f. may depend on T. If so, the following question arises: under what conditions on $\gamma(\lambda)$ and $\sigma(\lambda)$, and hence on $L(\lambda)$, does this l.c.f. converge to that prevailing in the PFSP case $\gamma(\lambda)\sigma(\lambda) = C'\sigma$? (We know that the product $C'\sigma$ determines the type of an PFSP.) Because of the identity that links $\sigma(\lambda)$ and $L(w)$ the two factors written between braces are identical (asymptotically, when $T \to \infty$); therefore the two halves of the l.c.f. yield the same condition on convergence. It is

$$\lim_{\lambda \to \infty} \gamma(\lambda)\sigma(\lambda) = C'\sigma.$$

In other words, the functions $1/\gamma(\lambda)$ and $\sigma(\lambda)$ must vary slowly with λ and asymptotically proportionally to each other.

The question concerning whether or not these conditions are also necessary has not been addressed yet.

Acknowledgments

In 1977-78, I studied semi-random PFSP with a one-dimensional t and a multi-dimensional F: early simulations for the second row of Table 1 were performed by M.R. Laff, and I conjectured that the closure of the set of values of $F(t)$ remains of Hausdorff-Besicovitch dimension $D = \delta$. Soon later, J. Hawkes proved this conjecture. In the mid-1980s, I studied an application of semi-random PFSP with a multi-dimensional t and one-dimensional F: early simulations performed by S. Lovejoy are reported in [5]. I also made conjectures concerning the random PFSP; in due time, they were proved in [2] and [4]. Earlier versions of this paper was discussed at length with R. Cioczek-Georges and M. Frame. Diagrams were drawn by H. Kaufman.

References

1. J. Barral and B.B. Mandelbrot: Probability Theory and Related Fields **124**, 409 (2002)
2. R. Cioczek-Georges and B.B. Mandelbrot: Stochastic Processes and Their Applications **60**, 1 (1995)
3. R. Cioczek-Georges and B.B. Mandelbrot: Stochastic Processes and their Applications **64**, 143 (1996)
4. R. Cioczek-Georges, B.B. Mandelbrot, G. Samorodnitsky and M. Taqqu: Bernoulli **1**, 201 (1995)
5. S. Lovejoy and B.B. Mandelbrot: Tellus **37A**, 209 (1985)
6. B.B. Mandelbrot: Zeitschrift für Wahrscheinlichkeitstheorie **22**, 145 (1972)
7. B.B. Mandelbrot: The Fractal Geometry of Nature (W.H. Freeman, New York, 1982)
8. B.B. Mandelbrot: In Lévy Flights and Related Phenomena in Physics (Nice, 1994). Ed. G. Zaslawsky, M. F. Schlesinger & U. Frisch (Lecture Notes in Physics) (Springer, New York, 1995)
9. B.B. Mandelbrot: Gaussian Self-Affinity and Fractals: Globality, the Earth, $1/f$, and R/S (Springer-Verlag, New York, 2002)
10. B.B. Mandelbrot: (to appear)
11. M.S. Taqqu and G. Samorodnitsky: Stable Non-Gaussian Processes: Stochastic Models with Infinite Variance (Chapman and Hall, New York, London, 1994)

Part II

Applications

Supra-diffusion

Michael F. Shlesinger

Physical Sciences Division, Office of Naval Research, 800 N. Quincy St., Arlington VA 22217, USA

Abstract. The standard random walk produces normal diffusion with a mean square displacement that grows linearly with time. We explore stochastic processes that under certain conditions produce supra-diffusion where the mean square displacement grows faster than linearly with time. Examples are presented from turbulent diffusion to random flights accelerating to relativistic velocities in a gravitational field.

1 Introduction

The original study of random walks involved jumps on a regular 1D lattice. Start at the origin, and then with equal probability make jumps of unit length to the right or left. After N jumps the probability of being M units to the right of the origin means that there were $(N+M)/2$ steps to the right and $(N-M)/2$ steps to the left. The probability $P(M,N)$ of this happening is

$$\frac{N!}{[(\frac{1}{2})(N+M)]![(\frac{1}{2})(N+M)]!}(\frac{1}{2})^N \tag{1}$$

This is also the probability of being M steps to the left of the origin as jumps to the right and left occur with equal probability. After N jumps the variance of the displacement will be proportional to N. One usually assumes that the number of jumps is proportional to the time elapsed. If the number of jumps grows at t^2 then so would the mean square displacement, This acceleration in the number of jumps is one way to have the diffusion constant increase in time. This is a trivial way to produce supra-diffusion,One can write a diffusion equation with a time dependent diffusion constant,

$$\frac{\partial P(x,t)}{\partial t} = D(t)\frac{\partial^2 P(x,t)}{\partial x^2} eq2) \tag{2}$$

In Fourier space, this diffusion equation has the solution

$$P(k,t) = -\exp\left[-k^2 \int_0^t D(\tau)\, d\tau\right] \equiv -\exp[-k^2 d(t)]$$

Transforming back to x, we have

$$P(x,t) = \frac{1}{\sqrt{4\pi d(t)}} \exp\left[-\frac{x^2}{4d(t)}\right] \tag{3}$$

with the second moment

$$\langle x^2(t) \rangle \propto d(t) \tag{4}$$

When $D(t)$ is a constant D, then $d(t) = Dt$, and we obtain the standard diffusion results. By assuming that the diffusion constant grows with time it is easy to obtain supra-diffusion. But this only shifts the problem to try to explain why the diffusion constant should grow with time. If the system is changing with time this might be a proper description. A possible case could be diffusion in laminar pipe flow. The flow has a parabolic velocity profile with flow going to zero at the pipe sides and following a parabolic profile with the maximum velocity at the center of the pipe. A drop of dye placed near the pipe boundary would diffuse towards the center where the velocity is larger. The spread of the dye would be faster and faster as the dye diffuses away from the boundary edge and towards the center. Super-ballistic diffusion has been studied in shear flows.[1]

For regular diffusion, although it is easy to solve for the Gaussian probability d/istribution, it is not trivial to describe the Brownian trajectory. The trajectory wiggles so much that the distance between any two points on the trajectory is infinity. This is because that the trajectory is a fractal. In fact the trail of the trajectory is two dimensional, so any 1D measure, like a length measure will be infinite. Also the derivative (velocity) of the trajectory is everywhere infinite. However, a discrete version of Brownian motion is very simple. It is just a random walk on a lattice with equal probability to jump to any nearest neighbour lattice site. We will present the random walk on a lattice with a probability distribution to jump a distance r in a time t. We will show how to write the random walk as a semi-Markov process and as an integral master equation with a coupled space-time memory. It is through the space-time coupled memory that we will be able to derive conditions for supra-diffusion.

2 Master Equations for Random Walks

A master equation (see (5) below) looks like a rate equation, but it is directly written for probabilities instead of for moments of probabilities.

$$\frac{\partial P(x,t)}{\partial t} = \sum_{x'} \int_0^t \phi(t') p(x') P(x - x', t - t')\, dt' - \int_0^t \phi(t') P(x, t - t')\, dt' \tag{5}$$

The quantity $p(x)$ is the probability for a jump of displacement x, and $\phi(t)$ has units of $1/t^2$. The meaning of $\phi(t)$ will be become clearer later, after we relate it to a probability waiting time density. Taking for (5) the Laplace (over time) and Fourier (over space) transforms we obtain

$$sP(k,s) - P(k, t=0) = \phi(s)p(k)P(k,s) - \phi(s)P(k,s)$$

and solving for the transformed probability gives

$$P(k,s) = \frac{1}{s + \phi(s) - \phi(s)p(k)} \tag{6}$$

Next, let us consider a random walk on a lattice where the walker waits for a random time, governed by a probability density $\Psi(t)$ for the waiting time between jumps[2]. The probability density $Q(x,t)$ to just arrive at a site x at time t is given by

$$Q(x,t) = \sum_{x'} \int_0^t Q(x-x', t-t') p(x') \psi(t') \, dt' + \delta_{x,0} \delta(t) \tag{7}$$

and the probability $P(x,t)$ to be at that site includes reaching the site at an earlier time $t-\tau$ and then not jumping away at least for a time τ

$$P(x,t) = \int_0^t Q(x, t-\tau)(1 - \int_0^\tau \psi(t') \, dt') \, d\tau \tag{8}$$

In Laplace-Fourier space,

$$P(k,s) = Q(k,s) \frac{1-\psi(s)}{s} = \frac{1}{1-p(k)\psi(s)} \frac{1-\psi(s)}{s} \tag{9}$$

If we set

$$\phi(s) = s\psi(s)/(1-\psi(s)) \tag{10}$$

then

$$\begin{aligned} P(k,s) &= \frac{1}{1-p(k)\phi(s)/(s+\phi(s))} \frac{1}{s+\phi(s)} \\ &= \frac{1}{s+\phi(s)-p(k)\phi(s)} \end{aligned}$$

and this is the same solution as in (6). This shows how to relate the memory in the random walk semi-Markov equation to the memory in the master equation such that both provide an equivalent description of the random process [3]. Markov means that the present jump only depends on where the jump originates, and not directly on previous jumps. Semi-Markov just means that the waiting time probability is taken into account. For long tail temporal memories, such that moments of the memory diverge, some re-write the master equation with a fractional derivative, instead of the memory term

3 Decoupled Memory: A Diffusive Case

Consider, the simple case of a random walk with equal probability steps to the left or right on a 1D lattice $p(x) = 1/2(\delta_{x,1} + \delta_{x,-1})$ so in Fourier space $p(k) = \cos(k)$, and choose $\psi(t) = \lambda \exp(-\lambda t)$ so in Laplace space $\psi(s) = \lambda/(\lambda + s)$. Then

$$P(k,s) = \frac{1}{\lambda(1-\cos(k))+s} \tag{11}$$

Inverse Laplace transforming gives

$$P(k,t) = \exp\left[-\lambda(1-\cos(k))t\right] \tag{12}$$

Since, $I_n(\lambda t) = \frac{1}{2\pi}\int_{-\pi}^{\pi}\cos(nk)\exp\left[-\lambda t\cos(k)\right]dk$, where I is the modified Bessel function, we can analytically calculate for integer x that,

$$P(x,t) = \exp(-\lambda t) I_x(\lambda t) \tag{13}$$

To get a better feel for this solution, look at the small k (large x) behavior in (12) to obtain

$$P(k,t) \approx \exp\left[-\frac{\lambda}{2}k^2 t\right]$$

This transforms into the Gaussian of (3) with $d(t) = Dt$ with $D = \lambda/2$. One problem with this solution is that for arbitrarily short times there is a finite probability for the random walker to be arbitrarily far away. This again shows the difficulty with the concept of velocity in Brownian motion, as the walker would need an arbitrarily large velocity to be arbitrarily far away in an arbitrarily short time. One way to solve this velocity problem is to introduce a coupled space-time memory into the random walk. This insures that the probability propagates out with a finite velocity and that it does not become instantaneously positive everywhere. We show this calculation in the next section.

4 Coupled Memory: A Diffusion Front

One obvious point to stress is that in (1) $P(M,N) = 0$ if $M > N$. The walker cannot go further than N units away in N unit steps. In the continuum limit leading to the diffusion equation, with D a constant, this condition of a front beyond which the probability is zero, is absent. At any time t, even at an infinitesimally small time, there will be a positive probability (3) for a diffusing particle to be arbitrarily far away from its starting point. For example, the Gaussian solution would over estimate the spread of pollution over a short time. A diffusion front can be obtained in a random walk with a coupled space-time memory such that if the particle possesses a velocity V then $P(x,t)$ will equal zero if $x > Vt$.

Let us return to the continuous-time random walk and introduce the coupled memory [4]

$$\Psi(x,t) \equiv p(x|t)\psi(t) \tag{14}$$

for the probability density of a single jump of displacement x taking a time t. This is given as the product of the conditional probability that the jump is of displacement x given that it is of duration t times the probability that the jump is of duration t. The solution to the random walk retains the same form of (9) except that the term $\psi(s)p(k)$ is replaced by $\Psi(k,s)$; a coupled memory.

Again, let us look at a simple case, with $p(x)$ allowing only nearest neighbour jumps with equal probability and,

$$p(x|t) = \delta(|x| - Vt) \tag{15}$$

where V, the particle velocity, is introduced into the problem. The delta function provides the kinematic relationship of $x = Vt$. Our discussion will have much in common with the discussion of the telegrapher's equation in ref. 5, Weiss' excellent book on random walks. For this random walk

$$\Psi(x,t) = \delta(|x| - Vt)\lambda \exp(-\lambda t) \tag{16}$$

Transforming to k-space gives $\lambda \cos(kVt) \exp(-\lambda t)$ and then taking the Laplace transform of (16) gives for the coupled memory

$$\Psi(k,s) = \frac{\lambda(s+\lambda)}{(s+\lambda)^2 + (kV)^2} \tag{17}$$

Using this result, the equation for $P(k,s)$ becomes,

$$P(k,s) = \frac{1}{(s+\lambda)} \frac{(s+\lambda)^2 + (kV)^2}{s^2 + s\lambda + (kV)^2} \tag{18}$$

If one takes the very long time long distance limit $(k, s \to 0)$ $P(k,s)$ becomes more familiar as

$$P(k,s) \approx \lambda/(s\lambda + (kV)^2)$$

which transforms back into a Gaussian,

$$P(x,t) \approx (4\pi Dt)^{-1/2} \exp(-\frac{x^2}{4Dt})$$

where $D = V^2/\lambda$. Let us transform (18) exactly back to x-space to calculate

$$P(x,s) = \frac{1}{(s+\lambda)} \frac{[(s+\lambda)^2] \exp(-|\frac{x}{V}|\sqrt{s^2+\lambda s}) + (s^2+\lambda s)\exp(-|\frac{x}{V}|\sqrt{s^2+\lambda s})}{2\sqrt{s^2+\lambda s}} \tag{19}$$

Let us again take the small s limit,

$$\lim_{s \to 0} P(k,s) \approx \frac{\lambda \exp(-|\frac{x}{V}|\sqrt{s^2+\lambda s})}{2\sqrt{s^2+\lambda s}} \tag{20}$$

Writing $\sqrt{s^2+\lambda s} = \sqrt{(s+\lambda/2)^2 - \lambda^2/4}$ and using identity,

$$L^{-1} \frac{\{\exp(-k\sqrt{s^2-a^2})\}}{\sqrt{s^2-a^2}} = I_0(a\sqrt{t^2-k^2})\Theta(t-|k|) \tag{21}$$

where the theta step function is one if $t > |k|$ and zero otherwise, we arrive at

$$P(x,t) \approx \exp(-\frac{\lambda t}{2}) I_0(\frac{\lambda}{2}(\sqrt{t^2 - \frac{x^2}{V^2}}))\Theta(Vt - |x|) \tag{22}$$

This is where the diffusion front at $x = Vt$ enters the equation, and the step function ensures that the argument of the modified Bessel function is non-zero. For Vt greater than the absolute value of x, the probability function becomes, at longer times, more diffusive-like inside of the diffusion front. This can be checked by writing for very large t that, $I_0(z) \approx \exp(z)/z$ where in our case $z = (\lambda t/2)\sqrt{1 - \frac{x^2}{V^2}} \approx \frac{\lambda t}{2}(1 - \frac{x^2}{2V^2t^2})$.

5 Long Random Flights with Constant Velocity

In the original work of Montroll and Weiss [2] on the continuous-time random walk the memory was decoupled into a probability waiting time density function and a probability jump distribution. The waiting time meant this was the time a random walker spent at a site between jumps. In the coupled memory picture the particle moves with velocity V, and the waiting time is the time of flight; i.e. the travel time to complete a jump. If we again consider a random walk with constant velocity, but now with a wide distribution of jump sizes characterized by their flight time probability density,

$$\psi(t) \propto t^{-1-\beta}, \beta > 0 \tag{23}$$

then for jumps to the right or left with equal probability, one can calculate for the second moment of $P(x,t)$ that [6,7]

$$\langle x^2(t) \rangle \propto \begin{cases} t^2, & \text{for } 0 < \beta < 1 \\ t^2/\ln(t), & \text{for } \beta = 1 \\ t^{2-(\beta-1)}, & \text{for } 1 < \beta < 2 \\ t\ln(t), & \text{for } \beta = 2 \\ t, & \text{for } \beta > 2 \end{cases} \tag{24}$$

The t^2 case corresponds to the mean flight time being infinite. For $\beta > 1$, but less than 2, the mean flight time is finite, but the mean square flight time is infinite. Only in the last case, for $\beta > 2$, does the mean square flight time become finite, and the standard linear growth in time occur. When the second moment of the flight length is infinite, and flight velocity is not considered, these are called Levy flights. When the flight duration (and flight velocity) is considered these called Levy walks or drives. The idea is that in a flight one only is concerned about the starting and end points. In a walk or a drive one also needs to include all the points along the path. Mathematicians studied Levy flights and physicists investigated Levy walk/drives. The mean squared displacement of a Levy flight is infinite, while this quantity for a Levy walk/drive is a function of time.

6 Random Flights in a Turbulent Fluid

Now we consider a case where the velocity is no longer a constant. This occurs in turbulent fluid flows. The idea is that one follows the separation of two buoyant particles in the flow and the longer their random flights the more they will encounter larger and larger vortices pulling them further away. The larger the vortex, the higher its energy and velocity. Effectively the relative diffusion constant for the separation of two particles grows in time. L. F. Richardson experimentally studied the phenomenon of turbulent diffusion and found that the mean squared displacement grows as t^3. For turbulent flow Kolmogorov hypothesized that the dissipation was a constant independent of scale R, i.e.

$$\text{dissipation} = \bar{\varepsilon} = \text{energy/time}$$

$$\overline{\varepsilon} = \frac{1}{2} m V^2 (V/R) \tag{25}$$

where $t = R/V$ has been used. To keep the dissipation independent of R we must have the Kolmogorov scaling of

$$V(R) \approx R^{1/3} \tag{26}$$

Since $R(t)$ scales as $V(R)\ t$, R will scale as

$$R \approx R^{1/3} t; R \approx t^{3/2} \tag{27}$$

If we go through the whole random walk formalism, it can be shown that[8]

$$\langle R^2(t) \rangle \approx \begin{cases} t^3, & \beta \leq 1/3 \\ t^{2+\frac{3}{2}(1-\beta)}, & 1/3 \leq \beta \leq 5/3 \\ t, & \beta \geq 5/3 \end{cases} \tag{28}$$

The fastest t^3 case occurs when the mean flight time is infinite. This result is called Richardson diffusion. The next case occurs when the mean flight time is finite, but the mean square flight time is infinite. The last case of linear growth occurs when the mean square flight time is finite.

7 Relativistic Turbulent Diffusion[9]

In our coupled memory random walk we will use $p(x|t) = \delta(x - f(t))$ where $f(t)$ is the function on the right-hand-side of the $x(t) = V(x)t$ equation. The function $f(t)$ can be very different for short times than for long times. The Kolmogorov conjecture for the mean dissipation $\overline{\varepsilon}$ in a relativistic setting is

$$(\gamma - 1) m_0 c^2 / t = (\gamma - 1) m_0 c^2 V/R = \overline{\varepsilon} \tag{29}$$

with

$$\gamma = \frac{1}{\sqrt{1 - (\frac{V}{c})^2}}$$

For small velocities compared to c, we have

$$\gamma - 1 = \frac{1}{\sqrt{1 - (\frac{V}{c})^2}} - 1 \tag{30}$$

$$\approx \frac{1}{2} (\frac{V}{c})^2$$

and this reduces to the previous classical case of $V(R)^3 \approx R$ yielding the t^3 result. For velocities approaching the speed of light c, perhaps in the turbulence of a supernova, we now return to (29) and use

$$\gamma - 1 \approx \gamma$$

therefore,

$$\frac{V^2}{1-\frac{V^2}{c^2}} \approx R^2$$

and solving for $\nu(R)$,

$$V^2(R) \approx \frac{R^2}{1+R^2/c^2}$$

Writing,

$$R^2 \approx V^2(R)t^2$$

$$\approx \frac{R^2}{1+R^2/c^2}t^2$$

which at long times gives that $R = ct$. When employed in the full random walk formalism for Levy walks of infinite mean duration,

$$\langle R^2(t)\rangle \approx (ct)^2 \tag{31}$$

in the asymptotically long time limit, as is proper in relativistic dynamics.

8 Accelerating Random Flights in a Gravitational Field

For a particle of rest mass m subject to a constant force mg, along the x-axis, has the relativistic displacement

$$x(t) = \frac{c^2}{g}\{[1+(g\frac{t}{c})^2]^{1/2} - 1\} \tag{32}$$

For short times,

$$x(t) \approx \frac{1}{2}gt^2 \tag{33}$$

the classical behavior of a particle falling in a gravitational field, and $\langle x^2(t)\rangle \;\alpha\; t^4$ for flights of infinite mean duration. For long times this behavior crosses over to,

$$x(t) \approx ct \tag{34}$$

and $\langle x^2(t)\rangle \;\alpha\; t^2$ illustrating that even for an accelerating particle the velocity cannot exceed the speed of light c.

9 Conclusions

One method of generalizing standard diffusion equations is to allow flights of a length and duration governed by probability distribution functions. This can be described mathematically with a coupled memory Markov process or equivalently as a coupled memory master equation. The concept of a flight duration involves the concept of a velocity. When the flight time probability distribution's second or first moment is infinite then supra-diffusion results. The limiting case of the mean squared displacement growing as t^2 is found when the velocity is a constant. Faster growth of the mean squared displacement is found when the physics leads to particle acceleration. However, relativistic effects always brings about a crossover to the t^2 case, otherwise acceleration passed the speed of light would occur. We have only discussed random processes, and several other cases are known that we have not yet mentioned. For example, the random breakup of micelles leads to faster diffusion as the pieces become smaller leading to a supra-diffusion [10].

A topic that we have not discussed is Levy flights in deterministic systems, including Hamiltonian systems. In these systems the orbits can be become stuck in a region of phase space. The trajectory can appear to be trapped in a fractally nested group of nearly integrable orbits. The orbit will eventually escape, but due to the sensitivity to initial conditions, the sticking times effectively possess a probability distribution function. The sticking can be in momentum space, so sticking is equivalent to a flight of constant momentum. The statistics of these complex trajectories are typically supra-diffusive. Examples include the Geisel map, the standard map, the web map, scattering in an egg crate potential, scattering in time dependent periodic potentials [15], trajectories in rotating flows [16]. Recent reviews of Levy flights and walks in random and deterministic systems can be found in refs. [11]-[14].

References

1. D. Ben Avraham, F. Leyvraz and S. Redner: Phys. Rev. A **45**, 2315 (1992)
2. E. W. Montroll and G. H. Weiss: J. Math. Phys. **6**, 167 (1965)
3. V. M. Kenkre, E. W. Montroll and M. F. Shlesinger: J.Stat. Phys. **9**, 45 (1973)
4. M. F. Shlesinger, J. Klafter, and Y. M. Wong: J. Stat. Phys. **27**, 499 (1982)
5. G. H. Weiss: *Aspects and Applications of the Random Walk* (North-Holland, Amsterdam 1994)
6. T. Geisel, J. Nierwetberg, and A. Zacherl: Phys. Rev. Lett. **54**, 616 (1985)
7. M. F. Shlesinger: Phys. Rev. Lett. **54**, 2551 (1985)
8. M. F. Shlesinger, B. West, and J. Klafter: Phys. Rev. Lett. **58**, 1100 (1987)
9. M. F. Shlesinger, J. Klafter, and G. Zumofen: Fractals **3**, 491 (1995)
10. A. Ott, J. P. Bouchaud, D. Langevin, and W. Urbach: Phys. Rev. Lett. **65**, 2201 (1990)
11. M. F. Shlesinger, G. Zaslavsky, and U. Frisch eds: *Levy flights and Related Topics in Physics* (Springer-Verlag, Berlin 1995)
12. J. Klafter, M. F. Shlesinger, and G. Zumofen: Physics Today **49**, 33 (1996)
13. G. Zaslavsky: *Physics of Chaos in Hamiltonian Systems* (Imperial College Press, London 1998)
14. R. Metzler, J. Klafter: Phys. Reports **339**, 1 (2000)
15. I. S. Aronson, M. I. Rabonovich, L. Sh. Tsimring: Phys. Lett. **A151**, 523 (1990)
16. T. H. Solomon, E. R. Weeks, and H. Swinney: Phys. Rev. Lett. **71**, 3975 (1993)

Fractional Diffusion Processes: Probability Distributions and Continuous Time Random Walk

Rudolf Gorenflo[1] and Francesco Mainardi[2]

[1] Department of Mathematics and Computer Science, Free University of Berlin, Arnimallee 3, D-14195 Berlin, Germany
[2] Dipartimento di Fisica, Università di Bologna, and INFN, Sezione di Bologna, Via Irnerio 46, I-40126 Bologna, Italy

Abstract. A physical-mathematical approach to anomalous diffusion may be based on generalized diffusion equations (containing derivatives of fractional order in space or/and time) and related random walk models. By the space-time fractional diffusion equation we mean an evolution equation obtained from the standard linear diffusion equation by replacing the second-order space derivative with a Riesz-Feller derivative of order $\alpha \in (0,2]$ and skewness θ ($|\theta| \le \min\{\alpha, 2-\alpha\}$), and the first-order time derivative with a Caputo derivative of order $\beta \in (0,1]$. The fundamental solution (for the Cauchy problem) of the fractional diffusion equation can be interpreted as a probability density evolving in time of a peculiar self-similar stochastic process. We view it as a generalized diffusion process that we call *fractional diffusion process*, and present an integral representation of the fundamental solution. A more general approach to anomalous diffusion is however known to be provided by the master equation for a continuous time random walk (CTRW). We show how this equation reduces to our fractional diffusion equation by a properly scaled passage to the limit of compressed waiting times and jump widths. Finally, we describe a method of simulation and display (via graphics) results of a few numerical case studies.

1 Introduction

It is well known that the fundamental solution (or *Green function*) for the Cauchy problem of the linear diffusion equation can be interpreted as a Gaussian (normal) probability density function (*pdf*) in space, evolving in time. All the moments of this *pdf* are finite; in particular, its variance is proportional to the first power of time, a noteworthy property of the *standard diffusion* that can be understood by means of an unbiased random walk model for the *Brownian motion.*

In recent years a number of master equations have been proposed for random walk models that turn out to be beyond the classical Brownian motion, see e.g. Klafter et al. [34]. In particular, evolution equations containing fractional derivatives have gained revived interest in that they are expected to provide suitable mathematical models for describing phenomena of anomalous diffusion, strange kinetics[1] and transport dynamics in complex systems. Recent references include e.g. [1,2,9,21,23,28,37,38,42,43,48,51,61].

The paper is divided as follows. In Sect. 2 we introduce our fractional diffusion equations providing the reader with the essential notions for the derivatives of fractional order (in space and in time) entering these equations. More precisely, we replace in the standard linear diffusion equation the second-order space derivative or/and the first-order time derivative by suitable *integro-differential* operators, which can be interpreted as a space or time derivative of fractional order $\alpha \in (0,2]$ or $\beta \in (0,1]$, respectively[2]. The space fractional derivative is required to depend also on a real parameter θ (the *skewness*) subjected to the restriction $|\theta| \le \min\{\alpha, 2-\alpha\}$. Then, in Sect. 3 we pay attention to the fact that the fundamental solutions (or *Green functions*) of our diffusion equations of fractional order in space or/and in time can be interpreted as spatial probability densities evolving in time, related to certain *self-similar* stochastic process. We view these processes as generalized (or *fractional*) diffusion processes to be properly understood through suitable random walk models that have been treated by us in previous papers, see e.g. [15,18–21,23]. In Sect. 4 we show how such evolution equations of fractional order can be obtained from a more general master equation which governs the so-called continuous time random walk (CTRW) by a properly scaled passage to the limit of compressed waiting times and jump widths. The CTRW structure immediately offers a method of simulation that we roughly describe in Sect. 5 where we also display graphs of a few numerical case studies. Finally, in Sect. 6, we point out the main conclusions and outline the direction for future work.

2 The Space-Time Fractional Diffusion Equation

By replacing in the standard diffusion equation

$$\frac{\partial}{\partial t} u(x,t) = \frac{\partial^2}{\partial x^2}\, u(x,t)\,, \quad -\infty < x < +\infty\,, \quad t \ge 0\,, \tag{1}$$

where $u = u(x,t)$ is the (real) field variable, the second-order space derivative and the first-order time derivative by suitable *integro-differential* operators, which can be interpreted as a space and time derivative of fractional order we obtain a generalized diffusion equation which may be referred to as the *space-time-fractional* diffusion equation. We write this equation as

$${}_tD_*^{\beta}\, u(x,t) = {}_xD_\theta^{\alpha}\, u(x,t)\,, \quad -\infty < x < +\infty\,, \quad t \ge 0\,, \tag{2}$$

[1] To the topic of strange kinetics a special issue (nowadays in press) of *Chemical Physics* is devoted where the interested reader can find a number of applications of fractional diffusion equations

[2] We remind that the term "fractional" is a misnomer since the order can be a real number and thus is not restricted to be rational. The term is kept only for historical reasons, see e.g. [17]. Our fractional derivatives are required to coincide with the standard derivatives of integer order as soon as $\alpha = 2$ (not as $\alpha = 1$!) and $\beta = 1$.

where the α, θ, β are real parameters restricted as follows

$$0 < \alpha \le 2\,, \quad |\theta| \le \min\{\alpha, 2-\alpha\}\,, \quad 0 < \beta \le 1\,. \tag{3}$$

In (2) ${}_xD_\theta^\alpha$ is the space-fractional *Riesz-Feller derivative* of order α and skewness θ, and ${}_tD_*^\beta$ is the time-fractional *Caputo derivative* of order β. The definitions of these fractional derivatives are more easily understood if given in terms of Fourier transform and Laplace transform, respectively. Generically, $u(x,t)$ is interpreted as a mass density or a probability density depending on the space variable x, evolving in time t.

In terms of the Fourier transform we have for the space-fractional *Riesz-Feller derivative*

$$\mathcal{F}\{ {}_xD_\theta^\alpha\, f(x); \kappa\} = -\psi_\alpha^\theta(\kappa)\, \widehat{f}(\kappa)\,, \quad \psi_\alpha^\theta(\kappa) = |\kappa|^\alpha\, \mathrm{e}^{i(\mathrm{sign}\,\kappa)\theta\pi/2}\,, \tag{4}$$

where $\widehat{f}(\kappa) = \mathcal{F}\{f(x);\kappa\} = \int_{-\infty}^{+\infty} \mathrm{e}^{+i\kappa x}\, f(x)\, dx$. In other words the symbol of the pseudo-differential operator ${}_xD_\theta^\alpha$ is required to be the logarithm of the characteristic function of the generic *stable* (in the Lévy sense) probability density, according to the Feller parameterization [12,13]. We note that the allowed region for the parameters α and θ turns out to be a diamond in the plane $\{\alpha\,,\,\theta\}$ with vertices in the points $(0,0)$, $(1,1)$, $(2,0)$, $(1,-1)$, that we call the *Feller-Takayasu diamond* [3]. For $\alpha = 2$ (hence $\theta = 0$) we have $\mathcal{F}\{ {}_xD_\theta^\alpha\, f(x);\kappa\} = -\kappa^2 = (-i\kappa)^2$, so we recover the standard second derivative. More generally for $\theta = 0$ we have $\mathcal{F}\{ {}_xD_\theta^\alpha\, f(x);\kappa\} = -|\kappa|^\alpha = -(\kappa^2)^{\alpha/2}$ so

$${}_xD_0^\alpha = -\left(-\frac{d^2}{dx^2}\right)^{\alpha/2}. \tag{5}$$

In this case we refer to the LHS of (5) as simply to the *Riesz fractional derivative* of order α. Assuming $\alpha \neq 1, 2$ and taking θ in its range one can show that the explicit expression of the *Riesz-Feller fractional derivative* obtained from (4) is

$${}_xD_\theta^\alpha\, f(x) := -\left[c_+(\alpha,\theta)\, {}_xD_+^\alpha + c_-(\alpha,\theta)\, {}_xD_-^\alpha\right] f(x)\,, \tag{6}$$

where, see [18],

$$c_+(\alpha,\theta) = \frac{\sin\left[(\alpha-\theta)\,\pi/2\right]}{\sin(\alpha\pi)}\,, \qquad c_-(\alpha,\theta) = \frac{\sin\left[(\alpha+\theta)\,\pi/2\right]}{\sin(\alpha\pi)}\,, \tag{7}$$

and ${}_xD_\pm^\alpha$ are Weyl fractional derivatives defined as

$${}_xD_\pm^\alpha\, f(x) = \begin{cases} \pm\dfrac{d}{dx}\left[{}_xI_\pm^{1-\alpha}\, f(x)\right], & \text{if} \quad 0 < \alpha < 1\,, \\[2ex] \dfrac{d^2}{dx^2}\left[{}_xI_\pm^{2-\alpha}\, f(x)\right], & \text{if} \quad 1 < \alpha < 2\,. \end{cases} \tag{8}$$

[3] Our notation for the stable distributions has been adapted from the original one by Feller. From 1998, see [18], we have found it as the most convenient among the others available in the literature, see e.g. [32,36,45,47,53,54,61]. Furthermore, this notation has the advantage that the whole class of the *strictly stable* densities is represented. As far as we know, the diamond representation in the plane $\{\alpha, \theta\}$ was formerly given by Takayasu in his 1990 book on *Fractals* [59].

In (8) the ${}_xI^{\mu}_{\pm}$ $(\mu > 0)$ denote the Weyl fractional integrals defined as

$$\begin{cases} {}_xI^{\mu}_{+}\, f(x) = \dfrac{1}{\Gamma(\mu)} \displaystyle\int_{-\infty}^{x} (x-\xi)^{\mu-1}\, f(\xi)\, d\xi \,, \\ {}_xI^{\mu}_{-}\, f(x) = \dfrac{1}{\Gamma(\mu)} \displaystyle\int_{x}^{+\infty} (\xi-x)^{\mu-1}\, f(\xi)\, d\xi \,. \end{cases} \quad (\mu > 0) \tag{9}$$

In the particular case $\theta = 0$ we get $c_+(\alpha, 0) = c_-(\alpha, 0) = 1/[2\cos(\alpha\pi/2)]$, and, by passing to the limit for $\alpha \to 2^-$, we get $c_+(2,0) = c_-(2,0) = -1/2$.

For $\alpha = 1$ we have

$${}_xD^1_{\theta}\, f(x) = \left[\cos(\theta\pi/2)\, {}_xD^1_0 + \sin(\theta\pi/2)\, {}_xD\right]\, f(x)\,, \tag{10}$$

where ${}_xD\, f(x) = \dfrac{d}{dx}\, f(x)$, and

$${}_xD^1_0\, f(x) = -\frac{d}{dx}\, [{}_xH\, f(x)]\,, \quad {}_xH\, f(x) = \frac{1}{\pi}\left(\int_{-\infty}^{+\infty} \frac{f(\xi)}{x-\xi}\, d\xi\right)\,. \tag{11}$$

In (11) the operator ${}_xH$ denotes the Hilbert transform and its singular integral is understood in the Cauchy principal value sense, see [20].

The operator ${}_xD^{\alpha}_{\theta}$ has been referred to as the *Riesz-Feller* fractional derivative since both Marcel Riesz and William Feller contributed to its definition[4].

Let us now consider the time-fractional *Caputo derivative*. Following the original idea by Caputo [6], see also [5,7,17,49], a proper time fractional derivative of order $\beta \in (0,1)$, useful for physical applications, may be defined in terms of the following rule for its Laplace transform[5]

$$\mathcal{L}\left\{{}_tD^{\beta}_{*}\, f(t); s\right\} = s^{\beta}\, \widetilde{f}(s) - s^{\beta-1}\, f(0^+)\,, \quad 0 < \beta < 1\,, \tag{12}$$

where $\widetilde{f}(s) = \mathcal{L}\{f(t); s\} = \int_0^{\infty} \mathrm{e}^{-st}\, f(t)\, dt$. Then the *Caputo fractional derivative* of $f(t)$ turns out to be defined as

$${}_tD^{\beta}_{*}\, f(t) := \frac{1}{\Gamma(1-\beta)} \int_0^t \frac{f^{(1)}(\tau)\, d\tau}{(t-\tau)^{\beta}}\,, \quad 0 < \beta < 1\,. \tag{13}$$

In other words the operator ${}_tD^{\beta}_{*}$ is required to generalize the well-known rule for the Laplace transform of the first derivative of a given (causal) function keeping the standard initial value of the function itself[6].

[4] Originally, in the late 1940's, Riesz [50] introduced the pseudo-differential operator ${}_xI^{\alpha}_0$ whose symbol is $|\kappa|^{-\alpha}$, well defined for any positive α with the exclusion of odd integer numbers, afterwards named the *Riesz potential.* The Riesz fractional derivative ${}_xD^{\alpha}_0 := -\, {}_xI^{-\alpha}_0$ defined by analytical continuation was generalized by Feller in his 1952 genial paper [12] to include the skewness parameter of the strictly stable densities.

[5] For our purposes we agree to take the Laplace parameter s real

[6] The reader should observe that the *Caputo* fractional derivative differs from the usual *Riemann-Liouville* fractional derivative which, defined as the left inverse of

The *space-time fractional diffusion* equation (2) contains as particular cases the *strictly space fractional diffusion* equation when $0 < \alpha < 2$ and $\beta = 1$, the *strictly time fractional diffusion* equation when $\alpha = 2$ and $0 < \beta < 1$, and the *standard diffusion equation* (1) when $\alpha = 2$ and $\beta = 1$.

For (2) we consider the Cauchy problem

$$u(x, 0^+) = \varphi(x), \quad x \in \mathbf{R}, \qquad u(\pm\infty, t) = 0, \quad t > 0, \tag{14}$$

where $\varphi(x)$ is a sufficiently well-behaved function. By its solution we mean a function $u^\theta_{\alpha,\beta}(x,t)$ which satisfies the conditions (14). By its Green function (or fundamental solution) we mean the (generalized) function $G^\theta_{\alpha,\beta}(x,t)$ which, being the formal solution of (2) corresponding to $\varphi(x) = \delta(x)$ (the Dirac delta function), allows us to represent the solution of the Cauchy problem by the integral formula

$$u^\theta_{\alpha,\beta}(x,t) = \int_{-\infty}^{+\infty} G^\theta_{\alpha,\beta}(\xi, t)\, \varphi(x - \xi)\, d\xi\,. \tag{15}$$

It is straightforward to derive from (2) and (14) the composite Fourier-Laplace transform of the Green function by taking into account the Fourier transform for the *Riesz-Feller* space-fractional derivative, see (4), and the Laplace transform for the *Caputo* time-fractional derivative, see (12). We have (in an obvious notation)

$$-\psi^\theta_\alpha(\kappa)\, \widehat{\widetilde{G^\theta_{\alpha,\beta}}}(\kappa, s) \;=\; s^\beta\, \widehat{\widetilde{G^\theta_{\alpha,\beta}}}(\kappa, s) - s^{\beta-1}, \tag{16}$$

so that

$$\widehat{\widetilde{G^\theta_{\alpha,\beta}}}(\kappa, s) = \frac{s^{\beta-1}}{s^\beta + \psi^\theta_\alpha(\kappa)}, \quad s > 0, \quad \kappa \in \mathbf{R}. \tag{17}$$

In the special case $\theta = 0$ we get

$$\widehat{\widetilde{G^0_{\alpha,\beta}}}(\kappa, s) = \frac{s^{\beta-1}}{s^\beta + |\kappa|^\alpha}, \quad s > 0, \quad \kappa \in \mathbf{R}. \tag{18}$$

---

the Riemann-Liouville fractional integral, is here denoted as ${}_tD^\beta\, f(t)$. We have, see e.g. [52],

$${}_tD^\beta\, f(t) := \frac{d}{dt}\left[\frac{1}{\Gamma(1-\beta)}\int_0^t \frac{f(\tau)\, d\tau}{(t-\tau)^\beta}\right], \quad 0 < \beta < 1.$$

It turns out that

$${}_tD^\beta_*\, f(t) \;=\; {}_tD^\beta\, \left[f(t) - f(0^+)\right] = {}_tD^\beta\, f(t) - f(0^+)\, \frac{t^{-\beta}}{\Gamma(1-\beta)}, \quad 0 < \beta < 1.$$

The *Caputo* fractional derivative, practically ignored in the mathematical treatises, represents a sort of regularization in the time origin for the *Riemann-Liouville* fractional derivative and satisfies the relevant property of being zero when applied to a constant. For more details on this fractional derivative (and its extension to higher orders) we refer the interested reader to Gorenflo and Mainardi [17].

By using the known scaling rules for the Fourier and Laplace transforms, we infer directly from (17) (without inverting the two transforms) the following noteworthy *scaling property* of the Green function,

$$G^{\theta}_{\alpha,\beta}(x,t) = t^{-\beta/\alpha}\, K^{\theta}_{\alpha,\beta}\left(x/t^{\beta/\alpha}\right) . \tag{19}$$

Here $x/t^{\beta/\alpha}$ acts as the *similarity variable* and $K^{\theta}_{\alpha,\beta}(\cdot)$ as the *reduced Green function*. Using (19) and the initial condition $G^{\theta}_{\alpha,\beta}(x,0^+) = \delta(x)$, we note that

$$\int_{-\infty}^{+\infty} G^{\theta}_{\alpha,\beta}(x,t)\, dx = \int_{-\infty}^{+\infty} K^{\theta}_{\alpha,\beta}(x)\, dx \equiv 1 . \tag{20}$$

In the case of the standard diffusion equation (1) the Green function is nothing but the Gaussian probability density with variance $\sigma^2 = 2t$, namely

$$G^{0}_{2,1}(x,t) = \frac{1}{2\sqrt{\pi}}\, t^{-1/2}\, \mathrm{e}^{-x^2/(4t)} . \tag{21}$$

In the general case, following the arguments by Mainardi, Luchko & Pagnini [38], we can prove that $G^{\theta}_{\alpha,\beta}(x,t)$ is still a probability density evolving in time. In the next section we summarise some results from [38].

3 The Green Function for Space-Time Fractional Diffusion

For the analytical and computational determination of the reduced Green function, from now on we restrict our attention to $x > 0$ because of the *symmetry relation* $K^{\theta}_{\alpha,\beta}(-x) = K^{-\theta}_{\alpha,\beta}(x)$. Mainardi, Luchko & Pagnini [38] have provided (for $x > 0$) the Mellin-Barnes integral representation

$$K^{\theta}_{\alpha,\beta}(x) = \frac{1}{\alpha x}\frac{1}{2\pi i}\int_{\gamma-i\infty}^{\gamma+i\infty} \frac{\Gamma(\frac{s}{\alpha})\,\Gamma(1-\frac{s}{\alpha})\,\Gamma(1-s)}{\Gamma(1-\frac{\beta}{\alpha}s)\,\Gamma(\rho\, s)\,\Gamma(1-\rho\, s)}\, x^{\,s}\, ds , \quad \rho = \frac{\alpha-\theta}{2\alpha} \tag{22}$$

where $0 < \gamma < \min\{\alpha, 1\}$. Following [38], we note that the Mellin-Barnes[7] integral representation allows us to construct computationally the fundamental solutions of (2) for any triplet $\{\alpha, \theta, \beta\}$ by matching their convergent and asymptotic expansions. Readers acquainted with Fox H functions can recognize in (22) the representation of a certain function of this class, see e.g. [28,41,52,56,58,61]. Unfortunately, as far as we know, computing routines for this general class of special functions are not yet available.

[7] The names Mellin and Barnes refer to the two authors, who in the first 1910's developed the theory of these integrals using them for a complete integration of the hypergeometric differential equation. We note that, as pointed out in [11] (Vol. 1, Ch. 1, §1.19, p. 49), these integrals were first used by S. Pincherle in 1888. For a revisited analysis of the pioneering work of Pincherle we refer to [39].

Let us now point out the main characteristics of the peculiar cases of *strictly space fractional diffusion* and *strictly time fractional diffusion*, for which the non-negativity of the corresponding reduced Green functions is known. For $\beta = 1$ and $0 < \alpha < 2$ (*strictly space fractional diffusion*) we have

$$K_{\alpha,1}^{\theta}(x) = L_{\alpha}^{\theta}(x) = \frac{1}{\alpha x}\frac{1}{2\pi i}\int_{\gamma-i\infty}^{\gamma+i\infty}\frac{\Gamma(s/\alpha)\,\Gamma(1-s)}{\Gamma(\rho\, s)\,\Gamma(1-\rho s)}\,x^{\,s}\,ds\,, \quad x>0\,, \tag{23}$$

with $0 < \gamma < \min\{\alpha, 1\}$, where $L_{\alpha}^{\theta}(x)$ denotes the class of the strictly stable (non-Gaussian) densities exhibiting heavy tails (with the algebraic decay $\propto |x|^{-(\alpha+1)}$) and infinite variance. For $\alpha = 2$ and $0 < \beta < 1$ (*strictly time fractional diffusion*) we have

$$K_{\alpha,1}^{\theta}(x) = \frac{1}{2}\,M_{\beta/2}(x) = \frac{1}{2x}\frac{1}{2\pi i}\int_{\gamma-i\infty}^{\gamma+i\infty}\frac{\Gamma(1-s)}{\Gamma(1-\beta s/2)}\,x^{\,s}\,ds\,, \quad x>0\,, \tag{24}$$

with $0 < \gamma < 1$, where $\frac{1}{2}\,M_{\beta/2}(x)$ denotes the class of the Wright-type densities exhibiting stretched exponential tails and therefore finite variance. The corresponding Green function evolves in time with the variance proportional to t^{β}. Mathematical details can be found in [38]; for further reading we refer to Schneider [56] for stable densities, and to Gorenflo, Luchko & Mainardi [16] for the Wright-type densities. For the special case $\alpha = \beta \le 1$ referred in [38] as *neutral diffusion*, we obtain from (22) an elementary (non-negative) expression

$$\begin{aligned} K_{\alpha,\alpha}^{\theta}(x) = N_{\alpha}^{\theta}(x) &= \frac{1}{\alpha x}\frac{1}{2\pi i}\int_{\gamma-i\infty}^{\gamma+i\infty}\frac{\Gamma(\frac{s}{\alpha})\,\Gamma(1-\frac{s}{\alpha})}{\Gamma(\rho\, s)\,\Gamma(1-\rho\, s)}\,x^{\,s}\,ds \\ &= \frac{1}{\alpha x}\frac{1}{2\pi i}\int_{\gamma-i\infty}^{\gamma+i\infty}\frac{\sin(\pi\,\rho\, s)}{\sin(\pi\, s/\alpha)}\,x^{\,s}\,ds \\ &= \frac{1}{\pi}\,\frac{x^{\alpha-1}\sin[\frac{\pi}{2}(\alpha-\theta)]}{1+2x^{\alpha}\cos[\frac{\pi}{2}(\alpha-\theta)]+x^{2\alpha}}\,, \end{aligned} \tag{25}$$

$x > 0\,,$

with $0 < \gamma < \alpha$, where $N_{\alpha}^{\theta}(x)$ denotes a peculiar class of densities exhibiting a power-law decay $\propto |x|^{-(\alpha+1)}$, which contains the well known (stable) Cauchy density (recovered for $\alpha = 1$ and $\theta = 0$).

For the generic case of *strictly space-time diffusion* ($0 < \alpha < 2$, $0 < \beta < 1$), including neutral diffusion, we can prove the non negativity of the corresponding reduced Green function in virtue of the identity, see [38],

$$K_{\alpha,\beta}^{\theta}(x) = \alpha\int_{0}^{\infty}\left[\xi^{\alpha-1}\,M_{\beta}\,(\xi^{\alpha})\right]\,L_{\alpha}^{\theta}\,(x/\xi)\,\frac{d\xi}{\xi}\,, \quad 0<\beta<1\,, \quad x>0\,. \tag{26}$$

Then, as a consequence of the previous discussion, for the *strictly space-time fractional* diffusion we obtain a class of probability densities (symmetric or non-symmetric according to $\theta = 0$ or $\theta \neq 0$) which exhibit heavy tails with an algebraic decay $\propto |x|^{-(\alpha+1)}$. Thus they belong to the domain of attraction of the Lévy stable densities of index α and can be referred to as *fractional stable densities*, according to a terminology proposed by Uchaikin [60].

In Fig. 1 we exhibit some plots of the probability densities provided by the reduced Green function for some "characteristic" values of the parameters α, β, and θ. These plots, taken from [38], were drawn for the values of the independent variable x in the range $|x| \leq 5$. To give the reader a better impression about the behaviour of the tails the logarithmic scale was adopted.

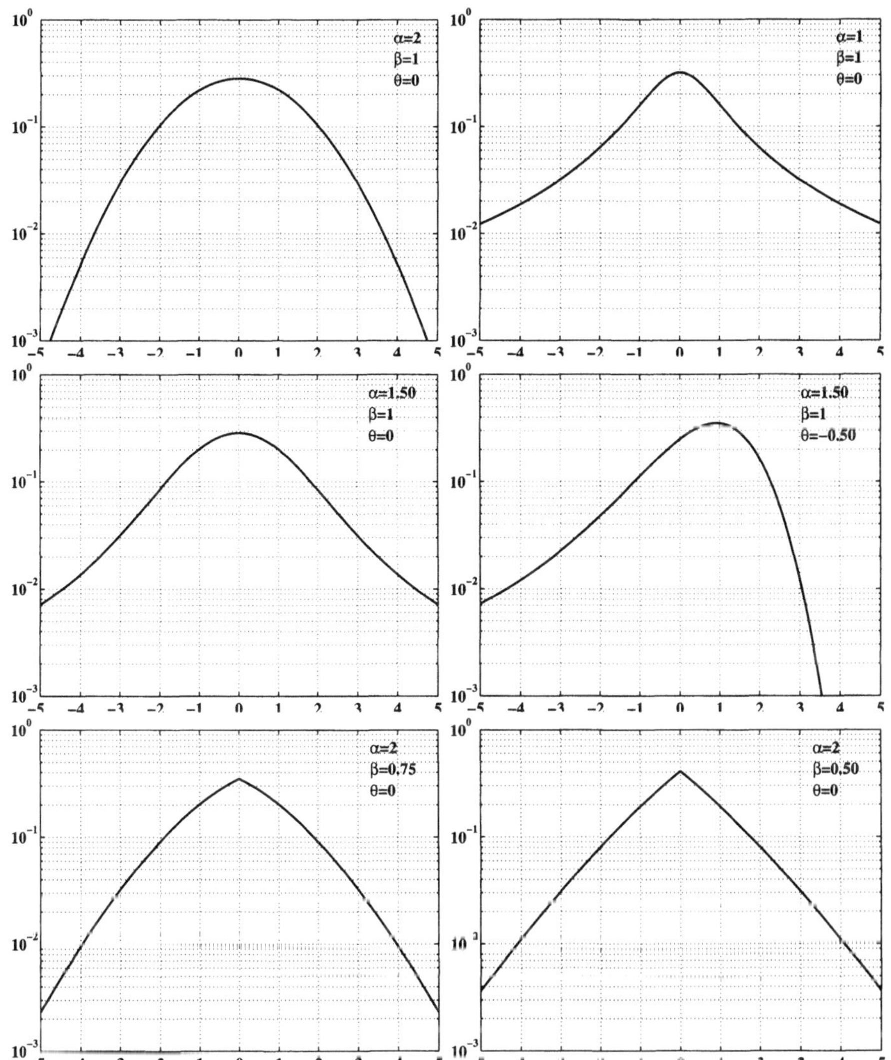

Fig. 1. Probability densities (reduced Green functions) for some values of the triplet $\{\alpha, \theta, \beta\}$

As for the stochastic processes governed by the above probability distributions we can expect the following. For the case of non-Gaussian stable densities we expect a special class of Markovian processes, called stable Lévy motions, which exhibit infinite variance associated to the possibility of arbitrarily large jumps (*Lévy flights*), whereas for the case of Wright-type densities we expect a class of stochastic non-Markovian processes, which exhibit a (finite) variance consistent with slow anomalous diffusion. Finally, for the case of fractional stable densities, the related stochastic processes are expected to possess the characteristics of the previous two classes. Indeed, they are non-Markovian (being $\beta < 1$) and exhibit infinite variance associated to the possibility of arbitrarily large jumps (being $\alpha < 2$). A way to realize (understand) all the above stochastic processes is to show sample paths and histograms of related random walk models. For random walks discrete both in space and in time we refer to our papers [15,21–23].

4 From CTRW to Fractional Diffusion

Here we show how the space-time fractional diffusion equation (2) can be obtained from the master equation for a *continuous time random walk* or, equivalently, from the master equation describing a cumulative *renewal process*, through an appropriate limit. As a matter of fact this limit will be carried out in the Fourier-Laplace domain, so the corresponding convergence is to be intended in a weak form, that is sufficient for our purposes. For the basic principles of *continuous time random walk* (simply referred to as CTRW), that was formerly introduced in Statistical Mechanics by Montroll and Weiss [46], see e.g. [30,45,47,63], of renewal processes, see e.g. [10,13,35,57].

The CTRW arises by a sequence of independently identically distributed (*iid*) positive random waiting times $T_1, T_2, \ldots$, each having a *pdf* $\psi(t)$, $t > 0$, and a sequence of *iid* random jumps $X_1, X_2, X_3, \ldots.$ in $\mathbf{R}$, each having a *pdf* $w(x)$, $x \in \mathbf{R}$. Setting $t_0 = 0$, $t_n = T_1 + T_2 + \ldots T_n$ for $n \in \mathbf{N}$, $0 < t_1 < t_2 < \ldots$, the wandering particle starts at point $x = 0$ in instant $t = 0$ and makes a jump of length X_n in instant t_n, so that its position is

$$x = 0 \quad \text{for} \quad 0 \le t < T_1 = t_1 ,$$

$$x = S_n = X_1 + X_2 + \ldots X_n , \quad \text{for} \quad t_n \le t < t_{n+1} .$$

An essential assumption is that the waiting time distribution and the jump width distribution are independent of each other. It is well known that this stochastic process is *Markovian* if and only if the waiting time *pdf* is of the form $\psi(t) = m \exp(-mt)$ with some positive constant m (*compound Poisson process*), see e.g. [13]. Then, by natural probabilistic arguments we arrive at the *master equation* for the spatial *pdf* $p(x,t)$ of the particle being in point x at instant t, see [25,40,55],

$$p(x,t) = \delta(x)\,\Psi(t) + \int_0^t \psi(t-t') \left[\int_{-\infty}^{+\infty} w(x-x')\,p(x',t')\,dx'\right] dt' , \qquad (27)$$

in which $\delta(x)$ denotes the Dirac generalized function, and, for abbreviation, $\Psi(t) = \int_t^\infty \psi(t')\,dt'$, is the probability that at instant t the particle is still sitting in its starting position $x = 0$. For this reason the function $\Psi(t)$ is usually referred to as the *survival probability*; in the *Markovian* case it reduces to the exponential function $\Psi(t) = \exp(-mt)$. Actually, $p(x,t)$ as containing a point measure, is a generalized *pdf*, but for ease of language we omit the qualification "generalized". Clearly, (27) satisfies the initial condition $p(x, 0^+) = \delta(x)$.

It is customary (and convenient for our purposes) to consider the master equation (27) in the Fourier-Laplace domain[8] where it reads

$$\widehat{\widetilde{p}}(\kappa, s) = \widetilde{\Psi}(s) + \widetilde{\psi}(s)\,\widehat{w}(\kappa)\,\widehat{\widetilde{p}}(\kappa, s)\,, \tag{28}$$

whence,

$$\widehat{\widetilde{p}}(\kappa, s) = \frac{\widetilde{\Psi}(s)}{1 - \widehat{w}(\kappa)\,\widetilde{\psi}(s)} = \frac{1 - \widetilde{\psi}(s)}{s}\,\frac{1}{1 - \widehat{w}(\kappa)\,\widetilde{\psi}(s)}\,. \tag{29}$$

We will henceforth assume that in our continuous time random walk the jump width *pdf* $w(x)$ is an even function ($w(x) = w(-x)$) and has a finite second moment (variance) or exhibits the asymptotic behaviour $w(x) \sim b\,|x|^{-(\alpha+1)}$ with some α, $0 < \alpha < 2$, for $|x| \to \infty$, and the waiting time *pdf* $\psi(t)$ has a finite first moment (mean) or exhibits the asymptotic behaviour $\psi(t) \sim c\,t^{-(\beta+1)}$ with some β, $0 < \beta < 1$, for $t \to \infty$, where b and c are positive constants.

Our aim is to derive from the master equation (27), by properly rescaling the waiting times and the jump widths and passing to the diffusion limit, the *space-time fractional diffusion equation.* By our derivation of (2) from (27) we de-mystify the often asked-for meaning of the time fractional derivative in the fractional diffusion equation. In plain words, the fractional derivatives in time as well as in space are caused by asymptotic power laws and well-scaled passage to the diffusion limit.

Scaling is achieved by making smaller all waiting times by a positive factor τ, and all jumps by a positive factor h. So we get the jump instants

$$t_n(\tau) = \tau T_1 + \tau T_2 + \ldots + \tau T_n \quad \text{for} \quad n \in \mathbf{N}\,, \tag{30}$$

and the jump sums,

$$S_0(h) = 0\,, \quad S_n(h) = hX_1 + hX_2 + \ldots + hX_n \quad \text{for} \quad n \in \mathbf{N}\,. \tag{31}$$

The reduced waiting times τT_n all have the *pdf* $\psi_\tau(t) = \psi(t/\tau)/\tau$, $t > 0$, and analogously the reduced jumps hX_n all have the *pdf* $w_h(x) = w(x/h)/h$, $x \in \mathbf{R}$. Readily we see

$$\widetilde{\psi}_\tau(s) = \widetilde{\psi}(s\tau)\,, \quad \widehat{w}_h(\kappa) = \widehat{w}(\kappa h)\,. \tag{32}$$

[8] It was in this domain that originally in 1965 Montroll and Weiss [46] derived their celebrated equation for the CTRW in Statistical Mechanics. However such equation can be derived by simply considering a random walk subordinated to a time renewal process as noted by us in [24] and by Baeumer and Meerschaert in [3]

Replacing in (27) $\psi(t)$ by $\psi_\tau(t)$, $\Psi(t)$ by $\Psi_\tau(t) = \int_t^\infty \psi_\tau(t')\,dt'$, $w(x)$ by $w_h(x)$, $p(x,t)$ by $p_{h,\tau}(x,t)$ we obtain the rescaled master equation which in the Fourier-Laplace domain reads as

$$\widehat{\widetilde{p_{h,\tau}}}(\kappa, s) = \frac{1 - \widetilde{\psi}_\tau(s)}{s} + \widetilde{\psi}_\tau(s)\,\widehat{w}_h(\kappa)\,\widehat{\widetilde{p_{h,\tau}}}(\kappa, s)\,, \tag{33}$$

whose solution is

$$\widehat{\widetilde{p_{h,\tau}}}(\kappa, s) = \frac{1 - \widetilde{\psi}_\tau(s)}{s}\,\frac{1}{1 - \widehat{w}_h(\kappa)\,\widetilde{\psi}_\tau(s)}\,. \tag{34}$$

To proceed further we assume the probability densities $w(x)$ and $\psi(t)$ of the jumps X_n and the waiting times T_n to meet the asymptotic conditions of the following Lemma 1 and Lemma 2, respectively, herewith recalled from [24] where the interested reader can find the proofs.

The first Lemma is a modified specialisation of Gnedenko's theorem in [14], see also [8]. It was already used by us, but not formally called a Lemma, in [20]. The second Lemma can be obtained by aid of a corollary in Widder's book [64].

Lemma 1

Assume $w(x) \ge 0$, $w(x) = w(-x)$ for $x \in \mathbf{R}$, $\int_{-\infty}^{+\infty} w(x)\,dx = 1$, and either

$$\sigma^2 := \int_{-\infty}^{+\infty} x^2\,w(x)\,dx < \infty \tag{35}$$

(relevant in the case $\alpha = 2$) or, with $b > 0$ and some $\alpha \in (0,2)$,

$$w(x) = (b + \epsilon(|x|))\,|x|^{-(\alpha+1)}\,. \tag{36}$$

In (36) assume $\epsilon(|x|)$ bounded and $O\left(|x|^{-\eta}\right)$ with some $\eta > 0$ as $|x| \to \infty$. Then, with a positive scaling parameter h and a scaling constant

$$\mu = \begin{cases} \dfrac{\sigma^2}{2}\,, & \text{if} \quad \alpha = 2\,, \\ \dfrac{b\,\pi}{\Gamma(\alpha+1)\,\sin(\alpha\pi/2)}\,, & \text{if} \quad 0 < \alpha < 2\,, \end{cases} \tag{37}$$

we have, for each fixed $\kappa \in \mathbf{R}$, the asymptotic relation

$$\widehat{w}(\kappa h) = 1 - \mu\,(|\kappa|\,h)^\alpha + o(h^\alpha) \quad \text{for} \quad h \to 0\,. \tag{38}$$

We note that (38) holds trivially if $\kappa = 0$ since $\widehat{w}(0) = 1$. □

Lemma 2

Assume $\psi(t) \ge 0$ for $t > 0$, $\int_0^\infty \psi(t)\,dt = 1$, and either

$$\rho := \int_0^\infty t\,\psi(t)\,dt < \infty \tag{39}$$

(relevant in the case $\beta = 1$), or, with $c > 0$ and some $\beta \in (0,1)$,

$$\psi(t) \sim c t^{-(\beta+1)} \quad \text{for} \quad t \to \infty\,. \tag{40}$$

Then, with a positive scaling parameter τ and a scaling constant

$$\lambda = \begin{cases} \rho\,, & \text{if} \quad \beta = 1\,, \\ \dfrac{c\,\Gamma(1-\beta)}{\beta}\,, & \text{if} \quad 0 < \beta < 1\,, \end{cases} \tag{41}$$

we have, for each fixed $s > 0$, the asymptotic relation

$$\widetilde{\psi}(s\tau) = 1 - \lambda\,(s\tau)^{\beta} + o(\tau^{\beta}) \quad \text{for} \quad \tau \to 0\,. \tag{42}$$

We note that (42) holds trivially if $s = 0$ since $\widetilde{\psi}(0) = 1$. □

Eq. (34) then becomes asymptotically

$$\widehat{\widetilde{p_{h,\tau}}}(\kappa, s) \sim \frac{\lambda\,\tau^{\beta}\, s^{\beta-1}}{\lambda\,\tau^{\beta}\, s^{\beta} + \mu\, h^{\alpha}\,|\kappa|^{\alpha}}\,, \quad \text{for} \quad h, \tau \to 0\,. \tag{43}$$

By imposing the *scaling relation*

$$\lambda\,\tau^{\beta} = \mu\, h^{\alpha}\,, \tag{44}$$

the asymptotics (43) yields

$$\widehat{\widetilde{p_{h,\tau}}}(\kappa, s) \to \frac{s^{\beta-1}}{s^{\beta} + |\kappa|^{\alpha}}\,. \tag{45}$$

Hence, in view of (18),

$$\widehat{\widetilde{p_{h,\tau}}}(\kappa, s) \to \widehat{\widetilde{G^{0}_{\alpha,\beta}}}(\kappa, s) \quad \text{for} \quad h, \tau \to 0\,, \tag{46}$$

under condition (44). Then, the asymptotic equivalence in the space-time domain between the master equation (27) after rescaling and the fractional diffusion equation (2) with $\theta = 0$ and the initial condition $u(x, 0^{+}) = \delta(x)$ is provided by the continuity theorem for sequences of characteristic functions after having applied the analogous theorem for sequences of Laplace transforms, see e.g. [13]. Therefore we have *convergence in law* or *weak convergence* for the corresponding probability distributions.

5 Simulations

By aid of the results of Sect. 4 we can produce approximate particle paths for space-time fractional diffusion in the spatially symmetric case $\theta = 0$ of (2). To this end, we require, for given α and β, a jump width *pdf* $w(x)$, obeying

Lemma 1, and a waiting time *pdf* $\psi(t)$, obeying Lemma 2. Natural choices are the corresponding symmetric stable density of index α, i.e. $w(x) = L_\alpha^0(x)$ $(0 < \alpha \le 2)$ and, following [40], the corresponding Mittag-Leffler type function

$$\psi(t) = -\frac{d}{dt} E_\beta(-t^\beta)\,, \quad 0 < \beta \le 1\,, \tag{47}$$

where

$$E_\beta(z) := \sum_{n=0}^{\infty} \frac{z^n}{\Gamma(\beta n + 1)}\,, \quad \beta > 0\,, \quad z \in \mathbf{C}\,, \tag{48}$$

denotes the (entire) transcendental function, known as the Mittag-Leffler function of order β [11] (Vol. 3, Ch 18, pp. 206-227). This function, which is a natural generalization of the exponential to which it reduces as $\beta = 1$, is playing a fundamental role in the applications of fractional calculus, see e.g. [17,37]. As has been shown in [40], see also [28,29], this choice of waiting-time density leads from the master equation (27) to the equation

$${}_tD_*^\beta\, p(x,t) = -p(x,t) + \int_{-\infty}^{+\infty} w(x-x')\, p(x',t)\, dx'\,, \quad p(x,0^+) = \delta(x)\,, \tag{49}$$

from which, by an appropriately scaled limiting process (analogous to that of Sect. 4), the fractional diffusion equation (2) with $u(x,0^+) = \delta(x)$, can be deduced, see [25]. Observe that (49) in the particular case $\beta = 1$ reduces to the well-known Feller-Kolmogorov equation for a compound Poisson process, in accordance with $E_1(-t) = \exp(-t)$.

Still some work must be invested in the inversions of the cumulative function $W(x) = \int_{-\infty}^{x} w(x')\, dx'$ and the survival probability $\Psi(t) = \int_t^\infty \psi(t')\, dt'$, which here is

$$\Psi(t) = E_\beta(-t^\beta)\,, \quad t \ge 0\,, \quad 0 < \beta \le 1\,. \tag{50}$$

In Fig. 2 we exhibit plots of $\Psi(t)$ versus time for some values of $\beta \in (0,1]$ from which we can get insight into the different behaviour for $0 < \beta < 1$ (fast decay for short times and slow decay for long times) and for $\beta = 1$ (exponential decay).

These inversions are required by the standard Monte-Carlo procedure of generating the corresponding jump-widths and waiting-times from $[0,1]$ - uniformly distributed (pseudo-) random numbers. A. Vivoli in his thesis [62] has described in detail how all this can be done and has carried out several case studies of which we show (here) three samples for CTRW's, just to convey a visual impression on the structure of such processes, see Fig. 3. In these samples we have $\alpha = 2$ so the jump density is a Gaussian, whereas $\beta = 1, 0.75, 0.50$.

Observe in Fig. 3 the striking contrast between the first graph and the other two. In the case $\beta = 1$ we have $\Psi(t) = \exp(-t)$ which results in long waiting times occurring rarely (the mean waiting time being finite!). So, we get a good approximation of Brownian motion, (2) reducing to (1). For $0 < \beta < 1$, however,

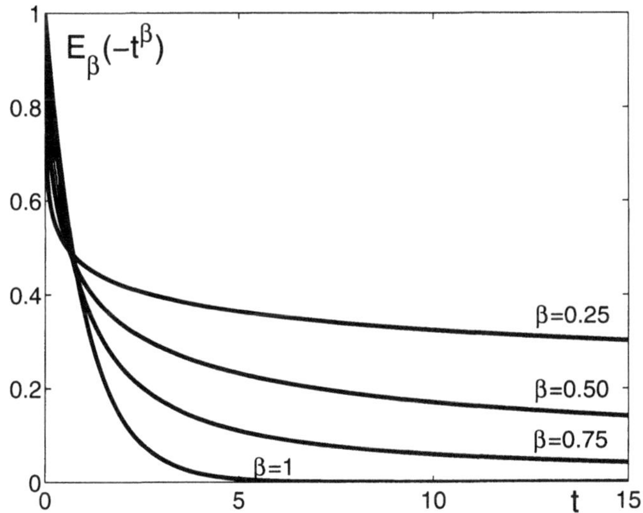

Fig. 2. The survival probability $\Psi(t) = E_\beta(-t^\beta)$ for $\beta = 0.25, 0.50, 0.75, 1$

the Mittag-Leffler function exhibits a power-law decay, namely

$$\Psi(t) = E_\beta(-t^\beta) \sim \frac{\sin(\beta\pi)}{\pi}\,\frac{\Gamma(\beta)}{t^\beta}, \quad t \to \infty. \tag{51}$$

As a consequence, we have a distinctly visible preponderance of long waiting times (the mean waiting time being infinite!).

As our emphasis in this paper is on waiting times (relevant in CTRW's) we should say that the essential aspect is the asymptotic behaviour for $t \to \infty$ of the corresponding probability densities, namely, according to Lemma 2, their decay like $c\,t^{-(\beta+1)}$ $(0 < \beta < 1)$ which implies for the survival probability a decay like $(c/\beta)\,t^{-\beta}$. This is true, of course, for the Mittag-Leffler waiting-time distributions used here, see (47) and (51). However, in the interest of easy inversion of $\Psi(t)$, it is advantageous to look for simpler suitable functions. One such function (which is more easily invertible) is

$$\Psi_*(t) = \frac{1}{1+\Gamma(1-\beta)t^\beta}, \quad t \geq 0, \quad 0 < \beta < 1, \tag{52a}$$

so that

$$\Psi_*(t)(t) \sim \frac{\sin(\beta\pi)}{\pi}\,\frac{\Gamma(\beta)}{t^\beta}, \quad t \to \infty. \tag{52b}$$

Happily this function shares with the function $\Psi(t) = E_\beta(-t^\beta)$ the desirable property of complete monotonicity in $t > 0$ [9]. Furthermore we note that the

[9] Complete monotonicity of a function $f(t)$, $t > 0$, means $(-1)^n \frac{d^n}{dt^n} f(t) \geq 0$, $(n = 0, 1, 2, \ldots)$, a characteristic property of $\exp(-t)$. For the Bernstein theorem this is

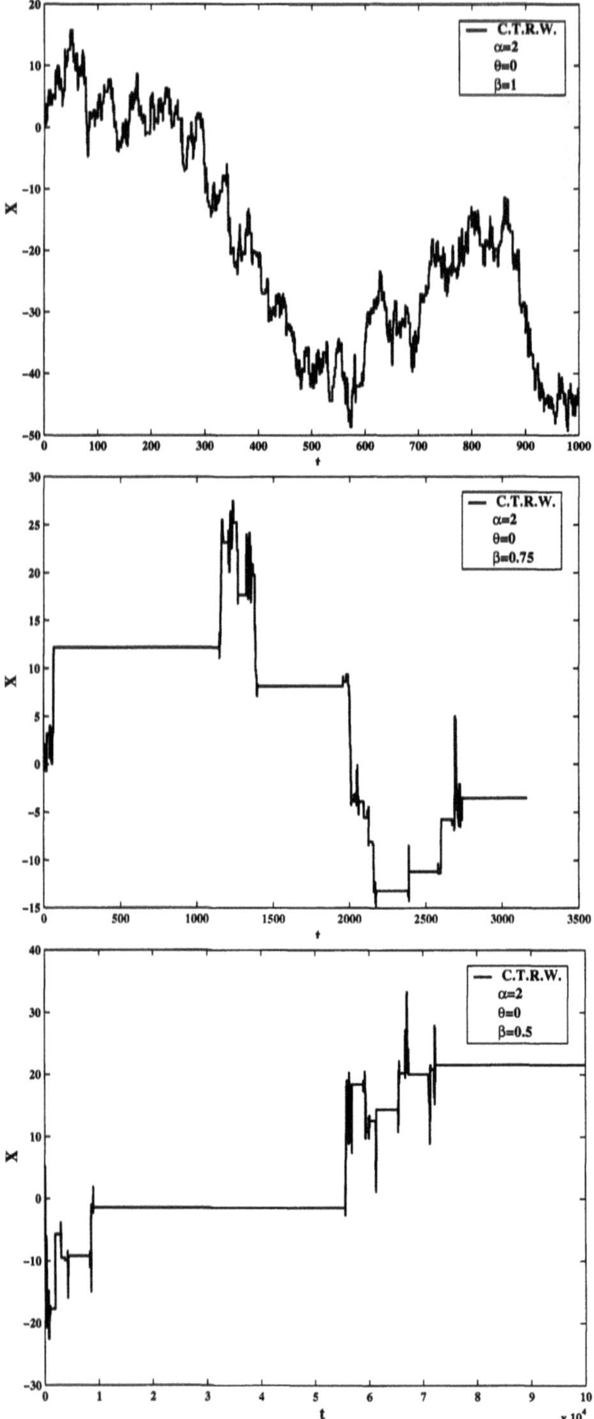

Fig. 3. Sample paths for CTRW's with $\alpha = 2$, $\theta = 0$ and $\beta = 1, 0.75, 0.50$ (from top to bottom)

functions $\Psi(t)$ and $\Psi_*(t)$ share the same order of asymptotics for $t \to 0^+$ (albeit with a different coefficient). In fact we find as $t \to 0^+$,

$$\Psi(t) = 1 - \frac{t^\beta}{\Gamma(1+\beta)} + o(t^\beta)\,, \quad \Psi_*(t) = 1 - \frac{\beta\,\pi}{\sin(\beta\pi)}\,\frac{t^\beta}{\Gamma(1+\beta)} + o(t^\beta)\,. \quad (53)$$

In a forthcoming paper [26] we will describe in more detail our methods of simulation and investigate their quality. In the interest of long-time (or, because of self-similarity, near-the-limit) simulations, it is highly desirable that such fast methods are developed.

6 Conclusions

In this paper we have surveyed the general theory of the one-dimensional space-time fractional diffusion equation and have presented representation of its fundamental solutions (the probability densities) in terms of Mellin-Barnes integrals. Then, we have outlined how, in the spatially symmetric case, this equation can be obtained by a limiting process from a master equation for a continuous time random walk via properly scaled compression of waiting times and jump widths. For the strictly space and/or time fractional cases ($\{0 < \alpha < 2\,,\, 0 < \beta < 1\}$), it suffices to assume asymptotic power laws $b\,|x|^{-(\alpha+1)}$ as $|x| \to \infty$ for the jump width *pdf* and $c\,t^{-(\beta+1)}$ as $t \to \infty$ for the waiting time *pdf*. For the compression factors h in space and τ in time we require a scaling relation of the kind $\lambda\,\tau^\beta = \mu\,h^\alpha$ where λ, μ are given positive constants. Here we have limited ourselves to show sample paths for some cases of the time fractional diffusion processes (the jump width *pdf* is Gaussian), referring for more comprehensive numerical studies to a forthcoming paper. The theory can be generalized to more than one space dimension and to non-symmetric jump *pdf*'s, likewise to probability distribution functions instead of densities for the jump widths and waiting times, but, in order to avoid too cumbersome notations and calculations, let us just hint here to such possibilities.

Acknowledgements

We are grateful to the Erasmus-Socrates project, to the INTAS project 00-0847, and to the Research Commissions of the Free University of Berlin and of the University of Bologna for supporting our work. F.M. is also grateful to the National Group of Mathematical Physics (G.N.F.M. - I.N.D.A.M.) and the National Institute of Nuclear Physics (I.N.F.N. - Sezione di Bologna) for partial support.

equivalent to the representability of $f(t)$ as (real) Laplace transform of a given non-negative (ordinary or generalized) function. For more information, see e.g. [4] (pp. 61-72), [13] (pp. 335-338), [31] (pp. 162-164) and [44].

References

1. E. Barkai, CTRW pathways to the fractional diffusion equation, *Chemical Physics* (2002), to appear
2. E. Barkai, R. Metzler, J. Klafter: From continuous-time random walks to the fractional Fokker-Planck equation, *Physical Review E* **61**, 132–138 (2000)
3. B. Baeumer, M.M. Meerschaert: Stochastic solutions for fractional Cauchy problems, *Fractional Calculus and Applied Analysis* **4**, 481-500 (2001)
4. C. Berg, G. Forst: *Potential Theory on Locally Compact Abelian Groups* (Springer, Berlin 1975)
5. P. Butzer, U. Westphal: 'Introduction to fractional calculus'. in: *Fractional Calculus, Applications in Physics*, ed. by R. Hilfer (World Scientific, Singapore 2000) pp. 1 - 85
6. M. Caputo: Linear models of dissipation whose Q is almost frequency independent, Part II *Geophys. J. R. Astr. Soc.* **13** 529–539 (1967)
7. M. Caputo, F. Mainardi: Linear models of dissipation in anelastic solids, *Riv. Nuovo Cimento* (Ser. II) **1**, 161–198 (1971)
8. A. V. Chechkin, V. Yu. Gonchar: A model for persistent Lévy motion, *Physica A* **277**, 312-326 (2000)
9. A.V. Chechkin, V.Yu. Gonchar: Linear relaxation processes governed by fractional symmetric kinetic equations, *JETP (Journal of Experimental and Theoretical Physics)* **91**, 635-651 (2000)
10. D.R. Cox: *Renewal Theory* (Methuen, London 1967)
11. A. Erdélyi, W. Magnus, F. Oberhettinger, F.G. Tricomi: *Higher Transcendental Functions*, Bateman Project, Vols. 1-3 (McGraw-Hill, New York 1953-1955)
12. W. Feller: On a generalization of Marcel Riesz' potentials and the semi-groups generated by them, *Meddelanden Lunds Universitets Matematiska Seminarium* (*Comm. Sém. Mathém. Université de Lund*). Tome suppl. dédié a M. Riesz, Lund (1952) 73-81
13. W. Feller: *An Introduction to Probability Theory and its Applications*, Vol. 2 (Wiley, New York 1971)
14. B.V. Gnedenko and A.N. Kolmogorov: *Limit Distributions for Sums of Independent Random Variables* (Addison-Wesley, Cambridge, Mass. 1954)
15. R. Gorenflo, G. De Fabritiis, F. Mainardi: Discrete random walk models for symmetric Lévy-Feller diffusion processes, *Physica A* **269**, 79–89 (1999)
16. R. Gorenflo, Yu. Luchko, F. Mainardi: Analytical properties and applications of the Wright function, *Fractional Calculus and Applied Analysis* **2**, 383-414 (1999)
17. R. Gorenflo, F. Mainardi: 'Fractional calculus: integral and differential equations of fractional order'. In: *Fractals and Fractional Calculus in Continuum Mechanics*, ed. by A. Carpinteri, F. Mainardi (Springer Verlag, Wien, 1997) pp. 223–276 [Reprinted in NEWS 010101, see `http://www.fracalmo.org`]
18. R. Gorenflo, F. Mainardi: Random walk models for space-fractional diffusion processes, *Fractional Calculus and Applied Analysis* **1**, 167–191 (1998)
19. R. Gorenflo, F. Mainardi: Approximation of Lévy-Feller diffusion by random walk, *Journal for Analysis and its Applications (ZAA)* **18**, 231-146 (1999)
20. R. Gorenflo, F. Mainardi: 'Random walk models approximating symmetric space-fractional diffusion processes'. In: J. Elschner, I. Gohberg and B. Silbermann (Editors), *Problems in Mathematical Physics*, ed. by J. Elschner, I. Gohberg, B. Silbermann (Birkhäuser Verlag, Basel 2001) pp. 120-145 [Series *Operator Theory: Advances and Applications*, No 121]

21. R. Gorenflo, F. Mainardi, D. Moretti, P. Paradisi: Time-fractional diffusion: a discrete random walk approach, *Nonlinear Dynamics* (2002), in press
22. R. Gorenflo, F. Mainardi, D. Moretti, G. Pagnini, P. Paradisi: Fractional diffusion: probability distributions and random walk models, *Physica A* **305**, 106-112 (2002)
23. R. Gorenflo, F. Mainardi, D. Moretti, G. Pagnini, P. Paradisi: Discrete random walk models for space-time fractional diffusion, *Chemical Physics* (2002), in press
24. R. Gorenflo and F. Mainardi: 'Non-Markovian random walk models, scaling and diffusion limits'. In: *Mini-Proceedings:e 2-nd MaPhySto Conference on Lévy Processes: Theory and Applications.* Dept. Mathematics, University of Aarhus, Denmark, 21-25 January 2002, ed. by O.E. Barndorff-Nielsen (Mathematical Physics and Stochastics Centre, Aarhus 2002) pp. 120-128 [see `http://www.maphysto.dk`, Publications, Miscellanea No. 22]
25. R. Gorenflo, F. Mainardi, E. Scalas, M. Raberto: 'Fractional calculus and continuous-time finance III: the diffusion limit'. In: *Mathematical Finance*, ed. by M. Kohlmann, S. Tang (Birkhäuser Verlag, Basel 2001) pp. 171-180
26. R. Gorenflo, F. Mainardi, E. Scalas, A. Vivoli: Continuous-time random walk models for fractional diffusion processes, in preparation.
27. I.S. Gradshteyn, I.M. Ryzhik: *Tables of Integrals, Series and Products* (Academic Press, New York 1980)
28. R. Hilfer: 'Fractional time evolution'. In: *Applications of Fractional Calculus in Physics*, ed. by R. Hilfer (World Scientific, Singapore, 2000) pp. 87-130
29. R. Hilfer, L. Anton: Fractional master equations and fractal time random walks, *Phys. Rev. E* **51**, R848-R851 (1995)
30. B.D. Hughes: *Random Walks and Random Environments, Vol. 1: Random Walks* (Clarendon Press, Oxford 1995)
31. N. Jacob, *Pseudo Differential Operators. Markov Processes*, Vol. 1: Fourier Analysis and Semigroups (Imperial College Press, London 2001)
32. A. Janicki, A. Weron: *Simulation and Chaotic Behavior of α–Stable Stochastic Processes* (Marcel Dekker, New York 1994)
33. J. Klafter, A. Blumen, M.F. Shlesinger: Stochastic pathway to anomalous diffusion, *Phys. Rev. A* **35**, 3081–3085 (1987)
34. J. Klafter, M. F. Shlesinger, G. Zumofen: Beyond Brownian motion, *Physics Today* **49**, 33–39 (1996)
35. M. Kotulski: Asymptotic distributions of continuous-time random walks: a probabilistic approach, *J. Stat. Phys.* **81**, 777–792 (1995)
36. P. Lévy: *Théorie de l'Addition des Variables Aléatoires*, 2nd edn. (Gauthier-Villars, Paris 1954)
37. F. Mainardi: 'Fractional calculus: some basic problems in continuum and statistical mechanics'. In: *Fractals and Fractional Calculus in Continuum Mechanics*, ed. by A. Carpinteri, F. Mainardi (Springer Verlag, Wien and New-York 1997) pp. 291–248 [Reprinted in NEWS 011201 `http://www.fracalmo.org`]
38. F. Mainardi, Yu. Luchko, G. Pagnini: The fundamental solution of the space-time fractional diffusion equation, *Fractional Calculus and Applied Analysis* **4**, 153-192 (2001) [Reprinted in NEWS 010401 `http://www.fracalmo.org`]
39. F. Mainardi, G. Pagnini: Salvatore Pincherle: the pioneer of the Mellin-Barnes integrals, *J. Computational and Applied Mathematics* (2002), to appear
40. F. Mainardi, M. Raberto, R. Gorenflo, E. Scalas: Fractional calculus and continuous-time finance II: the waiting-time distribution, *Physica A* **287**, 468–481 (2000)
41. A.M. Mathai, R.K. Saxena: *The H-function with Applications in Statistics and Other Disciplines* (New Delhi, Wiley Eastern Ltd 1978)

42. R. Metzler, J. Klafter: The random walk's guide to anomalous diffusion: a fractional dynamics approach, *Phys. Reports* **339**, 1-77 (2000)
43. R. Metzler, T.F. Nonnenmacher: Space- and time-fractional diffusion and wave equations, fractional Fokker-Planck equations, and physical motivation, *Chemical Physics* (2002), in press
44. K.S. Miller, S.G. Samko: Completely monotonic functions, *Integral Transforms and Special Functions* **12**, 389-402 (2001)
45. E.W. Montroll, M.F. Shlesinger: 'On the wonderful world of random walks'. In: *Nonequilibrium Phenomena II: from Stochastics to Hydrodynamics*, ed. by J. Leibowitz, E.W. Montroll (North-Holland, Amsterdam 1984) pp. 1-121 [Series *Studies in Statistical Mechanics*, Vol. XI]
46. E.W. Montroll, G.H. Weiss: Random walks on lattices II, *J. Math. Phys.* **6**, 167–181 (1965)
47. E.W. Montroll, B.J. West: 'On an enriched collection of stochastic processes'. In: *Fluctuation Phenomena*, ed. by E.W. Montroll, J. Leibowitz (North-Holland, Amsterdam 1979) pp. 61-175 [Series *Studies in Statistical Mechanics*, Vol. VII]
48. P. Paradisi, R. Cesari, F. Mainardi, F. Tampieri: The fractional Fick's law for non-local transport processes, *Physica A* **293**, 130-142 (2001)
49. I. Podlubny: *Fractional Differential Equations* (Academic Press, San Diego 1999)
50. M. Riesz: L'intégrales de Riemann-Liouville et le probléme de Cauchy, *Acta Math.* **81**, 1-223 (1949)
51. A. Saichev, G. Zaslavsky: Fractional kinetic equations: solutions and applications, *Chaos* **7**, 753-764 (1997)
52. S.G. Samko, A.A. Kilbas, O.I. Marichev: *Fractional Integrals and Derivatives: Theory and Applications* (Gordon and Breach, New York 1993)
53. G. Samorodnitsky, M.S. Taqqu: *Stable non-Gaussian Random Processes* (Chapman & Hall, New York 1994)
54. K. Sato: *Lévy Processes and Infinitely Divisible Distributions* (Cambridge University Press, Cambridge 1999)
55. E. Scalas, R. Gorenflo, F. Mainardi: Fractional calculus and continuous-time finance, *Physica A* **284**, 376-384 (2000)
56. W.R. Schneider: 'Stable distributions: Fox function representation and generalization'. In: *Stochastic Processes in Classical and Quantum Systems*, ed. by S. Albeverio, G. Casati, D. Merlini (Springer Verlag, Berlin-Heidelberg 1986) 497-511 [Lecture Notes in Physics, Vol. 262]
57. W.L. Smith: Renewal theory and its ramifications, *J. Roy. Statist. Soc. B* **20**, 243-284 (1958) [Discussion, pp. 284-302]
58. H.M. Srivastava, K.C. Gupta, S.P. Goyal: *The H-Functions of One and Two Variables with Applications* (South Asian Publishers, New Delhi and Madras 1982)
59. H. Takayasu: *Fractals in the Physical Sciences* (Manchester University Press, Manchester and New York 1990)
60. V.V. Uchaikin: private communication (2000)
61. V.V. Uchaikin, V.M. Zolotarev: *Chance and Stability. Stable Distributions and their Applications* (VSP, Utrecht 1999)
62. A. Vivoli, *Non-Gaussian Stochastic Processes and Their Applications*, Thesis for Degree in Physics, University of Bologna, March 2002, in Italian. [Supervisors: Prof. F. Mainardi and Prof. R. Gorenflo]
63. G.H. Weiss: *Aspects and Applications of Random Walks* (North-Holland, Amsterdam 1994)
64. D.V. Widder: *The Laplace Transform* (Princeton Univ. Press, Princeton 1946)

First Passage Distributions for Long Memory Processes

Govindan Rangarajan[1] and Mingzhou Ding[2]

[1] Department of Mathematics and Centre for Theoretical Studies, Indian Institute of Science, Bangalore 560 012, India
[2] Centre for Complex Systems and Brain Sciences, Florida Atlantic University, Boca Raton, FL 33431, USA

Abstract. We study the distribution of first passage time for Levy type anomalous diffusion. A fractional Fokker-Planck equation framework is introduced. For the zero drift case, using fractional calculus an explicit analytic solution for the first passage time density function in terms of Fox or H-functions is given. The asymptotic behaviour of the density function is discussed. For the nonzero drift case, we obtain an expression for the Laplace transform of the first passage time density function, from which the mean first passage time and variance are derived.

1 Introduction

Consider a stochastic process $X(t)$ with $X(0) = 0$. The first passage time (FPT) T to the point $X = a$ is defined as [1]

$$T = \inf\{t : X(t) = a\}.$$

For ordinary diffusion modeled by Brownian motion, where $\mathrm{Var}(X(t)) = 2Kt$, the exact density function of T is known and given by

$$f_\mu(t) = \frac{a}{(4\pi K t^3)^{1/2}} e^{-(a-\mu t)^2/4Kt}, \tag{1}$$

where μ denotes the drift, K is the diffusion coefficient and we have assumed $a > 0$ for ease of expression. Equation (1) is often referred to as the inverse Gaussian distribution and was first obtained by Schrodinger [2] and by Smoluchowski [3] in 1915. The zero drift case of $\mu = 0$ was considered earlier by Bachelier [4] around 1900 in the context of financial analysis. (See Seshadri [5] for an interesting account of their work and a general historical survey.)

The FPT problem finds applications in many areas of science and engineering [6–9]. A sampling of these applications is listed below.

- probability theory (study of Wiener process, fractional Brownian motion etc.)
- statistical physics (study of anomalous diffusion)
- neuroscience (analysis of neuron firing models)
- civil and mechanical engineering (analysis of structural failure)
- chemical physics (study of noise assisted potential barrier crossings)

- hydrology (optimal design of dams)
- financial mathematics (analysis of circuit breakers)
- imaging (study of image blurring due to hand jitter)

In this chapter we consider the FPT problem for a class of long memory processes. In particular, we study non Gaussian and non Markovian stochastic processes referred to as anomalous diffusions of the Levy type [10–12] where $\mathrm{Var}(X(t)) \sim t^{\gamma}$, $0 < \gamma < 2$, for large t. Generalizing earlier work for the zero drift case [13], our first result is the formulation of a fractional Fokker-Planck equation (FFPE) which describes Levy type anomalous diffusion with nonzero drift. Specializing to the zero drift case, we solve the FFPE under suitable initial and boundary value conditions using fractional calculus to obtain the FPT density function in terms of Fox or H-functions. We further prove that it is a valid probability density. For the nonzero drift case, we obtain the Laplace transform of the FPT density function. For $0 < \gamma < 1$, the Laplace transform is shown not to satisfy the completely monotone conditions [14] required for the Laplace transform of a valid probability density. Restricting ourselves to $1 \leq \gamma < 2$ we derive the mean and variance of the FPT density function. We show that for $\gamma = 1$, which corresponds to the ordinary diffusion, the inverse Laplace transform can be carried out explicitly and the result is (1). Finally, using properties of the H-functions, we obtain an asymptotic power law expression for the zero drift FPT density function.

2 Fractional Fokker-Planck Equation for Levy Type Anomalous Diffusion with Drift

Definition 1. Anomalous diffusion $X_{\gamma}(t)$ is a diffusive process with diffusion parameter γ ($0 < \gamma < 2$) where the mean $\mathbf{E}(X_{\gamma}(t)) = 0$ and the mean square displacement

$$\mathrm{Var}(X_{\gamma}(t)) \sim t^{\gamma} \quad \text{for large } t.$$

For $\gamma = 1$ we obtain ordinary diffusion. Subdiffusion corresponds to $0 < \gamma < 1$ and superdiffusion corresponds to $1 < \gamma < 2$.

Anomalous diffusion of the Levy type is a class of non Gaussian and non Markovian processes founded on the continuous time random walk (CTRW) where the waiting time comes from a renewal process and obeys certain power law distribution [11]. Let $\xi(y, u)$ denote the joint probability density between the jump size y and the waiting time u. It can be shown that, depending on the specific form of $\xi(y, u)$, the CTRW can produce both subdiffusive and superdiffusive processes as well as ordinary diffusion [11]. For example, consider

$$\xi(y, u) = \frac{1}{\sqrt{2\pi\sigma^2}} \exp[-y^2/2\sigma^2] \frac{(\alpha - 1)/\tau}{(1 + u/\tau)^{\alpha}}, \tag{2}$$

where y and u are independent with y being a Gaussian variable and u a Levy stable variable [15]. For $1 < \alpha < 2$, the corresponding CTRW is characterized by a subdiffusive process with $\gamma = \alpha - 1$, and for $\alpha \geq 2$, one gets ordinary diffusion with $\gamma = 1$. If, on the other hand, y and u are coupled through

$$\xi(y,u) = \frac{1}{2}\delta(u/\tau - |y|/\sigma)\frac{(\beta-1)/\tau}{(1+u/\tau)^{\beta}},$$

where $2 < \beta < 3$ and $\delta(\cdot)$ is the Dirac delta function, the CTRW describes a superdiffusive process with $\gamma = 4 - \beta$.

Definition 2. Levy type anomalous diffusion $X_\gamma(t)$ is the non-Gaussian stochastic process obtained by taking the generalized diffusion limit of the above CTRW. This limit is defined as: $\sigma^2 \to 0$, $\tau \to 0$ such that $K = \sigma^2/2\Gamma(1-\gamma)\tau^\gamma$ is a constant for a subdiffusive process and $K = (2-\gamma)\Gamma(\gamma-1)\sigma^2/2\tau^\gamma$ is a constant for a superdiffusive process. Here K is called the generalized diffusion constant.

Definition 3. Anomalous diffusion with drift μ is defined by

$$X_{\gamma,\mu}(t) = \mu t + X_\gamma(t).$$

Let the probability density function of a Levy type anomalous diffusive process with drift μ be denoted $p(x,t)$. Recently Metzler et al. [13] have formulated a fractional Fokker-Planck equation (FFPE), based on the renewal equation, which describes zero drift Levy type anomalous diffusive processes with $0 < \gamma \leq 1$. Generalizing this we obtain the following FFPE for the evolution of $p(x,t)$

$$p(x,t) - p(x,0) = K\ {}_0D_t^{-\gamma}\frac{\partial^2}{\partial x^2}p(x,t) - \mu\ {}_0D_t^{-1}\frac{\partial}{\partial x}p(x,t), \qquad (3)$$

where $0 < \gamma < 2$ and K is the generalized diffusion constant defined above. Here the Riemann-Liouville fractional integral operator ${}_0D_t^{-\gamma}$ is defined as [16,17]

$${}_0D_t^{-\gamma}p(x,t) = \frac{1}{\Gamma(\gamma)}\int_0^t dt'\ (t-t')^{\gamma-1}p(x,t'), \quad \gamma > 0,$$

with $\Gamma(z)$ being the gamma function [18]. The integration kernel on the right hand side accounts for the non Markovian memory effect stemming from the nonexponential waiting time distribution. From the above FFPE it is easily seen that

$$\mathbf{E}(X_{\gamma,\mu}(t)) = \mu t; \quad \mathrm{Var}(X_{\gamma,\mu}(t)) = 2Kt^\gamma/\Gamma(1+\gamma).$$

This demonstrates that our FFPE produces the correct time dependence for the mean and the variance.

To obtain the first passage time density function, we first solve (3) with the following initial and boundary value conditions:

$$p(\infty,t) = p(0,t) = 0\ (t > 0); \quad p(x,0) = \delta(x-a), \qquad (4)$$

where $x = a$ is the starting point of the diffusive process, containing the initial concentration of the distribution. We note that the above conditions are slightly different from the standard ones [19] used in such problems where $x = 0$ is the starting point but both sets of conditions are equivalent. The present formulation makes the subsequent derivation less cumbersome. Once $p(x,t)$ is known, the first passage time density function $f_{\gamma,\mu}(t)$ is given by [19]:

$$f_{\gamma,\mu}(t) = -\frac{d}{dt}\int_0^\infty p(x,t)dx. \tag{5}$$

3 FPT Density Function for Levy Type Anomalous Diffusion with Zero Drift

For a Levy type anomalous diffusive process with zero drift, the fractional Fokker-Planck equation for $p(x,t)$ reduces to

$$p(x,t) - p(x,0) = K\ {}_0D_t^{-\gamma}\frac{\partial^2}{\partial x^2}p(x,t). \tag{6}$$

We now state and prove

Theorem 1. *For Levy type anomalous diffusion with zero drift ($\mu = 0$) described by (6), the FPT density function is given by*

$$f_{\gamma,0}(t) = \frac{a\gamma}{2K^{1/2}t^{(2+\gamma)/2}}H_{1,1}^{1,0}\left(\frac{a}{(Kt^\gamma)^{1/2}}\middle|\begin{matrix}(1-\gamma/2,\gamma/2)\\(0,1)\end{matrix}\right),\quad a,t>0, \tag{7}$$

where $H(\cdot)$ denotes the Fox or H-function [20–22] to be described below.

Proof. Taking the Laplace transform of (6) with respect to time we get

$$q(x,s) - \frac{p(x,0)}{s} = \frac{K}{s^\gamma}\frac{\partial^2}{\partial x^2}q(x,s),$$

where $q(x,s)$ is the Laplace transform of $p(x,t)$ with respect to time. Here we have also applied the result [17] that the Laplace transform of ${}_0D_t^{-\gamma}p(x,t)$ is $q(x,s)/s^\gamma$. Using the initial condition $p(x,0) = \delta(x-a)$, the above equation can be rewritten as

$$\frac{\partial^2}{\partial x^2}q(x,s) - \frac{s^\gamma}{K}q(x,s) = -\frac{s^{\gamma-1}}{K}\delta(x-a).$$

The general solution of this equation satisfying all the boundary and initial conditions [cf. (4)] is given by

$$q(x,s) = \frac{s^{\gamma/2-1}}{2\sqrt{K}}\left[\exp(-s^{\gamma/2}|x-a|/\sqrt{K}) - \exp(-s^{\gamma/2}(x+a)/\sqrt{K})\right].$$

The inverse Laplace transform of $s^{\gamma/2-1}\exp(-|x|s^{\gamma/2})$ is known [23]. Writing the result using the H-function (which is more convenient for our purpose) we get

$$p(x,t) = \frac{1}{2(Kt^\gamma)^{1/2}} H_{1,1}^{1,0}\left(\frac{|x-a|}{(Kt^\gamma)^{1/2}} \middle| \begin{matrix} (1-\gamma/2,\gamma/2) \\ (0,1) \end{matrix} \right) - \frac{1}{2(Kt^\gamma)^{1/2}} H_{1,1}^{1,0}\left(\frac{x+a}{(Kt^\gamma)^{1/2}} \middle| \begin{matrix} (1-\gamma/2,\gamma/2) \\ (0,1) \end{matrix} \right). \tag{8}$$

Here, the Fox or H-function [20–22] has the following alternating power series expansion:

$$H_{p,q}^{m,n}\left(z \middle| \begin{matrix} (a_j,A_j)_{j=1,\dots,p} \\ (b_j,B_j)_{j=1,\dots,q} \end{matrix} \right) = \sum_{l=1}^{m} \sum_{k=0}^{\infty} \frac{(-1)^k z^{s_{lk}}}{k! B_l} \times \frac{\prod_{j=1,j\neq l}^{m} \Gamma(b_j - B_j s_{lk}) \prod_{r=1}^{n} \Gamma(1-a_r + A_r s_{lk})}{\prod_{u=m+1}^{q} \Gamma(1-b_u + B_u s_{lk}) \prod_{v=n+1}^{p} \Gamma(a_v - A_v s_{lk})}, \tag{9}$$

where $s_{lk} = (b_l + k)/B_l$ and an empty product is interpreted as unity. Further, m, n, p, q are nonnegative integers such that $0 \le n \le p$, $1 \le m \le q$; A_j, B_j are positive numbers; a_j, b_j can be complex numbers. For further discussions of the H-function, see Mathai [21]. A few key properties of H-functions are summarized in the Appendix.

Substituting (8) into (5) we have

$$f_{\gamma,0}(t) = -\frac{d}{dt}\left[\frac{1}{2(Kt^\gamma)^{1/2}} \int_0^\infty dx\ H_{1,1}^{1,0}\left(\frac{|x-a|}{(Kt^\gamma)^{1/2}} \middle| \begin{matrix} (1-\gamma/2,\gamma/2) \\ (0,1) \end{matrix} \right) \right] + \frac{d}{dt}\left[\frac{1}{2(Kt^\gamma)^{1/2}} \int_0^\infty dx\ H_{1,1}^{1,0}\left(\frac{x+a}{(Kt^\gamma)^{1/2}} \middle| \begin{matrix} (1-\gamma/2,\gamma/2) \\ (0,1) \end{matrix} \right) \right].$$

Defining $z = (x-a)/(Kt^\gamma)^{1/2}$, $z' = (x+a)/(Kt^\gamma)^{1/2}$, we obtain

$$\begin{aligned} f_{\gamma,0}(t) &= -\frac{d}{dt} \int_{-a/(Kt^\gamma)^{1/2}}^\infty dz\ H_{1,1}^{1,0}\left(|z| \middle| \begin{matrix} (1-\gamma/2,\gamma/2) \\ (0,1) \end{matrix} \right) + \frac{d}{dt} \int_{a/(Kt^\gamma)^{1/2}}^\infty dz'\ H_{1,1}^{1,0}\left(z' \middle| \begin{matrix} (1-\gamma/2,\gamma/2) \\ (0,1) \end{matrix} \right) \\ &= \frac{a\gamma}{2K^{1/2} t^{(2+\gamma)/2}} H_{1,1}^{1,0}\left(\frac{a}{(Kt^\gamma)^{1/2}} \middle| \begin{matrix} (1-\gamma/2,\gamma/2) \\ (0,1) \end{matrix} \right). \end{aligned}$$

This completes the proof. This result was obtained by us in earlier papers [24,25].

The above theoretical prediction for the full FPT density function were verified by numerically simulating the underlying CTRW process characterized by the probability density function $\xi(y,u)$ [cf. (2)]. The FPT density function obtained theoretically from Eq. (7) was compared with the FPT density function obtained numerically using 10 million realizations of the underlying stochastic process. We observed that the numerical simulation was in excellent agreement with the theoretical prediction.

Note that for ordinary diffusion ($\gamma = 1$), the expression for $f_{\gamma,0}(t)$ in (7) reduces to

$$f_{1,0}(t) = \frac{a}{2\sqrt{Kt^3}} \sum_{k=0}^{\infty} \frac{(-a/\sqrt{Kt})^k}{k!\Gamma(1/2 - k/2)}, \tag{10}$$

where we have used the alternating series expansion for the H-function given in (9). But [18]

$$\begin{aligned} \frac{1}{\Gamma(1/2 - k/2)} &= 0, \quad k \text{ odd}, \\ &= \frac{k!}{(-4)^{k/2}(k/2)!\sqrt{\pi}}, \quad k \text{ even}. \end{aligned}$$

Substituting this in (10) and letting $n = k/2$ we get

$$f_{1,0}(t) = \frac{a}{2\sqrt{Kt^3}} \sum_{n=0}^{\infty} \frac{(-a/\sqrt{Kt})^{2n}(2n)!}{(2n)!(-4)^n n!\sqrt{\pi}}.$$

Simplifying this we get the standard result [which agrees with (1) for $\mu = 0$]

$$\begin{aligned} f_{1,0}(t) &= \frac{a}{\sqrt{4\pi Kt^3}} \sum_{n=0}^{\infty} \frac{(-a^2/4\sqrt{Kt})^n}{n!} \\ &= \frac{a}{\sqrt{4\pi Kt^3}} \exp[-a^2/4Kt]. \end{aligned}$$

We next prove the following result.

Theorem 2. *$f_{\gamma,0}(t)$ is a proper probability density function.*

Proof. We know that $f_{\gamma,0}(t)$ is a proper probability density if and only if its Laplace transform $\phi_{\gamma,0}(s)$ satisfies the completely monotone condition [14], i.e.,

$$(-1)^n \frac{d^n \phi_{\gamma,0}(s)}{ds^n} \geq 0 \text{ for } s > 0,\ n = 1, 2, \ldots \quad \text{and } \phi_{\gamma,0}(0) = 1.$$

In our case,

$$\phi_{\gamma,0}(s) = e^{g(s)}, \tag{11}$$

where

$$g(s) = \frac{-a}{\sqrt{K}} s^{\gamma/2}. \tag{12}$$

To prove that $\phi_{\gamma,0}(s)$ is completely monotone, consider the following Lemma.

Lemma 2. *For $n \geq 1$, $(-1)^n \frac{d^n e^{g(s)}}{ds^n} \geq 0$ if $g^{(n)}(s)$ (the nth derivative of $g(s)$ with respect to s) is non-positive for n odd and non-negative for n even.*

Proof. We have

$$\frac{d^n e^{g(s)}}{ds^n} = \sum_{i=1}^{m} c_i [g^{(1)}]^{r_{i,1}} [g^{(2)}]^{r_{i,2}} \cdots [g^{(n)}]^{r_{i,n}} e^{g(s)},$$

where $r_{i,1}, r_{i,2}, \ldots, r_{i,n} \geq 0$, $c_i > 0$ and $r_{i,1} + r_{i,2} + \cdots + r_{i,n} \leq n$ for $i = 1, 2, \ldots, m$. Using standard differentiation rules from calculus, it is easily seen that $r_{i,1} + r_{i,3} + \cdots + r_{i,n}$ is odd for n odd and $r_{i,1} + r_{i,3} + \cdots + r_{i,n-1}$ is even for n even (where $i = 1, 2, \ldots, m$). This immediately gives us the required result thus completing the proof of Lemma 1.

Returning to the proof of Theorem 2, we have [cf. (12)]

$$g^{(n)}(s) = \frac{-a}{\sqrt{K}} \left(\frac{\gamma}{2}\right) \left(\frac{\gamma}{2} - 1\right) \cdots \left(\frac{\gamma}{2} - n + 1\right) s^{\gamma/2 - n}.$$

Since $0 < \gamma < 2$ and $s > 0$, $g^{(n)}(s)$ is non-positive for n odd and non-negative for n even. Combining this result with Lemma 1 and (11), we see that $(-1)^n \frac{d^n \phi_{\gamma,0}(s)}{ds^n} \geq 0$ for $n \geq 1$. From (11) we also see that $\phi_{\gamma,0}(0) = 1$. This completes the proof of Theorem 2.

Finally, we consider the asymptotic behaviour of the FPT density function for large values of t. Refer to (7). Let $z = a/(Kt^\gamma)^{1/2}$. It is known [21,27] that, for small z, $H_{1,1}^{1,0}(z) \sim |z|^{b_1/B_1} = 1$, since $b_1 = 0$ and $B_1 = 1$. Therefore, the FPT distribution $f(t)$, for large t, is characterized by the power law relation

$$f(t) \sim t^{-1-\gamma/2}, \quad t \to \infty, \tag{13}$$

which becomes the well known $-3/2$ scaling law for the ordinary Brownian motion. This power law behaviour has been observed earlier by Balakrishnan [28] for subdiffusive processes ($0 < \gamma < 1$) using a different method.

4 Laplace Transform of FPT Density Function for Levy Type Anomalous Diffusion with Drift

Theorem 3. *Given a Levy type anomalous diffusion with drift μ described by (3), the Laplace transform $\phi_{\gamma,\mu}(s)$ of its first passage time density function $f_{\gamma,\mu}(t)$ is given by*

$$\phi_{\gamma,\mu}(s) = \exp\left[-\frac{a\mu s^{\gamma-1}}{2K} - a\sqrt{\frac{\mu^2 s^{2\gamma-2}}{4K^2} + \frac{s^\gamma}{K}}\right]. \tag{14}$$

Proof. Taking the Laplace transform of Eq. (3) with respect to time we get

$$q(x,s) - \frac{p(x,0)}{s} = \frac{K}{s^\gamma} \frac{\partial^2}{\partial x^2} q(x,s) - \frac{\mu}{s} \frac{\partial}{\partial x} q(x,s),$$

where $q(x,s)$ is the Laplace transform of $p(x,t)$. Here we have again applied the result [17] that the Laplace transform of ${}_0D_t^{-\gamma}p(x,t)$ is $q(x,s)/s^\gamma$. The above equation can be rewritten as

$$\frac{\partial^2}{\partial x^2}q(x,s) + A\frac{\partial}{\partial x}q(x,s) + Bq(x,s) = -\frac{s^{\gamma-1}}{K}\delta(x-a), \tag{15}$$

where

$$A = -\frac{\mu s^{\gamma-1}}{K}; \quad B = -\frac{s^\gamma}{K}. \tag{16}$$

Since $K > 0$, we have

$$\lambda^2 \equiv A^2 - 4B = \frac{\mu^2 s^{2\gamma-2}}{K^2} + 4\frac{s^\gamma}{K} \geq 0.$$

Therefore two independent solutions of the homogeneous equation corresponding to (15) are given by [29]

$$q_1(x,s) = \exp[x(\lambda - A)/2]; \quad q_2(x,s) = \exp[x(-\lambda - A)/2].$$

Consequently, the general solution of (15) satisfying all the boundary and initial conditions [cf. (4)] is given by

$$q(x,s) = \frac{s^{\gamma-1}}{K\lambda}e^{-A(x-a)/2}[e^{-\lambda|x-a|/2} - e^{-\lambda(x+a)/2}]. \tag{17}$$

To obtain the Laplace transform of the FPT density function, we take the Laplace transform of (5) to get

$$\phi_{\gamma,\mu}(s) = -s\int_0^\infty dx\, q(x,s) + \int_0^\infty dx\, p(x,0).$$

Here we have used the fact that Laplace transform of $dp(x,t)/dt$ is given by [30] $sq(x,s) - p(x,0)$. Since $p(x,0) = \delta(x-a)$ [cf. (4)], we obtain

$$\phi_{\gamma,\mu}(s) = 1 - s\int_0^\infty dx\, q(x,s).$$

Substituting for $q(x,s)$ from (17), we get

$$\begin{aligned}\phi_{\gamma,\mu}(s) = 1 - \frac{s^\gamma}{K\lambda}\Bigg[&\int_0^a dx\, e^{-A(x-a)/2}e^{-\lambda(a-x)/2} \\ &+ \int_a^\infty dx\, e^{-A(x-a)/2}e^{-\lambda(x-a)/2}\Bigg] \\ &+ \frac{s^\gamma}{K\lambda}\int_0^\infty dx\, e^{-A(x-a)/2}e^{-\lambda(x+a)/2}.\end{aligned}$$

The integrals can be easily evaluated to finally give [upon using (16)]

$$\phi_{\gamma,\mu}(s) = \exp\left[-\frac{a\mu s^{\gamma-1}}{2K} - a\sqrt{\frac{\mu^2 s^{2\gamma-2}}{4K^2} + \frac{s^\gamma}{K}}\right].$$

This completes the proof of the theorem.

For $0 < \gamma < 1$, $\phi_{\gamma,\mu}(s)$ is not a completely monotone function [14] and hence is not a Laplace transform of a probability density function. For $1 \le \gamma < 2$, we have not been able to prove rigorously that $\phi_{\gamma,\mu}(s)$ is completely monotone. However, we have calculated the first hundred derivatives of $\phi_{\gamma,\mu}(s)$ using the symbolic manipulation program Mathematica and find that all of them satisfy the monotonicity condition. Hence we conjecture that $\phi_{\gamma,\mu}(s)$ is the Laplace transform of a probability density function for $1 \le \gamma < 2$. Henceforth, we restrict ourselves to this parameter range.

Corollary 1. *The mean first passage time* $\mathbf{E}(T)$ *for Levy type anomalous diffusion with* $1 \le \gamma < 2$ *and drift* $|\mu|$ *towards the barrier is given by*

$$\mathbf{E}(T) = \frac{a}{|\mu|}.$$

The variance is infinite for $\gamma > 1$ *whereas for* $\gamma = 1$ *(ordinary diffusion) it is given by the standard result*

$$\mathrm{Var}(T) = \frac{\sigma^2 a}{|\mu|^3}.$$

For drift away from the barrier, mean and variance are infinite for $1 \le \gamma < 2$.

Proof. First consider the case where the drift is towards the barrier. This implies that $\mu < 0$ since in our formulation the diffusive process starts at $x = a > 0$ and the barrier is at $x = 0$. In this case, the FPT density function can be written as [cf. (14)]

$$\phi_{\gamma,\mu}(s) = e^{ag(s)}, \tag{18}$$

where

$$g(s) = \frac{|\mu|s^{\gamma-1}}{2K} - \frac{|\mu|s^{\gamma-1}}{2K}\sqrt{1 + \frac{4Ks^{2-\gamma}}{\mu^2}}. \tag{19}$$

The mean first passage time is given by

$$\mathbf{E}(T) = -\frac{d\phi_{\gamma,\mu}(s)}{ds}\bigg|_{s=0}. \tag{20}$$

From Eqs. (18) and (19), we have

$$\frac{d\phi_{\gamma,\mu}(s)}{ds} = \left[\frac{|\mu|(\gamma-1)s^{\gamma-2}a}{2K}\left(1 - \sqrt{1 + \frac{4Ks^{2-\gamma}}{\mu^2}}\right) - \frac{(2-\gamma)a(1 + \frac{4Ks^{2-\gamma}}{\mu^2})^{-1/2}}{|\mu|}\right] e^{ag(s)}.$$

We need to find the limiting value of the above expression as $s \to 0$. First consider $e^{ag(s)}$. As $s \to 0$, we can expand the square root in Eq. (19) to give

$$g(s) = -\frac{s}{|\mu|} + \frac{Ks^{3-\gamma}}{|\mu|^3} - \cdots$$

Hence $g(s) \to 0$ as $s \to 0$. Consequently, $e^{ag(s)} \to 1$ as $s \to 0$. Performing similar expansions for the other terms in (21), we finally obtain [cf. (20)]

$$\mathbf{E}(T) = \frac{a}{|\mu|}.$$

Note that the mean first passage time is independent of γ.

The variance is obtained as follows:

$$\mathrm{Var}(T) = \mathbf{E}(T^2) - [\mathbf{E}(T)]^2. \tag{21}$$

Therefore we need to evaluate $\mathbf{E}(T^2)$. This is given by

$$\mathbf{E}(T^2) = \frac{d^2\phi_{\gamma,\mu}(s)}{ds^2}\Big|_{s=0}. \tag{22}$$

Now

$$\frac{d^2\phi_{\gamma,\mu}(s)}{ds^2} = \left[a\frac{d^2g(s)}{ds^2} + \left(a\frac{dg(s)}{ds}\right)^2\right]e^{ag(s)}. \tag{23}$$

Consider the first term. The second derivative of $g(s)$ is given by

$$\begin{aligned}\frac{d^2g(s)}{ds^2} = & \frac{|\mu|(\gamma-1)(\gamma-2)s^{\gamma-3}}{2K}\left(1-\sqrt{1+\frac{4Ks^{2-\gamma}}{\mu^2}}\right) \\ & - \frac{(\gamma-1)(2-\gamma)}{|\mu|s}\left(1+\frac{4Ks^{2-\gamma}}{\mu^2}\right)^{-1/2} \\ & + \frac{2K(2-\gamma)^2s^{1-\gamma}}{|\mu|^3}\left(1+\frac{4Ks^{2-\gamma}}{\mu^2}\right)^{-3/2}.\end{aligned}$$

Expanding all terms we get

$$\begin{aligned}\frac{d^2g(s)}{ds^2} = & -\frac{|\mu|(\gamma-1)(\gamma-2)s^{\gamma-3}}{2K}\sum_{n=2}^{\infty}\frac{(-1)^{n-1}1\cdot3\cdots(2n-3)}{2^n n!}\left(\frac{4Ks^{2-\gamma}}{\mu^2}\right)^n \\ & - \frac{(\gamma-1)(2-\gamma)}{|\mu|s}\sum_{n=1}^{\infty}\frac{(-1)^n 1\cdot3\cdots(2n-1)}{2^n n!}\left(\frac{4Ks^{2-\gamma}}{\mu^2}\right)^n \\ & + \frac{2K(2-\gamma)^2s^{1-\gamma}}{|\mu|^3}\sum_{n=0}^{\infty}\frac{(-1)^n 1\cdot3\cdots(2n+1)}{2^n n!}\left(\frac{4Ks^{2-\gamma}}{\mu^2}\right)^n.\end{aligned}$$

After considerable manipulation, this can be rewritten as

$$\begin{aligned}\frac{d^2g(s)}{ds^2} = & \frac{2K(2-\gamma)s^{1-\gamma}}{|\mu|^3}\sum_{n=0}^{\infty}\left[\frac{n(2-\gamma)+(3-\gamma)}{(n+2)}\right] \\ & \frac{(-1)^n 1\cdot3\cdots(2n+1)}{2^n n!}\left(\frac{4Ks^{2-\gamma}}{\mu^2}\right)^n.\end{aligned}$$

Thus the first term in (23) is given by

$$a\frac{d^2g(s)}{ds^2}e^{ag(s)} = \frac{2Ka(2-\gamma)s^{1-\gamma}}{|\mu|^3}e^{ag(s)}\sum_{n=0}^{\infty}\left[\frac{n(2-\gamma)+(3-\gamma)}{(n+2)}\right] \frac{(-1)^n 1\cdot 3\cdots(2n+1)}{2^n n!}\left(\frac{4Ks^{2-\gamma}}{\mu^2}\right)^n. \quad (24)$$

For $\gamma = 1$, the case of ordinary diffusion, we obtain from the above equation

$$a\frac{d^2g(s)}{ds^2}e^{ag(s)}\bigg|_{s=0} = \frac{\sigma^2 a}{|\mu|^3}.$$

Here we have also used the fact that $K = \sigma^2/2$ for $\gamma = 1$. On the other hand, for $1 < \gamma < 2$, as $s \to 0$ the prefactor multiplying the sum in (24) diverges whereas the sum itself is finite and bounded away from zero. Consequently, for $1 < \gamma < 2$ the first term in (23) diverges as $s \to 0$.

Next, we consider the second term in (23). We already know the limiting behaviour of this term for $1 \leq \gamma < 2$ as $s \to 0$ from our earlier analysis for mean first passage time, namely,

$$\left(\frac{dg(s)}{ds}\right)^2 e^{ag(s)}\bigg|_{s=0} = \frac{a^2}{\mu^2}.$$

Substituting the above results in (22), for $\gamma = 1$ we obtain

$$\mathbf{E}(T^2) = \frac{\sigma^2 a}{|\mu|^3} + \frac{a^2}{\mu^2}.$$

Therefore [cf. (21)]

$$\mathrm{Var}(T) = \sigma^2 a/|\mu|^3,$$

for $\gamma = 1$. For $1 < \gamma < 2$, the first term in (23) diverges as $s \to 0$ whereas the second term is finite. Therefore $\mathbf{E}(T^2)$ diverges. Hence the variance also diverges.

Finally, consider the case where the drift is away from the barrier. This implies that $\mu > 0$ in our formulation. Now the FPT density function can be written as in (18) where

$$g(s) = -\frac{|\mu|s^{\gamma-1}}{2K} - \frac{|\mu|s^{\gamma-1}}{2K}\sqrt{1+\frac{4Ks^{2-\gamma}}{\mu^2}}.$$

Performing the same analysis as above, it is easily seen that the mean and variance diverge for $1 \leq \gamma < 2$. This completes the proof.

Corollary 2. *For ordinary diffusion ($\gamma = 1$) with drift μ and diffusion constant K, the first passage time density function is given by*

$$f_{1,\mu}(t) = \frac{a}{\sqrt{4\pi K t^3}}\exp\left[-\frac{(a+\mu t)^2}{4Kt}\right], \quad a > 0, \quad t > 0.$$

Proof. For $\gamma = 1$, the Laplace transform of the first passage time density function reduces to [cf. Eq. (14)]

$$\phi_{1,\mu}(s) = \exp\left[-\frac{a}{2K}(-\mu - \sqrt{\mu^2 + 4Ks}\right]. \tag{25}$$

Now the inverse Laplace transform of $\exp(-\alpha\sqrt{s})$, $\alpha \geq 0$ is [30]

$$\frac{\alpha}{2\sqrt{\pi t^3}} \exp(-\alpha^2/4t).$$

Using this result, we can easily perform the inverse Laplace transform of (25) to obtain

$$f_{1,\mu}(t) = \frac{a}{\sqrt{4\pi K t^3}} \exp\left[-\frac{(a + \mu t)^2}{4Kt}\right], \quad a > 0, \quad t > 0.$$

We comment that, if the starting point of the diffusion is chosen at $x(0) = 0$, a negative μ will be used in the above equation and we recover the expected inverse Gaussian density, (1), for the FPT density function of a ordinary diffusion with drift. This completes the proof.

5 Summary

We have derived the explicit first passage time density function in terms of H-functions using a fractional Fokker-Planck equation formalism for zero drift Levy type anomalous diffusion. This was shown to be a proper probability density. For Levy type anomalous diffusion with drift and $1 \leq \gamma < 2$, the moment generating function (Laplace transform of the first passage time density function) was obtained. The mean first passage time was derived and shown to be independent of γ. The FPT density function for ordinary diffusion with drift was derived as a special case.

Acknowledgments

MD's work was supported by US Office of Naval Research and National Science Foundation. GR was supported by the Homi Bhabha Fellowship. GR is also associated with the Jawaharlal Nehru Centre for Advanced Scientific Research, Bangalore, India as a honorary faculty member.

Appendix: Properties of H-functions

The H-function has the following remarkable properties [21] which we will use later.

Property 1. The H-function is symmetric in the pairs $(a_1, A_1), \ldots, (a_n, A_n)$, likewise $(a_{n+1}, A_{n+1}), \ldots, (a_p, A_p)$; in $(b_1, B_1), \ldots, (b_m, B_m)$ and in (b_{m+1}, B_{m+1}), $\ldots, (b_q, B_q)$.

Property 2. Provided $n \geq 1$ and $q > m$,

$$H_{p,q}^{m,n}\left(z\left|\begin{matrix}(a_1,A_1), (a_2,A_2), & \cdots, & (a_p,A_p)\\ (b_1,B_1), \quad \cdots, & (b_{q-1},B_{q-1}), & (a_1,A_1)\end{matrix}\right.\right)$$
$$= H_{p-1,q-1}^{m,n-1}\left(z\left|\begin{matrix}(a_2,A_2), \cdots, & (a_p,A_p)\\ (b_1,B_1), \cdots, & (b_{q-1},B_{q-1})\end{matrix}\right.\right).$$

Property 3. Provided $m \geq 2$ and $p > n$,

$$H_{p,q}^{m,n}\left(z\left|\begin{matrix}(a_1,A_1), \quad \cdots, & (a_{p-1},A_{p-1}), & (b_1,B_1)\\ (b_1,B_1), (b_2,B_2), & \cdots, & (b_q,B_q)\end{matrix}\right.\right)$$
$$= H_{p-1,q-1}^{m-1,n}\left(z\left|\begin{matrix}(a_1,A_1), \cdots, (a_{p-1},A_{p-1})\\ (b_2,B_2), \cdots, \quad (b_q,B_q)\end{matrix}\right.\right).$$

Property 4.

$$H_{p,q}^{m,n}\left(z\left|\begin{matrix}(a_j,A_j)_{j=1,\dots,p}\\ (b_j,B_j)_{j=1,\dots,q}\end{matrix}\right.\right) = H_{q,p}^{n,m}\left(\frac{1}{z}\left|\begin{matrix}(1-b_j,B_j)_{j=1,\dots,q}\\ (1-a_j,A_j)_{j=1,\dots,p}\end{matrix}\right.\right).$$

Property 5. For $k > 0$,

$$\frac{1}{k}H_{p,q}^{m,n}\left(z\left|\begin{matrix}(a_j,A_j)_{j=1,\dots,p}\\ (b_j,B_j)_{j=1,\dots,q}\end{matrix}\right.\right) = H_{p,q}^{m,n}\left(z^k\left|\begin{matrix}(a_j,kA_j)_{j=1,\dots,p}\\ (b_j,kB_j)_{j=1,\dots,q}\end{matrix}\right.\right).$$

Property 6.

$$z^\rho H_{p,q}^{m,n}\left(z\left|\begin{matrix}(a_j,A_j)_{j=1,\dots,p}\\ (b_j,B_j)_{j=1,\dots,q}\end{matrix}\right.\right) = H_{p,q}^{m,n}\left(z\left|\begin{matrix}(a_j+\rho A_j,A_j)_{j=1,\dots,p}\\ (b_j+\rho B_j,B_j)_{j=1,\dots,q}\end{matrix}\right.\right).$$

References

1. G. R. Grimmet and D. R. Stirzaker: *Probability and Random Processes* (Oxford University Press, New York 1994)
2. E. Schrödinger: Physikalische Zeitschrift **16**, 289 (1915)
3. M. V. Smoluchowski: Physikalische Zeitschrift **16**, 318 (1915)
4. L. Bachelier: Annales des Sciences de l'Ecole Superieure **17**, 21 (1900)
5. V. Seshadri: *The Inverse Gaussian Distribution* (Clarendon, Oxford 1993)
6. H. Risken: *The Fokker-Planck Equation* (Springer-Verlag, Berlin 1989)
7. C. W. Gardiner: *Handbook of Stochastic Methods* (Springer-Verlag, Berlin 1997)
8. H. C. Tuckwell: *Introduction to Theoretical Neurobiology*, Vol. 1 & 2 (Cambridge University Press, Cambridge 1988)
9. Y. K. Lin and G. Q. Cai: *Probabilistic Structural Dynamics* (McGraw-Hill, New York 1995)
10. J.-P. Bouchaud and A. Georges: Phys. Rep. **195**, 12 (1990)
11. M. F. Shlesinger, J. Klafter and Y. M. Wong: J. Stat. Phys. **27**, 499 (1982)
12. J. Klafter, A. Blumen and M. F. Shlesinger: Phys. Rev. A **35**, 3081 (1987)
13. R. Metzler, E. Barkai, and J. Klafter: Phys. Rev. Lett. **82**, 3563 (1999)
14. W. Feller: *An Introduction to Probability Theory and Applications* Volume 2 (Wiley, New York 971)

15. G. Samorodnitsky and M. S. Taqqu: *Stable Non-Gaussian Random Processes* (Chapman & Hall, New York 1994)
16. K. B. Oldham and J. Spanier: *The Fractional Calculus* (Academic, New York 1974)
17. K. S. Miller and B. Ross: *An Introduction to the Fractional Calculus and Fractional Differential Equations* (Wiley, New York 1993)
18. I. S. Gradshteyn and I. M. Ryzhik: *Tables of Integrals, Series, and Products* (Academic, New York 1965)
19. D. R. Cox and H. D. Miller: *Theory of Stochastic Processes* (Methuen & Co., London 1965)
20. C. Fox: Trans. Am. Math. Soc. **98**, 395 (1961)
21. A. M. Mathai and R. K. Saxena: *The H-function with Applications in Statistics and Other Disciplines* (Wiley Eastern, New Delhi 1978)
22. H. M. Srivastava, K. C. Gupta, and S. P. Goyal: *The H-functions of One and Two Variables with Applications* (South Asian, New Delhi 1982).
23. I. Podlubny: *Fractional Differential Equations* (Academic Press, San Diego 1999)
24. G. Rangaran and M. Ding: Phys. Lett. A **273**, 322 (2000)
25. G. Rangaran and M. Ding: Phys. Rev. E **62**, 120 (2000)
26. M. Evans, N. Hastings and B. Peacock: *Statistical Distributions* (Wiley & Sons, New York 1993)
27. B. L. J. Braaksma: Compos. Math. **15**, 239 (1964)
28. V. Balakrishnan: Physica A **132**, 569 (1985)
29. A. D. Polyanin and V. F. Zaitsev: *Handbook of Exact Solutions for Ordinary Differential Equations* (CRC Press, Boca Raton 1995)
30. A. Erdelyi: *Tables of Integral Transforms.* Volume 1 (McGraw-Hill, New York 1954).

Non-Gaussian Statistics and Anomalous Diffusion in Porous Media

Pabitra N. Sen

Schlumberger-Doll Research, Ridgefield, CT 06877-4108, USA

"Can one hear the shape of a drum?", M.Kac (1966)

Abstract. Two time displacement correlation functions $< x(t)x(0) >$ grow linearly with time for ordinary diffusion, but for the case of anomalous diffusion it shows more complex time dependence. Anomalous diffusion is abundant in fluid dynamics with complex flow, generally chaotic and turbulent flow. In porous media this behavior is seen even in absence of flow, i.e. pure diffusion – as well as for hydrodynamic dispersion – molecular transport at low Reynold's number. These involve motion of molecules restricted by structures. We describe how transport of complex fluids contained within complex matrices can be used to explore the physics of complex porous media with their multiple characteristic time and length scales, varying over many orders of magnitude.

1 Introduction

A keystone of modern statistical physics is the theory of Brownian motion, random walk and diffusion. In the coarse-graining paradigm, first introduced by Einstein [1], for times much larger than a characteristic time τ_c, that, in turn, is much larger than times for underlying molecular collisions, the distribution of the displacement becomes a Gaussian whose width, given by the mean square displacement $< [x(t) - x(0)]^2 >\sim t^\gamma$, grows linearly with time ($\gamma = 1$) for ordinary diffusion. Anomalous diffusion ($\gamma \neq 1$) is abundant in fluid dynamics with complex flow, generally chaotic and turbulent flow.

The goal of this article is to show that there are numerous occasions where geometrical restriction to diffusion and flow can give rise to "anomalous" behavior – the resulting distribution is not a Gaussian and the mean square displacement does not grow linearly with time. This happens even in simple smooth geometries and in absence of any exotic waiting time distribution or Levy statistics. Restricted diffusion appears in porous media. Porous media are ubiquitous: examples can be taken from biology (nutrient transport, perfusion, membrane function, blood flow), catalysis, foodstuff, materials (polymer networks, self organizing materials), and geology (fluid movement in hydrocarbon reservoirs, ground water migration, contamination).

2 Deviation from Gaussian

The simplest example of effect of restriction comes from random walks with reflecting and absorbing boundary conditions [2] at a wall. These show non-

Gaussian distributions. There are at least a couple ways of deriving these results: either solving the diffusion equation with a reflecting (or absorbing, or, more general partially absorbing) boundary condition or by looking at random walks [2], and counting the paths that get bounced back (or absorbed) at a wall located at $x = 0$. This is basically the method of images. The probability $G(x, t; x_1, 0)\Delta x$ that a particle starting at x_1 at time $t = 0$ is found between x and $x + \Delta x$, after time t (the coarse graining distance $\Delta x >> l$, the unit step size) is

$$\begin{aligned} G(x,t;x_1,0) &= \frac{1}{\sqrt{4\pi D_0 t}} e^{-\frac{(x-x_1)^2}{4D_0 t}}, \quad -\infty \le x \le \infty, \ \textit{unbounded} \\ &= \frac{1}{\sqrt{4\pi D_0 t}} \left[e^{-\frac{(x-x_1)^2}{4D_0 t}} + e^{-\frac{(x+x_1)^2}{4D_0 t}} \right], \quad 0 \le x \le \infty, \\ &\qquad \textit{reflecting wall} \\ &= \frac{1}{\sqrt{4\pi D_0 t}} \left[e^{-\frac{(x-x_1)^2}{4D_0 t}} - e^{-\frac{(x+x_1)^2}{4D_0 t}} \right], \quad 0 \le x \le \infty, \\ &\qquad \textit{absorbing wall} \end{aligned} \tag{1}$$

These three behaviors are shown in Fig. 1 Clearly the distributions with a wall are far from being Gaussian. The deviation from Gaussian is more and more pronounced for initial and final positions that are close to the wall – where the walkers can feel the wall.

The mean square displacement is given by the integrals over x and x_1 of the above propagator (1) times the displacement squared $(x - x_1)^2$ times the initial density of walkers $\rho(x_1)$.

$$< (x - x_1)^2 > = 2D(t)\, t = \int_0^\infty \int_0^\infty dx\, dx_1\, (x - x_1)^2\, \rho(x_1)\, G(x, t; x_1, 0) \tag{2}$$

For normalization, it is convenient to take, instead of the semi-infinite space to the right of the wall, a large, but finite domain $0 < x < L_S$ and sprinkle the walkers uniformly with density $1/L_S$, so $\rho(x_1) = 1/L_S$. In the limit $L_S >> L_D = \sqrt{2D_0 t}$, i. e. at early times, we find,

$$\begin{aligned} D(t) &= \frac{< (x - x_1)^2 >}{2t} = D_0, \ \textit{unbounded} \\ &= D_0 \left[1 - \frac{4\sqrt{D_0\, t}}{3\sqrt{\pi}\, L_S} \right], \ \textit{reflecting wall} \\ &= D_0 \left[1 - \frac{4\, S \sqrt{D_0\, t}}{3\, V \sqrt{\pi}} \right] \end{aligned} \tag{3}$$

This result is for a single wall. Note that mean-square displacement is changed by the presence of a wall – it acquires a correction that depends on L_D/L_S.

In 3-dimension

$$D(t) = \frac{< (x - x_1)^2 > + < (y - y_1)^2 > + < (z - z_1)^2 >}{6t}$$

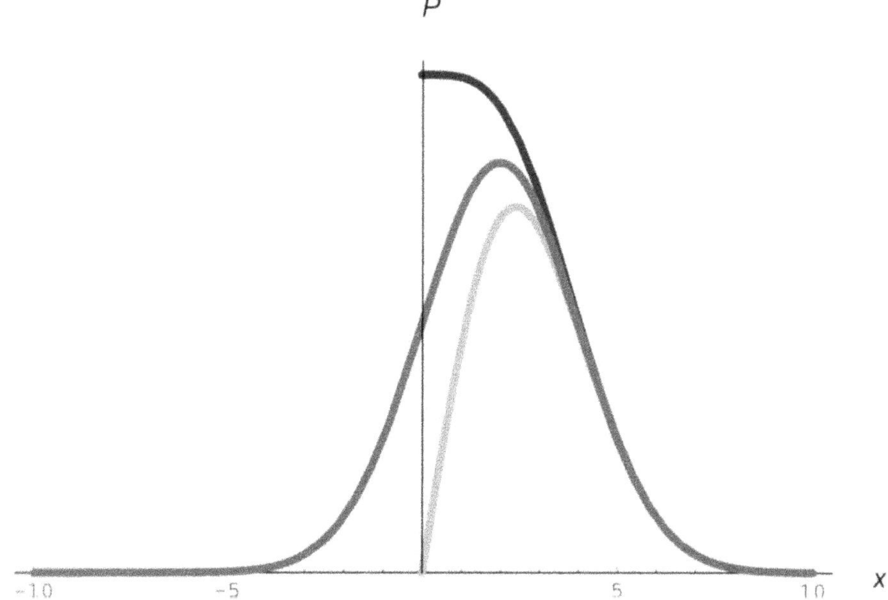

Fig. 1. The probability density $G(x, t; 2, 0)$ for a particle starting initially at $x = 2$ seen after a duration t such that $L_D = \sqrt{2D_0 t} = 2$. The top and the bottom curves are for the reflecting and the absorbing boundary conditions respectively at a wall located at the origin. The middle curve is for free unrestricted diffusion in absence of the wall.

$$= D_0, \; unbounded$$

$$= D_0 \left[1 - \frac{4\, S \sqrt{D_0\, t}}{9\, V \sqrt{\pi}}\right] \tag{4}$$

An easy way to derive the above result is to evaluate the time derivative of the mean-square displacement –

$$\begin{aligned} \frac{\partial < (x - x_1)^2 >}{\partial t} &= \int_0^\infty \int_0^\infty dx\, dx_1\, (x - x_1)^2 \,\rho(x_1)\, \frac{\partial G(x, t; x_1, 0)}{\partial t} \\ &= D_0 \int_0^\infty \int_0^\infty dx\, dx_1\, (x - x_1)^2 \,\rho(x_1)\, \frac{\partial^2 G(x, t; x_1, 0)}{\partial x^2} \end{aligned} \tag{5}$$

Here we have used the diffusion equation to replace the time derivative of the propagator $G(x, t; x_1, 0)$. Next, integrating by parts, and using the reflecting boundary condition that

$$\frac{\partial G(x, t; x_1, 0)}{\partial x}|_{x=0} = 0 \tag{6}$$

gives (3).

Mark Kac's a landmark paper titled "Can one hear the shape of a drum?" [3] is a wonderful exposition of the mathematics connected with the inverse problem of extracting geometrical information from non-invasive experiments. For a tutorial on later developments, see [4]. Kac gives the essential physical insights needed to analyze the problem of restricted diffusion.

Kac considered diffusion inside a closed boundary ("drum") with absorbing boundary conditions. He showed that the spectral function, which is a sum over all eigenvalues λ_n of the diffusion operator with the absorbing boundary condition

$$\int d\vec{x} G(\vec{x}, t; \vec{x}, 0) = \sum_n e^{-\lambda_n t}, \tag{7}$$

has information on geometry.

Initially, the walkers do not know about the presence of walls nor the shape or size of the container. This he called the principle of not feeling the wall. Later, at short enough time, only a small fraction of walkers, that start out within a diffusion length $(D_0 t)^{1/2}$, from the wall, will sense the presence of the boundary. Initially a (smooth) wall of any shape will appear flat (just as the earth appears to us), and it is not until later the walkers will find out about the curvature of the wall. As time goes on, more and more geometrical features will be known to the walkers.

These ideas can be extended to give a heuristic argument for the influence of wall on $D(t)$ in (3) (or its extensions to include curvature). In absence of the interface, $D(t) = D_0$, the diffusion constant of the bulk fluid. At early times, the number of the walkers that have sensed the wall is $(D_0 t)^{1/2} S$, where S is the total surface area (open or closed pore). If V_p is the pore volume that contains these walkers, then a fraction $1 - (D_0 t)^{1/2} S/V_P$ of walkers have not felt the wall and have a mean square displacement same as that of the free walkers $2D_0 t$. The remaining fraction $(D_0 t)^{1/2} S/V_P$ is hindered. The total mean square displacement is given by the sum of the two fractions, and thus the measured diffusion coefficient starts decaying as $D(t) = D_0(1 - \textit{constant}\ (D_0 t)^{1/2} S/V_P + \ldots)$. The value of the constant is given in (4). Some exact asymptotic results, including effects of curvature, were obtained in [5].

In the long time limit, in a closed pore, with the reflecting walls, mean-square displacement will be bounded by the size L_S of the pore. In one dimension,

$$\begin{aligned} < (x - x_1)^2 > &\to \frac{L_S^2}{6} \\ D(t) \quad &\to \frac{L_S^2}{12t}, \end{aligned} \tag{8}$$

and the diffusion coefficient will approach to zero as $(L_S/L_D)^2$with $L_D = \sqrt{2D_0 t}]$, i. e. as $1/t$ at long times.

The factor of 6 in (8) comes from the fact that the rms displacement for some walkers is much less than the box size.

When the walkers are confined between two reflecting walls, the propagator can be obtained by summing the infinite number of images (as between mirrors in a barber-shop) instead of a single image as in (1) . Thus, the propagator will be given by a sum of an infinite number of Gaussians. By using the well known

Poisson's summation trick, one can show that the sum can be represented in terms of the eigenfunctions of the diffusion operator.

The propagator can be expressed in terms of the eigenfunctions of the diffusion operator [7]

$$D_0 \nabla^2 C(\boldsymbol{x}) = \frac{\partial C(\boldsymbol{x})}{\partial t}, \tag{9}$$

$$D_0 \hat{n} \cdot \nabla C(\boldsymbol{x}) + \rho C(\boldsymbol{x})|_{\Sigma} = 0, \tag{10}$$

with the appropriate initial conditions. The expansion in terms of eigenvalues reads as

$$G(\boldsymbol{x}, \boldsymbol{y}; t) = \sum_{n=0}^{\infty} \psi_n(\boldsymbol{x}) \psi_n(\boldsymbol{y}) e^{-\lambda_n t}. \tag{11}$$

The eigenfunctions $\psi_n(\boldsymbol{r})$ and eigenvalues λ_n are the normalized solutions of the equations

$$D_0 \nabla^2 \psi_n(\boldsymbol{x}) = -\lambda_n \psi_n(\boldsymbol{x}), \tag{12}$$

$$D_0 \hat{n} \cdot \nabla \psi_n(\boldsymbol{x}) + \rho \psi_n(\boldsymbol{x})|_{\Sigma} = 0, \tag{13}$$

and form an orthogonal complete set

$$\int_V d\boldsymbol{x}\, \psi_n(\boldsymbol{x}) \psi_m(\boldsymbol{x}) = \delta_{n,m}, \tag{14}$$

with the completeness relation for the eigenfunctions $\psi_n(\boldsymbol{x})$

$$\sum_{n=0}^{\infty} \psi_n(\boldsymbol{x}) \psi_n(\boldsymbol{y}) = \delta(\boldsymbol{x} - \boldsymbol{y}). \tag{15}$$

For particles confined between two reflecting walls,one at $x = 0$ and the other at $x = L_S$, $\rho \equiv 0$, the eigenfunctions are simple cosines $\psi_n(x) \sim \cos(n\pi x / L_S)$, and $\lambda_n = D_0 n^2 \pi^2 / L_S^2$. By using the well known Poisson's summation trick [7], one can show that the sum (11) can be represented as an infinite series of Gaussians arising out of reflection between "barber-shop mirrors".

For diffusion coefficient, a simple integration, gives:

$$< (x - x_1)^2 >= L_S^2 \left[\frac{1}{6} - 16 \sum_{m=0}^{\infty} \frac{1}{(2m+1)^4 \pi^4} e^{-(2m+1)^2 \pi^2 \frac{L_D^2}{2L_S^2}} \right] \tag{16}$$

Note that at sufficiently short-time, one needs to sum over an infinite number of eigen-modes to get asymptotic behavior of (3). The long time behavior $< (x - x_1)^2 > \sim L_S^2 / 6$ sets in exponentially fast.

3 Diffusion in Connected Porous Media

In connected porous media, there are some directions which are open and some that are restricted. Imagine walking in a labyrinth – you may run into a dead end in the direction you are walking, but then if you take a turn, you may keep walking. A collection of capillary tube is often used to model porous media as they offer a combination of restriction and openness. A collection of fractures in otherwise non-porous matrix can also serve to illustrate the main idea how the diffusion coefficient is reduced in a connected porous medium and its time dependence.

Consider a single slab shaped pore, like a smooth fracture in a non-porous solid. The walkers moving perpendicular to the pore-walls (along x-axis) will be confined but in the other two direction (y, z) the mean square displacement will continue to grow linearly with time as in free diffusion

$$\begin{aligned} < (\mathbf{r} - \mathbf{r}_1)^2 > &= < (x - x_1)^2 > + < (y - y_1)^2 > + < (z - z_1)^2 > \\ &\rightarrow \frac{L_S^2}{6} + 4D_0 t \\ \frac{D(t)}{D_0} &= \frac{< (\mathbf{r} - \mathbf{r}_1)^2 >}{6D_0 t} \rightarrow \frac{1}{\alpha} + \frac{L_S^2}{18L_D^2}, \end{aligned} \tag{17}$$

Here $\alpha = 3/2$. One can easily extend this result to a collection of randomly oriented fractures.

4 Time Dependent Diffusion Coefficient in a Disordered Medium

So far we have discussed the short and long time asymptotics of the time dependence of the diffusion coefficient in simple models. Now let us turn our attention to porous media. Consider the case of an interconnected pore space containing Brownian particles of uniform concentration, confined by a convoluted reflecting boundary as in Fig. 2. This might be a model for water in a sedimentary rock.

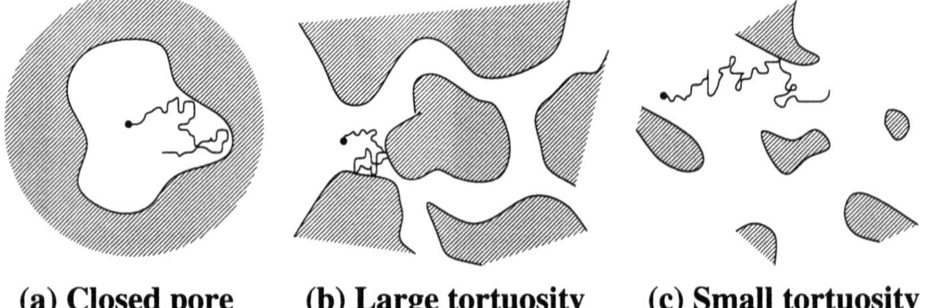

Fig. 2. Illustrations of a closed pore and two well connected porous media and some random walks. In the long time limit, the diffusion coefficient will eventually be zero in the closed pore, but, in connected systems, it will have a finite value which is reduced from its free value by a factor α, known as the tortousity.

There is a close resemblance between diffusion equation and Schrödinger's equation [7] and many formulations, particularly, the perturbation theory are similar. In quantum mechanics, the problem of electron motion in the presence of randomly placed scatterers is often treated with a multiple scattering formulation. This was an outcome of formulations for band structure in crystalline solids, like the KKR or the pseudo-potential theory, where the atomic potential cause scattering of outer electrons running though the solid. In diffusion equation, one is generally faced with an initial-boundary value problem where a Laplace transform is more convenient.

Now let us imagine the the solid phase of a porous medium as a set of scatterers. For diffusion in presence of a set of scatterers, the Green's function can be developed in a multiple scattering expansion, see for example, references cited in [8].

The motion of fluid particles in the pore space is governed by the diffusion equation. The multiple scattering formulation gives

$$C(\mathbf{r},s) = G_0 C_0(\mathbf{r}) + \sum_{i=1}^{N} G_0 t_i C_i(\mathbf{r},s) \tag{18}$$

$$C_i(\mathbf{r},s) = G_0 C_0(\mathbf{r}) + \sum_{j(\neq i)=1}^{N} G_0 t_j C_j(\mathbf{r},s) \tag{19}$$

Here G_0 is the unperturbed Green's function, in the free-space,

$$G_0 = \frac{1}{s - D_0 \nabla^2}, \tag{20}$$

and t_i is the t-matrix for scattering from the $i-$th scatterer. In this formulation, t_i takes into account scattering to all orders by an individual scatterer.

Iterating (18) and (19), and then averaging the t-operators over configurations of scatterers leads to a series for the configurationally averaged Green's function, $\langle G \rangle$,

$$\langle C(\mathbf{r},s)\rangle \equiv \langle G\rangle C_0(\mathbf{r}) = G_0 C_0(\mathbf{r}) + \sum_{i}^{N} G_0 \langle t_i \rangle G_0 C_0(\mathbf{r})$$
$$+ \sum_{i}^{N} \sum_{j\neq i}^{N} G_0 \langle t_i G_0 t_j \rangle G_0 C_0(\mathbf{r}) + \cdots \tag{21}$$

Here the angle brackets indicate averages over the configurations of the scatterers.

A simple model of a fluid saturated porous medium is a dilute suspension of non-porous material suspended in a fluid, like Fig. (2,c). A closed form expression for the diffusion coefficient in a dilute suspension of reflecting spheres, which is valid for all times, is given in [8]. This result is in agreement with general long and short time asymptotic results. In particular, the diffusion coefficient approaches its final asymptote as

$$\frac{D(t)}{D_0} \to \frac{1}{\alpha} + \frac{\beta}{D_0 t} - \frac{\gamma}{(D_0 t)^{3/2}} + \cdots \tag{22}$$

For a general porous media, the coefficients α, β, and γ depend on the details of the pore geometry, but, unlike the short time asymptotics, cannot be predicted as a part of some general theory. However, the functional form (22) is thought to be universal.

The coefficient α is a dimensionless number that defines the dc limit of diffusion and conductivity, and in porous media, it is the tortuosity. The tortuosity plays an important role in various transport processes in porous media – ranging from fourth sound in superleaks to electrical conduction [9]. There is an **exact** calculation by Zwanzig [10] of α in a periodic 2D channel with refleting walls.

In the petroleum industry, tortuosity plays a central role in the estimation of hydrocarbon reserves. Via Einstein relation, the electrical conductivity and diffusion coefficients are related by the volume fraction of the conducting phase. Conductivity of a rock containing oil and salt water equals water conductivity times the volume fraction of water divided by α. Thus, α is needed to estimate the "reserve" – the volume fraction of oil – from measured electrical conductivity of rock with intersticial fluids.

The second most important parameter of the rock is its permeability. This describes how much of oil will flow under a pressure gradient. This depends on pore-sizes, among other things. Recall that the Poiseulli flow rate in a pipe depends on the square of the cross-section. However, α contains no information about the length scale of the pore space. To determine the length scale one requires at least the knowledge of β, which is the first dimensionful parameter in the long time asymptotic form above.

Complex porous media in nature, like naturally occurring sedimentary rocks, (Fig. 3) have numerous length scales associated with them. Determining these is an active area of research.

5 Effect of Relaxation

Diffusion in the presence of spherical traps has been an active field, dating back to Smoluchowski in 1916 [11]. In chemistry and chemical engineering, this model is used as a paradigm of diffusion-limited chemical reaction.

In nature, the boundary condition that applies most often, is the partially absorbing boundary condition. This boundary condition implies that the normal flux (given by the normal derivative of the density of walkers at the pore-wall times D_0) is related to the density of walkers at the wall (see references cited in [6]) times a relaxation velocity ρ. For reflecting boundary condition, ρ is zero and for absorbing boundary it is infinite.

$$6tD(t \mid \rho) = \langle r^2(t) \rangle_s = \frac{\int (\mathbf{r} - \mathbf{r}')^2 G(\mathbf{r}, \mathbf{r}', t) d\mathbf{r} d\mathbf{r}'}{\int G(\mathbf{r}, \mathbf{r}', t) d\mathbf{r} d\mathbf{r}'} \tag{23}$$

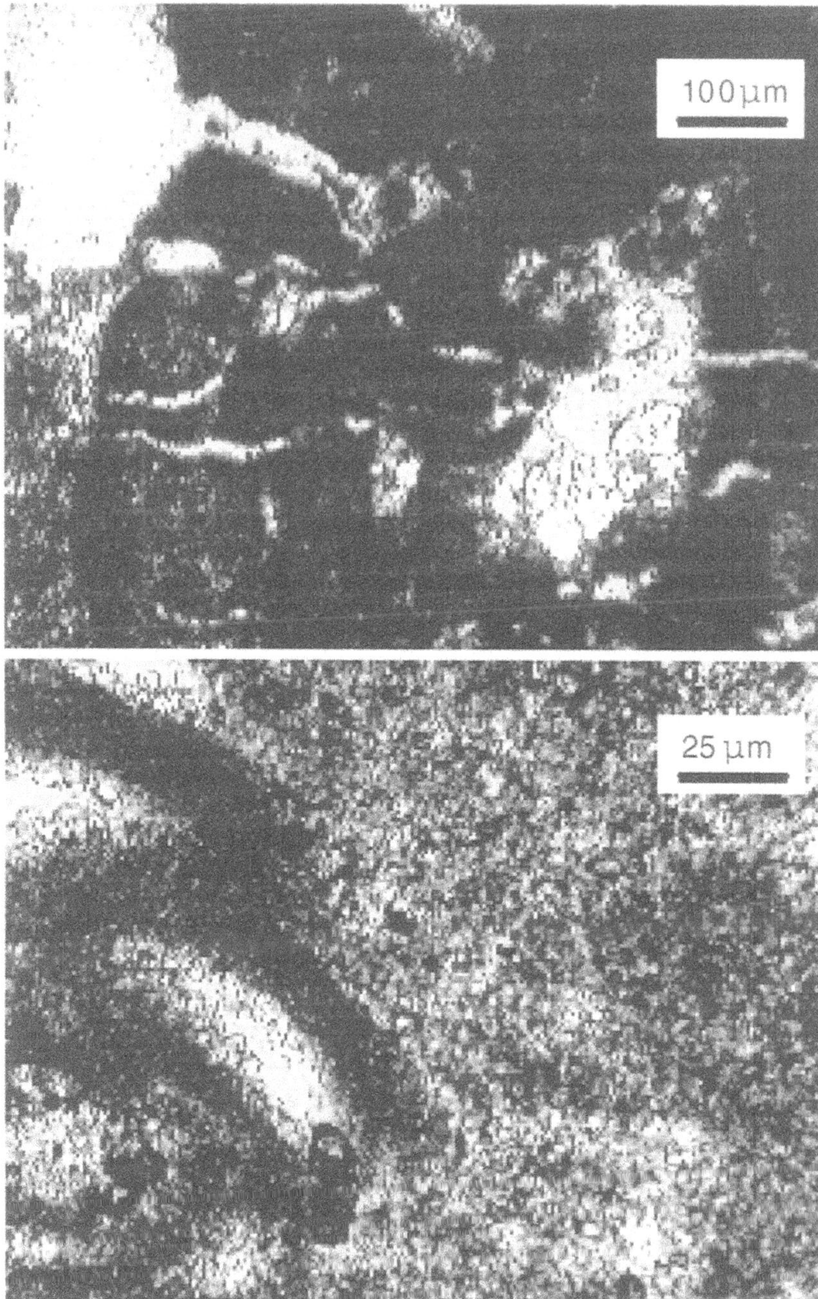

Fig. 3. Optical micrograph of a thin section of a carbonate rock at two magnifications. The pores were injected with a blue epoxy (white hole on top left corner is where expoxy got ground out). Grains scatter the transmitted light and appear black. White parts denote calcite crystals. The grains themselves are microporous, but their pores are sufficiently small (about $2\mu m$) such that epoxy could not penetrate.

The analog of (1) for a single wall can be obtained many different ways. One way is is to use the Laplace representation for finite ρ

$$\widetilde{G}(x_<, x_>, s) = \frac{e^{-\kappa x_>}}{\kappa D_0}\left[\frac{\kappa D + \rho}{\kappa D - \rho} e^{-\kappa x_<} + e^{\kappa x_<}\right], \ \kappa = \sqrt{\frac{D_0}{s}} \tag{24}$$

After performing the integration, we expand in powers of $\sqrt{D_0/s}$. Laplace transforming, and dividing throughout by $M(t) = 1 - \rho t S/V_P + O(t^{3/2})$, we have

$$\frac{D(t)}{D_0} = 1 - \frac{4\sqrt{D_o t}S}{9V_P\sqrt{\pi}} + \frac{\rho t S}{6V_P} + O((D_0 t)^{3/2}) \tag{25}$$

The problems with non-zero, but finite, ρ are rather complicated, especially for open structures. We do not have expressions for time dependent diffusion coefficient that combines restriction with partially reflecting boundary condition in open structures. A computation of the long time diffusion coefficient is available [12] for a 3D periodic array of spheres with non zero but small values of $\rho a/D_0$, where a is the sphere radius. For larger values of this parameter, no results are available, even numerically. Disentangling the complexity due to combined effects of diffusion and decay remains a major hurdle.

6 Dispersion

Dispersion - the spreading of molecules due to advection and molecular diffusion within a flow field at low Reynold's number - is a fundamental problem in hydrodynamics and provides an example of non-standard statistics. Taylor first addressed dispersion for the case of tracers spreading in pipe flow [13]. The probability distribution of displacements in the long time and large displacement regime is expected to become a Taylorian, i. e. its dynamics are given by a coarse-grained advection diffusion equation giving rise to a Gaussian distribution whose center moves at the average velocity, and whose mean square displacement σ^2 is given by a time-independent dispersion coefficient $D = \sigma^2/2t$.

Dispersion in porous media is considerably more complex and has been of interest because of its practical importance. The stream-lines of Stokes flow ($R_e \ll 1$) in a porous medium provide a quenched random system in which molecules hop between stream lines by random walks. In this regard the dispersion in complex geometries, in a slow flow regime, is analogous to turbulent diffusion. de Gennes [14] pointed out that the Taylorian picture is "upset" in a porous-medium when stagnating zones are present. In porous media dead-ends may also give rise to anomalous trapping and release times [14], leading to a deviation from a Taylorian [15].

7 Conclusion

Dispersion and diffusion are important in their application in diverse fields ranging from biological perfusion, chemical reactors to soil remediation and oil recovery. I have no space to describe these application nor the experiments that use the ideas described here. I apologize for not being able to mention many important work, such as Donsker and Varadhan's[16].

As this is a tutorial, it is appropriate to urge the reader to go to the "fountain head" for joy and inspiration. Brownian motion papers by Einstein [1] are extremely elegant, far better than any text-book, so is Chandrasekhar's RMP article [2]. Kac's paper on "Can one hear the shape of a drum" [3] reflects his personality and is full of joy – has a tutorial flavor. The book by Harald Crámer is full of insight and I end my tutorial with a quote from Crámer [19]:

"There is a famous remark by Lippman (quoted by Poncarétin Calcul des probabilité, Second Ed., Paris 1912) to the effect that 'everybody believes in the law of errors, the experimenters because they think it is a mathematical theorem, the mathematicians because they think it is an experimental fact'. – It seems appropriate to comment that both parties are perfectly right, provided that their belief is not too absolute: mathematical proof tells us that, *under certain qualifying conditions*, we are justified in expecting a normal distribution, while statistical experience shows that, in fact distributions are often *approximately normal*".

Acknowledgement

I thank Lukasz Zielinski for Fig. and pointing out several typographical errors.

References

1. A. Einstein: *Investigations on the Theory of the Brownian Movement.* (Dover, New York 1956)
2. S. Chandrasekhar: Rev. Mod. Phys **15**, 1 (1943) Reprinted in *Selected Papers on Noise and Stochastic Processes.* (N.Wax, Dover, New York 1954)
3. M. Kac: American Mathematical Monthly **73** 1 (1966)
4. S. J. Chapman: American Mathematical Monthly 124 (1995)
5. P. P. Mitra, P. N. Sen and L. M. Schwartz: Phys. Rev. B **47**, 8565 (1993)
6. P. P. Mitra and P. N. Sen: Phys. Rev. B **45**, 143 (1992)
7. P. M. Morse and H. Feshbach: *Methods of Theoretical Physics.* (McGraw-Hill, New York 1953)
8. T. de Swiet and P. N. Sen: J. Chem. Phys **100**, 5597 (1994)
9. D. L. Johnson and P. N. Sen: Phys. Rev.B. **24** 2486 (1981); D. L. Johnson, L. M. Schwartz and J. Koplick: Phys. Rev. Lett. **57**, 2564 (1986); R. J. S. Brown: Geophysics **45** 1269 (1980)
10. R. Zwanzig: Physica **117A**, 277 (1983)
11. M.V. Smoluchowski: Phys. Z. **17** 557 (1916)
12. P. N. Sen, L. M. Schwartz, P. P. Mitra and B. I. Halperin: Phys. Rev. **B49**, 215 (1994)
13. G. I. Taylor: Proc. R. Soc. London **219**, 186 (1953)
14. P-G. deGennes: J. Fluid Mech **136**, 189 (1983)
15. Ulrich M. Scheven and Pabitra. N. Sen: Phys. Rev. Letts., **89**, 254501-1 (2002)
16. M. D. Donsker and S. R. S. Varadhan: Comm. Pure Appl. Math. **28** 525 (1975)
17. P. Grassberger and I. Procaccia: J. Chem. Phys. **77** 6281 (1982)

18. R. F. Kayser and J. B. Hubbard: Phys. Rev. Lett. **51** 79 (1983); R. F. Kayser and J .B . Hubbard: J. Chem. Phys. **80** 1127 (1984)
19. Harald Crámer: *Mathematical Methods of Statistics.* (Princeton University Press, Princeton 1958)

Directed Transport in AC-Driven Hamiltonian Systems

S. Denisov, J. Klafter, and M. Urbakh

School of Chemistry, Tel-Aviv University, Tel-Aviv 69978, Israel

Abstract. We study the mechanism of current rectification in Hamiltonian systems driven by an external periodic force with broken time reversal symmetry. We show that directed transport arises due to breaking the symmetry of flights near regular islands. In the framework of the continuous time random walk (CTRW) approach we construct a generalized asymmetric flights model and derive an expression for the current in terms of the characteristics of the relevant islands. The broken-symmetry strategy allows to manipulate the transport properties of both individual particles and statistical ensembles of particles.

1 Introduction

The *ratchet* idea allows to obtain directionality in the absence of a global bias or a gradients, by using breaking time or/and spatial symmeties [1]. Inspired by the interest in the physical principles of microbiological motility, the phenomenon of directed transport produced by symmetry breaking covers now a broad class of physical systems, such as Josepshon junctions, cold atom systems and more (see [1]).

Dynamical systems can be classified according to the strength of their dissipation, given by the damping rate. In the limit of strong dissipation we obtain overdamped dynamics [2]. This situation corresponds to motion at microscales, i. e. molecular motors and Brownian rectifiers [1]. For intermediate strength of dissipation we get into the range of underdamped systems where inertial dynamics dominates. The mechanisms of current generation in such systems have been studied in detail [3].

In the limit of zero-dissipation, one deals with Hamiltonian systems, which, in contrast to overdamped and underdamped dynamics, is much less understood. It has been shown in [4] that breaking time or/and spatial symmetries leads to the appearance of a strong DC current in the chaotic layer of an AC driven Hamiltonian system. Recently, an approach has been proposed that allows to estimate the mean value of a DC current [5].

While the necessary conditions for the appearance of directionality in Hamiltonian systems can be formulated in terms of symmetry breaking (SB) relations [4], the direction and value of the current are determined by the specific mechanisms. Hamiltonian dynamics are characterized by a rich spectrum of behaviors, ranging from regular to anomalous motions, related to the mixed structure of the phase space [6]. This mixed structure of the phase space causes the failure of the simple Brownian description of Hamiltonian kinetics [7,8]. Examples for this failure are, among others,the appearance of anomalous diffusion with a degree anomaly of which depends on the statistical properties of the flights [7,8], and strongly nonuniform mixing [9].

In this contribution we demonstrate that directionality in AC-driven Hamiltonian systems emerges due to symmetry breaking of a mixed phase space structure [10]. Namely, DC current is induced by breaking of the symmetry of Lévy flights which are obtained by regular islands embedded in the chaotic layer (these are basically Lévy walks due to presence of velocity [7]) [11]. Thus, as in the case of Hamiltonian anomalous diffusion [8,12], also the direction and magnitude of the current are determined by the structure of regular islands. We show how the understanding of current rectification mechanisms opens new ways to manipulate classical Hamiltonian systems [13].

2 AC-Driven Hamiltonian: The Model

As an example, we consider a relatively simple model, which describes a particle moving in a spatially periodic potential $U(x)$, $U(x+L) = U(x)$, under the influence of a time-periodic, zero-mean force, $E(t)$, $E(t+T) = E(t), \langle E(t) \rangle = 0$ [5]:

$$H = \frac{p^2}{2} + U(x) - xE(t) \ , \ \ddot{x} = -\frac{\partial U(x)}{\partial x} + E(t) \ , \tag{1}$$

where L and T are the spatial and temporal periods.

The Hamiltonian in (1), with a three dimensional phase space, is generically nonintegrable and is thus characterized by the presence of a stochastic layer, which replaces the separatrix of the corresponding integrable nondriven system. We are interested in the dynamical properties of trajectories in the main stochastic layer, which determine the transport characheristics.

The ergodic properties of the stochastic layer allow to apply symmetry considerations which can be used to predict whether the trajectories have a zero average velocity and therefore no current [4]. Note that symmetry breaking is a necessary but not a sufficient condition for the appearance of a DC current. The idea behind the symmetry analysis is to find all the symmetries of (1) which leave the equations of motion invariant, but change the sign of the velocity $\dot{x}$. Since such symmetry operations map a given trajectory onto another one, by making sure that both trajectories belong to the same stochastic layer, and recalling the ergodic properties of the layer, we may immediately conclude that the average velocity in each trajectory from the chaotic area will be exactly zero.

For the system in (1) the possible symmetries are [4]:

$$\hat{S}_1 : x \to -x, t \to t + \frac{T}{2}, \quad \{F(-x+x_0) = -F(x+x_0), \quad E(t+T/2+t_0) = -E(t+t_0))\}, \tag{2}$$

$$\hat{S}_2 : x \to x, t \to -t, \quad \{E(-t+t_0) = E(t+t_0)\}, \tag{3}$$

where $F(x) = -\frac{\partial U(x)}{\partial x}$, and x_0 and t_0 are some appropriate argument shifts. The conditions in the brackets are the requirements for the periodic functions

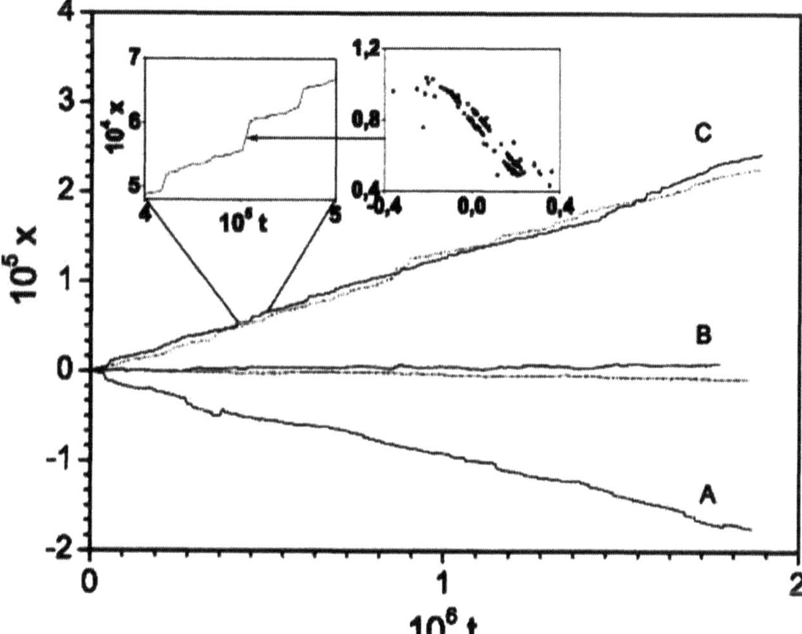

Fig. 1. $x(t)$ versus t for different values of the third harmonic amplitude E_3 ($E_1 = 0.2$, $E_2 = -0.4$, $\phi_2 = 0.4$, $\phi_3 = 0$): $E_3 = 0$, $J \approx -0.1$ (case A); $E_3 = 0.143$, $J = 0$ (case B); $E_3 = 0.3$, $J \approx 0.12$ (case C). Left upper inset: zooming of $x(t)$ for the case C. Right inset shows the Poincare section which corresponds to the longest flight in the blown up part of trajectory. The dashed curve corresponds to a filtered trajectory without flights (see text) and the dotted one presents the result of the CTRW simulation

$U(x)$ and $E(t)$. The way to break the symmetries is to choose functions $U(x)$ and $E(t)$ which violate the listed requirements. In general, we deal here with a broader definition of *ratchet* transport, which is obtained by breaking the reflection symmetry in space, or the reflection and shift symmetries in time. Unlike the case of dissipative systems [2,3], here breaking the symmetry of the potential $U(x)$ only is not enough to obtain a current. This is possible, however, by a proper choice of $E(t)$ alone. Thus, it turns out not to be necessary to break the reflection symmetry of the periodic potential $U(x)$. Of course one can break all symmetries in time and space simultaneously, which may lead to a quantitative change of the results.

We consider the case of a simple potential $U(x) = -cos(2\pi x)$ and a driving force which contains three harmonics:

$$E(t) = E_1 cos(\omega t) + E_2 cos(2\omega t + \phi_2) + E_3 cos(3\omega t + \phi_3). \tag{4}$$

Here and in the following we choose $T = 2\pi$ and $\omega = 1$. In order to get a current we must break both symmetries, $\hat{S}_1$ and $\hat{S}_1$, (2)–(3). For $E_2 = E_3 = 0$ both symmetries, $\hat{S}_1$ and $\hat{S}_2$, are present and the total current equals zero. For

$E_2 = 0$, symmetry $\hat{S}_1$ is present and we have again zero current. For $E_3 = 0$ and $E_2 \neq 0$ all symmetries in (2)–(3) are broken, except for the specific values $\phi_2 = k\pi$, $k = 0, 1, 2...$, and, in general, we can expect a nonzero current in the system in (1). In Fig. 1 we show the time dependence of the coordinate $x(t)$ for several values of E_3 and fixed nonzero values of E_1, E_2. The initial conditions are chosen to be within the main stochastic layer. An interesting effect results from the variation of E_3: the increase of E_3 from 0 to 0.3 leads to current reversal, which is the result of a nonlinear interaction of the harmonics. Thus, for some value of E_3 between $E_3 = 0$ and $E_3 = 0.3$, the average velocity in the layer should be equal to zero. This is the case for $E_3 \approx 0.143$. Thus, while all symmetries (2,3) are broken, we find zero current as the result of a balance between the microscopic dynamical mechanisms.

The dynamics in the stochastic layer appears to be complex. The insets in Fig. 1(a) show that in parts the particle moves in seemingly free ballistic flights of different time lengths. A Poincarè map of such a flight shows that the trajectory in the stochastic layer sticks to regular islands.

3 The Role of Regular Islands

The nonlinear Hamiltonian system (1) has a mixed phase space, which contains a main stochastic layer and regular islands [6]. These islands are impermeable to chaotic trajectories and, at a first glance, can be excluded from the phase space. This is, however, impossible because of the complex structure of the boundaries between the chaotic and these regular regions. Close to islands the chaotic layer contains hierarchical sets of *cantori*, which form bottleneck-like barriers through which orbits can penetrate. Due to the cantori structure, a trajectory can be trapped for a long time *near* the corresponding island. The sticking effect leads to the appearance of long regular flights which alternate with chaotic motion. For a nonzero winding number v these phases correspond to long unidirectional flights. The case of $v = 0$ corresponds to a localized rotating motion [6]. Thus, islands with $v \neq 0$ form a set of ballistic channels in the chaotic layer, and sticking to such an island corresponds to trapping of a particle in the corresponding channel. This would lead us later to Lévy walks [7,8].

A possible way for obtaining a nonzero current can be realized through correlated modes of motion, where the characteristic time of the correlation decay is much longer than the period of external driving T. In the stochastic layer such modes can be associated with ballistic flights only [10]. Thus the flights play a crucial role in the process of current rectification inside the chaotic layer. Another possibility is that the purely diffusive dynamics inside the stochastic layer also contributes to a directed current. We will show that this possibility is excluded in the cases under consideration.

Breaking the symmetry of ballistic flights implies that there is an asymmetry of resonance structures, and that the value of the current should provide a quantitative measure of this asymmetry. For the analysis of the system's dynamics we used the propagator $P(x,t)$, i. e. the probability density for a particle to be

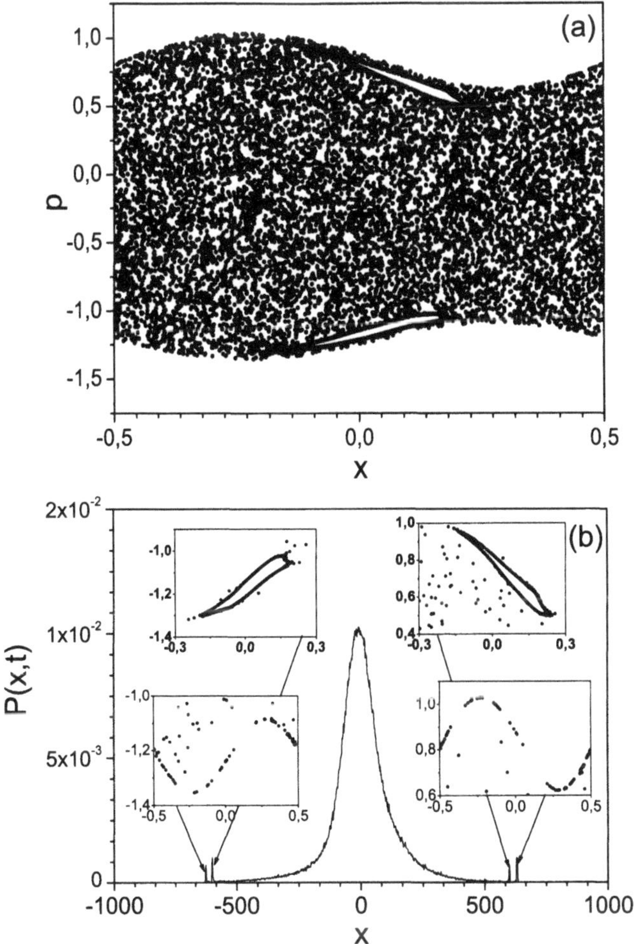

Fig. 2. (a) Poincare section and (b) the propagator for a fixed time ($t = 100T$) for $E_3 = 0.143$ (case B). The insets show Poincarè sections correspondinto flights near different ballistic islands

at x at time t [14]. In Figs. 2b and 3b we show the propagator for a fixed time $t = 100T$ which was obtained using a long chaotic trajectory with the duration $10^7 T$. This trajectory, due to the ergodicity condition, should cover the stochastic layer uniformly. The propagator structure has two prominent features: a central "bell shaped" part and several sharp asymmetric peaks in the tails. These peaks correspond to flights which a particle performs when it sticks to ballistic resonances. The peak location is determined exactly by the corresponding resonance winding number. It is easy to identify all relevant ballistic channels using the propagator for a given time and stroboscopic Poincarè section (see Figs. 2a and 3a.). For clarity we emphasize that also the central part of the propagator at a given time contains contributions from flights. In other words, the propagator

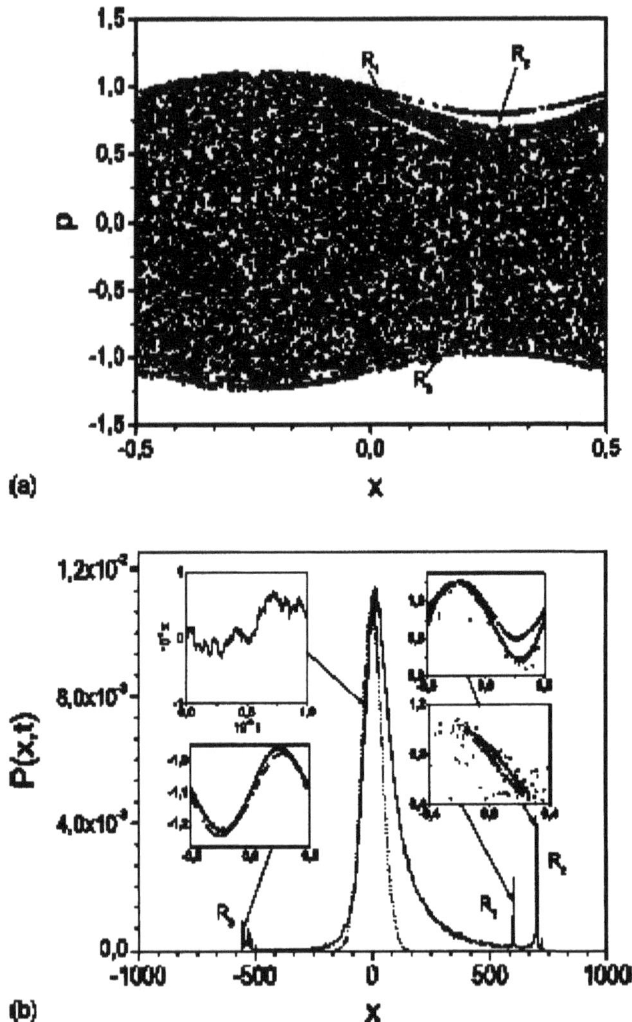

Fig. 3. Same as in Fig. 2 for $E_3 = 0.3$ (case C). Dashed line corresponds to the propagator for a dynamical process without flights in the ballistic channels R_1 , R_2 and R_3 (see text)

has a quite interesting dependence on the fixed time t and it allows to observe the relevant ballistic channels which contribute for the chosen fixed time t.

For $E_3 = 0.3$, the resonance structure has a well pronounced asymmetry, which becomes evident in the asymmetry of the propagator. There are two ballistic channels for positive direction: R_1 with winding number $v_1 = 6/T \approx 0.955$ and R_2 with $v_2 = 7/T \approx 1.115$ and only one for negative direction, R_3 with winding number $v_3 = -22/4T \approx -0.876$. As a result we observe a positive net current.

Now we show that a nonzero current is indeed fully determined by the sticking to ballistic channels. We simply eliminate all ballistic flights from the dynamics of any trajectory by using a velocity gated technique [11,13]. We take into account the existence of channels R_1 , R_2 and R_3. After successive waiting times of $10T$ we test whether the particle displacement corresponds to flight at a given ballistic channel with a known winding number and within an uncertainty of 5%. If we apply this filtering to a single trajectory, we observe practically zero current $J \approx -0.005$ (see dotted line in Fig. 1). The resulting reduced propagator is shown as a dashed line in Fig. 3. It is clearly symmetrical. Thus, we conclude that within an uncertainty of 5% purely diffusive motion does not contribute to the induced current. In fact we argue that even these possible 5% are due to smaller resonances and other ballistic channels embedded in the stochastic layer.

In contrast to the previous case, the propagator for $E_3 = 0.143$ is symmetric, which corresponds to zero current. This is the consequence of the fully symmetric resonance structure for which the resonances with opposite winding numbers compensate each other. Thus, in this case the zero current value is the result of a *dynamical* symmetry rather than a *geometrical* asymmetry.

4 Generalized Asymmetric CTRW-Model

In order to quantify our result that the observed current results from ballistic flights (corresponding to long correlations), while random diffusion (corresponding to fast decay of correlations) gives no contribution to the total current, we simulate the dynamics by a sequence of alternating processes: *flights* (sticking to ballistic channels) and an unbiased *random walk* (chaotic diffusion in random area). It is reasonable to assume that there is no correlation between flights because successive flights are usually separated by a diffusive component. Under these assumptions, the dynamics can be effectively modelled within the CTRW formalism [15] as a generalized asymmetric flight process [16].

Assume that there are N relevant different regular islands with winding numbers $v_i, i = 1, ..., N$. Every resonance is characterized by a probability distribution function (PDF) of sticking times $\psi_i(t)$ and a probability of sticking events p_i. The random walk phase is characterized by the PDF $\psi_c(t)$. Using the CTRW scheme (see Appendix for details) we obtain the following expression for the current:

$$J = \frac{\sum_{i=1}^{N} p_i v_i \langle t_i \rangle}{\sum_{i}^{N} p_i \langle t_i \rangle + \langle t_c \rangle}, \tag{5}$$

where $\langle t_i \rangle = \int_0^\infty t\psi_i(t)\mathrm{d}t$ and $\langle t_c \rangle = \int_0^\infty t\psi_c(t)\mathrm{d}t$. These first moments are finite due to the Kac theorem on the finiteness of recurrence times in Hamiltonian systems [17].

The contribution from a single flight near the i-th island is given by probability density to be in a ballistic mode with velocity v_i for a time chosen from $\psi_i(t)$,

$$\Psi_i(x,t) = \delta(x - v_i t)\psi_i(t) \tag{6}$$

and, correspondingly, the contribution of the random diffusion is

$$\Psi_c(x,t) = \frac{1}{\sqrt{\pi Dt}} exp(-\frac{x^2}{Dt})\psi_c(t), \tag{7}$$

The *total* propagator for the time $t = MT$ is a convolution of the single motion events, (6) and (7):

$$P(x,MT) = Q\sum_{n=1}^{M}\sum_{k=1}^{M-n}\cdots\sum_{l=1}^{M-n-\ldots-g}\Psi_c(x,(M-n-\ldots-l)T)\Psi_1(x,nT)\ldots\Psi_N(x,lT) \tag{8}$$

where the number of sums equals to the number of ballistic channels N, and Q is a normalization constant.

Let us consider in detail the case of $E_3 = 0.3$, for which there are three relevant ballistic channels, R_1, R_2 and R_3. Since the winding numbers of all resonances are known, we can separate those parts of the trajectories, which correspond to the particle flights in a certain channel. This has been done numerically by identifying an elementary flight using a velocity gate technique [11,13]. We use here the same procedure as used above for filtering. The corresponding sticking time PDF's for resonances R_1 and R_2 are shown in Fig. 4. Sticking to the resonance R_3 is extremely rare, and the characteristic time spent by the particle in this channel is much smaller than the characteristic time for channels R_1 and R_2. Thus the influence of R_3 is negligible.

Both PDF's for R_1 and R_2 have power-like tails : $\psi_{1,2} \sim t^{-\alpha_{1,2}}$, $\alpha_1 \approx -2.6$, $\alpha_2 \approx -2.4$. Thus, we are confronted with the appearance of Lévy walks. Resonance R_1 has the most dominant contribution to the directed transport with a sticking probability $p_1 = 0.94$ and its mean sticking time is $\langle t_1 \rangle \approx 286$. Resonance R_2 has a sticking probability $p_2 = 0.06$ and a mean sticking time $\langle t_2 \rangle \approx 1673$. Thus, although the particle gets relatively seldom into the ballistic channel R_2, it spends there a time which is almost one order of magnitude longer then the characteristic sticking time for channel R_1. The resonance R_2 shows up with a stronger anomalous character than R_1, and therefore this resonance determines the asymptotics of the global diffusion in the stochastic layer. The evolution of the corresponding mean square displacement $(\langle x^2 \rangle - \langle x \rangle^2) \sim t^\gamma$ is shown as an inset in Fig. 4b. The diffusion has a strong anomalous character with a characteristic exponent $\gamma \approx 1.6$, which is in agreement with the exponent α_2 ($\gamma = 4 - \alpha_2$)[15]. Numerical analysis of the random walk PDF $\psi_c(t)$ displays a Poissonian distribution:

$$\psi_c(t) = \frac{1}{\tau_c} exp(-\frac{t}{\tau_c}) \tag{9}$$

with a time constant $\tau_c \approx 2642$, which is equal to the mean time of a the random diffusion process. Using the numerical values of the relevant parameters, (5) for the current gives $J_{num} \approx 0.118$, which is very close to the result of direct numerical integration of (1) (see Fig. 1a).

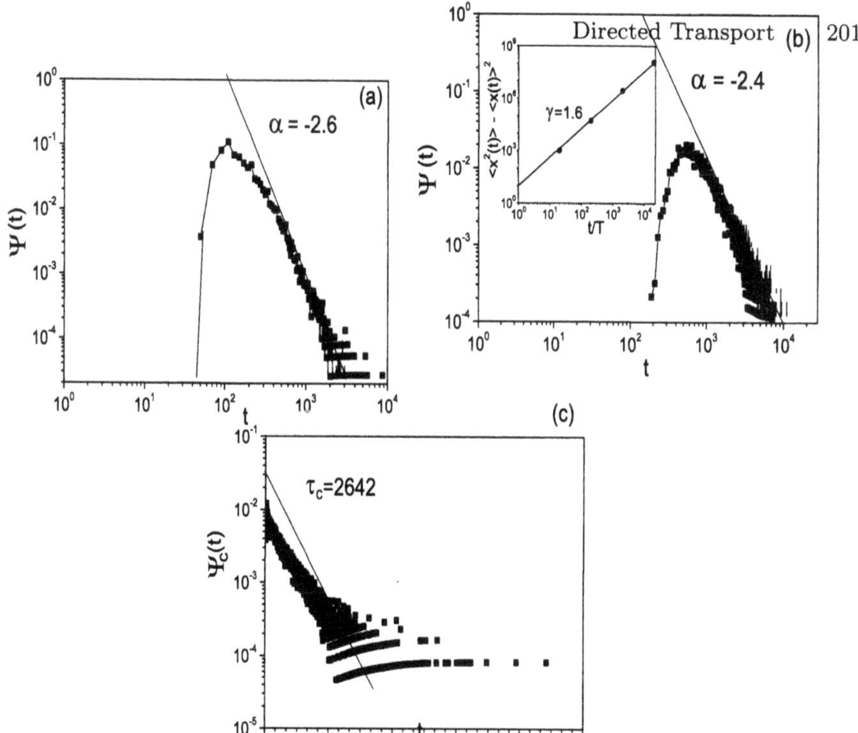

Fig. 4. The numerically obtained sticking time PDF for: (a) R_1, (b) R_2, and (c) random walk phase for $E_3 = 0.3$ (see text). Inset in (b) shows the mean square evolution for the global diffusion into the chaotic layer

Using the exponents α_1 and α_2, we simulate the dynamics within the chaotic layer by a CTRW process with the following PDF distributions [18]:

$$\psi_{1,2} = \begin{cases} 0, & t < t_c^{1,2} \\ A_{1,2} t^{-\alpha_{1,2}}, & t \geq t_c^{1,2}, \end{cases} \qquad (10)$$

where $t_c^1 = 115$, $t_c^2 = 530$, and the mean time $\langle t \rangle = t_c \frac{\alpha - 1}{\alpha - 2}$. In order to model the diffusional part we used a Langevin equation with a reasonable diffusion coefficient [19] and the Poisson distribution (9) for the duration time of chaotic walking. The obtained trajectory is very close to the real trajectory (dotted line in Fig. 1a.).

5 Temporary Symmetry Breaking Action: Manipulation of Systems

In this section we show how understanding the SB mechanism for directed motion leads to a new way to manipulate classical Hamiltonian systems. One can easily imagine a situation where remote control is required, or when a global bias is not desirable. Here we demonstrate how the SB approach helps to manipulate both single particles and statistical ensembles of particles. In the case of a statistical ensemble, SB opens new possibilities for handling some fraction of the particles, a process which cannot be performed using standard bias techniques.

Let us consider another example of driven Hamiltonian systems, namely, a particle which moves in the periodic nonlinear potential $U(x) = cos(x)$ under the influence of a periodic train of delta-kicks with an amplitude E_1 and period T [6]:

$$H_s(p,x,t) = \frac{p^2}{2} + E_1 cos(x) \sum_{n=-\infty}^{\infty} \delta(t - nT). \tag{11}$$

The Hamiltonian in (11) is symmetric with respect to time and space reversal transformation $\{t \to -t, x \to -x\}$, so a particle, whose dynamics obey (11), performs a diffusive motion with zero drift. The symmetry can be broken by switching on a second source of kicks which are shifted, in time and space domains, with respect to the first one:

$$\begin{aligned} H(p,x,t) &= H_s(p,x,t) + \\ &\Phi(t, t_{on}, t_{off}) E_2 cos(x + \phi) \sum_{l=-\infty}^{\infty} \delta(t - nT + \tau), \end{aligned} \tag{12}$$

where ϕ and τ are spatial and temporal shift constants, $\Phi(t, t_{on}, t_{off}) = \Theta(t - t_{on})\Theta(t_{off} - t)$ is a square pulse function, and t_{on} and t_{off} are the switching on and switching off times. The duration time for the SB force is therefore $t_{SB} = t_{off} - t_{on}$. Notice that this SB method is different from the usually used two-harmonics ratchet approach [1].

Let us start from the case of a permanent SB ($t_{on} = -\infty$, $t_{off} = \infty$). For nonequidistant kicks, $\tau \neq mT/2$, and $\phi \neq s\pi$ $(m, s = ... - 1, 0, 1, ...)$, all relevant symmetries are broken and we fulfill the necessary conditions for the appearance of a current [4]. The evolution of the Hamiltonian in (12) can be described by a pair of consecutive maps in position x and momentum p:

$$\begin{aligned} x'_{n+1} &= x_n + p_n \tau, \\ p'_{n+1} &= p_n + U'(x_{n+1}); \end{aligned} \tag{13}$$

$$\begin{aligned} x_{n+1} &= x'_{n+1} + p'_{n+1}(T - \tau), \\ p_{n+1} &= p'_{n+1} + \Phi(t, t_{on}, t_{off}) U'(x_{n+1} + \phi), \end{aligned} \tag{14}$$

where (13) correspond to kicks from the main source, and (14) correspond to kicks from the SB source.

Switching on the second kicking source with amplitude E_2 and phase shifts ϕ and τ results in asymmetric overlapping of the main chaotic layer with the layer of ballistic islands (see Fig. 5). This overlap produces a current. Current inversion (mirroring the layers overlap) can be obtained by a simple shift inversion $\phi \to -\phi$ or $\tau \to T - \tau$.

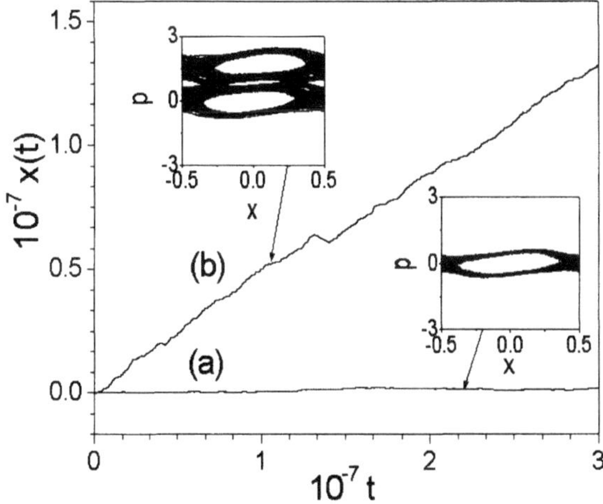

Fig. 5. Dependence of $x(t)$ versus t for (a) the symmetric Hamiltonian, (11) ($E_1 = 0.24$, $T = 0.6$) and (b) the Hamiltonian with the additional SB source (2), (12) ($E_1 = 0.24$, $T = 0.6$, $E_2 = 0.11$, $\phi = 0.8$, $\tau = 0.4$). Insets show the corresponding Poincarè sections

For the analysis of the dynamics we again use the propagator $P(x, t)$. In Fig. 6 we show the propagator for time $t = 1000T$ obtained by averaging over 10^5 trajectories, starting in a chaotic area of the main layer. The peaks in the propagator correspond to flights which a particle performs when it sticks to ballistic islands. The locations of the peaks are determined by the corresponding winding numbers. From the structure of the propagator it is clearly seen that the large scale particle displacements, when compared to the period of the potential $U(x)$, are a result of the flights. Directed particle drift stems from asymmetry in the structure of the ballistic islands with positive and negative winding numbers. Moreover, most of the contribution to the particle's transport in the positive direction comes directly from the main ballistic islands with $v = 1$ (see inset in Fig. 6).

Based on the above, we expect that controlling the manifold overlap in phase space, one can control the directed transport by tuning the value of the velocity. We now describe a possible way to manipulate a particle through SB during a finite time interval t_{SB}. In order to do this we use two features of the system: (i) the possibility to temporarily remove the barriers in phase space (formed by invariant KAM tori) between different invariant manifolds, and (ii) the sticky nature of the regular islands. Namely, one can remove the barriers from the phase space during a time interval t_{SB} and then restore them. This can be viewed as an act of a *demon*. Here the demon removes the barrier ("opens a door") at time t_{on} and restores the barrier ("closes the door") at time t_{off}. Due to the stickness property, the information required for the control the particle is its velocity only. This means that the "door" closes when the velocity of the particle is close to a desired winding number. The most efficient manipulation can be achieved using the "stickiest" islands, which are present in both Hamiltonians, the symmetric

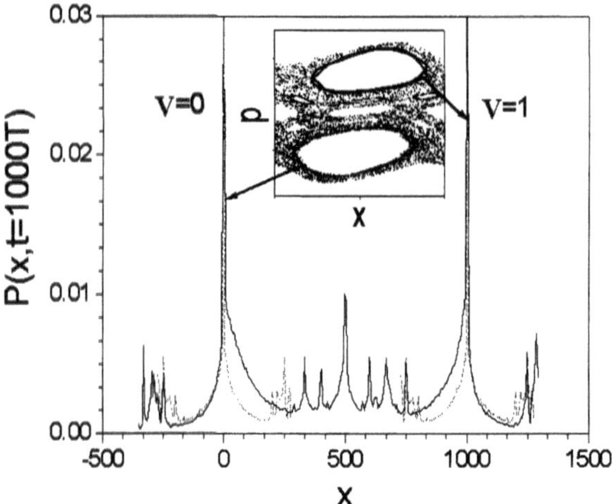

Fig. 6. The propagator for a fixed time ($t = 1000T$) for (a) the symmetric Hamiltonian, (11)(dotted line) and (b) the Hamiltonian with constant SB, (12) (solid line). Inset displays sticky islands which correspond to the main peaks in the propagator shape. Parameters same as in Fig. 5

one, H_s, in (11), and the SB one, H, in (12). In this way the time duration t_{SB} needed for the manipulation decreases, and the accuracy of the procedure increases. Below we outline an example of the SB strategy of our demon.

Let us start from the situation in which a particle is located inside the main stochastic layer. If the demon wants to move the particle in the positive direction then the particle must be shifted into the upper ballistic layer. In this case, the demon must switch on the second source of delta-kicks, that leads to overlapping of the main layer with the upper one. Then demon has to follow the velocity of the particle. When, for duration of about $10T$, the velocity is close enough to the winding number of stickiest island, it indicates that the particle is trapped in the upper layer with a high probability. At this stage the demon switches off the second source. Now the particle remains locked inside this layer and moves approximately with a constant velocity in the positive direction. After some time, when the particle reaches a required region in space and the demon would like to stop it, he should again switch on the SB source and follow the velocity. If the particle velocity is close enough to zero then it means that the particle sticks to a localized resonance and has returned to the main chaotic layer. The demon now switches off the second source and the particle is locked back inside the main layer. The mean energy returns to its value before the SB action.

In Fig. 7 we show the realization of the SB procedure for the system in (13)–(14). We check the displacement of the particle after each time step 10T. If this displacement is close enough to $v_i t_{contr}$ (about 10%), we take it as a sign that the particle is near the corresponding island. The direction of the motion is defined by the value of the time shift τ of the second source.

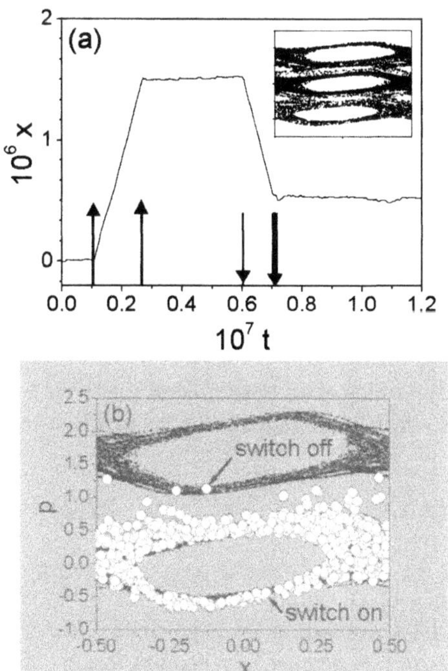

Fig. 7. A realization of the control approach. (a) Trajectory for the Hamiltonian in (12). The SB source acts for time windows marked by the bars. The widths of the bars equal the duration of the SB action. The arrows point to the resonance involved in the overlap (upper and lower ones correspondingly). The time phase shift is $\tau = 0.4$ for the two first pulses and $\tau = 0.2$ for the last ones. Inset shows the Poincarè section for the manipulation period. (b) The Poincarè section (white circles) of the system in (12) after every time step $t = 10T$ (see text for details) during the first SB pulse. The parameters as in Fig. 5

The mean time needed for the demon's SB action can be evaluated using the distribution of times between consecutive sticking events in the case of a permanent SB action, (13)–(14) [7,8]. The time needed to moves particle to the flying mode can be estimated from the first moment of the PDF for the ballistic island. For the same parameters as in Fig. 7 this procedure gives $\simeq 35t_{contr}$. The time needed for return the particle back to the nondrift diffusive mode can be estimated from the first moment of the PDF for the central localized island. This gives $\simeq 48t_{contr}$.

The SB strategy can be also used in the case of a statistical ensemble of particles. In this case, the SB can change populations of particles on the different manifolds through the control of the KAM-tori barriers. Let us consider the example of a continuous system with a Hamiltonian, which describes the motion of particles in a standing wave with a modulating amplitude [6]:

$$H_s(p, x, t) = \frac{p^2}{2} + E_1 \cos(x) \cos^2(\omega t). \tag{15}$$

Such a Hamiltonian system has been realized in atomic optics experiments probing motion in a wave produced by a laser field [20]. Here we investigate the classical limit.

Following steps analogous to those applied above, we can brake all relevant symmetries by switching on a second standing wave, shifted with respect to the main one:

$$H(p,x,t) = H_s(p,x,t)$$
$$+\Phi(t,t_{on},t_{off})E_2 cos(x+\phi)cos^2(\omega t+\tau), \tag{16}$$

where ϕ and τ are spatial and temporal shift constants and E_2 is the amplitude of the second standing wave.

We consider the dynamics of an ensemble of particles with an initial Maxwellian distribution in p, and homogeneous in x, inside one spatial period of the potential

$$\rho(p,x,0) = \frac{1}{2\pi}\sqrt{\frac{\beta}{2\pi}}\mathrm{e}^{-\frac{\beta}{2}p^2}\Theta(x)\Theta(2\pi - x), \tag{17}$$

with $\beta = 10$.

Under the influence of the main standing wave, $E_1\cos(x)\cos^2(\omega t)$, the ensemble performs diffusive spreading with no drift. Now we show that using SB for a finite duration t_{SB}, i. e. a pulse of a second force, (16), it is possible to chip off a small fraction of the particles from the an initial "cloud" and transport it in a preassigned direction. Namely, a small fraction (compared to the initial ensemble) of particles can be locked into the manifold with a nonzero drift. After switching off the pulse of the second standing wave the particles move in the prescribed direction with a velocity of the corresponding manifold. This is a kind of tweezers, which chip off some fraction of particles. In Fig. 8 we show an example of the realization of this strategy. SB induces an overlap of the main chaotic layer with the thin upper ballistic layer. The number of chipped particles can be controlled by variation of t_{SB}. For example, for $t_{SB} = 10T$, the chipped fraction is about 3% and for $t_{SB} = 20T$ is about 7%. It should be mentioned that this manipulation can not be achieved by standard technique using an external bias, since this will lead to the total displacement of the ensemble only. The SB approach, proposed here, provides therefore a new possibility to perform a manipulation with statistical ensembles using zero-mean external fields.

6 Conclusion

Directed transport inside the main chaotic layer is determined by breaking the symmetry of Lévy flights inside ballistic channels, which are generated due to the presence of resonance islands with nonzero winding numbers. This suggests a rather simple algorithm for estimating and controlling the current in the system. Namely, using the Poincare section and the propagator $P(x,t)$, one can identify the relevant resonance islands and calculate their winding numbers. Then, by

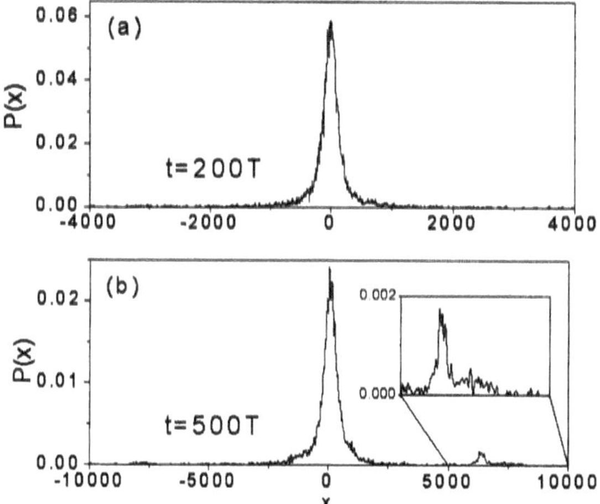

Fig. 8. The spatial distribution of an ensemble of particle $N = 10^4$ (see text) for the Hamiltonian in (15) ($E_1 = 0.5$, $E_2 = 0.1$, $T = 2\pi$, $\phi = 1.2$, $\tau = 0.8$) (a) just before and (b) after the action of the SB force ($t_{on} = 200T$, $t_{off} = 220T$). The inset is an enlargement of the additional peak which corresponds to the chipped fraction of the directed particles

changing the system parameters one can vary the resonance structure by opening and closing relevant ballistic channels.

We have found that the nonlinear interaction between different harmonics of the driving force $E(t)$ can trigger a nontrivial current inversion in Hamiltonian systems. This change of sign is related to the appearance and disappearance of regular islands. We have shown also that SB approach to Hamiltonian kinetic provides a new tool for manipulating and directing dynamical systems.

Acknowledgment

The authors thank the support of INTAS project 00-0847 (JK) and of the Israel Science Foundation, project 573/00-2 (MU).

Appendix

Let $P(x,t)$ be a propagator for a random walker position for large times. The first momemt of x is obtained from the Fourier transform of $P(x,t)$:

$$i\frac{\partial P(k,t)}{\partial k}|_{k=0} = < x(t) > \tag{18}$$

We now introduce some relevant definitions.

The probability density that a flight event has a distance x and duration t:

$$\xi(x,t) = \sum_{i=1}^{N} p_i \delta(x - v_i t)\psi_i(t)$$

The probability that the particle has moved a distance x in time t in a single flight (not necessarily stopping):

$$\Gamma(x,t) = \sum_{i=1}^{N} p_i \delta(x - v_i t) \int_t^\infty \psi_i(\tau) d\tau$$

The probability that the particle performs a random walk for at least time t (and remains walking):

$$\Phi(t) = \int_\infty^t \psi_c(\tau) d\tau.$$

The probability of just starting a random walk at (x,t):

$$Y(x,t) = \frac{1}{2}\delta(t)\delta(x) + \int_{-\infty}^{\infty} dx' \int_0^t dt' Z(x',t')\xi(x-x',t-t'),$$

where $Z(x,t)$ describes the probability of just starting a flight at (x,t):

$$Z(x,t) = \frac{1}{2}\delta(t)\delta(x) + \int_0^t dt' Y(x',t')\psi_c(t-t').$$

The final equation for $P(x,t)$ is therefore

$$P(x,t) = \int_0^t dt' \Phi(t-t') Y(x,t') + \int_0^\infty dx' \int_0^t \Gamma(x-x',t-t') Z(x',t'), \quad (19)$$

The integral equation (A-2) can be solved by Fourier transforming space and Laplace transforming time,

$$P(k,s) = \{s^{-1}[1-\psi_s(s)]\}[\frac{\frac{1}{2}+\frac{1}{2}\xi(k,s)}{1-\xi(k,s)\psi_s(s)}] + [\sum_{i=1}^{N} p_i \lambda_i][\frac{\frac{1}{2}+\frac{1}{2}\psi_s(s)}{1-\xi(k,s)\psi_s(s)}],$$

where $\xi(k,s) = \sum_{i=1}^{N} p_i \psi_i(s+jkv_i)$, $\lambda_i(k,s) = p_i(s+jkv_i)^{-1}[1-\psi_i(s+jkv_i)]$, and j is the imaginary unit.

Thus,

$$\langle x \rangle = \frac{(1+\psi_c)\sum_{i=1}^{N} p_i v_i (1-\psi_i(s))}{2s^2(1-\xi\psi_c)}. \quad (20)$$

All functions $\psi_i(t)$ and $\psi_c(t)$ must have a finite first moment (because of the Kac theorem about finitness of reccurence times in Hamiltonian systems [17]). Thus, for small s (that corresponds to $t \to \infty$):

$$\psi_i(s) = 1 - s\langle t_i \rangle + ..., \psi_c(s) = 1 - s\langle t_s \rangle + ... \ . \quad (21)$$

Using expansion (A-4) we obtain from (A-3):

$$\langle x(s) \rangle = \frac{\sum_{i=1}^{N} p_i v_i \langle t_i \rangle}{\sum_i^N p_i \langle t_i \rangle + \langle t_c \rangle} \frac{2 - s\langle t_c \rangle}{2s^2} \approx \frac{\sum_{i=1}^{N} p_i v_i \langle t_i \rangle}{\sum_i^N p_i \langle t_i \rangle + \langle t_c \rangle} \frac{1}{s^2}$$

This leads to

$$\langle x(t)\rangle = \frac{\sum_{i=1}^{N} p_i v_i \langle t_i\rangle}{\sum_i^N p_i \langle t_i\rangle + \langle t_c\rangle} t,$$

and, finally, we obtain for the asymptotic current:

$$J = \frac{\sum_{i=1}^{N} p_i v_i \langle t_i\rangle}{\sum_i^N p_i \langle t_i\rangle + \langle t_c\rangle}. \tag{22}$$

References

1. P. Reimann: Phys. Rep. **361**, 57 (2002)
2. H. Risken: *The Fokker Planck Equation.* (Springer, Berlin, 1984)
3. P. Hänggi and R. Bartussek, in: *Nonlinear Physics of Complex Systems*, Lecture Notes in Physics, **476**, ed. by J. Parisi, S. C. Muller, and W. Zimmermann (Springer, Berlin 1996) p.294; P. Jung, J. G. Kissner, and P. Hänggi: Phys. Rev. Lett. **76**, 3436 (1996); J. L. Mateos: Phys. Rev. Lett. **84**, 258 (2000)
4. S. Flach, O. Yevtushenko, and Y. Zolotaryuk: Phys. Rev. Lett. **84**, 2358 (2000)
5. T. Dittrich, R. Ketzmerick, M.-F. Otto and H. Schanz: Ann. Phys. (Leipzig) **9**, 755 (2000); H. Schanz, M.-F. Otto, R. Ketzmerick, and T. Dittrich: Phys. Rev. Lett **87** 07601 (2001)
6. R. Z. Sagdeev, D. A. Usikov, and G. M. Zaslavsky: *Nonlinear physics : from the pendulum to turbulence and chaos.* (Harwood Academic Publ. 1992)
7. M. F. Shlesinger, G. M. Zaslavsky and J. Klafter: Nature **363**, 31 (1993); J. Klafter, M. F. Shlesinger, and G. Zumofen: Physics Today **49**, 33 (1996)
8. J. Klafter and G. Zumofen: Phys. Rev. E **49**, 4873 (1994)
9. S. Wiggins: *Chaotic transport in dynamical systems.* (Springer, Berlin 1992)
10. S. Denisov, S. Flach: Phys. Rev. E **64**, 056236 (2001)
11. S. Denisov, J. Klafter, M. Urbakh, and S. Flach, Physica D (*in press*)
12. T. Geisel, A. Zacherl, and G. Radons: Phys. Rev. Lett. **59** 2503 (1987)
13. S. Denisov, J. Klafter, M. Urbakh, *submitted*
14. There is a difference between the propagator $P(x,t)$ and function $\rho(p,x,t)$ disscused in Sec.2: $\rho(p,x,t)$ takes into account trajectories weighted over the *whole* phase space (including resonance islands and KAM-tori), whereas the propagator $P(x,t)$ includes only trajectories initiated at the chaotic layer
15. G. Zumofen, J. Klafter, and A. Blumen: Phys. Rev. E **47**, 2183 (1993)
16. E. R. Weeks and H. L. Swinney: Phys. Rev. E **57** 4915 (1998)
17. N. Kac: *Statistical Independence in Probability, Analysis, and Number Theory.* (Mathematical Association of America, Oberlin, OH, 1959)
18. For generation of random variable x with distribution (11) we have used the random variable ξ with uniform distribution in unit interval and transformation $x = t_c \xi^{-\frac{1}{\alpha-1}}$
19. We choose the value $D = 1.6$, which is comparable with the chaotic diffusion rate in the system, (1)
20. J.R. Robinson et al: Phys. Rev. Lett. **76**, 3304 (1996); F. L. Monte et al: Phys.Rev.Lett. **75**, 4598 (1995); B. G. Klappauf, W. H. Oskay, D. A. Steck, and M. G. Raizen, Phys. Rev. Lett. **81**, 4044 (1998)

Patterns and Correlations in Economic Phenomena Uncovered Using Concepts of Statistical Physics

H.E. Stanley[1], P. Gopikrishnan[1], V. Plerou[1], and M.A. Salinger[2]

[1] Center for Polymer Studies and Department of Physics, Boston University, Boston, MA 02215 USA
[2] Department of Finance and Economics, School of Management, Boston University, Boston, MA 02215 USA

Abstract. This paper discusses some of the *similarities* between work being done by economists and by physicists seeking to find "patterns" in economics. We also mention some of the *differences* in the approaches taken and seek to justify these different approaches by developing the argument that by approaching the same problem from different points of view, new results might emerge. In particular, we review two such new results. Specifically, we discuss the two newly-discovered scaling results that appear to be "universal", in the sense that they hold for widely different economies as well as for different time periods: (i) the fluctuation of price changes of any stock market is characterized by a probability density function (PDF), which is a simple power law with exponent $\alpha + 1 = 4$ extending over 10^2 standard deviations (a factor of 10^8 on the y-axis); this result is analogous to the Gutenberg-Richter power law describing the histogram of earthquakes of a given strength; (ii) for a wide range of economic organizations, the histogram that shows how size of organization is inversely correlated to fluctuations in size with an exponent ≈ 0.2. Neither of these two new empirical laws has a firm theoretical foundation. We also discuss results that are reminiscent of phase transitions in spin systems, where the divergent behavior of the response function at the critical point (zero magnetic field) leads to large fluctuations. We discuss a curious "symmetry breaking" for values of Σ above a certain threshold value Σ_c; here Σ is defined to be the local first moment of the probability distribution of demand Ω – the difference between the number of shares traded in buyer-initiated and seller-initiated trades. This feature is qualitatively identical to the behavior of the probability density of the magnetization for fixed values of the inverse temperature.

1 Introduction to Patterns in Economics

One prevalent paradigm in economics is to marry finance with mathematics, with the fruit of this marriage the development of models. In physics, we also develop and make use of models or, as they are sometimes called, "artificial worlds," but many physicists are fundamentally empirical in their approach to science – indeed, some physicists never make reference to models at all (other than in classroom teaching situations). This empirical approach has led to advances when theory has grown out of experiment. One such example is the understanding of phase transitions and critical phenomena [3]. Might this "empirics first" physics paradigm influence the way physicists approach economics?

Our group's approach to economic questions has been to follow the paradigm of critical phenomena, which also studies complex systems comprised of many interacting subunits, i.e., to first examine the empirical facts as thoroughly as possible, and search for any "patterns".

That at least *some* economic phenomena are described by "scaling patterns" has been recognized for over 100 years since Pareto investigated the statistical character of the wealth of individuals by modeling them using the scale-invariant distribution

$$f(x) \sim x^{-\alpha}, \tag{1}$$

where $f(x)$ denotes the number of people having income x or greater than x, and α is an exponent that Pareto estimated to be 1.5 [1]. Pareto noticed that his result was universal in the sense that it applied to nations *"as different as those of England, of Ireland, of Germany, of the Italian cities, and even of Peru"*. A physicist would say that the universality class of the scaling law (1) includes all the aforementioned countries as well as Italian cities, since by definition two systems belong to the same universality class if they are characterized by the same exponents.

In the century following Pareto's discovery, the twin concepts of scaling and universality have proved to be important in a number of scientific fields [2,3]. A striking example was the elucidation of the puzzling behavior of systems near their critical points. Over the past few decades it has come to be appreciated that the scale-free nature of fluctuations near critical points also characterizes a huge number of diverse systems also characterized by strong fluctuations. This set of systems includes examples that at first sight are as far removed from physics as is economics. For example, consider the percolation problem, which in its simplest form consists of placing blue pixels on a fraction p of randomly-chosen plaquettes of a yellow computer screen (Fig. 1). A remarkable fact is that the largest connected component of blue pixels magically spans the screen at a threshold value p_c. This purely geometrical problem has nothing to do with the small branch of physics called critical point phenomena. Nonetheless, the fluctuations that occur near $p = p_c$ are scale free and functions describing various aspects of the incipient spanning cluster that appears at $p = p_c$ are described by power laws. Indeed, the concepts of scaling and universality provide the conceptual framework for understanding this geometry problem.

It is becoming clear that almost any system comprised of a large number of interacting units has the potential of displaying power law behavior. Since economic systems are in fact comprised of a large number of interacting units has the potential of displaying power law behavior, it is perhaps not unreasonable to examine economic phenomena within the conceptual framework of scaling and universality [2–7]. We will discuss this topic in detail below.

So having embarked on a path guided by these two theoretical concepts, what does one do? Initially, critical phenomena research – guided by the Pareto principles of scaling and universality – was focused finding which systems display scaling phenomena, and on discovering the actual values of the relevant exponents. This initial empirical phase of critical phenomena research proved vital, for only by carefully obtaining empirical values of exponents such as α could scientists learn which systems have the same exponents (and hence belong to the same *universality class*). The fashion in which physical systems partition into disjoint universality classes proved essential to later theoretical developments such as the renormalization group [5] – which offered some insight into the reasons why scaling and universality seem to hold; ultimately it led to a better understanding of the critical point.

Similarly, the initial research in economics guided by the Pareto principles has largely been concerned with establishing which systems display scaling phenomena, and with measuring the numerical values of the exponents with sufficient accuracy that one can begin to identify universality classes if they exist. Economics systems differ from often-studied physical systems in that the number of subunits are considerably smaller in contrast to macroscopic samples in physical systems that contain a huge number of interacting subunits, as many as Avogadro's number 6×10^{23}. In contrast, in an economic system, one initial work was limited to analyzing time series comprising of order of magnitude 10^3 terms, and nowadays with high frequency data the standard, one may have 10^8 terms. Scaling laws of the form of (1) are found that hold over a range of a factor of $\approx 10^6$ on the x-axis [8–10].

2 Classic Approaches to Finance Patterns

As do economists, physicists view the economy as a collection of interacting units. This collection is complex; everything depends on everything else. The interesting problem is: *how* does everything depend on everything else? Physicists are looking for robust empirical laws that will describe – and theories that will help understand – this complex interaction.

To a physicist, the most interesting thing about an economic time series – e.g., the S&P 500 stock average index – is that it is dominated by fluctuations. If we make a curve of the values of the S&P 500 over a 35-year period, we see a fluctuating signal. Statistical physicists are particularly interested in fluctuating signals. The nature of this fluctuation immediately suggests to a physicist a model that was developed 100 years ago by Bachelier: the *biased random walk* [11].

A one-dimensional random walk is a drunk with a coin and a metronome. At each beat of the metronome, the coin is flipped – heads means one step to the right, tails one step to the left. If we look at our S&P 500 plot placed alongside a graph of a one-dimensional *biased* random walk – it is biased because it uses a "biased coin" that has a slight tendency to go up rather than down – we see a reasonable visual similarity. In fact, many economic pricing models – e.g., Black and Scholes – use this biased random walk.

Still there are certain points in the S&P 500 plot – such as October 19, 1987 ("Black Monday"), or the 15 percent drop over the week following the events of 11 September 2001 – that are not mirrored anywhere in the biased random walk model. Nowhere do we see a drop anywhere near the 30 percent drop of Black

Monday. This could not occur in a biased random walk – the probability that a walk will move two steps in the same direction is p^2, three steps is p^3, and so on – so the probability of many steps in the same direction is exponentially rare, and virtually impossible.

Then how do we quantify these S&P 500 fluctuations? We begin by graphing the values of the fluctuations as a function of time. We place the plot of the empirical data next to the predictions of Bachelier's model. The fluctuations in the model are normalized by one standard deviation. Note that the biased random walk has a probability density function (PDF) that is a Gaussian, so the probability of having more than five standard deviations is essentially zero – you can see that a line drawn at five standard deviations is outside the range of the fluctuations.

If we normalize the empirical data we see a difference. A line drawn at five standard deviations is *not* outside the range of the fluctuations – there are many "shocks" that exceed five standard deviations. A bar placed on the positive side at five standard deviations also has 30 or 40 hits – fluctuations that equal or exceed five standard deviations in the positive direction. Some, such as Black Monday, are more than 34 standard deviations. The exponential of $(-1/2)(34)^2$ is approximately $10^{-267/2}$.

Because big economic shocks affect the economy around the world ("everything depends on everything else"), the possibility of an economic "earthquake" is one that we must take seriously. Big changes in stocks affect not only people with large amounts, but also those on the margins of society. One person's portfolio collapse is another's physical starvation; e.g., literal starvation in some areas was one result of the recent Indonesian currency collapse.

Another example is the recent Merriwether LTCM (Long Term Capital Management) collapse, caused in part by the use of models that do not take into account these catastrophic rare events. Thus there are many reasons we physicists might be interested in understanding economic fluctuations.

3 Patterns in Finance Fluctuations

One topic we physicists are interested in is symmetry. An example of traditional symmetry is sodium chloride. One can displace the lattice an amount equal to exactly two lattice constants and the configuration will remain the same. One can rotate it 90 degrees, or invert it, and the configuration will remain the same. Not only are these properties fascinating to mathematicians, they are also very relevant to solid state physics. This simple symmetry and the mathematics and physics that are built on it have led to extremely useful inventions, e.g., the transistor. The framework for our approach to systems with many interacting subunits is something that is usually called "scale invariance." These systems vary greatly from systems that do have scales.

We are all familiar with algebraic equations such as

$$x^2 = 4, \tag{2}$$

and we know the solution is a number, ± 2. Most of us are also familiar with functional equations, which are statements not about relations between numbers, but about the functional form of a function $f(x)$. Algebraic equations have solutions that are numbers, but functional equations have solutions that are functional forms. Power law functions are the solutions of certain functional equations of the form

$$f(\lambda x) = \lambda^p f(x). \tag{3}$$

In a functional equation, the converse also holds, i.e., every function that is of this power-law form also obeys this functional equation. This applies to a large number of contexts, in particular, to physical systems that have been tuned to be near critical points. An example is a binary mixture of two fluids in which the temperature has been tuned to be a special value called the critical temperature. At that temperature, there occur fluctuations in density in the binary mixture that extend over all length scales up to and including the wavelength of light. If you shine a flashlight on a tube of the binary mixture, you see an eerie glow – because the density fluctuations are so big in spatial extent, they become comparable to the wavelength of the light that is interacting with them. When that occurs, you see something that is visible – "critical opalescence." The same conceptual framework that describes this system appears to be able to describe economic systems.

Newcomers to the field of scale invariance often ask why a power law does not extend "forever" as it would for a mathematical power law of the form $f(x) = x^{-\alpha}$. This legitimate concern is put to rest by by reflecting on the fact that power laws for natural phenomena are not equalities, but rather are asymptotic relations of the form $f(x) \sim x^{-\alpha}$. Here the tilde denotes *asymptotic equality.* Thus $f(x)$ is not "approximately equal to" a power law so the notation $f(x) \approx x^{-\alpha}$ is inappropriate. Similarly, $f(x)$ is not proportional to a power law, so the notation $f(x) \propto x^{-\alpha}$ is also inappropriate. Rather, asymptotic equality means that $f(x)$ becomes increasingly like a power law as $x \to \infty$. Moreover, crossovers abound in financial data, such as the characteristic crossover from power law behavior to simple Gaussian behavior as the time horizon Δt over which fluctuations are calculated increases; such crossovers are characteristic also of other scale-free phenomena in the physical sciences [3], where the Yule distribution often proves quite useful.

For reasons of this sort, standard statistical fits to data are inappropriate, and often give distinctly erroneous values of the exponent α. Rather, one reliable way of estimating the exponent α is to form successive slopes of pairs of points on a log-log plot, since these successive slopes will be monotonic and converge to the true asymptotic exponent α.

The scale-invariance symmetry involved here is just as much a symmetry as the translational invariance symmetry in sodium chloride. We do not know how useful this scale-invariance symmetry will ultimately prove to be. Over the past 30 years physicists have used the theme of scale-invariance symmetry to understand systems near their critical points. Previous to this period of time, this class of problems was one no one could solve: there were many, many length scales, not just one. The length scales could run from one nearest-neighbor spacing out to approximately 5,000 (approximately the wavelength of light). The elements that make up this system are molecules that interact only over a short range – almost entirely with nearest neighbors. But this nearest-neighbor interaction propagates a small amount of torque through the system of nearest-neighbor interactions, so that the entire system is affected.

This is beginning to sound like economics, in which "everything affects everything else," And in economics, the first thing a physicist would do is look for the correlations. If we look at a graph of the autocorrelation function, we see a measure of the quantity G, which is a price change over some time horizon Δt. If we look at how G is now correlated with G at a time τ later, we measure that quantity as a function of τ, and as the size of τ increases, the correlation decreases. It is remarkable that this decrease happens in a regular fashion. How do we interpret this decrease? If we put the autocorrelation function in logarithmic units and the time lag in linear units, we see that the data fall on an approximate straight line This means that the function is decaying exponentially, which means it does indeed have a characteristic scale [12–14]. So the autocorrelation function is not scale invariant. This differs from systems near their critical points in which the autocorrelation functions are scale invariant.

The decay time in this finance example is short (4 minutes), so one cannot easily "make money" on these correlations [12,13]. A little less well-known is the measure of the volatility [13,14]. One way to quantify volatility is to replace G (the price change) with the absolute value of G. The data now are not at all linear on log-linear paper, but they are linear on log-log paper. And, of course, a power-law $y = x^p$ is linear on log-log paper, because $\log y = p \log x$. The slope of the log-log plot p is the value of the exponent. These exponents turn out to be fundamental quantities. In this case, $p = -0.3$. The data are straight from about 200 minutes out to about 10^5 minutes – a range of almost 1000. With the data graphed,one can see the approximate region in which the data are straight – the data are not straight over all regions. Qualitatively, we have known for a long time that there are long-range correlations in the volatility, e.g., volatility "clustering" and "persistence," but this graph helps quantify this known empirical fact.

If we cannot find an ordinary correlation near a critical point, we must try something else. For example, we might simply dump all of our data "on the floor." After we do that, the data no longer have time ordering nor do they have long-range or short-range power-law correlations in the volatility of the autocorrelation function itself. Now we pick the data points up off the floor and make a histogram. Mandelbrot did this in 1963 with 1000 data points – a tiny number by today's standards – for cotton-price fluctuations [2]. He concluded that those data were consistent with a Lévy distribution, i.e., a power-law distribution in that histogram – a so-called "fat tail."

In 1995, Mantegna and Stanley decided to test this result using data with Δt shorter than the daily data available in 1963 [12]. We used approximately 1 million data points: three orders of magnitude greater than Mandelbrot's data set. Instead of Mandelbrot's daily returns on cotton prices, we had returns as frequent as 15 seconds on the S&P 500 index. We found that on a log-linear plot (i) the histogram of the G data points for the S&P 500 clearly is not a Bachelier/Black-Scholes Gaussian, and (ii) although the *center* of the histogram agrees fairly well with Mandelbrot's Lévy distribution, it begins to disagree after a few standard deviations. This disagreement led us to develop a class of mathematical processes called *truncated* Lévy distributions – which has attracted the attention of a number of mathematicians, who have carried this concept far further than we could [15–20].

What about "universality," the notion in statistical physics that many laws seem to be remarkably independent of details? A physics example is that dramatically different materials behave exactly the same near their respective critical points. Binary mixtures, binary alloys, ferromagnets, even biological systems that involve switching, all behave the same way. An analog of this universality appears in economics. For example, Skjeltorp [21] did a study that utilized the Mantegna approach. Instead of 1,500,000 points from the S&P 500 (15-second intervals spread over six years), Skjeltorp did a parallel study for the Norwegian stock exchange and got almost exactly the same result.

We assumed that the reason we saw the truncated Lévy distribution while Mandelbrot did not was because we had more data – by three orders of magnitude. Gopikrishnan et al. recently acquired a data set three orders of magnitude larger still (of order 10^9) – one that records *every* transaction of every stock. They found that when their data were graphed on log-log paper, the result was linearity [22–25]. This is the log of the cumulative distribution, the same quantity Mandelbrot plotted for cotton. But where Mandelbrot's straight line had a slope of about 1.7 (well inside the Lévy regime, which stops at slope 2.0), Gopikrishnan's straight line has a slope of ≈ 3.0 (far outside the limit for a Lévy distribution). The fact that these data are approximately linear over two orders of magnitude means that fluctuations that are as much as 100 standard deviations are still conforming to the same law that describes the smaller fluctuations. This is reminiscent of the Gutenberg-Richter law that describes earthquakes [26–28]. Thus it would seem that these very rare events, which are conventionally treated as totally unexpected and unexplainable, have a precise probability describable by the same law that describes much more common events. These rare events occur with a frequency 8 orders of magnitude less than the common, everyday event.

This means that Mandelbrot's results for cotton (10^3 points) are at total odds with Gopikrishnan's results for the S&P 500 (10^9 points). Why this difference? Is it simply because Mandelbrot did not have enough data to draw reliable conclusions? Or do commodities intrinsically have fatter tails? In recent work with data from British Petroleum, it appears that commodity data may have a slightly smaller slope – consistent with the possibility that perhaps there is not one universal behavior for everything, but at least two separate universal behaviors – one for commodities and one for equities [29]. This smaller slope is still above 2, so the commodity data are not in the Lévy regime.

4 Patterns Resembling "Diffusion in a Tsunami Wave"

Over this past year, we and our collaborators have been trying to understand these exponents using procedures similar to those used in critical phenomena, e.g., we relate one exponent to another and call the relation a scaling law, or we derive some microscopic model.

In particular, there appears to be an intriguing analog not with the classic diffusion process studied in 1900 by Bachelier, [11] but rather with a generalization called anomalous diffusion. It is plausible that classical diffusion does not describe all random motion. The Brownian motion seen in the behavior of a grain of pollen in relatively calm water becomes something quite different if the grain of pollen is in a tsunami wave. The histograms would certainly be perturbed by a tsunami. A tsunami is an apt verbal metaphor for such economic "earthquakes" as the Merriwether disaster, so why not explore the stock market as an example of anomalous diffusion?

In one-dimensional classic diffusion, a particle moves at constant velocity until it collides with something. One calculates, e.g., the end position of the particle, and (of course) finds a Gaussian. Within a fixed time interval Δt, one might calculate a histogram for the number of collisions $p(N)$, and also find a Gaussian. And if one calculates a histogram of the variance W^2, one also finds a Gaussian. The fact that these are relatively narrow Gaussians means that there is a characteristic value, i.e., the width of that Gaussian, and that this is the basis for classical diffusion theory.

The corresponding quantity in the stock market to the displacement x is the price. At each transaction there is a probability that the price will change, and after a given time horizon there is a total change G. We've seen the histogram of G values – the cumulative obeyed an inverse cubic law, and therefore the pdf, by differentiation, obeys an inverse quartic law.

What about these histograms? Apparently no one had calculated these previously. Plerou et al. set about using the same data analyzed previously for G to calculate the histograms of N and W^2. They also found power laws – not Gaussians, as in classic diffusion. That means there is no characteristic scale for the anomalous diffusion case (there is a characteristic scale for the classic diffusion case) and for an obvious reason. If you are diffusing around in a medium – such as the "economic universe" we live in – in which the medium itself is changing, then the laws of diffusion change and, in particular, they adopt this scale-free form. Further, the exponents that describe $p(N)$ and $p(W^2)$ appear [30,31] to be the analogs of exponents in critical phenomena in the sense that they seem to be related to one another in interesting ways.

5 Patterns Resembling Critical Point Phenomena

Before concluding, we ask what sort of understanding could eventually develop if one takes seriously the power laws that appear to characterize finance fluctuations. It is tempting to imagine that there might be analogies between finance and known physical processes displaying similar scale-invariant fluctuations. For example, if one measures the wind velocity in turbulent air, one finds intermittent fluctuations that display some similarities with finance fluctuations [32]. However these similarities are not borne out by quantitative analysis – e.g., one finds non-Gaussian statistics, and intermittency, for both turbulence fluctuations and stock price fluctuations, but the time evolution of the second moment and the shape of the probability density functions are different for turbulence and for stock market dynamics [33,34].

More recent work pursues a rather different analogy, phase transitions in spin systems. Stock prices respond to fluctuations in demand, just as the magnetization of an interacting spin system responds to fluctuations in the magnetic field. Periods with large number of market participants buying the stock imply mainly positive changes in price, analogous to a magnetic field causing spins in a magnet to align. Recently, Plerou et al. [35] addressed the question of how stock prices respond to changes in demand. They quantified the relations between price change G over a time interval Δt and two different measures of demand fluctuations: (a) Φ, defined as the difference between the number of buyer-initiated and seller-initiated trades, and (b) Ω, defined as the difference in number of shares traded in buyer and seller initiated trades. They find that the conditional expectations $\langle G\rangle_{\Phi}$ and $\langle G\rangle_{\Omega}$ of price change for a given Φ or Ω are both concave. They find that large price fluctuations occur when demand is very small – a fact which is reminiscent of large fluctuations that occur at critical points in spin systems, where the divergent nature of the response function leads to large fluctuations. Their findings are reminiscent of phase transitions in spin systems, where the divergent behavior of the response function at the critical point (zero magnetic field) leads to large fluctuations. Further, Plerou et al. [36] find a curious "symmetry breaking" for values of Σ above a certain threshold value Σ_c; here Σ is defined to be the local first moment of the probability distribution of demand Ω, the difference between the number of shares traded in buyer-initiated and seller-initiated trades. This feature is qualitatively identical to the behavior of the probability density of the magnetization for fixed values of the inverse temperature.

Since the evidence for an analogy between stock price fluctuations and magnetization fluctuations near a critical point is backed up by quantitative analysis of finance data, it is legitimate to demand a theoretical reason for this analogy. To this end, we discuss briefly one possible theoretical understanding for the origin of scaling and universality in economic systems. As mentioned above, economic systems consist of interacting units just as critical point systems consist of interacting units. Two units are correlated in what might seem a hopelessly complex fashion – consider, e.g., two spins on a lattice, which are correlated regardless of how far apart they are. The correlation between two given spins on a finite lattice can be partitioned into the set of all possible topologically linear paths connecting these two spins – indeed this is the starting point of one of the solutions of the two-dimensional Ising model (see Appendix B of [3]). Since correlations decay exponentially along a one-dimensional path, the correlation between two spins would at first glance seem to decay exponentially. Now it is a mathematical fact that the total number of such paths grows exponentially with the distance between the two spins – to be very precise, the number of paths is given by a function which is a product of an exponential and a power law. The constant of the exponential *decay* depends on temperature while the constant for the exponential *growth* depends only on geometric properties of the system [3]. Hence by tuning temperature it is possible to achieve a threshold temperature where these two "warring exponentials" just balance each other, and a previously negligible power law factor that enters into the expression for the number of paths will dominate. Thus power law scale invariance emerges as a result of canceling exponentials, and universality emerges from the fact that the interaction paths depend not on the interactions but rather on the connectivity. Similarly, in economics, two units are correlated through a myriad of different correlation paths; "everything depends on everything else" is the adage expressing the intuitive fact that when one firm changes, it influences other firms. A more careful discussion of this argument is presented, not for the economy but for the critical phenomena problem, in Ref. [5].

Finally, a word of humility with respect to our esteemed economics colleagues is perhaps not inappropriate. Physicists may care passionately if there are analogies between physics systems they understand (like critical point phenomena) and economics systems they do not understand. But why should anyone else care? One reason is that scientific understanding of earthquakes moved ahead after it was recognized [26,27] that extremely rare events – previously regarded as statistical outliers regarding for their interpretation a theory quite distinct from the theories that explain everyday shocks – in fact possess the identical statistical properties as everyday events (e.g., all earthquakes fall on the same straight line on an appropriate log-log plot). Similarly, if it were ever to be convincingly demonstrated that phenomena of interest in economics and finance possess the analogous property, then the challenge will be to develop a coherent understanding of financial fluctuations that incorporates not only everyday fluctuations but also those extremely rare "financial earthquakes."

6 Cross-Correlations in Price Fluctuations of Different Stocks

We know that a stock price does not vary in isolation from other stock prices, but that stock prices are correlated. This fact is not surprising because we know that "in economics everything depends on everything else." How do we *quantify* these cross-correlations of one stock with another? If we take the G values of four companies out of the 1000 that we have studied – corresponding to the shrinking or growing of each of these four companies in, say, a 30-minute interval. How does the behavior of these four companies during that half-hour interval affect your response to their price activity? If two of the companies were Pepsi and Coke, there would probably be some correlation in their behaviors.

In order to quantify these cross-correlations, we begin by calculating a cross-correlation matrix. If we have 1000 firms, we have a 1000×1000 matrix $\mathbf{C}$ each element C_{ij} which is the correlation of firm i with firm j. This large number of elements (1 million) does not frighten a physicist with a computer. Eugene Wigner applied random matrix theory 50 years ago to interpret the complex spectrum of energy levels in nuclear physics [37–48]. We do exactly the same thing, and apply random matrix theory to the matrix $\mathbf{C}$. We find that certain eigenvalues of that 1000×1000 matrix deviate from the predictions of random matrix theory, which has not eigenvalues greater than an upper bound of ≈ 2.0. Furthermore, the content of the eigenvectors corresponding to those eigenvalues correspond to well-defined business sectors. This allows us to define business sectors without knowing anything about the separate stocks: a Martian who cannot understand stock symbols could nonetheless identify business sectors [47,49].

7 Patterns in Firm Growth

In the economy, each firm depends on every other firm, and the interactions are not short-ranged nor are they of uniform sign. For example, Ford Motor Company is in trouble because they have been selling their Explorer vehicle with extremely unsafe tires – and the price of their stock is going down. Prospective buyers purchase General Motors cars instead. There is thus a *negative* correlation between the stock prices of the two companies. But now General Motors needs to hire more workers to make a larger number of cars, and the McDonald's near the assembly plant has many more customers at lunchtime – a *positive*. Sometime later the situation may change again. So we can say that the "spins" all interact with one another, and that these interactions change as a function of time.

Nevertheless, the general idea of a critical phenomenon seems to work. If the firms were spread out in a kind of chain, the correlations among them would decay exponentially. Instead, the firms interact with each other much the same way that subunits in critical phenomena interact with each other. This fact motivated a study carried out about five years ago by a group of physicists interacting with an economist [10,50,51]. They calculated the fluctuations in business firms from one year to the next. They found that if they broke the fluctuations into bins by size a tent-shaped distribution function was produced for each day of trading. The width of the tent was narrower for large firms than the width of the tent for small firms. This is not surprising, since a small firm has a potential to grow or shrink much more rapidly than a larger firm. When the widths of these tent-shaped distribution functions were plotted on log-log paper as a function of histogram size, the decreasing function turns out to be a straight line – corresponding to a power-law behavior in that function. The exponent in that power-law is ≈ 0.2. The linearity extends over a number of decades, indicating that the data collapse onto a single plot irrespective of scale.

Recent attempts to make models that reproduce the empirical scaling relationships suggest that significant progress on understanding firm growth may be well underway [52–55], leading to the hope of ultimately developing a clear and coherent "theory of the firm". One utility of the recent empirical work is that now any acceptable theory must respect the fact that power laws hold over typically six orders of magnitude; as Axtell put the matter rather graphically: *"the power law distribution is an unambiguous target that any empirically accurate theory of the firm must hit"* [8].

8 Universality of the Firm Growth Problem

Takayasu et al. have demonstrated that the above results are universal by moving outside the realm of US economies and studying firm behavior in other parts of the world [56].

Buldyrev et al. have shown that organizations (such as business firms) that are organized like trees will fluctuate in size [51]. The hierarchical structure is set up so that instructions from the top of the hierarchy propagate down to the branching lower levels of the structure. Within that structure is a disobedience factor – those lower down do not always obey the directives handed down from those above them. This factor is, of couse, crucial to the survival of the system. If employees always did only and exactly what they were told, any small mistake put into the system by a manager would grow and do an increasing amount of damage as it propagated through the expanding tree structure of the organization. On the other hand, the probability of an instruction being disobeyed cannot be one – or chaos would result. The propensity to disobey can be neither infinitesimal nor unity. The "obeying probability" needs to settle at a point at which the organization can maintain both its integrity and self-corrective flexibility. And the behavior of the exponent describing this probability is very similar to the behavior of critical exponents.

This result is fairly robust, not only as far as business firm fluctuations are concerned, but also in the size of countries. Lee et al. extend the same analysis used for business firms to countries – and with the same exponent [57]. Data can therefore be graphed on the same curve both for firms and for countries – where country size is measured by GDP.

We can see a similar pattern in the funding of university-based research. We researchers compete for research money the same way business firms compete for customers. Plerou et al. analyzed the funding of research groups over a 17-year period in the same way fluctuations in firm size were analyzed [58]. The results were very similar with the data collapsing onto the same curve.

As a final example, we mention the case of fluctuating bird populations in North America. In this case the exponent is 0.35 instead of 0.2. But, nevertheless, there seems to be some kind of property of organizations that we do not understand well [59].

9 "Take-Home Message"

So – what have we learned? First, that the approach we have emphasized is an empirical approach where one first seeks to uncover features of the complex economy that are challenges to understand. We find that there are two new universal scaling models in economics: (i) the fluctuation of price changes of any stock market is characterized by a PDF which is a simple power law with exponent $\alpha+1=4$ that extends over 10^2 standard deviations (a factor of 10^8 on the y-axis); (ii) for a wide range of economic organizations, the histogram that shows how size of organization is inversely correlated to fluctuations in size with an exponent $\beta \approx 0.2$.

Neither of these two new laws has a firm theoretical foundation. This situation parallels the situation in the 1960s when the new field of critical phenomena also did not have a firm theoretical foundation for its new laws, but was awaiting the renormalization group. It is my hope that some of you in this room will rise to the challenge and try to find a firm theoretical foundation for the structure of the empirical laws that appear to be describing (i) finance fluctuations, and (ii) economic organizations.

Acknowledgements

We wish to thank our collaborators, L. A. Amaral, D. Canning, X. Gabaix, R. N. Mantegna, K. Matia, M. Meyer, and B. Rosenow from whom we have learned a great deal. We also thank NSF for financial support.

References

1. V. Pareto: *Cours d'Economie Politique*, (Lausanne and Paris 1897)
2. B. B. Mandelbrot: J. Business **36**, 394 (1963)
3. H. E. Stanley: *Introduction to Phase Transitions and Critical Phenomena* (Oxford University Press, Oxford 1971)
4. H. Takayasu [ed]: *Empirical Science of Financial Fluctuations: The Advent of Econophysics* (Springer, Berlin 2002)
5. H. E. Stanley: Rev. Mod. Phys. **71**, S358 (1999)
6. R. N. Mantegna and H. E. Stanley: *An Introduction to Econophysics: Correlations and Complexity in Finance*, (Cambridge University Press, Cambridge 2000)
7. J. P. Bouchaud: Quantitative Finance, **1**, 105 (2001)
8. R. L. Axtell: Science **293**, 1818 (2001)
9. M. H. R. Stanley, S. V. Buldyrev, S. Havlin, R. Mantegna, M. A. Salinger and H. E. Stanley: Econ. Lett. **49**, 453 (1996)
10. M. H. R. Stanley, L. A. N. Amaral, S. V. Buldyrev, S. Havlin, H. Leschhorn, P. Maass, M. A. Salinger and H. E. Stanley: Nature **379**, 804 (1996)
11. L. Bachelier: *Théorie de la spéculation* Ph.D. thesis in mathematics, Annales Scientifiques de l'Ecole Normale Supérieure *III-17*, 21 (1900)
12. R. N. Mantegna and H. E. Stanley: Nature **376**, 46 (1995)
13. Y. Liu, P. Gopikrishnan, P. Cizeau, M. Meyer, C. K. Peng and H. E. Stanley: Phys. Rev. E **60**, 1390 (1999)
14. Z. Ding, C. W. J. Granger, and R. F. Engle: J. Empirical Finance **1**, 83 (1993)
15. R. N. Mantegna and H. E. Stanley: Phys. Rev. Lett. **73**, 2946 (1994)
16. B. Podobnik, P. Ch. Ivanov, Y. Lee, A. Chessa, and H. E. Stanley: Europhysics Letters **50**, 711 (2000)

17. R. N. Mantegna and H. E. Stanley: *Lévy Flights and Related Topics in Physics* edited by M. F. Shlesinger, G. M. Zaslavsky, and U. Frisch (Springer, Berlin 1995) pp. 300–312
18. R. N. Mantegna and H. E. Stanley: Physica A **254**, 77 (1998)
19. B. Podobnik, P. Ch. Ivanov, Y. Lee, and H. E. Stanley: Europhysics Letters **52**, 491 (2000)
20. P. Ch. Ivanov, B. Podobnik, Y. Lee, and H. E. Stanley: *Truncated Lévy Process with Scale-Invariant Behavior* [Proc. NATO Advanced Research Workshop on Application of Physics in Economic Modeling, Prague 8–10 February 2001] Physica A **299**, 154 (2001)
21. J. A. Skjeltorp: Physica **283**, 486 (2001)
22. T. Lux: Appl. Finan. Econ **6**, 463 (1996)
23. P. Gopikrishnan, M. Meyer, L. A. N. Amaral, and H. E. Stanley: Eur. Phys. J. B **3**, 139 (1998)
24. V. Plerou, P. Gopikrishnan, L. A. N. Amaral, M. Meyer, and H. E. Stanley: Phys. Rev. E **60**, 6519 (1999)
25. P. Gopikrishnan, V. Plerou, L. A. N. Amaral, M. Meyer, and H. E. Stanley: Phys. Rev. E **60**, 5305 (1999)
26. B. Gutenberg and C. F. Richter: *Seismicity of the Earth and Associated Phenomenon* (2nd Edition, Princeton University Press, Princeton 1954)
27. D. L. Turcotte: *Fractals and Chaos in Geology and Geophysics*, (Cambridge Univ. Press, Cambridge 1992)
28. J. B. Rundle, D. L. Turcotte, and W. Klein: *Reduction and Predictability of Natural Disasters* (Addison-Wesley, Reading MA 1996)
29. K. Matia, L. A. N. Amaral, S. Goodwin, and H. E. Stanley: Phys. Rev. E Rapid Communications **66**, 045103 (2002)
30. V. Plerou, P. Gopikrishnan, L. A. N. Amaral, X. Gabaix, and H. E. Stanley: Phys. Rev. E (Rapid Communications) **62**, 3023 (2000)
31. P. Gopikrishnan, V. Plerou, X. Gabaix, and H. E. Stanley: Phys. Rev. E (Rapid Communications) **62**, 4493 (2000)
32. S. Ghashgaie, W. Breymann, J. Peinke, P. Talkner and Y. Dodge: Nature, **381**, 767 (1996)
33. R. N. Mantegna and H. E. Stanley: Nature **383**, p.587 (1996)
34. R. N. Mantegna and H. E. Stanley: Physica A **239**, 255 (1997)
35. V. Plerou, P. Gopikrishnan, X. Gabaix, and H. E. Stanley: Phys. Rev. E **66**, 027104-1 (2002)
36. V. Plerou, P. Gopikrishnan, and H. E. Stanley: Nature, **420**, cond-mat/0111349 (2002)
37. M. L. Mehta: *Random Matrices* (Academic Press, Boston 1991)
38. T. Guhr, A. Muller-Groeling, and H. A. Weidenmüller: Phys. Reports **299**, 189 (1998)
39. E. P. Wigner: Ann. Math. **53**, 36 (1951)
40. E. P. Wigner: Proc. Cambridge Philos. Soc. **47**, 790 (1951)
41. E. P. Wigner: *Conference on Neutron Physics by Time-of-flight* (Gatlinburg, Tennessee 1956) pp. 59-70
42. M. L. Mehta and F. J. Dyson: J. Math. Phys. **4**, 713 (1963)
43. F. J. Dyson: Revista Mexicana de Fisica **20**, 231 (1971)
44. A. M. Sengupta and P. P. Mitra: Phys. Rev. E **60**, 3389 (1999)
45. L. Laloux, P. Cizeau, J.-P. Bouchaud and M. Potters: Phys. Rev. Lett. **83**, 1469 (1999)

46. V. Plerou, P. Gopikrishnan, B. Rosenow, L. A. N. Amaral, and H. E. Stanley: Phys. Rev. Lett. **83**, 1471 (1999)
47. P. Gopikrishnan, B. Rosenow, V. Plerou, and H. E. Stanley: Phys. Rev. E Rapid Communications **64**, 035106-1 (2001)
48. V. Plerou, P. Gopikrishnan, B. Rosenow, L.A.N. Amaral, T. Guhr, and H. E. Stanley: Phys. Rev. E **65**, 066126 (2002)
49. B. Rosenow, V. Plerou, P. Gopikrishnan, and H. E. Stanley: Europhysics Letters **59**, 500 cond-mat/0111537 (2002)
50. L. A. N. Amaral, S. V. Buldyrev, S. Havlin, H. Leschhorn, P. Maass, M. A. Salinger, H. E. Stanley, and M. H. R. Stanley: J. Phys. I France **7**, 621 (1997)
51. S. V. Buldyrev, L. A. N. Amaral, S. Havlin, H. Leschhorn, P. Maass, M. A. Salinger, H. E. Stanley, and M. H. R. Stanley: J. Phys. I France **7**, 635 (1997)
52. L. A. N. Amaral, S. V. Buldyrev, S. Havlin, M. A. Salinger and H. E. Stanley: Phys. Rev. Lett. **80** 1385 (1998)
53. J. Sutton: *The variance of firm growth rates: the 'scaling' puzzle*, (working paper, London School of Economics 2001)
54. G. Bottazzi, G. Dosi, M. Riccaboni and F. Pammolli: *The scaling of growth processes and the dynamics of business firms* (Economics Department at University of Siena and Pisa San' Anna School of Advanced Studies working paper, based on Invited Talk at the 56th ESEM Econometric Society meeting, Lausanne, 26-29 Aug 2001)
55. H. E. Stanley, L. A. N. Amaral, P. Gopikrishnan, V. Plerou and M. A. Salinger: *Scale invariance and universality in economic phenomena* (working paper, based on Invited Talk at the 56th ESEM Econometric Society meeting, Lausanne 26-29 Aug 2001)
56. H. Takayasu and K. Okuyama: Fractals **6**, 67 (1998)
57. Y. Lee, L. A. N. Amaral, D. Canning, M. Meyer, and H. E. Stanley: Phys. Rev. Letters **81**, 3275 (1998)
58. V. Plerou, L. A. N. Amaral, P. Gopikrishnan, M. Meyer, and H. E. Stanley: Nature **400**, 433 (1999)
59. T. Keitt and H. E. Stanley: Nature **393**, 257 (1998)

Semiparametric Modeling of Stochastic and Deterministic Trends and Fractional Stationarity

Jan Beran[1], Yuanhua Feng[1], Günter Franke[2], Dieter Hess[2], and Dirk Ocker[3]

[1] Department of Mathematics and Statistics
[2] Department of Economics
University of Konstanz, D-78457 Konstanz, Germany
[3] Schweizer Verband der Raiffeisenbanken, CH-9001 St.Gallen

Abstract. The distinction between stationarity, difference stationarity, deterministic trends as well as between short- and long-range dependence has a major impact on statistical conclusions, such as confidence intervals for population quantities or point and interval forecasts. *SEMIFAR* models introduced by [6] provide a unified approach that allows for simultaneous modelling of and distinction between deterministic trends, difference stationarity and stationarity with short- and long-range dependence. In this paper, recent results on the *SEMIFAR* models are summarized and their potential usefulness for economic time series analysis is illustrated by analyzing several commodities, exchange rates, the volatility of stock market indices and some simulated series. Predictions combine stochastic prediction of the random part with functional extrapolation of the deterministic part.

1 Introduction

Many economic time series exhibit apparent local or global 'trends'. A large number of methods for dealing with trends under specific assumptions are described in the literature (see e.g. standard time series books, such as [16]; [37]. Essentially, models for trends can be classified as either (1) deterministic or (2) stochastic. A deterministic trend is described by a deterministic function $g(t)$, whereas a stochastic trend is generated by a purely stochastic nonstationary process such as random walk, (fractional) Brownian motion or an integrated ARIMA process. As a third possibility, local "spurious" trends can be generated by stationary processes with long-range dependence, such as stationary fractional ARIMA models. Statistical inference about population quantities and statistical forecasts are greatly influenced by our decision about the type of the 'trend' generating mechanism. For instance, for a stationary series, forecasts of a conditional expected value converge to the sample mean, with increasing forecasting horizon, and the width of forecast intervals is asymptotically constant. In contrast, for difference stationary series, forecasts converge to the last observation and the width of forecast intervals diverges to infinity. Forecasts for time series with a deterministic trend require reliable trend extrapolation which can usually not be trusted beyond a small forecasting horizon. On a finer scale, the rate at which forecast intervals converge to the asymptotic width (for stationary processes) or diverge to infinity (for difference stationary processes) depends on the fractional differencing parameter (see Sect. 4).

In practical applications, it is often very difficult to find the "right" model and, in particular, to decide whether a series is stationary, has a deterministic or stochastic trend, or whether there may be long-range correlations. In fact, often, a combination of these may be present. To resolve this problem, [6] introduced the so-called *SEMIFAR* (semiparametric fractional autoregressive) models. These models provide a unified data-driven semiparametric approach that allows for simultaneous modelling of and distinction between deterministic trends, stochastic trends and stationary short- and long-memory components. Within the given framework, the approach helps the data analyst to decide which components are present in the observed data. In this paper, recent results on *SEMIFAR* models ([6], [9], [8]) are summarized and their application to economic time series is discussed.

Briefly speaking, a *SEMIFAR* model is a fractional stationary or non-stationary autoregressive model with a nonparametric deterministic trend. This extends Box-Jenkins ARIMA models ([10]), by using a fractional differencing parameter $d > -0.5$, and by including a nonparametric trend function g. The trend function can be estimated, for example, by kernel smoothing. The parameters may be estimated by an approximate maximum likelihood method introduced in [5]. Note in particular that, with this method the integer differencing parameter is also estimated from the data. A data-driven algorithm for estimating *SEMIFAR* models, which is a mixture of these two approaches, was introduced in [6]. Clearly, as any statistical method, the analysis by *SEMIFAR* models has to be accompanied by appropriate subject-specific considerations.

The paper is organized as follows. The model is defined in Sect. 2. Estimation issues are discussed in Sect. 3, especially nonparametric estimation of the trend and the method for estimating the parameters characterizing the stochastic component of the process. Forecasting with *SEMIFAR* models is described in Sect. 4. The application of *SEMIFAR* models to economic time series is discussed in Sect. 5. In particular, we discuss modelling and forecasting commodities and exchange rates, and modelling the volatility of stock market indices. Also, four simulated series are analyzed to illustrate the usefulness of the method for cases where the answer is known. (A broader simulation study is reported in [6]). Some final remarks are given in Sect. 6.

2 The Model

2.1 Definition

A *SEMIFAR* model is a Gaussian process Y_i with an existing smallest integer $m \in \{0, 1\}$ such that

$$\phi(B)(1-B)^{\delta}\{(1-B)^{m}Y_i - g(t_i)\} = \epsilon_i, \tag{1}$$

where $t_i = (i/n)$, $\delta \in (-0.5, 0.5)$, g is a smooth function on $[0,1]$, B is the backshift operator, $\phi(x) = 1 - \sum_{j=1}^{p} \phi x^{j}$ is a polynomial with roots outside the unit circle and ϵ_i $(i = ..., -1, 0, 1, 2, ...)$ are iid zero mean normal with var $(\epsilon_i) = \sigma_\epsilon^2$. Here, the fractional difference $(1-B)^\delta$ ([25], [30]) is defined by

$$(1-B)^\delta = \sum_{k=0}^{\infty} b_k(\delta) B^k \tag{2}$$

with

$$b_k(\delta) = (-1)^k \frac{\Gamma(\delta+1)}{\Gamma(k+1)\Gamma(\delta-k+1)} . \tag{3}$$

2.2 Intuitive Explanation of the Definition

The motivation for this definition can be summarized as follows: We wish to have a model that may be decomposed into an arbitrary deterministic (possibly zero) trend and a random component that may be stationary or difference stationary. Moreover, short-range and long-range dependence as well as antipersistence should be included. Here, long-range dependence is defined as follows (see, e.g. [35]; [14]; [27]; [31]; [4] and references therein): A stationary process Y_i with autocovariances $\gamma(k) = \text{cov}(Y_i, Y_{t+k})$ is said to have long-range dependence, if the spectral density $f(\lambda) = (2\pi)^{-1} \sum_{k=-\infty}^{\infty} \exp(ik\lambda)\gamma(k)$ has a pole at the origin of the form

$$f(\lambda) \sim c_f |\lambda|^{-\alpha} \ (|\lambda| \to 0) \tag{4}$$

for a constant $c_f > 0$ and $\alpha \in (0,1)$, where " $\sim$ " means that the ratio of the left and right hand sides converges to one. In particular, this implies that, as $k \to \infty$, the autocovariances $\gamma(k)$ are proportional to $k^{\alpha-1}$ and hence their sum is infinite. On the other hand, a stationary process is called antipersistent, if (4) holds with $\alpha \in (-1,0)$. This implies that the sum of all autocovariances is zero, i.e. $\sum_{k=-\infty}^{\infty} \gamma(k) = 0$. Note that for usual short-memory processes, such as stationary ARMA processes, (4) holds with $\alpha = 0$, and the autocovariances sum up to a nonzero finite value.

The reason for including long-memory and anti-persistence is that for traditional ARIMA models an extreme choice has to be made between taking no or the first difference. The result of this dichotomy is that for many data sets, taking no difference is not enough (i.e. the series seems nonstationary), but taking the first difference leads to overdifferencing. The latter often results in a large negative lag-one correlation for the differenced data. To avoid this and to model slowly decaying correlations, [30]) and [25] introduced fractional ARIMA processes. However, there, the differencing parameter d is restricted to the stationarity region $(-1/2, 1/2)$. In a direct extension, [5] defines an arbitrary differencing parameter $d > -1/2$ such that $(1-B)^m Y_t$ is a stationary fractional ARIMA(p, δ, q) process, $m = [d+1/2]$ is the integer part of $d+1/2$ and $\delta = d-m$. This corresponds to (1) with a constant function $g \equiv \mu$. Since the integer differencing parameter m assumes integer values only and the fractional differencing parameter δ is in $(-1/2, 1/2)$, both differencing parameters can be recovered uniquely from the 'overall differencing parameter' $d = m + \delta$. If $d > 1/2$, then we have a nonstationary fractional ARIMA process. It should be noted, in particular, that this parameterisation allows for maximum likelihood estimation of d. Thus not only δ, but also m can be estimated from the data and confidence intervals can be given for both differencing parameters (see [5]).

Finally, SEMIFAR models extend the definition of fractional ARIMA models with arbitrary d by including an arbitrary deterministic trend function g. (For simplicity only cases with $q = 0$ (i.e. no moving average terms) are considered. An extension to $q > 0$, which may be called 'SEMIFARIMA models', is obvious.) The definition of SEMIFAR models includes all the desired cases mentioned above. In particular, setting $\delta = 0$ and $g(t) = \mu$, we obtain classical Box-Jenkins ARIMA models. For $g = 0$, and $m = 0$ we have stationary fractional ARIMA models as defined in [30] and [25].

More specifically, for SEMIFAR models, $Z_i = \{(1 - B)^m Y_i - g(t_i)\}$ is a stationary fractional autoregressive process. Thus, the spectral density of Z_i is proportional to $|\lambda|^{-2\delta}$ at the origin so that the process $\{(1 - B)^m Y_i - g(t_i)\}$ has long-memory if $\delta > 0$, antipersistence if $\delta < 0$ and short memory if $\delta = 0$. (1) generalizes stationary fractional AR-processes to the nonstationary case, including difference stationarity and deterministic trend. The following special cases are thus included in (1):

(a) Y_t =no deterministic trend + stationary process with short- or long-range dependence, or antipersistence;
(b) Y_t = deterministic trend + stationary process with short- or long-range dependence, or antipersistence;
(c) Y_t =no deterministic trend + difference-stationary process, whose first difference has short- or long-range dependence, or antipersistence;
(d) Y_t = deterministic trend + difference-stationary process, whose first difference has short- or long-range dependence, or antipersistence.

Simulated time series for these special cases are shown in Fig. 1, where Figs. 1a to 1d correspond to case (a) trough (d), respectively. A full description of the models used in figures 1a to d is given in Sect. 5.2.

2.3 Some Economic Motivation

Since the estimation of the SEMIFAR-model is purely data-driven, there exists a danger that the estimated model is inconsistent with economic reasoning. If this happens to be true, then the estimated model and the economic reasoning are called into question. In the following, we will briefly discuss some economic models which can explain short and long-term dependence in time series of prices of commodities and financial securities.

The implications of pricing models necessarily depend on the assumptions made. Many models assume perfect markets and perfectly rational economic agents. A basic requirement for any viable model is that it precludes arbitrage. A market can be arbitrage-free only if all prices for state-contingent claims are positive and finite. Let S_t be the price of some security at date t. For example, consider a stock whose price may be considered the risk-adjusted present

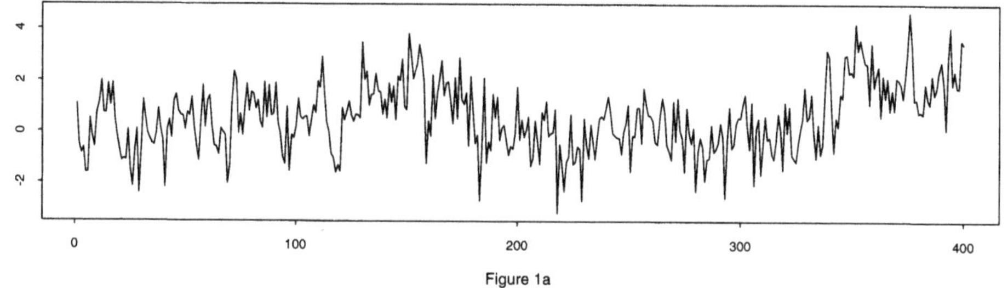

Figure 1a

Second simulated series

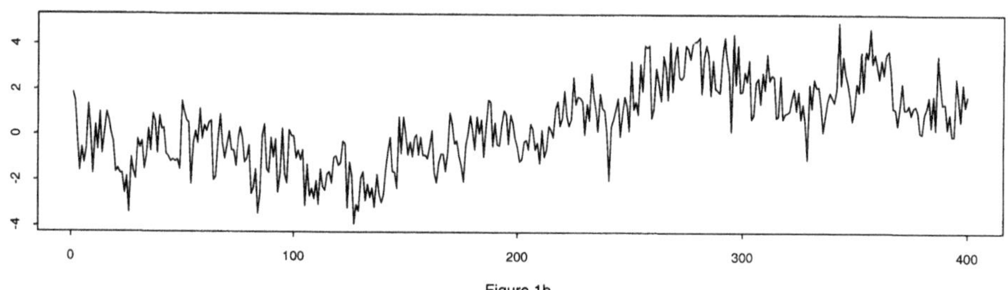

Figure 1b

Third simulated series

Figure 1c

Fourth simulated series

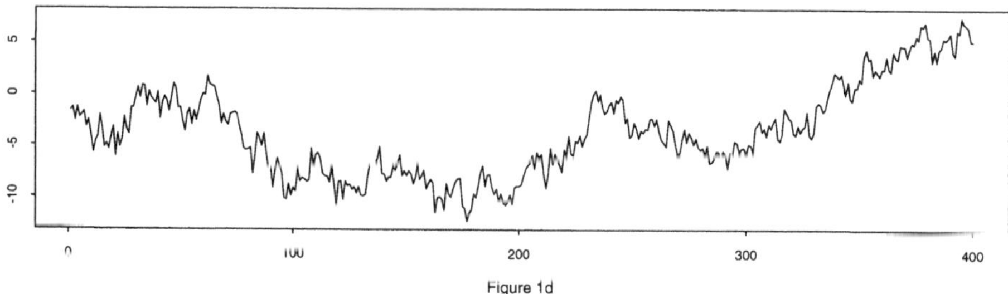

Figure 1d

Fig. 1. Simulated series

value of future dividends. Suppose that there exist exogenous shocks at date 0 which increase (reduce) all future dividends d_t, $(t = 1, 2, \ldots)$ by finite amounts. Then, given sufficiently low discount rates, the stock price would change by an infinite amount. Hence, the market would not satisfy the no-arbitrage requirement ([34]). As Mandelbrot points out, fractional Brownian motions, which are typical stochastic models with long-range dependence or antipersistence (in the increment process), do not rule out these cases.

Another violation of the no-arbitrage requirement is obtained if the short run-autocorrelation of price changes is very high or very low. Then an observed price change would permit an almost riskless forecast of the price change over the next time period which then could be arbitraged against a risk-free asset. [39] also proves the existence of arbitrage opportunities in fractional Brownian motions, but he also shows that a slight modification of the model suffices to rule out these opportunities. Note, in particular, that according to Mandelbrot's definition, arbitrage exists for all long-memory processes whereas this is not the case according to Rogers' definition. Thus, the answer to the question whether arbitrage is possible depends on which definition of arbitrage is used.

Even if arbitrage opportunities do not exist, the economist wonders how any short or long range dependencies in price series might be explained. [40] has shown that prices must follow a random walk in a risk neutral world with a non-random risk-free interest rate. For simplicity, consider an asset with an exogenously given random price S_T at date T. Then in a risk neutral world with homogeneous expectations of economic agents, the forward price S_t^f of the asset at date t equals $E_t[S_T]$, i.e. the conditional expectation of S_T. The forward price eliminates, by definition, the discounting effects of the risk-free interest rates. Since the conditional expectation of S_T follows a random walk without drift, any dependencies in forward price changes are ruled out.

In a risk-averse world with a frictionless complete market, there exists a unique forward pricing kernel $\phi_{t,T}(S_T)$ at date t $(t < T)$, by which the forward price S_t^f can be derived. We have $E[\phi_{t,T}(S_T)] = 1$ and $S_T^f = E_t[\phi_{t,T}(S_T)S_T]$. Then there still exist cases in which dependencies in forward price changes do not exist. Suppose, for example, that $E_t(S_T)$ follows a standard geometric Brownian motion without drift. Then $ln(S_{t+1}^f/S_t^f)$ follows a standard geometric Brownian motion with drift if and only if the forward pricing kernel has constant elasticity $\eta_{t,T}$, i.e. if $dln\phi_{t,T}/d\ln S_T = \eta_{t,T}, \forall t$ ([20]). Now suppose that the elasticity depends on S_T, holding the current forward price S_0^f constant. Suppose that $d\eta_{t,T}/dS_T < 0$ which may be thought of as "declining relative risk aversion of the market". Then the variance of the forward price S_t^f increases and the log returns $\ln(S_{t+1}^f/S_t^f)$ are negatively autocorrelated. In the case of increasing elasticity of the pricing kernel, the variance of the forward price declines relative to the constant elasticity case, but autocorrelation of log returns still is negative. The intuition behind this result is straightforward. Whenever the forward price S_t^f is higher or lower, relative to the constant elasticity case which implies zero autocorrelation, then given S_0^f and the distribution of S_T, a lower return in one period must be compensated by a higher return in the residual period. This implies short and long-range negative autocorrelation. Hence, in this framework, antipersistence may exist whereas it is difficult to argue in favor of positive autocorrelation.

Of course, real markets are not perfect. Introducing asymmetric information broadens the spectrum of return processes. Insiders, for example, attempt to exploit their information privilege by strategic trading which leads to a gradual price adjustment and, thus, short range positive autocorrelation of returns. The same autocorrelation is to be expected in the case of positive feedback trading. Then agents observe a price increase and place additional buy orders since they expect a further price increase. Finally, if large investors buy or sell consecutively small portions of a rather illiquid security, this induces positive autocorrelation.

[34] suggests that economic agents have a finite foresight horizon. This may imply various dependencies in returns. If agents, for example, ignore effects of a shock on a corporation's profits beyond some horizon, then these effects will gradually be taken into consideration and generate long-range dependent price changes. Alternatively, if agents naively extrapolate growth rates of profits over very long time horizons, this extrapolation error will gradually be corrected with corresponding gradual price changes. Such behavior might explain the well documented winner-loser effect which states that stocks with high returns over the last years tend to generate low returns over the next years and vice versa. Also cyclical macroeconomic factors tend to generate cyclical stock price behavior given either a short foresight horizon or naive extrapolative behavior. Otherwise it would be useless to distinguish between cyclical and noncyclical stocks.

Similar considerations apply to commodities for which short and long range-contracts are traded. The economic analysis of commodity prices becomes more complicated since durability of commodities and side effects of storing commodities summarized in convenience yields come into play.

Finally, long memory in aggregated indices may be a result of aggregation. As was shown, for instance, in [23], adding up a large number of time series can lead (asymptotically) to a series with long-range dependence, even if the individual series do not exhibit any long memory.

Yet, adding these real world aspects to purely theoretical models should not be understood as providing unlimited freedom to all kinds of short and long-range dependencies in time series of prices. The ultimate purpose of research is to find out price processes that are observable with sufficient reliability and are grounded on solid economic reasoning. *SEMIFAR* models provide a rather general class of models to do the empirical job. The economists should use the empirical insights for developing sensible economic models.

3 Estimation of *SEMIFAR* Models – A Review

Estimation of *SEMIFAR* models includes (1) nonparametric estimation of the trend component and (2) estimation of the parameters characterizing the stochastic component. This section summarizes theoretical results on the proposed kernel estimator of the trend function and approximate maximum likelihood estimator of the parameters without proofs. See [6] and [8] for details. A data-driven algorithm for estimating the whole model is also briefly described.

3.1 Kernel Estimation of the Trend Function

The problem of estimating g from data given by

$$Y_i = g(t_i) + X_i \tag{5}$$

has been considered by various authors for the case where the error process X_t is stationary with (i) short-range dependence, i.e. (4) holds with $\alpha = 0$ (see e.g. [13], [2], [26] and [29]) or (ii) long-range dependence, i.e. $0 < \alpha < 1$ (see e.g. [26]; [15] and [38]). For *SEMIFAR* models defined by (1), the cases (i) and (ii) are obtained by setting $m = 0$ and $\delta = \alpha/2 = 0$ (case (i)), or $m = 0$ and $\delta \in (0, 1/2)$ (case (ii)) respectively. For $m = 1$, the same is true for the first difference $Y_i - Y_{i-1}$. (Note, however, that for *SEMIFAR* models, $m \in \{0, 1\}$ is an unknown parameter!) In addition to cases (i) and (ii), definition 1 also includes the antipersistent case, i.e. $\delta < 0$ so that the spectral density f of Y_i (or $Y_i - Y_{i-1}$ respectively) converges to zero at the origin. The theorem below extends previous results on kernel estimation to the anti-persistent case, and gives formulas for the mean squared error and the optimal bandwidth that are valid for the whole range $\delta \in (-0.5, 0.5)$.

For estimating g by kernel smoothing, symmetric polynomial kernels of the form $K(x) = \{\sum_{l=0}^{r} \alpha_l x^{2l}\}\mathbb{I}_{\{|x|\leq 1\}}$ (see e.g. [21]) will be used. If (5) holds, then, for a given bandwidth $b > 0$ and $t \in [0, 1]$, the kernel estimate of g is defined by

$$\hat{g}(t) = K_b \diamond y(n) = \frac{1}{nb} \sum_{i=1}^{n} K(\frac{t - t_i}{b}) Y_i \tag{6}$$

where $y(n) = (Y_1, ..., Y_n)$. Let $n_0 = [nt]$, $n_1 = [nb]$ and $0 < \Delta < 0.5$, the following notations will be used:

$$V_n(\theta, b) = (nb)^{-1-2\delta} \sum_{i,j=n_0-n_1}^{n_0+n_1} K\left(\frac{t - t_i}{b}\right) K\left(\frac{t - t_j}{b}\right) \gamma(i - j), \tag{7}$$

$$I(g'') = \int_{\Delta}^{1-\Delta} [g''(t)]^2 dt \tag{8}$$

and

$$I(K) = \int_{-1}^{1} x^2 K(x) dx. \tag{9}$$

The following result is obtained under the assumption that (5) holds and that g is at least twice continuously differentiable (see [6] for the proof).

Theorem 1. *Let $b_n > 0$ be a sequence of bandwidths such that $b_n \to 0$ and $nb_n \to \infty$. Then, under the stated assumptions and δ in (1) in the interval (-0.5,0.5), we have*

(i) Bias:

$$E[\hat{g}(t) - g(t)] = b_n^2 \frac{g''(t)I(K)}{2} + o(b_n^2) \tag{10}$$

uniformly in $\Delta < t < 1 - \Delta$;

(ii)

$$\lim_{n \to \infty} V_n(\theta, b_n) = V(\theta) \tag{11}$$

where $0 < V(\theta) < \infty$ is a constant;

(iii) Variance:

$$(nb_n)^{1-2\delta} \operatorname{var}(\hat{g}(t)) = V(\theta) + o(1) \tag{12}$$

uniformly in $\Delta < t < 1 - \Delta$;

(iv) IMSE: The integrated mean squared error in $[\Delta, 1 - \Delta]$ is given by

$$\int_{\Delta}^{1-\Delta} E\{[\hat{g}(t) - g(t)]^2\} dt = IMSE_{asympt}(n, b_n) + o(\max(b_n^4, (nb_n)^{2\delta-1}))$$

$$= b_n^4 \frac{I(g'')I^2(K)}{4} + (nb_n)^{2\delta-1} V(\theta) + o(\max(b_n^4, (nb_n)^{2\delta-1})) \tag{13}$$

(v) Optimal bandwidth: *The bandwidth that minimizes the asymptotic IMSE is given by*

$$b_{opt} = C_{opt}\, n^{(2\delta-1)/(5-2\delta)} \tag{14}$$

where

$$C_{opt} = C_{opt}(\theta) = \left(\frac{(1 - 2\delta)V(\theta)}{I(g'')I^2(K)} \right)^{1/(5-2\delta)}. \tag{15}$$

Similar results can be obtained for kernel estimates of derivatives of g. For instance, the second derivative can be estimated by $\hat{g}''(t) = n^{-1}b^{-3} \sum K((t_j - t)/b)Y_j$ where K is a symmetric polynomial kernel such that $\int K(x)dx = 0$ and $\int K(x)x^2 dx = 2$. By analogous arguments, the optimal bandwidth is then of the order $O(n^{(2\delta-1)/(9-2\delta)})$.

Simple explicit formulas for $V(\theta)$ can be given for $\delta = 0$ and $\delta > 0$ as follows (see e.g. [26]):

$$V(\theta) = 2\pi c_f \int_{-1}^{1} K^2(x)dx, \ (\delta = 0), \tag{16}$$

$$V(\theta) = 2c_f \Gamma(1 - 2\delta) \sin \pi\delta \int_{-1}^{1} \int_{-1}^{1} K(x)K(y)|x - y|^{2\delta-1} dx dy, \ (\delta > 0). \tag{17}$$

In order to obtain a similar formula for $\delta < 0$, at a point x let $K(y) = \sum_{l=0}^{r} \beta_l(x)(x - y)^l =: K_0(x) + K_1(x - y)$, where $K_0(x) = \beta_0(x)$, $K_1(x - y) = \sum_{l=1}^{r} \beta_l(x)(x - y)^l$. Then we have (see [8])

$$V(\theta) = 2c_f \Gamma(1 - 2\delta) \sin(\pi\delta) \int_{-1}^{1} K(x) \times$$

$$\left\{ \int_{-1}^{1} K_1(x - y)|x - y|^{2\delta-1} dy - \int_{|y|>1} K_0(x)|x - y|^{2\delta-1} dy \right\} dx \tag{18}$$

for $\delta < 0$. For the box-kernel (i.e. $r = 0$), formulas (16), (17) and (18) give the same result

$$V = \frac{2^{2\delta} c_f \Gamma(1-2\delta)\sin(\pi\delta)}{\delta(2\delta+1)} \tag{19}$$

with $V(0) = \lim_{\delta\to 0} V(\delta) = \pi c_f$ (see corollary 1 in [6]).

3.2 Maximum Likelihood Estimation

The maximum likelihood estimation proposed by [5] for a constant function $g = \mu$ can be carried over directly to *SEMIFAR* models with time-deterministic trend functions (see [6]). In particular, from the 'overall differencing parameter' $d = m+\delta$ both, the discrete differencing parameter m and the fractional differencing parameter δ can be recovered uniquely, since m can take on integer values only and δ is in $(-1/2, 1/2)$. Moreover, this parameterization allows for maximum likelihood estimation of d (and thus of δ *and* m) along with the autoregressive parameters and the trend function. Moreover, inference about the autoregressive parameters takes into account that m and δ were not known a priori.

Let $\theta^o = (\sigma^2_{\epsilon,o}, d^o, \phi^o_1, ..., \phi^o_p)^T = (\sigma^2_{\epsilon,o}, \eta^o)^T$ be the true unknown parameter vector in (1) where $d^o = m^o + \delta^o$, $-1/2 < \delta^o < 1/2$ and $m^o \in \{0,1\}$. Then

$$\begin{aligned}\phi(B)(1-B)^{\delta^o}\{(1-B)^{m^o}Y_i - g(t_i)\} &= \sum_{j=0}^{\infty} a_j(\eta^o)B^j[c_j(\eta^o)Y_i - g(t_i)]\\ &= \sum_{j=0}^{\infty} a_j(\eta^o)[c_j(\eta^o)Y_{i-j} - g(t_{i-j})],\end{aligned}$$

where the coefficients a_j and a_jc_j are obtained by matching the powers in B. Hence, Y_i admits an infinite autoregressive representation

$$\sum_{j=0}^{\infty} a_j(\eta^o)[c_j(\eta^o)Y_{i-j} - g(t_{i-j})] = \epsilon_i. \tag{20}$$

Let b_n $(n \in N)$ be a sequence of positive bandwidths such that $b_n \to 0$ and $nb_n \to \infty$ and define $\hat{g}(t_i) = \hat{g}(t_i; m)$ by

$$\hat{g}(t_i; 0) = K_{b_n} \diamond y(n), \tag{21}$$

and

$$\hat{g}(t_i; 1) = K_{b_n} \diamond Dy(n), \tag{22}$$

with $Dy(n) = (Y_2 - Y_1, Y_3 - Y_2, ..., Y_n - Y_{n-1})$. Consider now ϵ_i as a function of η. For a chosen value of $\theta = (\sigma^2_\epsilon, m+\delta, \phi_1, ..., \phi_p)^T = (\sigma^2_\epsilon, \eta)^T$, denote by

$$e_i(\eta) = \sum_{j=0}^{i-m-2} a_j(\eta)[c_j(\eta)Y_{i-j} - \hat{g}(t_{i-j}; m)] \tag{23}$$

the (approximate) residuals and by $r_i(\theta) = e_i(\eta)/\sqrt{\theta_1}$ the standardized residuals. Assuming that $\{\epsilon_i(\eta^o)\}$ are independent zero mean normal with variance $\sigma^2_{\epsilon,o}$, an approximate maximum likelihood estimator of θ^o is obtained by maximizing the approximate log-likelihood

$$l(Y_1, ..., Y_n; \theta) = -\frac{n}{2}\log 2\pi - \frac{n}{2}\log \sigma_\epsilon^2 - \frac{1}{2}n^{-1}\sum_{i=m+2}^{n} r_i^2 \tag{24}$$

with respect to θ and hence by solving the equations

$$\dot{l}(Y_1, ..., Y_n; \theta) = 0 \tag{25}$$

where $\dot{l}$ is the vector of partial derivatives with respect to θ_j $(j = 1, ..., p+2)$. More explicitly, $\hat{\eta}$ is obtained by minimizing

$$S_n(\eta) = \frac{1}{n}\sum_{i=m+2}^{n} e_i^2(\eta) \tag{26}$$

with respect to η and setting

$$\hat{\sigma}_\epsilon^2 = \frac{1}{n}\sum_{i=m+2}^{n} e_i^2(\hat{\eta}). \tag{27}$$

The result in [5] can be extended to *SEMIFAR* models [6]:

Theorem 2. *Let $\hat{\theta}$ be the solution of (26) and (27), and define $\theta^o_* = (\sigma^2_{\epsilon,o}, \eta^o_*)^T = (\sigma^2_{\epsilon,o}, \delta^o, \eta^o_2, ..., \eta^o_{p+1})^T$. This means that, $\theta^o_2 = d = m^o + \delta^o$ is replaced by $\theta^o_{2,*} = \delta^o$. Then, as $n \to \infty$,*

(i) $\hat{\theta}$ converges in probability to the true value θ^o;
(ii) $n^{\frac{1}{2}}(\hat{\theta} - \theta^o)$ converges in distribution to a normal random vector with mean zero and covariance matrix

$$\Sigma = 2D^{-1} \tag{28}$$

where

$$D_{ij} = (2\pi)^{-1}\left[\int_{-\pi}^{\pi} \frac{\partial}{\partial\theta_i}\log f(x)\frac{\partial}{\partial\theta_j}\log f(x)dx\right]|_{\theta=\theta^o_*}. \tag{29}$$

It should be noted that in theorem 2, both, the fractional differencing parameter δ and the integer differencing parameter m are estimated from the data. The asymptotic covariance matrix does not depend on m. Theorem 2 can be generalized to the case where the innovations ϵ_i are not normal, and satisfy suitable moment conditions.

Theorem 2 is derived under the assumption that the order $p = p_o$ of the autoregressive polynomial in (1) is known. In practice p_o needs to be estimated by applying a suitable model choice criterion. It can be shown, however, that consistency properties of model choice criteria, such as the BIC ([41]; [1]) and the HIC ([28]), are analogous to the case of stationary short-memory autoregressive processes ([6]):

Theorem 3. *Under the assumptions of theorem 2, let p_o be the true order of the polynomial ϕ in (1) and define*

$$\hat{p} = \arg\min\{AIC_\alpha(p); p = 0, 1, ..., L\} \tag{30}$$

where L is a fixed integer, $AIC_\alpha(p) = n \log \hat{\sigma}_\epsilon^2(p) + \alpha \cdot p$ and $\hat{\sigma}_\epsilon^2(p)$ is the maximum likelihood estimate of the innovation variance $\sigma_{\epsilon,o}^2$ using a SEMIFAR model with autoregressive order p. Moreover, define $\hat{\theta}$ by (26) and (27) with p set equal to $\hat{p}$. Suppose furthermore that α is at least of the order $O(2c \log\log n)$ for some $c > 1$. Then the results of theorem 2 hold.

Combining Theorems 1 through 3, It is straightforward to obtain from confidence intervals for the unknown parameter vector θ and the unknown trend function g, as well as for testing hypotheses about θ and g. Note, in particular, that the integer differencing parameter m is also estimated by maximum likelihood ($\hat{m}$ is equal to the integer part of $\hat{d} + 1/2$).

3.3 Estimation of the Whole Model

For estimating the whole model one needs a semiparametric data-driven algorithm combining the two estimation methods described above. An algorithm for the case where g is assumed to be equal to a constant μ is given in [5]. A data-driven algorithm for estimating the *SEMIFAR* model with a general trend function g was proposed by Beran in 1997 in the original, unpublished paper on the *SEMIFAR* model. What follows is a brief description of this algorithm.

The algorithm makes use of the fact that d is the only additional parameter, in addition to the autoregressive parameters, so that a systematic search with respect to d can be made. This algorithm can be adapted to the case where g is an unknown function, by replacing $\hat{\mu}$ by a kernel estimate of g. The optimal bandwidth can be estimated by an iterative plug-in method similar to the one in [29] and [38]. These authors consider the case of stationary errors, i.e. m is known to be equal to zero. The algorithm in [38] is as follows:

1. an initial bandwidth is defined;
2. a preliminary estimate of g is computed and subtracted from the observations;
3. the relevant parameters of the error process are estimated from the residuals;
4. the bandwidth is updated.

Steps 2 to 4 are repeated until the change in the bandwidth is below a predefined threshold. This algorithm has been extended to fitting *SEMIFAR* models ([6]). A detailed study on the consistency, rates of convergence and comparison of different iterative algorithms for *SEMIFAR* fitting will be given in a forthcoming paper.

4 *SEMIFAR* Forecasting

This section describes out-of-sample predictions of *SEMIFAR* processes. Let $Y_1, ..., Y_n$ be observations generated by a *SEMIFAR* model of order p with parameter vector $\theta = (\sigma_\epsilon^2, d, \phi_1, ..., \phi_p)^T$ (where $d = m + \delta$). The aim is to predict a future observation Y_{n+k} for some $k \in \{1, 2, 3, ...\}$. Denote by X_i a zero mean fractional AR process of order p with parameter vector $\theta_* = (\sigma_\epsilon^2, \delta, \eta_2, ..., \eta_{p+1})^T$, and define $t_{n+k} = (n + k)/n = t_n + k/n$. Then

$$Y_{n+k} = \mu(t_{n+k}) + U_{n+k} \tag{31}$$

with

$$\mu(t_{n+k}) = g(t_{n+k}), \quad U_{n+k} = X_{n+k} \tag{32}$$

if $m = 0$, and

$$\mu(t_{n+k}) = Y_n + \sum_{j=1}^{k} g(t_{n+j}), \quad U_{n+k} = \sum_{j=1}^{k} X_{n+j} \tag{33}$$

if $m = 1$. Thus, to predict Y_{n+k} from $Y_1, ..., Y_n$, two problems need to be solved:

1. *extrapolation* of the function $\mu(t)$ to $t = t_{n+k}$;
2. *prediction* of the stochastic component U_{n+k}.

4.1 Extrapolation of the Trend Function

Since for *SEMIFAR* models only general regularity conditions on g are imposed, the deterministic trend $g(t)$ may behave in an arbitrary way in the future. This is in contrast to parametric trend models. However, we may obtain the predictions of $\hat{g}(t_{n+j})$ for $j \in \{1, 2, ..., k\}$, for instance by a local constant or a local linear extension of $\hat{g}(t_n)$. $\hat{\mu}(t_{n+k})$ is obtained by inserting $\hat{g}(t_{n+k})$ in (32) or $\hat{g}(t_{n+j})$ for $j \in \{1, 2, ..., k\}$ in (33) (see [9]).

4.2 Prediction of the Stochastic Component

Note that $X_i = U_i = Y_i - g(t_i)$ for $m = 0$, and $X_i = U_i - U_{i-1} = Y_i - Y_{i-1} - g(t_i)$ for $m = 1$. Let $\gamma(k) = \text{cov}(X_i, X_{i+k})$ denote the autocovariances of X_i. Using the mean square criterion, the best linear predictor of U_{n+k} based on $Y_1, ..., Y_n$ is defined by $\hat{U}_{n+k} = \beta_{opt}^T X(n)$ where $X(n) = (X_1, ..., X_n)^T$ and the vector $\beta_{opt} = (\beta_1, ..., \beta_n)^T$ minimizes the mean squared prediction error $MSE = E[(U_{n+k} - \hat{U}_{n+k})^2]$. The values of β_{opt} and the corresponding optimal mean squared prediction error MSE_{opt} are given by ([9])

Theorem 4. *For all integers $r, s > 0$, define*

$$\gamma_r^{(s)} = [\gamma(r + s - 1), \gamma(r + s - 2), ..., \gamma(r)]^T, \tag{34}$$

$$\tilde{\gamma}_k^{(n)} = \sum_{j=1}^{k} \gamma_j^{(n-1)}, \tag{35}$$

and denote by $\Sigma_n = [\gamma(i-j)]_{i,j=1,\dots,n}$ *the covariance matrix of* $X(n)$*. Then, the following holds.*

i) If $m = 0$,

$$\beta_{opt} = \Sigma_n^{-1} \gamma_k^{(n)}, \tag{36}$$

$$MSE_{opt} = \gamma(0) - [\gamma_k^{(n)}]^T \Sigma_n^{-1} [\gamma_k^{(n)}]; \tag{37}$$

ii) If $m = 1$,

$$\beta_{opt} = \Sigma_n^{-1} \tilde{\gamma}_k^{(n)}, \tag{38}$$

$$MSE_{opt} = \sum_{s=-(k-1)}^{k-1} (k - |s|)\gamma(s) - [\tilde{\gamma}_k^{(n)}]^T \Sigma_n^{-1} [\tilde{\gamma}_k^{(n)}]. \tag{39}$$

Note in particular that, as $k \to \infty$, the MSE tends to a finite constant in the case of a stationary stochastic component ($m = 0$), whereas it diverges to infinity in the case of a nonstationary stochastic component ($m = 1$). More specifically we have ([9])

Corollary 3. *Define* $c_f = \lim_{\lambda \to 0} |\lambda|^{2\delta} f(\lambda)$ *where* f *is the spectral density of* X_i*, and let*

$$\nu(\delta) = \frac{2\Gamma(1 - 2\delta) \sin \pi\delta}{\delta(2\delta + 1)} \tag{40}$$

for $0 < |\delta| < 0.5$ *and* $\nu(0) = \lim_{\delta \to 0} \nu(\delta) = 2\pi$*. Then, as* $k \to \infty$*, the following holds:*

i) If $m = 0$,

$$MSE_{opt} \to \gamma(0) = \mathrm{var}(X_i); \tag{41}$$

ii) If $m = 1$,

$$MSE_{opt} \sim c_f \nu(\delta) k^{1+2\delta}. \tag{42}$$

Note in particular that, for $m = 1$ and $\delta < 0$, the MSE_{opt} diverges to infinity at a slower rate than in the case of a random walk (with $\delta = 0$). Similarly, for $m = 1$ and $\delta > 0$, the MSE_{opt} diverges faster to infinity.

4.3 Prediction Intervals

Results in theorem 4 and corollary 1 can be used to obtain prediction intervals for Y_{n+k} with $k \geq 1$. For known values of g and θ a $100(1-\alpha)$ percent prediction interval for Y_{n+k}, is given by

$$\hat{Y}_{n+k} \pm z_{\alpha/2} \sqrt{MSE_{opt}} \tag{43}$$

where $\hat{Y}_{n+k} = \mu(t_{n+k}) + \beta_{opt}^T X(n)$ and the values of β_{opt} and MSE_{opt} are obtained from theorem 1. If g and θ are estimated, the quantities in (43) are replaced by the corresponding estimated quantities.

5 Examples

In this section we provide some insight into the empirical validity of the *SEMIFAR* models by analyzing some price series and some index volatility series. Moreover, some simulation exercises demonstrate the model's capacity to find out the true properties of a time series.

5.1 Commodities and Exchange Rates

The data (Figs. 2a to d) include daily spot prices for copper (between January 2, 1997, and September 2, 1998, n=421), a monthly price series for cocoa beans (between January 1971 and September 1996, n=310), and two daily nominal exchange rates (between September 17, 1997, and August 4, 1998, n=221). The currencies are the Swiss Franc (chf) and the European Currency Unit (xeu). The data are expressed in US dollars per unit of the corresponding series. The log-transformation (natural logarithm) was applied to each series.

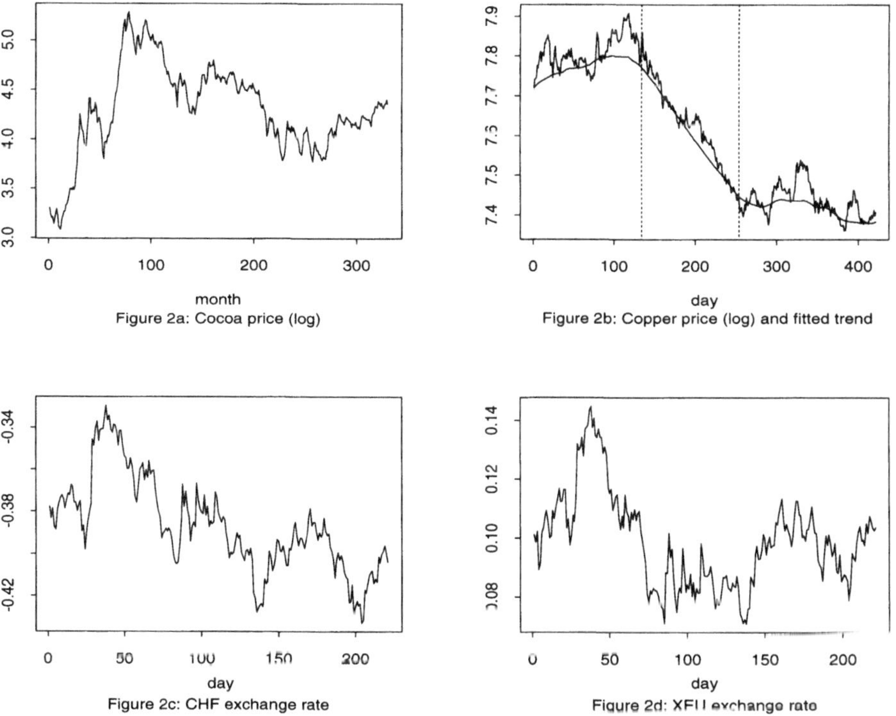

Fig. 2. Monthly prices for cocoa beans (Jan. 1971 - Sept. 1996, n=310), daily prices for copper (Jan. 2, 1997 - Sept. 2, 1998, n=421), daily nominal exchange rates for the Swiss Franc (log(USD/CHF)) and the European Currency Unit (log(USD/XEU)) (Sept. 17, 1997 - Aug. 4, 1998, n=221).

First, we fit *SEMIFAR* models to the observed series. Note in particular that, instead of continuously compounded returns (first difference in natural logarithm of the closing price for consecutive trading days/months), the *original series of observed* (log-)prices is considered. Thus, in contrast to the traditional approach, it is not assumed *a priori* that the first integer difference has to be taken to make the series stationary. Instead, the possibilities of stationarity, difference stationarity, deterministic trend, short memory, long memory and antipersistence are left open. It is then decided based on the data which combination of these components may be present.

There has been some discussion in the recent literature about possible unit root behaviour or long memory in financial time series. In view of this, it is interesting to see which hypothesis may be supported by fitting *SEMIFAR* models. Table 1 summarizes the essential features of the fitted models. The corresponding 95%-confidence intervals are given in brackets. The models were selected using the BIC.

Table 1. Estimation results

series	$\hat{d}$	95%-c.i. d	$\hat{\phi}_1$	95%-c.i. ϕ_1	significant trend
cocoa	.897	[.682,1.112]	.394	[.142, .646]	no
copper	.780	[.705,.855]	-	-	yes
chf	.913	[.810,1.016]	-		no
xeu	.870	[.767,.973]	-	-	no

The estimated value of d and the confidence intervals suggest that all series are nonstationary ($d > 1/2$). In addition, the unit roots hypothesis ($d = 1$) can not be rejected for cocoa and chf. On the other hand, for copper and the European Currency Unit, $d = 1$ is not contained in the 95%-confidence interval. Thus, for these data, taking the first (integer) difference would lead to overerdifferencing. Furthermore, there is substantial short-term dependence in the cocoa series in form of a strong AR(1) term.

Since in all cases the estimated value of m was one, testing the presence/absence of a deterministic trend can be done by testing $H_o : g \equiv 0$ against $H_a : g \not\equiv 0$. (Note that for $m = 1$, g is the trend function for the first difference.) The only series where H_o was rejected (at the 5% level) was copper. As one may expect (at least a posteriori), for this series, a significant trend is detected due to the relatively long descent in the middle part of the observed time period. The starting and end point of the time interval where $\hat{g}$ exceeded the critical bound are marked in Fig. 2b by two vertical lines. Note in particular that fitting a global linear trend would not be appropriate here. For the other three series, apparent local trends do not persist long enough, and can therefore be 'explained' as purely stochastic.

The satisfactory fits of the models are demonstrated by the q-q-plots and correlograms of the residuals in Figs. 3 and 4. Slight departure from normality (for the residuals) can be noticed for the exchange rate data. (Note, however,

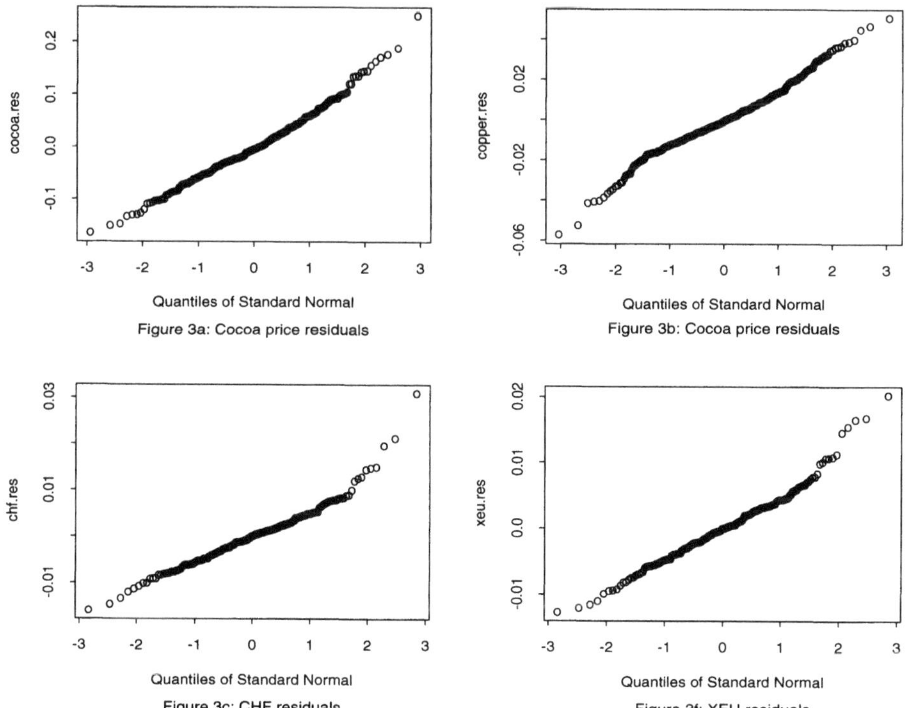

Fig. 3. Normal probability plots of SEMIFAR-residuals for the examples in Figs. 2a through d.

that normality of ϵ_t is not required for the theoretical results described above to hold.) Also, there is no strong evidence for ARCH (autoregressive conditional heteroskedasticity) errors in the correlograms of the squared residuals (Figs. 4e through h).

Second, we explore the reliability of forecasts. The $k-$steps ahead out-of-sample forecasts and 95%- and 99%-forecast intervals for $k = 1, 2, ..., 20$, using constant extrapolation of g, are displayed in Fig. 5. Overall, every future value was inside the 95% prediction interval. Observe also the weak US dollar in the exchange rate data during the last quarter of the period under consideration. Despite this sudden development, the future values were within the 95% prediction intervals. It should also be noted that, for $1/2 < \hat{d} < 1$, the width of forecast intervals diverges to infinity at a slower rate than under the unit root hypothesis $d = 1$. Thus, shorter forecast intervals are obtained than with unit root models, such as a random walk. For a detailed discussion see [9]. Clearly, as always with forecasting, sudden extreme structural changes in the behaviour of the data that have not occurred in the past cannot be foreseen (except perhaps with the help of additional information).

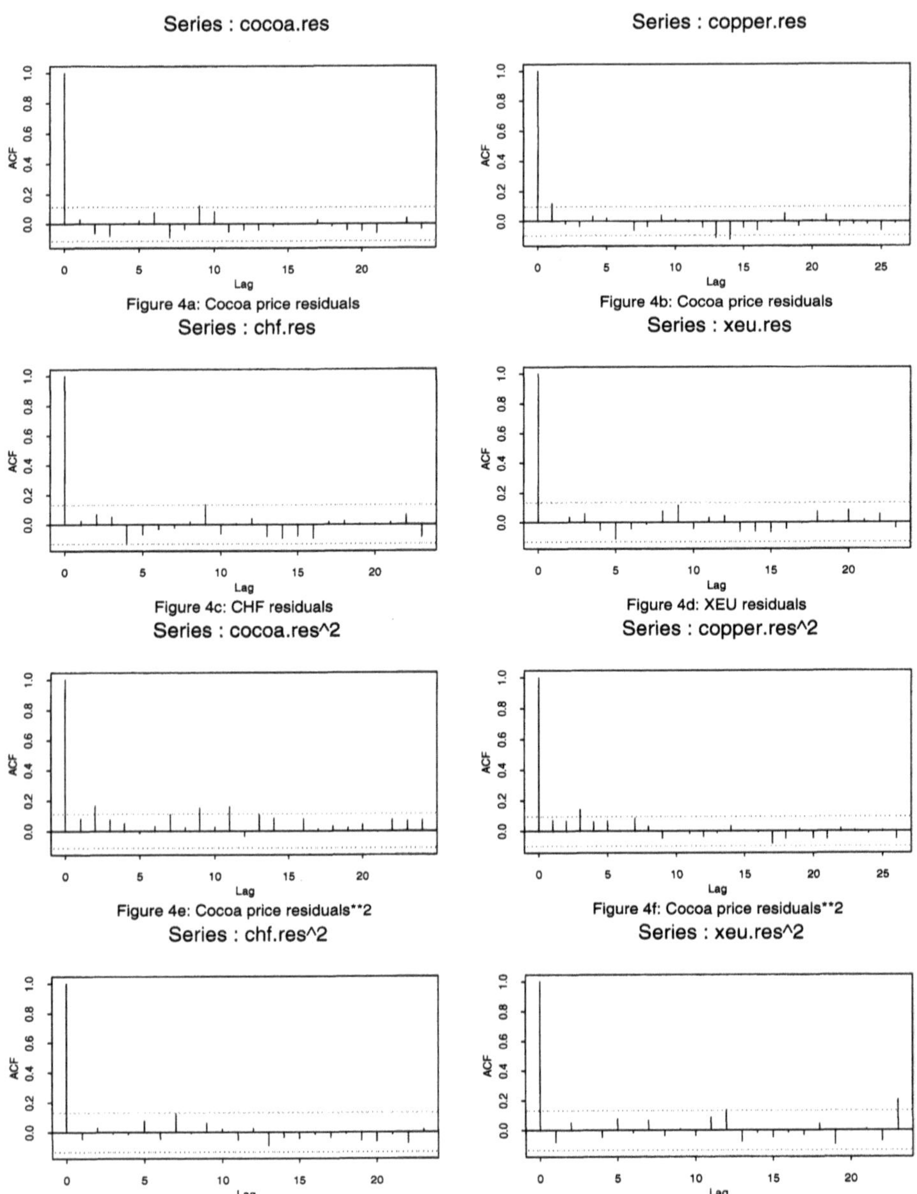

Fig. 4. Autocorrelations of SEMIFAR-residuals (figures 4a through d) and of the squared residuals (figures 4e through h) for the examples in figures 2a through d.

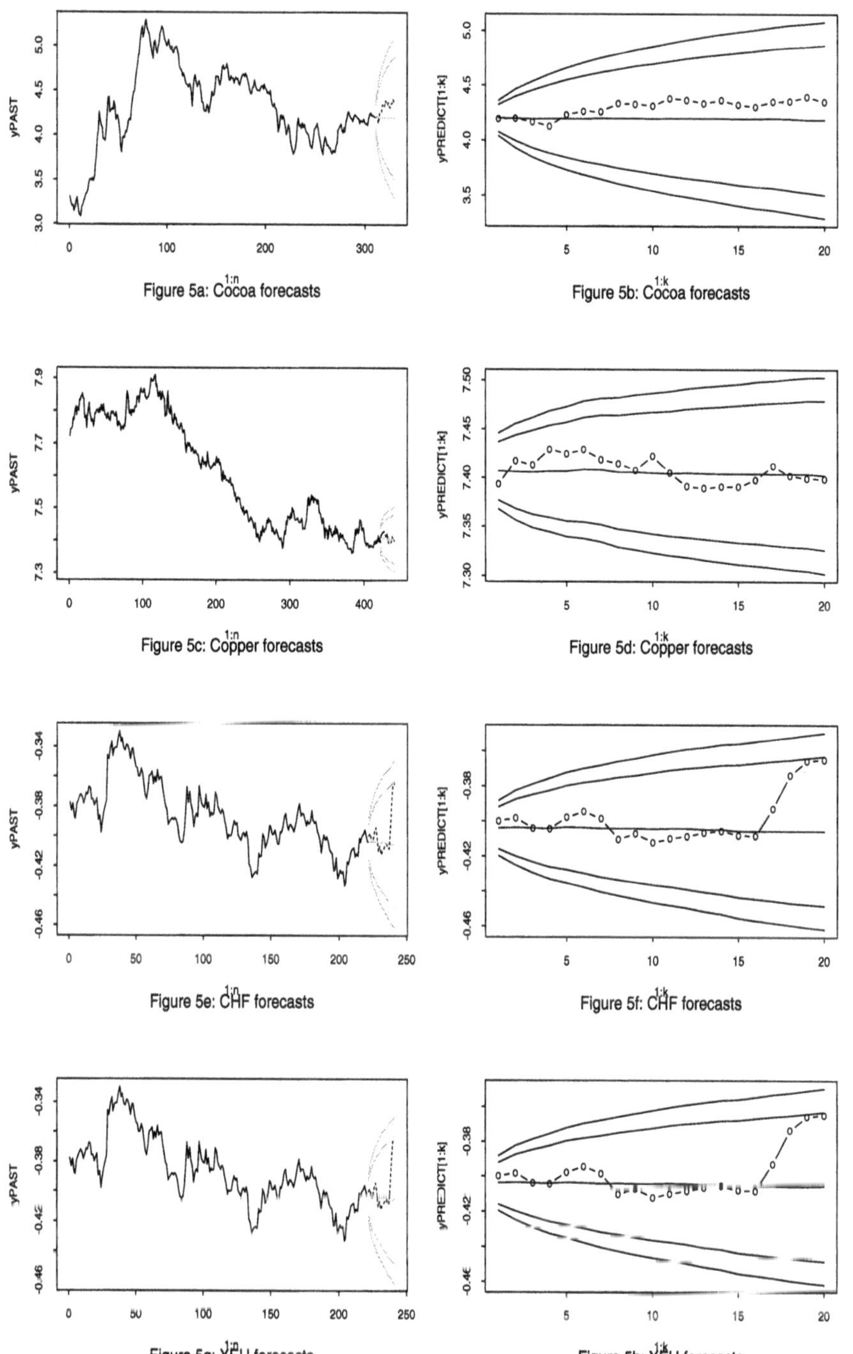

Fig. 5. Observed values with k-step ahead SEMIFAR forecasts and 95%- and 99%-forecast intervals for the examples in figures 2a through d. Figures 5b, d, f and h display close-ups of the forecasts and forecast intervals in figures 1a, c, e and g.

5.2 Volatility of Stock Market Indices

Figure 6a shows daily values of the DAX and the FTSE300 between January 2, 1992 and November 10, 1995 (weekdays only, excluding holidays). The first differences are given in Fig. 6b. Let I_t be the original index. To study volatility, we analyze the transformed absolute differences $Y_t = |I_t - I_{t-1}|^{\frac{1}{4}}$. The reason for taking the fourth root of the increments is that the marginal distribution of the resulting series is very close to normal (see the normal probability plots in Figs. 6c and d). A similar transformation approach is used, for instance, by [18], [24] and [24]. Ding and Granger found long range dependence in several volatility series that were defined in a similar way. The correlograms of Y_t in Figs. 6e and f do indeed indicate slowly decaying autocorrelations.

Applying the *SEMIFAR* method yields $\hat{p} = 0$ for both series, $\hat{d} = -0.02$ $[-0.07, 0.03]$ for the DAX and $\hat{d} = 0.05$ $[0.003, 0.100]$ for the FTSE300. In both cases, a significant deterministic trend is found. Fig. 7 shows the two Y_t series with the fitted trends and upper and lower 5% critical limits for testing significance of the trends. The result indicates that there are relatively long periods where volatility is high/low systematically for both series. This extends and is comparable to results by Ding and Granger in the following sense. For stationary long-memory processes, long-term behaviour is determined by the fractional parameter d. SEMIFAR models include, apart from d, a deterministic (and essentially arbitrary) trend function as an additional building block that can 'explain' long-term fluctuations. A smooth deterministic function can be interpreted as an even stronger (and more systematic) degree of temporal dependence than stationarity with slowly decaying correlations. The significant trends fitted to the volatility series of DAX and FTSE300 thus indicate that there may be even stronger 'long memory' in volatility than suggested by a stationary model with long-range dependence.

A more sophisticated analysis of volatility may be obtained by applying GARCH-type extensions of *SEMIFAR* models to the original series I_t. The mathematical theory necessary for such extensions is the subject of current research. For fractional GARCH models that do not include deterministic trend functions see e.g. [3], [17], [24], [32]. In particular, [32] extend the maximum likelihood method of [5] to fractional GARCH models.

5.3 Simulated Examples

In this subsection *SEMIFAR* models are fitted to some simulated series. The series ($n = 400$) are shown in Figs. 1a through d, which are:

Figure 1a: $Y_i = X_i$ where X_i is a fractional autoregressive process of oder $p_0 = 0$ with $d^0 = 0.4$.

Figure 1b: $Y_i = g(t) + X_i$ where X_i is a fractional autoregressive process of oder $p_0 = 0$ with $d^0 = 0.4$ (but not the same realization as in figure 1a) and $g(t) = 1.75 * (1/(1 + e^{4-8t}) - \sin(2\pi t))$.

Figure 1c: $Y_i - Y_{i-1} = X_i$ where X_i is a fractional autoregressive process of oder $p_0 = 0$ with $d^0 = -0.3$.

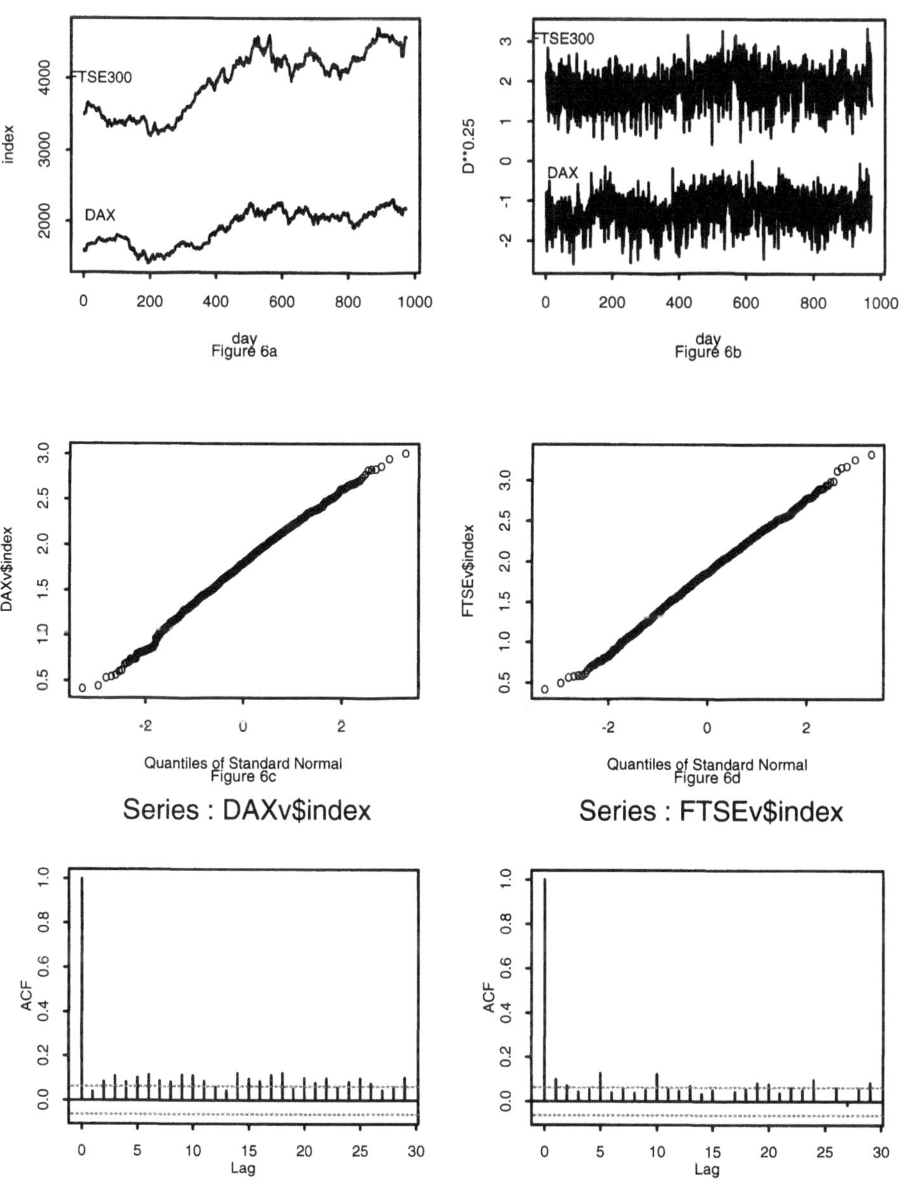

Fig. 6. Daily DAX and FTSE300 values I_t (Fig, 6a), first difference (Fig. 6b), normal probability plots of $Y_t = |I_t - I_{t-1}|^{\frac{1}{4}}$ (Figs. 6c,d) and autocorrelations of Y_t (Figs. 6e,f).

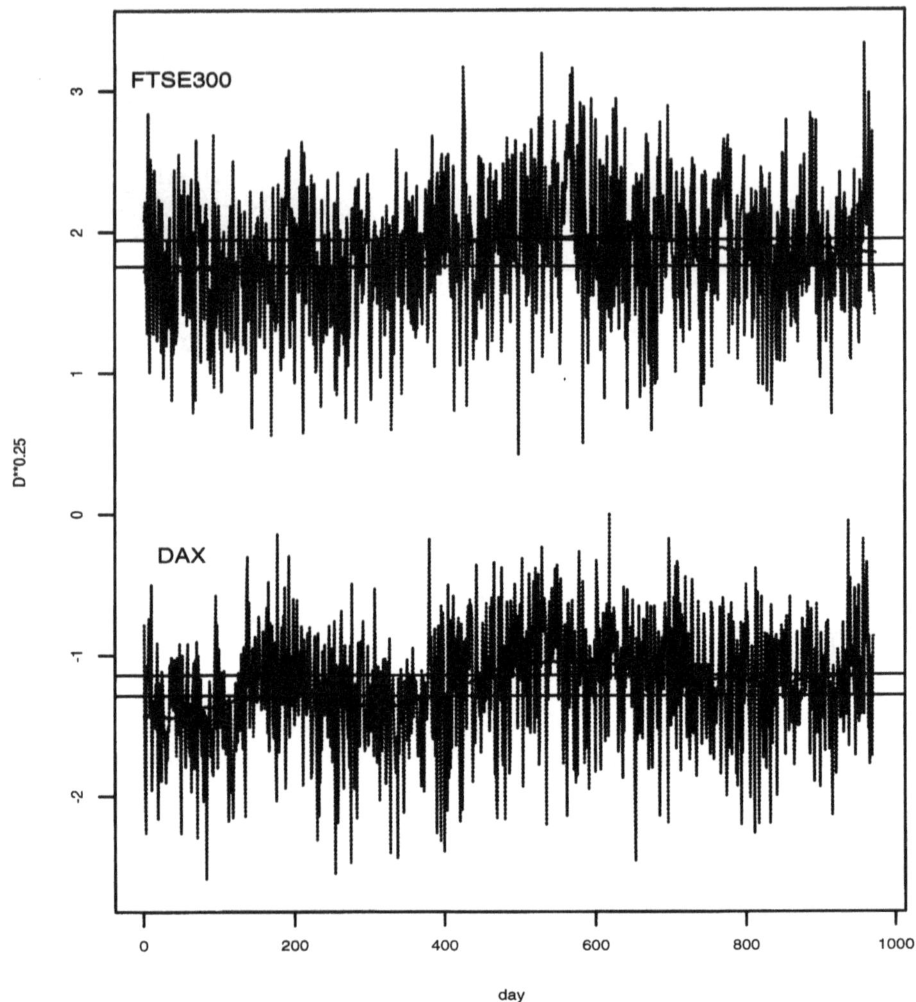

Fig. 7. Trends fitted by the SEMIFAR method to $Y_t = |I_t - I_{t-1}|^{\frac{1}{4}}$ where $I_t = DAX$ and FTSE300 respectively. Also given are the 5% rejection limits for testing where the trend is significant.

Figure 1d: $Y_i - Y_{i-1} = g(t) + X_i$ where X_i is the same fractional autoregressive process as in Fig. 1c and $g(t) = 0.2 * (t - 0.5)$.

All of these simulated series were generated by S-Plus with the "error" series X_i generated by the function *arima.fracdiff.sim*. Since a visual assessment of the time series plots appears to be difficult, it is interesting to see in how far the proposed method provides better information. The estimates $\hat{p}$ and $\hat{\eta} = \hat{d}$ (because of $p_0 = 0$) together with 95%-confidence intervals, obtained by fitting *SEMIFAR* models for $p = 0, 1, 2, 3, 4, 5$ and choosing p based on the BIC, are

Table 2. Estimates of p_o, d^o and $m^o = [d^o + 0.5]$ for the four simulated examples in Figs. 1a through 1d. The true values of p_o, d^o and m^o are given in brackets. Also given are the 95%-confidence intervals for d^o and the results of the testing on whether there is a significant trend g in the data.

Figure	$\hat{p}(p_o)$	$\hat{m}(m^o)$	$\hat{d}(d^o)$	95%-C.I. for d^o	testing on g
Fig. 1a	0(0)	0(0)	0.425(0.4)	[0.348, 0.502]	not significant
Fig. 1b	0(0)	0(0)	0.329(0.4)	[0.252, 0.406]	significant
Fig. 1c	0(0)	1(1)	0.764(0.7)	[0.687, 0.841]	not significant
Fig. 1d	0(0)	1(1)	0.762(0.7)	[0.685, 0.839]	significant

given in Table 2. Also given are the 95%-confidence intervals for $d^o = [d^o + 0.5]$ and the results of the testing whether there is a significant trend g in the data.

The values of $\hat{m}$ and $\hat{p}$ are correct for all four series. Thus, in particular, the method yields the correct answer to the question whether differencing is needed, i.e. whether the observed series has a stochastic trend component. Moreover, the estimates $\hat{\eta}$ are very close to the true values and the true values are always in the confidence intervals. Similarly, regarding the presence of a deterministic trend component, the results give correct indications. Hence the proposed models provide a way to distinguish stochastic trends, deterministic trends, long- and short memory or mixtures of these. It can be expected that more refined smoothing methods, such as local bandwidth choice (see e.g. [11]), may lead to even better estimates of g. This will be pursued elsewhere.

5.4 Comparison Between SEMIFAR and AR

In this section a brief comparison between the *SEMIFAR* model and the well known AR model will be made using the four examples in Sect. 5.1. Using the S-PLUS function *arima.mle* and the AIC criterion, an AR(2) model $y_t = 0.3392 y_{t-1} - 0.0896 y_{t-2} + \epsilon_t$ with $\hat{\sigma}^2 = 0.00402$ was obtained for the **cocoa** data. The AR model obtained for the **xeu** data was $y_i = -0.1178 y_{t-1} + \epsilon_t$ of order 1 with $\hat{\sigma}^2 = 0.000027$. The other two data sets **copper** and **chf** were shown to be white noises. The ratios between the widths of prediction intervals for the k-step forecasting obtained be the fitted *SEMIFAR* model and the AR one are given in Table 3.

Note in particular that, in 'stationary versus unit root' approaches, a decision has to be made between $d = 0$ and $d = 1$. A wrong decision has an extreme impact on forecast intervals, since the width of forecast intervals is asymptotically

Table 3. Ratios of prediction intervals by SEMIFAR model and AR one

k	1	2	3	4	5	6	7	8	9	10	15	20
cocoa	1.00	0.97	0.98	0.99	0.99	0.99	0.98	0.98	0.97	0.97	0.94	0.92
copper	0.98	0.88	0.82	0.77	0.74	0.71	0.69	0.67	0.66	0.64	0.59	0.56
chf	1.00	0.96	0.93	0.91	0.89	0.88	0.87	0.86	0.86	0.85	0.82	0.80
xeu	1.00	0.99	0.96	0.94	0.92	0.91	0.89	0.88	0.87	0.86	0.82	0.79

constant for $d = 0$ whereas it diverges to infinity at the rate $\sqrt{k}$ for $d = 1$. In contrast, for FARIMA models, prediction intervals are of order $O(k^{\tau/2})$ with τ varying in a continuous range, including $\tau = 0$ and $\tau = 1$ as special cases. The value of $\tau = max\{0, 2d - 1\}$ is estimated from the data by maximum likelihood and the extreme decision between $O(1)$ and $O(k^{.5})$ is avoided. As a result, prediction intervals are better adapted to the observed data, and often shorter if there is antipersistent. This is, in particular, often the case for foreign exchange rates. Consider for example the results in [9] on forecasting nominal exchange rates. There, the most dramatic improvement was achieved for the British Pound. Already for $k = 20$, the average interval was shorter by a factor of about 0.7, while the coverage probability of the interval appeared to be correct. Similar results were obtained in a recent PhD. thesis of [36], who found, in comparison, shorter prediction intervals for eight (out of eight) nominal foreign exchange rates. Many of them were shorter by a factor clearly smaller than 0.9 for $k = 20$ (see [36]). Further evidence of antipersistence in financial time series can also be found in [36]. In contrast to foreign exchange rates (and commodities), [36] found that traditional [10] ARIMA forecast intervals are typically too optimistic (i.e. too short) if the degree of persistence is strong, such as for nominal stock market indices.

6 Final Remarks

In this paper, we summarized recent results on so-called *SEMIFAR* models for time series that incorporate stochastic trends, deterministic trends, long-range dependence and short-range dependence. The potential usefulness of this model for economic time series analysis is illustrated by several data examples. In particular, the proposed method helps the data analyst to answer the question which of these components are present in the observed series. How well the different components can be distinguished depends on the specific process and, in particular, on the shape of the trend function. Therefore, in order that the proposed method is effective in general, the observed series must not be too short. In cases where one has sufficient a priori knowledge about the type of trend (e.g. linear, exponential etc.), parametric trend estimation is likely to provide more accurate results. This can be done simply by replacing the general function g in Definition (1) by the corresponding parametric function.

Further refinements of the method, such as local polynomial fitting of g, local bandwidth choice (see e.g. [11]), bootstrap confidence intervals, faster algorithms (see [22]) or other smoothing methods, etc., will be worth pursuing in future. Also, various extensions of *SEMIFAR* models are possible. For instance, as for classical ARIMA models, stochastic seasonal components can be included by multiplying the left hand side of (1) by a polynomial $\phi_{seas}(B) = \sum \phi_{j,seas} B^{sj}$ where $s \in N$ is the seasonal period. Other extensions, such as inclusion of parametric and nonparametric explanatory variables, other seasonal components and nonlinearities in the stochastic part of the process, are the subject of current research.

Acknowledgements

This research was supported in part by an NSF grant to MathSoft, Inc. (Seattle), and by the Center of Finance and Econometrics, University of Konstanz, Germany. The data for the exchange rates were obtained from the Web-site of PACIFIC (Policy Analysis Computing & Information Facility in Commerce) at the University of British Columbia, Vancouver, Canada; the prices of copper is from the homepage of the London Metal Exchange, and the cocoa price series from the ICCO's (International Cocoa Organization) Web-page. We would like to thank the authors of these data sets for making their data publicly available. Finally, we would like to thank Dr. Elke M. Hennig (Citibank, Frankfurt) for the stock market series.

References

1. H. Akaike: Biometrika **66**, 237 (1979)
2. N. S. Altman: Journal of the American Statistical Association **85**, 749 (1990)
3. R. T. Bailie, T. Bollerslev and H. O. Mikkelsen: Journal of Econometrics **74**, 3 (1996)
4. J. Beran: *Statistics for long-memory processes* (Chapman and Hall, New York 1994)
5. J. Beran: Journal of the Royal Statistical Society, series B **57**, 695 (1995)
6. J. Beran: *SEMIFAR models – a semiparametric fractional framework for modelling trends, long-range dependence and nonstationarity* (Preprint, University of Konstanz 1999)
7. J. Beran, R. J. Bhansali and D. Ocker: *On unified model selection for stationary and nonstationary short and long-memory autoregressive processes* (To appear in Biometrika 1998)
8. J. Beran and Y. Feng: *Local polynomial fitting with long-memory, short-memory and antipersistent errors*, The Ann. Instit. Statist. Math. (in press) (2002)
9. J. Beran and D. Ocker: J. Statist. Plann. Infer. **80**, 137 (1999)
10. G. E. Box and G. M. Jenkins: *Time Series Analysis: Forecasting and Control* (Holden Day, San Francisco 1976)
11. M. Brockmann: Journal of the American Statistical Association **88**, 1302 (1993)
12. Y. W. Cheung: Journal of Business and Economic Statistics **11**, 93 (1993)
13. S. T. Chiu: Statistics and Probability Letters **8**, 347 (1989)
14. D. R. Cox: 'Long-range dependence: a review'. In: *Statistics: An Appraisal, Proceedings 50th Anniversary Conference*, ed. by H. A. David and H. T. David (The Iowa State University Press 1984) 55
15. Csörgö, S. J. Mielniczuk. Annals of Statistics **23**, 1000 (1995)
16. P. J. Diggle: *Time Series - a biostatistical introduction* (Oxford University Press, Oxford 1990)
17. Z. Ding and C. W. J. Granger: Journal of Econometrics **73**, 185 (1996)
18. Z. Ding, C. W. J. Granger and R. F. Engle: Journal of Empirical Finance **1**, 83 (1993)
19. W. M. Fong and S. Ouliaris: Journal of Applied Econometrics **10**, 255 (1995)
20. G. Franke, R. Stapleton and M. G. Subrahmanyan: *When are options overpriced? The Black-Shcholes model and alternative characterisations of the pricing kernel* (Forthcoming in European Finance Review, 1999)

21. T. Gasser and H. G. Muller: 'Kernel estimation of regression functions'. In: *Smoothing Techniques for Curve Estimation*, ed. by T. Gasser and M. Rosenblatt, Lecture Notes in Mathematics, (Springer, New York 1979) 757, pp.23–68
22. T. Gasser, A. Kneip, W. Köhler: Journal of the American Statistical Association **86**, 643 (1991)
23. C. W. J. Granger: Journal of Econometrics **14**, 227 (1980)
24. C. W. J. Granger and Z. Ding: Journal of Econometrics **73**, 61 (1996)
25. C. W. J. Granger and R. Joyeux: Journal of Time Series Analysis **1**, 15 (1980)
26. P. Hall and J. Hart: Stochastic Processes and Their Applications **36**, 339 (1990)
27. F. R. Hampel: 'Data analysis and self-similar processes'. *Proceedings of the 46th Session of ISI* (Tokyo, Book 4 1987) pp. 235–254
28. E. J. Hannan and B. G. Quinn: Journal of the Royal Statistical Society, series B **41**, 190 (1979)
29. E. Herrmann, T. Gasser, and A. Kneip: Biometrika **79**, 783 (1992)
30. J. R. M. Hosking: Biometrika **68**, 165 (1981)
31. H. Künsch: Proc. First World Congress of the Bernoulli Society, Tashkent Vol **1**, 67 (1986)
32. S. Q. Ling and W. K. Li: JASA **92**, 1184 (1997)
33. C. Y. Liu, and J. He: Journal of Finance **36**, 773 (1991)
34. B. B. Mandelbrot: The Review of Economics and Statistics **53**, 225 (1971)
35. B. B. Mandelbrot: *The fractal geometry of nature* (Freeman, New York 1983)
36. D. Ocker: Stationary and nonstationary fractional ARIMA models - model choice, forecasting, aggregation and intervention. Unpublished PhD thesis, University of Konstanz (1999)
37. M. B. Priestley: *Spectral Analysis and Time Series* (Academic Press, London 1981)
38. B. K. Ray and R. S. Tsay: Biometrika **84**, 791 (1997)
39. L. C. G. Rogers: Mathematical Finance **7**, 95 (1997)
40. P. A. Samuelson: Industrial Management Review **VI**, 41 (1965)
41. G. Schwarz: Annals of Statistics **6**, 461 (1978)

Interaction Models for Common Long-Range Dependence in Asset Prices Volatility

Gilles Teyssière

GREQAM & CORE, Centre de la Vieille Charité, F-13002 Marseille, France

Abstract. We consider a class of microeconomic models with interacting agents which replicate the main properties of asset prices time series: non-linearities in levels and common degree of long-memory in the volatilities and co-volatilities of multivariate time series. For these models, long-range dependence in asset price volatility is the consequence of swings in opinions and herding behavior of market participants, which generate switches in the heteroskedastic structure of asset prices. Thus, the observed long-memory in asset prices volatility might be the outcome of a change-point in the conditional variance process, a conclusion supported by a wavelet anaysis of the volatility series. This explains why volatility processes share only the properties of the second moments of long-memory processes, but not the properties of the first moments.

1 Long-Range Dependence in Finance

Asset prices time series are characterized by several features: leptokurtic distribution, nonlinear variations, volatility clustering, unit roots in the conditional mean, and strong dependence in the volatility. These empirical features have been documented in [38,39], [48], [9], [22,23], [3] among others.

Daily prices P_t are modeled by martingale processes, i.e., $E(P_{t+1}|I_t) = P_t$, where I_t denotes the information set available at time t. This property is termed as 'Efficient Market Hypothesis', the content of I_t defining the type of market efficiency considered, see e.g., Fama [13]. As a consequence, the returns $R_t = \log(P_t/P_{t-1})$ are uncorrelated and unpredictable.

However, the power transformation $|R_t|^\delta$ displays strong dependence, the degree of which is the highest for $\delta = 1$. This empirical feature, termed as 'Taylor effect' [48], motivated the use of the class of long-memory volatility models introduced by Robinson [45], and developed in [22], [10], [18] and other works.

This statistical univariate approach was incomplete, as a multivariate analysis, pioneered by Teyssière [50,51], revealed that several time series share a common degree of strong dependence in their conditional variances and covariances. This regularity suggested the presence of a common structural model generating these features.

Furthermore, the series $|R_t|^\delta$ differ from standard long-range dependent, henceforth LRD, processes: while the autocorrelation function and the spectrum of the series $|R_t|^\delta$ display a LRD-type behavior, the series $|R_t|^\delta$ are not trended unlike standard LRD processes. Recent works, see [12], [34], [25], [32], considered the change-point problem for volatility processes, as the class of non-homogenous stochastic variance processes is also able to match the empirical properties of asset prices returns.

These empirical results motivated further research for devising structural microeconomic models explaining these features. Kirman and Teyssière [31,29,30] produced a model, based on microeconomic models with interacting agents, which generates these empirical properties of asset prices.

This chapter is organized as follows. Section 2 reviews some statistical methods used for testing for long-range dependence and for the presence of a change-point in the volatility process. Section 3 presents the class of microeconomic models generating the empirical property of common long-range dependence in multivariate asset price volatility. Simulation results for our models are given in Sect. 4.

2 Long-Range Dependent vs. Change-Point Processes

A stationary process Y_t is called a stationary process with long-memory if its autocorrelation function, henceforth ACF, $\rho(k)$ has asymptotically the following hyperbolic rate of decay, see [2], [20], [21], [26], [46]:

$$\rho(k) \sim L(k)k^{2d-1} \quad \text{as } k \to \infty, \tag{1}$$

where $L(k)$ is a slowly varying function,[1] and $d \in (0, 1/2)$ is the long-memory parameter which governs the slow rate of decay of the ACF and then parsimoniously summarizes the degree of long-range dependence of the series. Equivalently, the spectrum $f(\lambda)$ of a long-memory process can be approximated in the neighborhood of the zero frequency as

$$f(\lambda) \sim G\lambda^{-2d}, \quad \text{as } \lambda \to 0^+, \quad 0 < G < \infty. \tag{2}$$

2.1 Statistical Inference

Since the statistical characteristics of volatility processes are more complex than the ones of standard parametric long-memory processes, we resort in this study to semiparametric statistical tools which require mild assumptions on the process generating the data, henceforth DGP.

Several tests for stationarity against long-range dependent alternatives have been proposed by Lo [37], Kwiatkowski et al. [36], and Giraitis, Kokoszka, Leipus and Teyssière [15,16]. These statistics are based on the partial sum process $S_k = \sum_{t=1}^{k}(Y_t - \bar{Y})$ and the assumption that under the null hypothesis of stationarity, the standardized partial sum process satisfies a functional central limit theorem. Lo [37] considered the standardized range of S_k, i.e.,

$$R/S(q) = \frac{1}{\hat{s}_T(q)}\left[\max_{1\le k\le T} S_k - \min_{1\le k\le T} S_k\right] = \frac{\hat{R}_T}{\hat{s}_T(q)}. \tag{3}$$

[1] A function $L(k)$, $k \geq 0$, is called slowly varying function if $L(\lambda k)/L(k) \to 1$ as $k \to \infty$, $\forall \lambda > 0$.

Kwiatkowski et al. [36] considered the standardized second moment of S_k:

$$KPSS(q) = \frac{1}{T^2 \hat{s}_T^2(q)} \sum_{k=1}^{T} S_k^2 = \frac{\hat{M}_T}{T\hat{s}_T^2(q)}, \tag{4}$$

while Giraitis et al. [16] considered the standardized variance of S_k:

$$V/S(q) = \frac{1}{T^2 \hat{s}_T^2(q)} \left[\sum_{k=1}^{T} S_k^2 - \frac{1}{T} \left(\sum_{k=1}^{T} S_k \right)^2 \right] = \frac{\hat{V}_T}{T\hat{s}_T^2(q)}, \tag{5}$$

where $\hat{s}_T^2(q)$ is the heteroskedastic and autocorrelation consistent variance estimator, see [44]:

$$\hat{s}_T^2(q) = T^{-1} \sum_{i=1}^{T} (Y_i - \bar{Y})^2 + 2 \sum_{i=1}^{q} \omega_i(q) \hat{\gamma}_i \quad \text{with} \quad \omega_i(q) = 1 - \frac{1}{q+1}, \tag{6}$$

where the sample auto-covariances $\hat{\gamma}_i$ at lag i account for the possible short-range dependence up to the qth order.

Under the null hypothesis of no long-range dependence, the R/S statistic has the following asymptotic distribution: $T^{-1/2} R/S(q) \xrightarrow{d} \max_{0 \le t \le 1} W^0(t) - \min_{0 \le t \le 1} W^0(t)$, i.e., the range of a Brownian bridge $W^0(t) = W(t) - tW(1)$, on the unit interval, $KPSS(q) \xrightarrow{d} U_{KPSS} = \int_0^1 (W^0(t))^2 dt$, while $V/S(q) \xrightarrow{d} U_{V/S} = \int_0^1 (W^0(t))^2 dt - (\int_0^1 W^0(t) dt)^2$. The V/S statistic is less sensitive to the choice of the truncation order q than the R/S statistic, and is more powerful than the KPSS statistic. Furthermore, $E(U_{KPSS}) = 1/6$, $V(U_{KPSS}) = 1/45$, while $E(U_{V/S}) = 1/12$ and $V(U_{V/S}) = 1/360$. The smaller variance of the random variable V/S might explain its superior power for small samples.

The R/S statistic has been used by Mandelbrot and his co-authors, see [40], for estimating the degree of long-range dependence d. Define $\hat{s}_T^2 = \hat{s}_T^2(0)$, then $\hat{s}_T^2 \to \text{Var}(Y)$. Since

$$S_k = \sum_{j=1}^{k} (Y_j - EY_j) - \frac{k}{T} \sum_{j=1}^{T} (Y_j - EY_j), \tag{7}$$

and

$$\frac{1}{T^{1/2+d}} \sum_{j=1}^{[Tt]} (Y_j - EY_j) \overset{D[0,1]}{\to} C\, W_{1/2+d}(t), \tag{8}$$

where C is a positive constant, and $\overset{D[0,1]}{\longrightarrow}$ means weak convergence in the space $D[0,1]$ endowed with Skorokhod topology. Then

$$\frac{\hat{R}_T}{T^{1/2+d}} \xrightarrow{d} C \left[\max_{0 \le t \le 1} W^0_{1/2+d}(t) - \min_{0 \le t \le 1} W^0_{1/2+d}(t) \right], \tag{9}$$

$W^0_{1/2+d}(t)$ being the fractional Brownian bridge, defined as

$$W^0_{1/2+d}(t) = W_{1/2+d}(t) - tW_{1/2+d}(1). \tag{10}$$

Thus,

$$\frac{1}{T^{1/2+d}} \frac{\hat{R}_T}{\hat{s}_T} \xrightarrow{d} \frac{C\left[\max_{0\le t\le 1} W^0_{1/2+d}(t) - \min_{0\le t\le 1} W^0_{1/2+d}(t)\right]}{\mathrm{Var}(Y)^{1/2}}, \tag{11}$$

Equation (11) constitutes a theoretical foundation for the R/S estimator. Taking logarithms of both sides yields the heuristic identity:

$$\log\left(\hat{R}_T/\hat{s}_T\right) \approx \left(\frac{1}{2}+d\right)\log T + \mathbf{constant}, \quad \text{as } T \to \infty, \tag{12}$$

Denote $\hat{d}_{R/S} = (\log(\hat{R}_T/\hat{s}_T)/\log T) - 1/2$, then $\hat{d}_{R/S} - d = O_P(1/\log T)$. Thus, $1/2 + d$ can be interpreted as the slope of a regression line of $\log(\hat{R}_T/\hat{s}_T)$ on $\log T$.

Giraitis, Kokoszka, Leipus and Teyssière [17] suggested to extend this principle to the KPSS and the V/S statistics. By (8)

$$\frac{\hat{M}_T}{T^{1+2d}} \xrightarrow{d} C^2 \int_0^1 \left[W^0_{1/2+d}(t)\right]^2 dt. \tag{13}$$

Define $\hat{d}_{KPSS} = (\log(\hat{M}_T^{1/2}/\hat{s}_T)/\log T) - 1/2$, we get $\hat{d}_{KPSS} - d = O_P(1/\log T)$. Thus, the slope of the regression line of $\log(\hat{M}_T^{1/2}/\hat{s}_T)$ on $\log T$ estimates $d + 1/2$. Similarly, the regression of $\log(\hat{V}_T^{1/2}/\hat{s}_T)$ on $\log T$ estimates $d + 1/2$. Setting $\hat{d}_{V/S} = (\log(\hat{V}_T^{1/2}/\hat{s}_T)/\log T) - 1/2$, we get $\hat{d}_{V/S} - d = O_P(1/\log T)$.

The technical details of the implementation of these 'pox-plot' estimators are described in [2] and [17]. These semiparametric estimators have a few drawbacks. There is no formal asymptotic theory for them, and they have the slow rate of convergence of order $\log(T)$. For that reason, we complete the empirical study of the long-range dependent properties of our microeconomic model by considering another semi-parametric estimator of the degree of long-range dependence proposed by Robinson [47], which is the discrete version of the Whittle approximate maximum likelihood estimator in the spectral domain. This estimator, suggested by Kunsch [35], is based on the mild assumption (2) of the spectrum $f(\lambda)$ of a long-memory process in the neighborhood of the zero frequency. The consequences of a misspecification of the functional form of the spectrum in the Whittle estimator are avoided with this local approximation. After concentrating in G, the estimator is given by:

$$\hat{d} = \arg\min_d \left\{ \ln\left(\frac{1}{m}\sum_{j=1}^{m} \frac{I(\lambda_j)}{\lambda_j^{-2d}}\right) - \frac{2d}{m}\sum_{j=1}^{m} \ln(\lambda_j) \right\}, \tag{14}$$

where $I(\lambda_j)$ is the periodogram estimated for the range of Fourier frequencies $\lambda_j = \pi j/T, j = 1, \ldots, m \ll [T/2]$, the bandwidth parameter m tends to infinity with T, but more slowly since $1/m + m/T \to 0$ as $T \to \infty$. Under appropriate conditions, which include the existence of a moving average representation and the differentiability of the spectrum near the zero frequency, this estimator has the following distribution independent of the value of d:

$$\sqrt{m}(\hat{d} - d) \sim N(0, 1/4). \tag{15}$$

Furthermore, this estimator is robust to the presence of conditional heteroskedasticity of general form, and an optimal bandwidth with the same robustness properties does exist under mild assumptions, see [24].

2.2 Long-Memory Volatility Models

The clustering of the variations of asset returns can be modeled by the class of Generalized Autoregressive Conditional Heteroskedastic GARCH(1,1), processes, see [7] and [48], defined as:

$$R_t = \mu + \varepsilon_t, \quad \sigma_t^2 = \omega + \beta\sigma_{t-1}^2 + \alpha\varepsilon_{t-1}^2, \quad \varepsilon_t \sim N(0, \sigma_t^2), \tag{16}$$

with $\omega > 0$, and α, $\beta \geq 0$. It has been empirically found that for large samples the sum of the estimated parameters $\hat{\alpha} + \hat{\beta}$ was close to one, the restricted model being an Integrated GARCH(1,1), henceforth IGARCH(1,1) see [11], defined as:

$$R_t = \mu + \varepsilon_t, \quad \sigma_t^2 = \omega + \beta\sigma_{t-1}^2 + (1-\beta)\varepsilon_{t-1}^2, \quad \varepsilon_t \sim N(0, \sigma_t^2), \tag{17}$$

which can be written as an ARCH process

$$R_t = \mu + \varepsilon_t, \quad \sigma_t^2 = \omega + \psi(L)\varepsilon_t^2, \quad \varepsilon_t \sim N(0, \sigma_t^2), \tag{18}$$

the coefficients of the lag polynomial $\psi(L)$ sum to one but decrease exponentially to zero. For the class of IGARCH processes, the shocks of the innovations ε_t on the level of the conditional variance σ_τ^2 have a strong persistence $\forall \tau > t$, which is not consistent with what is empirically observed. Thus, the occurrence of IGARCH(1,1) processes can be considered as a large sample artefact of a more complex phenomenon.

The IGARCH process is generalized with the class of long-memory ARCH, henceforth LM-ARCH, processes introduced by Robinson [45], and defined as:

$$R_t = \mu + \varepsilon_t, \quad \sigma_t^\delta = \omega + \psi(L)\varepsilon_t^\delta, \quad \varepsilon_t \sim N(0, \sigma_t^2), \tag{19}$$

where $\psi(L) = \sum_{i=1}^{\infty} \psi_i L^i$ is an infinite order lag polynomial the coefficients of which are positive and have asymptotically the following hyperbolic rate of decay $\psi_j = O\left(j^{-(1+d)}\right)$, and $\delta > 0$ is a parameter. Unlike IGARCH(1,1) processes, the persistence of the variations of the innovations on the volatility decays slowly. However, there is no stationary solutions to the equations defining a long-memory ARCH process, see e.g., [19], [28], [14], the only exception being the long-memory linear ARCH process introduced by Giraitis, Robinson and Surgailis [18]. Granger and Ding [22] and other authors considered the occurrence of long-range dependence in asset price volatilities.

2.3 Multivariate Analysis

The multivariate properties of volatility processes can be analyzed by considering the 'co-volatility' processes. The volatility processes associated with a conditional mean process R_t can be represented by its absolute value $|R_t|$ or the squared returns process R_t^2. Thus, the co-volatility of the bivariate processes $(R_{1,t}, R_{2,t})$ can be represented by the processes $\sqrt{|R_{1,t}R_{2,t}|}$ or $R_{1,t}R_{2,t}$, although only the first process is positive. Empirical evidence on asset price series, e.g., FX rates reported on Table 1, has shown that several time series share a common degree of long-range dependence in their volatilities and co-volatilities.

Table 1. Estimation of the fractional degree of integration for the series of absolute returns on Pound-Dollar $|R_{1,t}|$, Deutschmark-Dollar $|R_{2,t}|$, squared returns $R_{1,t}^2$, $R_{2,t}^2$, and the co-volatilities $\sqrt{|R_{1,t}R_{2,t}|}$ and $R_{1,t}R_{2,t}$ for the period April 1979 - January 1997. We use here the Gaussian estimator defined in (14). Asymptotic S.E. $(2\sqrt{m})^{-1}$ are between parentheses.

m	$\lvert R_{1,t}\rvert$	$\lvert R_{2,t}\rvert$	$\sqrt{\lvert R_{1,t}R_{2,t}\rvert}$
$T/4$	0.2385 (0.0147)	0.2312 (0.0147)	0.2413 (0.0147)
$T/8$	0.3071 (0.0207)	0.3219 (0.0207)	0.3230 (0.0207)
$T/16$	0.4113 (0.0293)	0.4073 (0.0293)	0.4393 (0.0293)
m	$R_{1,t}^2$	$R_{2,t}^2$	$R_{1,t}R_{2,t}$
$T/4$	0.1569 (0.0147)	0.1478 (0.0147)	0.1397 (0.0147)
$T/8$	0.2312 (0.0207)	0.2119 (0.0207)	0.2073 (0.0207)
$T/16$	0.2770 (0.0293)	0.2787 (0.0293)	0.2952 (0.0293)

A multivariate analysis of long-range dependent volatility processes can be carried by considering the parametric framework of the class of multivariate long-memory ARCH processes, introduced by Teyssière [50,51], and defined as:

$$\boldsymbol{R_t} = \boldsymbol{m}(\boldsymbol{R_t}) + \boldsymbol{\varepsilon_t}, \quad \boldsymbol{\varepsilon_t} \sim \text{i.i.d.}\, \boldsymbol{N}(\boldsymbol{0}, \boldsymbol{\Sigma_t}), \tag{20}$$

where $\boldsymbol{m}(\boldsymbol{R_t})$ denotes the vector regression function, $\boldsymbol{\varepsilon_t}$ is a n-dimensional vector of Gaussian error terms with conditional covariance matrix $\boldsymbol{\Sigma_t}$. The typical element $s_{ij,t}$ of $\boldsymbol{\Sigma_t}$ being either

$$s_{ij,t} = \frac{\omega_{ij}}{1-\beta_{ij}(1)} + \left(1 - \frac{(1-\phi_{ij}L)(1-L)^{d_{ij}}}{1-\beta_{ij}L}\right)\varepsilon_{i,t}\varepsilon_{j,t} \quad i,j = 1,\ldots,n, \tag{21}$$

or

$$s_{ij,t} = \sum_{k=1}^{\infty} \frac{B(p_{ij}+k-1, d_{ij}+1)}{B(p_{ij}, d_{ij})} \varepsilon_{i,t-k}\varepsilon_{j,t-k}, \quad i,j = 1,\ldots,n, \tag{22}$$

i.e., both conditional variances and covariances are modeled as LM-ARCH processes, which differ by different parameterizations: (21) is termed as fractionally integrated GARCH, see [1], while (22) defines the long-memory ARCH devised by Ding and Granger [10]. This class of multivariate LM-ARCH models has a few restrictions: the conditions on the parameters insuring that the matrix Σ_t is positive definite have to be implemented numerically in the estimation procedure. Furthermore, the number of parameters increases quickly with the dimension of the vector process, so that so far only three-dimensional models have been estimated, see [51]. However, empirical estimation results have shown that the conditional variances and covariances of several asset prices returns share the same degree of long-memory, an interesting property which stimulated further research producing the theoretical models presented later in this chapter.

2.4 Change-Point Processes

Volatility processes differ from standard long-range dependent processes: while long-range dependent time series exhibit local trends, the proxy of volatility processes, e.g., the absolute returns $|R_t|$ or the squared returns R_t^2 do not contain such a trend. Figure 1 below displays the absolute value of returns on the FTSE 100 index, which is not trended, although the estimated degree of long-memory with Robinson's [47] Gaussian estimator yields $d = 0.33$.

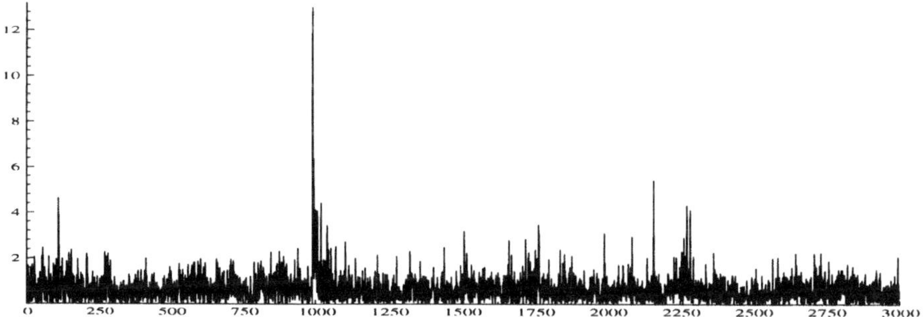

Fig. 1. Absolute returns on the FTSE 100 index.

Mikosch and Stărică [42,43] have shown that the ACF of the absolute value of a non-homogenous GARCH(1,1) process, i.e., a GARCH(1,1) process with changing coefficients, has a hyperbolic rate of decay which resembles the one of a long-range dependent process. We consider as example the following change-point GARCH(1,1) process defined as:

$$y_t = \mu + \varepsilon_t, \quad \sigma_t^2 = \omega + \beta\sigma_{t-1}^2 + \alpha\varepsilon_{t-1}^2, \quad \varepsilon_t \sim N(0, \sigma_t^2), \tag{23}$$

where the parameters ω, β and α change as follows:

DGP 1: a GARCH(1,1) process with change point in the middle of the sample, such that the unconditional variance $\sigma^2 = \omega/(1 - \alpha - \beta)$ remains unchanged ($\sigma^2 = 0.25$)

$$\begin{aligned} &\omega = 0.1, \quad \beta = 0.3, \quad \alpha = 0.3 \text{ for } t = 1, \dots, [T/2] \\ &\omega = 0.15, \quad \beta = 0.25, \quad \alpha = 0.15 \text{ for } t = [T/2] + 1, \dots, T \end{aligned} \tag{24}$$

DGP 2: a GARCH(1,1) process with change in the middle of the sample, with change in the unconditional variance of the process,

$$\omega = 0.1, \quad \beta = 0.3, \quad \alpha = 0.3 \text{ for } t = 1, \dots, [T/2] \quad (\sigma^2 = 0.25) \tag{25}$$
$$\omega = 0.15, \quad \beta = 0.65, \quad \alpha = 0.25 \text{ for } t = [T/2]+1, \dots, T \quad (\sigma^2 = 1.5) \tag{26}$$

DGP 3: a smooth transition GARCH(1,1) process, such that the parameters $\omega(t)$, $\beta(t)$ and $\alpha(t)$ change smoothly, i.e.,

$$\begin{aligned} \omega(t) &= 0.1 + 0.05F(t, [T/2]), \quad \beta(t) = 0.3 + 0.35F(t, [T/2]), \\ \alpha(t) &= 0.3 - 0.05F(t, [T/2]), \quad \gamma = 0.05, \end{aligned} \tag{27}$$

where $F(t,k) = (1 + \exp(-\gamma(t-k)))^{-1}$, γ is a strictly positive parameter which governs the smoothness of the change. If γ becomes very large, this DGP reduces to DGP 2.

Table 2. Tests for long-range dependence on the absolute value of GARCH process with change-point in the middle of the sample. $T = 500$. Test size 5%.

	DGP 1			DGP 2			DGP 3		
q	KPSS	V/S	R/S	KPSS	V/S	R/S	KPSS	V/S	R/S
0	0.2015	0.2770	0.2995	1.0000	1.0000	1.0000	1.0000	1.0000	1.0000
1	0.1470	0.1912	0.1785	1.0000	1.0000	1.0000	1.0000	1.0000	1.0000
2	0.1203	0.1443	0.1218	1.0000	1.0000	1.0000	1.0000	1.0000	1.0000
5	0.0874	0.0918	0.0601	1.0000	0.9998	0.9996	1.0000	0.9994	0.9991
10	0.0735	0.0674	0.0356	0.9993	0.9981	0.9891	0.9994	0.9979	0.9858
20	0.0632	0.0470	0.0188	0.9945	0.9819	0.8596	0.9978	0.9817	0.8285
30	0.0567	0.0325	0.0076	0.9846	0.9274	0.4844	0.9930	0.9361	0.4112

Table 2 displays the simulation results of the various tests for long-range dependence in the absolute returns generated by the change-point GARCH processes defined above. Similar results are obtained when considering the series of squares of a non-homogeneous GARCH(1,1) processes. Thus, the tests proposed in [37], [36] and [16] can wrongly detect the presence of long-range dependence in the volatility process, while the true DGP is a non-homogeneous GARCH process with a non-constant unconditional variance. However, when the unconditional variance is constant, the power of these tests tends to their size, a statistical property which is also observed for change-point tests, see Kokoszka and Teyssière [32]. We observe that the R/S statistic is more sentitive to the truncation order q than the other statistics. Furthermore, Fig. 2 below shows absolute returns of a series generated by DGP 2. Although standard tests and estimators detect the presence of long-range dependence this series is not trended. The class of non-homogeneous GARCH(1,1) processes is also appropriate for fitting asset prices returns.

There is a substantial literature on change-point processes, interested readers are referred to [5] and [8] for complete surveys. Most of these tests are concerned

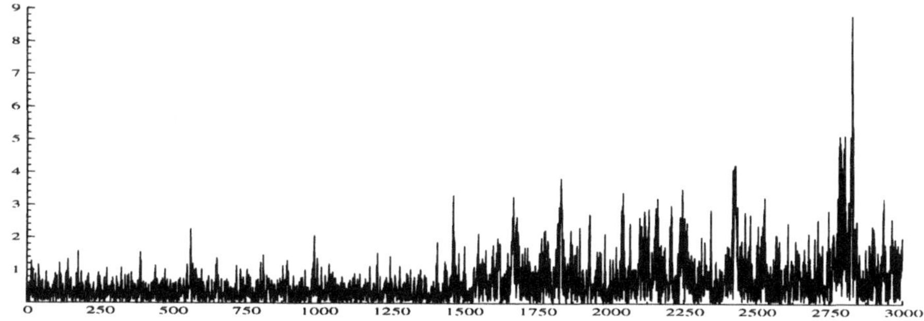

Fig. 2. Absolute value of the realization of a change-point GARCH process.

with change-point in the conditional mean processes, while we are interested here in conditional variance processes, although one can use, without theoretical foundations, these change-point tests for conditional mean processes to the volatilities and co-volatility proxy processes, i.e., $R_{1,t}^2$, $|R_{1,t}|$, $R_{1,t}R_{2,t}$, and $\sqrt{|R_{1,t}R_{2,t}|}$. We consider in this survey the tests for change point in conditional variance, proposed by Kokoszka and Leipus [34], Horváth, Kokoszka and Teyssière [25] and Kokoszka and Teyssière [32].

Kokoszka and Leipus [34] proposed a CUSUM based estimator for change-point in the class of ARCH(∞) processes at unknown time t. This estimator is defined by:

$$\hat{t} = \min\left\{t : |C_t| = \max_{1\leq j\leq T} |C_j|\right\}, \tag{28}$$

where

$$C_t = \frac{t(T-t)}{T^2}\left(t^{-1}\sum_{j=1}^{t} R_j^2 - \frac{1}{T-t}\sum_{j=t+1}^{T} R_j^2\right). \tag{29}$$

Horvath et al. [25] proposed several tests for change-point in ARCH sequences, based on the empirical process of squared residuals. Berkes and Horváth [4] analyzed the empirical process of squared residuals for GARCH(p,q) sequences. According to Kokoszka and Teyssière [32], some of these asymptotic tests work well when considering the squared residuals for GARCH(1,1) sequences although bootstrap tests have always the correct size and are then more reliable.

We consider here a GARCH(1,1) model fitted on the simulated returns, i.e.,

$$R_t = \mu + \varepsilon_t, \quad \varepsilon_t \sim N(0, \sigma_t^2), \quad \sigma_t^2 = \omega + \beta\sigma_{t-1}^2 + \alpha\varepsilon_{t-1}^2, \tag{30}$$

and we denote by $\hat{\varepsilon}_t^2$ the sequence of squared standardized residuals for this GARCH(1,1) model.

The first statistic is a Kolmogorov-Smirnov type statistic. For $1 \leq k \leq T$, define

$$\hat{T}(k,t) = \sqrt{T}\,\frac{k}{T}\left(1 - \frac{k}{T}\right)\left|\hat{F}_k(t) - \hat{F}_k^*(t)\right|, \tag{31}$$

with

$$\hat{F}_k(t) = \frac{1}{k}\#\{i \le k : \hat{\varepsilon}_i^2 \le t\}, \quad \hat{F}_k^*(t) = \frac{1}{T-k}\#\{i > k : \hat{\varepsilon}_i^2 \le t\}. \tag{32}$$

The K-S statistic is defined as

$$\hat{M} = \sup_{0 \le t \le \infty} \max_{1 \le k \le T} |\hat{T}(k,t)|. \tag{33}$$

According to [32], correct inference is obtained by using bootstrap based inference. Horvath et al. [25] proposed also a Cramér-von Mises statistic:

$$\hat{B} = \int_0^1 \left\{ \frac{1}{T} \sum_{i=1}^{T} [\hat{T}([Ts], \hat{\varepsilon}_i^2)]^2 \right\} ds. \tag{34}$$

The distribution function of B can be derived from Blum, Kiefer and Rosenblatt [6]. Kokoszka and Teyssière [32] have shown that this asymptotic test provides correct inference.

3 Interaction Models

The class of models considered here differ from standard microeconomic models as we consider that agents are heterogeneous and do not act independently on the markets, but their beliefs and actions are affected by the predominant opinion among market participants. Keynes pointed out that individuals trades are concerned about what 'market sentiment' is rather than about fundamental values. We consider here equilibrium models, thus we rule out the case of the intra-day prices, which are not equilibrium prices but result from the content of book orders.

If the markets are efficient, the expected price $E(P_{t+1})$ of an asset at time $t+1$ conditional on the information set I_t is given by:

$$E(P_{t+1}|I_t) = P_t. \tag{35}$$

In our model, agents do not consider markets to be efficient and assume that they can predict the next price P_{t+1}. Chartists assume that the exchange rate P_{t+1} is a convex linear function of the previous prices, i.e.,

$$E^c(P_{t+1}|I_t) = \sum_{j=0}^{M^c} h_j P_{t-j}, \quad \text{with} \quad \sum_{j=0}^{M^c} h_j = 1, \tag{36}$$

where h_j, $j = 0, \ldots, M^c$ are constants, M^c is the memory of the chartists, while fundamentalists forecast the next price as:

$$E^f(P_{t+1}|I_t) = \bar{P}_t + \sum_{j=1}^{M^f} \nu_j (P_{t-j+1} - \bar{P}_{t-j}), \tag{37}$$

where ν_j, $j = 1, \ldots, M^f$ are positive constants, representing the degree of reversion to the fundamentals, M^f is the memory of the fundamentalists. This series of 'fundamentals' $\bar{P}_t$, which can be thought as the price if it were only to be explained by a set of relevant variables, is assumed to follow a random walk:

$$\bar{P}_t = \bar{P}_{t-1} + \varepsilon_t, \quad \text{with} \quad \varepsilon_t \sim N(0, \sigma_\varepsilon^2). \tag{38}$$

Individuals i, $i = 1, \ldots, N$ have a standard mean-variance utility function:

$$U(W_{t+1}^i) = E(W_{t+1}^i) - \lambda V(W_{t+1}^i), \tag{39}$$

where λ denotes the risk aversion coefficient, $E(.)$ and $V(.)$ denote the expectation and variance operators. Agents have the possibility of investing at home in a risk free asset or investing abroad in a risky asset.

Denote by ρ_t the foreign interest rate, by d_t^i the demand by the ith individual for foreign currency, and by r the domestic interest rate. The exchange rate P_t and the foreign interest rate ρ_t are considered by agents as independent random variables, with

$$\rho_t \sim N(\rho, \sigma_\rho^2) \quad \text{with} \quad \rho_t > r. \tag{40}$$

Hence, the cumulated wealth of individual i at time $t+1$, W_{t+1}^i is given by:

$$W_{t+1}^i = (1 + \rho_{t+1}) P_{t+1} d_t^i + (W_t^i - P_t d_t^i)(1 + r). \tag{41}$$

Thus, we have:

$$E(W_{t+1}^i | I_t) = (1 + \rho) E^i(P_{t+1} | I_t) d_t^i + (W_t^i - P_t d_t^i)(1 + r), \tag{42}$$

$$V(W_{t+1}^i | I_t) = (d_t^i)^2 \zeta_t \quad \text{where} \quad \zeta_t = V\left(P_{t+1}(1 + \rho_{t+1})\right). \tag{43}$$

Demand d_t^i is found by maximizing utility. First order condition gives

$$(1 + \rho) E^i(P_{t+1} | I_t) - (1 + r) P_t - 2\zeta_t \lambda d_t^i = 0, \tag{44}$$

where $E^i(.|I_t)$ denotes the expectation of an agent of type i. Let k_t be the proportion of fundamentalists at time t, the market demand is:

$$d_t = \frac{(1 + \rho)\left(k_t E^f(P_{t+1}|I_t) + (1 - k_t) E^c(P_{t+1}|I_t)\right) - (1 + r) P_t}{2\zeta_t \lambda}. \tag{45}$$

Now consider the exogenous supply of foreign exchange X_t, then the market is in equilibrium if aggregate supply is equal to aggregate demand, i.e., $X_t = d_t$, which gives

$$P_t = \frac{1 + \rho}{1 + r}\left(k_t E^f(P_{t+1}|I_t) + (1 - k_t) E^c(P_{t+1}|I_t)\right) - \frac{2\zeta_t \lambda X_t}{1 + r}. \tag{46}$$

We assume that $2\zeta_t \lambda X_t / (1 + \rho) = \gamma \bar{P}_t$. If $M^f = M^c = 1$, then the equilibrium price is given by

$$P_t = \frac{k_t - \gamma}{A} \bar{P}_t - \frac{k_t \nu_1}{A} \bar{P}_{t-1} + \frac{(1 - k_t) h_1}{A} P_{t-1}, \tag{47}$$

with

$$A = \frac{1+r}{1+\rho} - (1-k_t)h_0 - k_t\nu_1. \tag{48}$$

Thus, when k_t jumps from zero to one, our so called 'Havana-India' model resembles a change-point process in the conditional mean. Since the process k_t is likely to take all values between 0 and 1, it is of interest to study the effects of the evolution of the process k_t on the occurrence of long-range dependence in the volatility of the series generated by the microeconomic model.

We consider a multivariate extension of this model, i.e., the joint modeling of a bivariate process $(P_{1,t}, P_{2,t})$. Both exchange rates depend on a pair of foreign interest rates $(\rho_{1,t}, \rho_{2,t})$. Our bivariate model then becomes:

$$\begin{pmatrix} P_{1,t} \\ P_{2,t} \end{pmatrix} = \begin{pmatrix} \frac{k_t-\gamma}{A_1}\bar{P}_{1,t} - \frac{k_t\nu_{1,1}}{A_1}\bar{P}_{1,t-1} + \frac{(1-k_t)h_{1,1}}{A_1}P_{1,t-1} \\ \frac{k_t-\gamma}{A_2}\bar{P}_{2,t} - \frac{k_t\nu_{2,1}}{A_2}\bar{P}_{2,t-1} + \frac{(1-k_t)h_{2,1}}{A_2}P_{2,t-1} \end{pmatrix}, \tag{49}$$

with

$$A_i = \frac{1+r}{1+\rho_i} - (1-k_t)h_{i,0} - k_t\nu_{i,1}. \tag{50}$$

We assume that the bivariate process of fundamentals $(\bar{P}_{1,t}, \bar{P}_{2,t})$ displays some form of positive correlation, i.e.,

$$\begin{aligned} \begin{pmatrix} \bar{P}_{1,t} \\ \bar{P}_{2,t} \end{pmatrix} &= \begin{pmatrix} \bar{P}_{1,t-1} \\ \bar{P}_{2,t-1} \end{pmatrix} + \begin{pmatrix} \varepsilon_{1,t} \\ \varepsilon_{2,t} \end{pmatrix}, \\ \begin{pmatrix} \varepsilon_{1,t} \\ \varepsilon_{2,t} \end{pmatrix} &\sim N\left[\begin{pmatrix} 0 \\ 0 \end{pmatrix}, \begin{pmatrix} \sigma_{1,1}^2 & \sigma_{1,2} \\ \sigma_{1,2} & \sigma_{2,2}^2 \end{pmatrix}\right], \quad \sigma_{1,2} > 0. \end{aligned} \tag{51}$$

In the simulation study, we set $\sigma_{1,2}$ so that the coefficient of correlation between the two processes $\varepsilon_{1,t}$ and $\varepsilon_{2,t}$ is equal to 0.75. This choice has been motivated by the estimation results in [50] for the bivariate long-memory ARCH processes, where the coefficient of correlation in the conditional covariance matrix $\boldsymbol{\Sigma_t}$ has been found equal to 0.75. As we will see in Sect. 4, the assumption of a positive correlation is crucial if we are interested in the co-volatility processes $\sqrt{|R_{1,t}R_{2,t}|}$ and $R_{1,t}R_{2,t}$: in that case these co-volatility processes have exactly the same degree of long-memory as the processes $|R_{1,t}|$, $|R_{2,t}|$ and $R_{1,t}^2$, $R_{2,t}^2$ respectively, in accordance with the empirical findings of [50]. In [29], we assume $\sigma_{1,2} = 0$, and simulation results are less satisfactory than the current ones, as the degree of LRD for the series $\sqrt{|R_{1,t}R_{2,t}|}$ is slightly higher than the ones for the series $|R_{1,t}|$, $|R_{2,t}|$.

We also assume that the process k_t is the same for both markets, i.e., the proportion of fundamentalists is the same for both currencies. This assumption is consistent with the one that fundamentals for both series are correlated, i.e., both FX markets are linked. This is a reasonable assumption if we consider that both currencies belong to the same 'target-zone', see [12].

We consider here several types of processes for $\{k_t\}_{t=1}^T$. The first one is the epidemiologic process introduced by Hans Föllmer and used in [31,29,30], where agents interact and communicate their beliefs on the next period forecast through Föllmer's epidemiologic process.

Let N be the total number of agents and ϑ_t be the number of agents with a fundamentalist forecast at time t. We assume that pairs of agents meet at random and that the probability that the first agent is converted to the opinion of the second one is equal to $(1-\delta)$. Furthermore, each agent can independently change his opinion with probability ξ, so that the process is not trapped in the extremes, i.e., agents are either all chartists or all fundamentalists.

Given that the state of the process is summarized by the value of ϑ_t, its evolution is defined by the following transition matrix:

$$\Pr(\vartheta, \vartheta+1) = \left(1 - \frac{\vartheta}{N}\right)\left(\xi + (1-\delta)\frac{\vartheta}{N-1}\right), \tag{52}$$

$$\Pr(\vartheta, \vartheta-1) = \frac{\vartheta}{N}\left(\xi + (1-\delta)\frac{N-\vartheta}{N-1}\right), \tag{53}$$

$$\Pr(\vartheta, \vartheta) = 1 - \Pr(\vartheta, \vartheta+1) - \Pr(\vartheta, \vartheta-1). \tag{54}$$

For this epidemiologic process, the proportion of fundamentalists and the forecasts of agents does not depend on the past performance of forecasts functions. For that reason, we can consider a diffusion process for k_t which depends on the accuracy of the forecast functions in the recent periods: the probability of choosing a particular forecast function depends on its comparative performance over the competing forecast function. We can use Theil's [52] U statistic as measure of forecast accuracy over the last M periods:

$$U_M^j = \sqrt{\frac{M^{-1}\sum_{l=1}^{M} w_l \left(P_{t-l} - E^j(P_{t-l}|I_{t-1-l})\right)^2}{M^{-1}\sum_{l=1}^{M} w_l P_{t-l}^2}}, \quad j \in \{c, f\}, \quad \sum_l w_l = 1, \tag{55}$$

M being the learning memory of agents, the weights $w_l, l = 1, \ldots, M$ representing the relative importance of the forecast errors at time $t-l$. We choose here an exponential choice function $g^j(\cdot)$ for the forecast function $E^j(\cdot)$ defined by:

$$g^j(t) = \exp(-\Upsilon U_M^j), \quad \Upsilon > 0, \quad j \in \{c, f\}, \tag{56}$$

the parameter Υ is called the "intensity of choice". At time t, agents will chose with probability $\pi^f(t)$ the fundamentalist forecast function, where

$$\pi^f(t) = \frac{g^f(t)}{g^f(t) + g^c(t)}, \tag{57}$$

the probability of choosing the chartist forecast function is $\pi^c(t) = 1 - \pi^f(t)$. For the bivariate process, the probability of choosing the fundamentalist forecast function is given by averaging the two choice functions for both markets.

Let ϑ_t/N be the proportion of fundamentalists resulting from either the epidemiologic process or the learning process. We assume that agents observe this proportion with error, i.e., agent i observe $k_{i,t}$ defined as:

$$k_{i,t} = \frac{\vartheta_t}{N} + \varepsilon_{i,t} \quad \text{with} \quad \varepsilon_{i,t} \sim N(0, \sigma_\vartheta^2). \tag{58}$$

If agent i observe $k_{i,t} \geq 0.5$, then he will make a fundamentalist forecast, otherwise he will make a chartist forecast. The proportion k_t of agents making a fundamentalist forecast is then given by:

$$k_t = N^{-1} \# \left\{ i : k_{i,t} \geq \frac{1}{2} \right\}. \tag{59}$$

For the epidemiologic case, the herding behavior of the process k_t into the extremes depends on ξ and σ_ϑ, while it depends on Υ and σ_ϑ for the process based on the forecasts accuracy. For both processes, the parameter σ_ϑ measures the accuracy of observation of the proportion of fundamentalists; see (58). If σ_ϑ becomes smaller, the prevailing opinion is observed with more accuracy, which results in massive swings of opinion.

As we will see in the next section, these parameters govern the level of long-range dependence in the volatility of the simulated returns.

4 Simulation Study

We simulated 10.000 replications of our microeconomic models. We considered samples of 1500 observations. The models generates the empirical properties of asset prices returns. The series of asset returns R_t do not display dependence, the average estimated value for d is $\hat{d} = 0.002$ for the series R_t. When the sample size increases from 750 to 1500, the estimated value for d increases from $\hat{d} = 0.20$ to $\hat{d} = 0.28$. When estimating the parameters of a GARCH processes on the series of 750 observations, we get $\hat{\alpha} = 0.04$ and $\hat{\beta} = 0.74$, while for the series of 1500 observations, $\hat{\alpha} = 0.055$ and $\hat{\beta} = 0.88$: the model replicates the empirical property of occurrence of IGARCH processes when the sample size increases, see [30]. The occurrence of long-range dependence in asset prices volatility might be the consequence of several changes in regime in the price process.

The level of long-range dependence d of the simulated processes increases when we reduce the value of σ_ϑ, i.e., when the proportion of fundamentalists is observed with more accuracy: in that case the process k_t herds into the extremes. The level of long-range dependence is linked to the swings in the predominant opinion which make the price process defined by (47) switching between two regimes.

The assumption of a positive correlation between the fundamentals proved to be important. In [29], we assume that there is no correlation between the two processes $(\varepsilon_{1,t}, \varepsilon_{2,t})$, i.e., $\sigma_{1,2} = 0$. As a consequence, the estimated level of long-range dependence in the co-volatility process $\sqrt{|R_{1,t}R_{2,t}|}$ was slightly higher than the one of the volatility processes $|R_{1,t}|$ and $|R_{2,t}|$. Furthermore, for this uncorrelated setting, the co-volatility process $R_{1,t}R_{2,t}$ does not display

Table 3. Tests for long-range dependence on the absolute value of Simulated returns, R_t, absolute returns $|R_t|$ and squared returns R_t^2. $T = 1500$. Test size 5%.

	R_t, $P(d=0)$			$\|R_t\|$, $P(d>0)$			R_t^2, $P(d>0)$		
q	KPSS	V/S	R/S	KPSS	V/S	R/S	KPSS	V/S	R/S
0	0.9369	0.9358	0.9343	0.9460	0.9772	0.9720	0.9440	0.9733	0.9739
1	0.9389	0.9376	0.9369	0.9408	0.9739	0.9687	0.9349	0.9674	0.9655
2	0.9395	0.9408	0.9388	0.9375	0.9687	0.9642	0.9323	0.9648	0.9609
3	0.9402	0.9486	0.9414	0.9343	0.9635	0.9622	0.9271	0.9609	0.9557
4	0.9395	0.9512	0.9421	0.9297	0.9616	0.9609	0.9226	0.9577	0.9531
5	0.9388	0.9577	0.9453	0.9271	0.9590	0.9590	0.9219	0.9551	0.9512
10	0.9375	0.9629	0.9512	0.9161	0.9486	0.9473	0.9076	0.9408	0.9421
15	0.9395	0.9603	0.9531	0.8946	0.9375	0.9388	0.8875	0.9284	0.9336
20	0.9369	0.9649	0.9557	0.8777	0.9265	0.9271	0.8719	0.9167	0.9232
25	0.9395	0.9649	0.9557	0.8595	0.9031	0.9161	0.8491	0.8907	0.9083
30	0.9375	0.9681	0.9551	0.8270	0.8823	0.9024	0.8296	0.8615	0.8927

Table 4. Gaussian estimates of d for the bivariate series of simulated absolute returns $|R_{1,t}|$, $|R_{2,t}|$, $\sqrt{|R_{1,t}R_{2,t}|}$. (Monte Carlo S.E. in parenthesis.) $T = 1500$. m_{opt} denotes the optimal bandwidth.

m	$\|R_{1,t}\|$	$\|R_{2,t}\|$	$\sqrt{\|R_{1,t}R_{2,t}\|}$
m_{opt}	0.2771 (0.1127)	0.2766 (0.1124)	0.2793 (0.1130)
84	0.2984 (0.0965)	0.2978 (0.0975)	0.3008 (0.0978)
108	0.2660 (0.0880)	0.2653 (0.0873)	0.2670 (0.0883)
132	0.2421 (0.0811)	0.2411 (0.0814)	0.2432 (0.0821)
156	0.2244 (0.0753)	0.2232 (0.0757)	0.2250 (0.0755)

Table 5. Gaussian estimates of d for the bivariate series of simulated squared returns $R_{1,t}^2$, $R_{2,t}^2$, $R_{1,t}R_{2,t}$. (Monte Carlo S.E. in parenthesis.) $T = 1500$. m_{opt} denotes the optimal bandwidth.

m	$R_{1,t}^2$	$R_{2,t}^2$	$R_{1,t}R_{2,t}$
m_{opt}	0.2582 (0.1076)	0.2592 (0.1096)	0.2121 (0.1009)
84	0.2851 (0.0939)	0.2852 (0.0956)	0.2462 (0.0928)
108	0.2534 (0.0855)	0.2541 (0.0857)	0.2176 (0.0835)
132	0.2304 (0.0788)	0.2308 (0.0793)	0.1978 (0.0775)
156	0.2137 (0.0723)	0.2130 (0.0736)	0.1822 (0.0706)

any long-range dependence. With the assumption that $\sigma_{1,2} > 0$, the simulated co-volatility process $R_{1,t}R_{2,t}$ displays long-memory, the degree of which is close to the one of the series $R_{1,t}^2$ and $R_{2,t}^2$, as empirically observed, see Tables 4 and 5.

From Table 6, we can see that the V/S and R/S 'pox-plot' estimation results do not differ too much from the ones provided by the Gaussian estimator [47].

We report here simulation results for the CVM and K-S change-point tests. Interested readers are referred to [31,30] for the performance of the test by Kokoszka and Leipus [34]. Given that the asymptotic K-S test does not have the correct size, we resort to bootstrap based inference for this test, the number

Table 6. "Pox-plot" estimates of d based on the squared returns series R_t^2. (Monte Carlo S.E. in parenthesis.) $T = 1500$.

	R/S estimate of d	V/S estimate of d	KPSS estimate of d
d	0.2506 (0.0791)	0.2604 (0.0993)	0.3324 (0.1182)

of bootstraps B is set to 399 for all replications. For a test of size 5%, the CVM test rejects 22.97% of the times the null hypothesis of no change-point, while the K-S test rejects this null hypothesis 20.58% of the times. When interpreting these results, we have to keep in mind that these tests have been devised for processes with a single change-point in the conditional variance, and that we apply them to the first-difference of non-standard conditional mean processes, which can have multiple changes in regime.

Figure 3 displays the absolute value of a series of simulated returns generated by the model. This series resembles the series of absolute returns on asset prices, i.e., it does not have a trend, although the ACF of this series displayed LRD-type behavior, see Fig. 4. In Kokoszka and Teyssière [32] and Kirman and Teyssière [31], we used the wavelet estimator by Veitch and Abry [53] for estimating the degree of LRD of several asset prices volatilities and the volatility process generated by our model. Wavelet analysis is of interest as this multi-

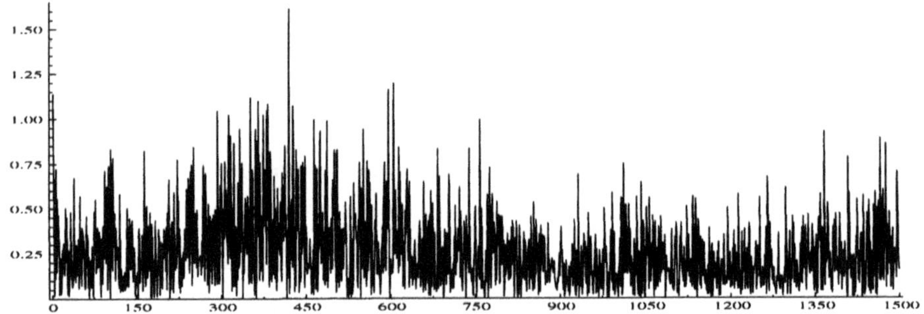

Fig. 3. Absolute value of a series of returns produced by the microeconomic model.

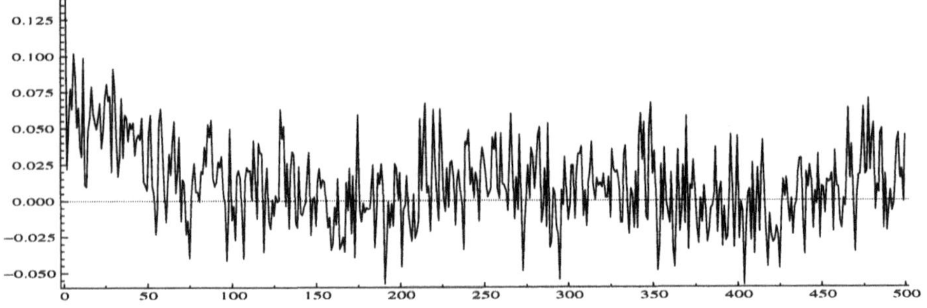

Fig. 4. ACF of the absolute value of a series of returns produced by the model.

resolution analysis is unaffected by changes in the location parameter of a time series and is then able to distinguish between genuine long-range dependence and spurious long-range dependence caused by changes in regimes.

For both real data and series simulated by our model, we observed that the estimated degree of LRD with the wavelet estimator is far lower than the one obtained with the Whittle estimator. The degree of LRD for the absolute returns $|R_{1,t}|$ on British Pound-US dollar drops from 0.41 when estimated with the local Whittle estimator to 0.12 when estimated with the wavelet estimator, while the degree of LRD for the absolute returns $|R_{2,t}|$ on German Deutschmark-US dollar falls from 0.40 to 0.12 respectively. We observe the same changes for the degrees of LRD for the other empirical and simulated volatility and co-volatility processes, i.e., $R^2_{1,t}$, $R^2_{2,t}$, $\sqrt{|R_{1,t}R_{2,t}|}$ and $R_{1,t}R_{2,t}$. Furthermore, for several series, the confidence intervals for the wavelet estimates often contain the value zero. Our microeconomic models are then able to generate most of the empirical dependence properties of daily asset returns.

Acknowledgments

I wish to thank Patrice Abry, Liudas Giraitis, Lajos Hor-váth, Piotr Kokoszka, Alan Kirman, Davar Koshnevisan, Remigijus Leipus, Murad Taqqu, and participants in the International Conference on "Long-Range Dependent Stochastic Processes and their Applications" in Bangalore, January 2002, and to the Stochastics Seminar at the Department of Mathematics of the University of Utah, May 2002, for useful discussions and comments. The final version of this paper has been mainly written in April–May 2002, during my visit to the Department of Mathematics and Statistics of Utah State University, which I thank as well as Piotr Kokoszka for their hospitality.
This text presents research results of the Belgian Program on Interuniversity Poles of Attraction initiated by the Belgian State, Prime Minister's Office, Science Policy Programming. The scientific responsibility is assumed by the author.

References

1. R. Baille, T. Bollerslev, H. Mikkelsen: J. Econometrics **74,** 3 (1996)
2. J. Beran: *Statistics for Long-Memory Processes.* (Chapman and Hall, New York 1994)
3. J. Beran, D. Ocker: J. Business. and Eco. Statist. **19,** 103 (2001)
4. I. Berkes, L. Horváth: Limit Results for the Empirical Process of Squared Residuals in GARCH Models. Stoch. Process. Appl., forthcoming (2003)
5. M. Besseville, I. V. Nikifirov: *Detection of Abrupt Changes: Theory and Applications.* (Prentice Hall, Upper Saddle River 1993)
6. J. R. Blum, J. Kiefer, M. Rosenblatt: Ann. Math. Statist. **32,** 485 (1961)
7. T. Bollerslev: J. Econometrics. **31,** 307 (1986)
8. M. Csörgo, L. Horváth: *Limit Theorems in Change-Point Analysis.* (Wiley, New York 1997)
9. M.M. Dacorogna, U.A. Müller, R.J. Nagler, R.B. Olsen, O. V. Pictet: J. International Money and Finance **12,** 413 (1993)

10. Z. Ding, C.W.J. Granger: J. Econometrics **73,** 185 (1996)
11. R.F Engle, T. Bollerslev: Econometric Rev. **5l,** 1 (1986)
12. R.F. Engle, Y-F. Gau: Conditional Volatility of Exchange Rates under a Target Zone. Preprint, University of California San Diego (1997)
13. E.F. Fama: J. Business **38,** 34 (1965)
14. L. Giraitis, D. Surgailis: Stoch. Process. Appl. **100,** 275 (2002)
15. L. Giraitis, P.S. Kokoszka, R. Leipus, G. Teyssière: J. Econometrics **112,** 265 (2003)
16. L. Giraitis, P.S. Kokoszka, R. Leipus, G. Teyssière: On the Power ot R/S-Type Tests under Contiguous and Semi Long Memory Alternatives. Acta Math. Applicandae, forthcoming (2002)
17. L. Giraitis, P.S. Kokoszka, R. Leipus, G. Teyssière: Statist. Inference Stoch. Process. **3,** 113 (2000)
18. Giraitis, L., P.M. Robinson, D. Surgailis: Ann. Appl. Prob. **10,** 1002 (2000)
19. L. Giraitis, P.S. Kokoszka, R. Leipus: Econometric Theory **16** , 3 (2000)
20. C.W.J. Granger: J. Econometrics. **14,** 227 (1980)
21. C.W.J. Granger, R. Joyeux: J. Time-Ser. Anal. **1,** 15 (1980)
22. C.W.J. Granger, Z. Ding: Annales d'Économie et de Statistique **40,** 67 (1995)
23. C.W.J. Granger, Z. Ding: J. Econometrics **73,** 61 (1996)
24. M. Henry: J. Time-Ser. Anal. **22,** 293 (2001)
25. L. Horváth, P.S. Kokoszka, G. Teyssière: Ann. Statist. **29,** 445 (2001)
26. J.R.M. Hosking: Biometrika **68,** 165 (1981)
27. H.E. Hurst: Trans. Amer. Soc. Civil Engineers **116,** 770 (1951)
28. V. Kazakevičius, R. Leipus: Econometric Theory **18,** 1 (2002)
29. A. Kirman, G. Teyssière: Studies in Nonlinear Dynamics and Econometrics **5**, 281 (2002)
30. A. Kirman, G. Teyssière 'Bubbles and Long Range Dependence in Asset Prices Volatilities,' In: *Equilibrium, Markets and Dynamics. Essays in Honour of Claus Weddepohl.* ed. by C.H. Hommes, R. Ramer, C. Withagen. (Springer Verlag 2002) pp. 307–327.
31. A. Kirman, G. Teyssière: Testing for Bubbles and Change-Points. Preprint, GREQAM & CORE (2001)
32. P.S. Kokoszka, G. Teyssière: Change-Point Detection in GARCH Models: Asymptotic and Bootstrap Tests. Preprint, Utah State University & CORE (2002)
33. P.S. Kokoszka, R. Leipus: 'Detection and Estimation of Changes in Regime.' In: *Long-Range Dependence: Theory and Applications.* ed. by M.S. Taqqu, G. Oppenheim, P. Doukhan (Birkhauser, 2002) pp 325–337.
34. P.S. Kokoszka, R. Leipus: Bernoulli **6,** 513 (2000)
35. H.R. Künsch: 'Statistical Aspects of Self-Similar Processes'. In: *Proceedings of the First World Congress of the Bernoulli Society,***1**, ed. by Yu. Prohorov, V.V. Sazanov (VNU Science Press, Utrecht 1987) pp. 67–74.
36. D. Kwiatkowski, P.C.B. Phillips, P. Schmidt, Y. Shin: J. Econometrics **54,** 159 (1992)
37. A.W. Lo: Econometrica **59,** 1279 (1991)
38. B.B. Mandelbrot: J. Business **36,** 384 (1963)
39. B.B. Mandelbrot: *Fractals and Scaling in Finance: Discontinuity, Concentration, Risk.* (Springer Verlag, 1997)
40. B.B. Mandelbrot, M.S. Taqqu: 'Robust R/S Analysis of Long-run Serial Correlation.' In *42nd Session of the International Statistical Institute, Manila, Book* **2**, pp. 69–99.
41. B.B. Mandelbrot, A. Fisher, L. Calvet: A Multifractal Model of Asset Returns. Preprint, Yale University (1997)

42. T. Mikosch, C. Stărică: Change of Structure in Financial Time Series, Long Range Dependence and the GARCH Model. Preprint, University of Groningen (1999)
43. T. Mikosch, C. Stărică: Non-stationarities in Financial Time Series: The Long Range Dependence and the IGARCH Effects. Preprint, Chalmers University (2002)
44. W.K. Newey, K.D. West: Econometrica **55,** 703 (1987)
45. P.M. Robinson: J. Econometrics **47,** 67 (1991)
46. P.M. Robinson: Time Series with Strong Dependence. In: *Advances in Econometrics, Sixth World Congress*, ed. by C.A. Sims (Cambridge University Press 1994) pp. 47–95.
47. P.M. Robinson: Ann. Statist. **23,** 1630 (1995)
48. S.J. Taylor: *Modelling Financial Time Series.* (Wiley, New York 1986)
49. G. Teyssière: Double Long-Memory Financial Time Series. Preprint, GREQAM (1996)
50. G. Teyssière: Modelling Exchange Rates Volatility with Multivariate Long-Memory ARCH Processes. Under revision for the J. Business and Econ. Statist., (1997)
51. G. Teyssière: 'Multivariate Long-Memory ARCH Modelling for High Frequency Foreign Exchange Rates.' In *Proceedings of the HFDF-II Conference*, (Olsen & Associates 1998)
52. H. Theil: *Economic Forecast and Policy.* (North Holland, Amsterdam, 1961)
53. D. Veitch, D., P. Abry: IEEE Trans. Information Theo. **45,** 878 (1999)

Long Memory and Economic Growth in the World Economy Since the 19th Century

Gerald Silverberg[1] and Bart Verspagen[2]

[1] III and MERIT, University of Maastricht, The Netherlands
[2] ECIS, Eindhoven University of Technology, The Netherlands

Abstract. We discuss the relevance of long memory for the investigation of long-term economic growth and then briefly review the state-of-the-art of statistical estimators of long memory on small samples and their application to economic datasets. We discuss theoretical mechanisms for long memory such as cross-sectional heterogeneity. We argue that this endogeneity should be explained endogenously and not simply assumed. Evolutionary models of growth appear to offer one natural explanation of such heterogeneity. Using the Maddison (1995) [1] data on 16 countries starting in 1870, supplemented by more recent data down to the year 2001, we then apply different estimators to test the hypothesis of long memory on individual country GDP and GDP per capita. These estimators are Beran's FGN nonparametric test based on an approximate Whittle ML estimator, Robinson's semiparametric log periodogram regressor, Sowell's parametric ML ARFIMA estimator and the ML FAR estimator. The results are mixed and somewhat ambiguous between methods. Moving from the nonparametric to the parametric methods (i.e., controlling for short memory) we find less evidence of long memory. We find that Robinson's semiparametric method also suffers from severe sensitivity to the cutoff parameters. We compare our results with those of Michelacci and Zaffaroni [2] and criticize their methodology. We conclude that the lack until now of a single test that deals successfully with all known problems (small sample bias, short memory contamination, specification error, parameter sensitivity) precludes the formulation of a definitive statement about long memory in economic growth.

1 Introduction

One of the most hotly debated topics in macroeconomics in recent years has been the nature of fluctuations in the growth process of aggregate output in the longer term. Traditionally, economists have conceived of the growth process as consisting of a deterministic trend (such as exponential growth) on which is superimposed either deterministic cyclical fluctuations (a decidedly minority view in the meantime) or a stable, mean-reverting stochastic process of ARMA type. Since Nelson and Plosser's seminal 1982 paper [3] this view has been challenged by purported evidence of a unit root in univariate time series such as GDP, i.e., essentially a random walk component, which puts into question the notion of any underlying deterministic trend. However, this knife-edge distinction between a unit root and a near unit root but stationary process has proven difficult to maintain empirically, especially since the main test employed, the Dickey-Fuller test, has been shown to have limited power against a range of alternative hypotheses.

The main empirical feature about which most authors agree consists of the observation of persistence in aggregate time series, meaning concretely the slower decay of the autocorrelation function than would be expected of a pure ARMA process (which is geometric in the lag). An equivalent way of looking at this is in the frequency domain, where it was already observed by Granger in 1966 [4] that most macroeconomic variables display a typical spectral shape, apparently diverging to infinity at zero frequency and declining uniformly at high frequencies. Time series with this property show periods of all lengths and have no characteristic time scale, and are often referred to as aperiodic cycles. The random walk itself is one example of this class of processes, but in fact it is only a special case of a family of such processes which can be parameterized by their degree of fractional integration. Given that agreement holds about persistence, one can enlarge the test hypothesis from unit root or not to the question of the degree of fractional integration of the process.

A second question naturally arises as to what the underlying mechanism in the economy could be which would produce this typical pattern. One path is via dynamic aggregation of an economy with cross-sectional heterogeneity ([5], [6], [7], [2]). Another, somewhat related approach, invokes a dynamic, evolutionary framework of technology diffusion which naturally results in cross-sectional heterogeneity ([8], [9], [10]). We will describe this approach in more detail in Sect. 4.

Why are persistence and long memory of interest in the analysis of long-term patterns of growth and fluctuations? For one thing, it is intermediate between a relatively unstructured stochastic world in which the present is just the summation of unrelated random events in the past (a random walk), and a rigidly predictable deterministic cycle or trend with relatively negligible, mean-reverting stochastic disturbances. It preserves the notion of even the distant past continuing to influence the present in a somewhat law-like fashion, allowing the future to be structured while remaining shrouded in a haze of uncertainty. It also allows for developments at a range of time scales, with what appear to be trends at one time scale being revealed to be parts of irregular cycles at longer time scales. Moreover, many natural phenomena display long memory, such as the flows of rivers, which originally led Hurst in 1951 [11] to investigate the matter statistically.

Economists have been interested in long memory since the original work of Mandelbrot in the 1960s on price series. Since then attention has concentrated on long memory in asset prices and foreign exchange rates, and within the last fifteen years, on real national output measures as well. In the present paper we propose to use recent refinements in technique due to Beran [12] and Sowell [13], and the more extensive national income data set of Maddison [1] to examine both long memory in national times series. The paper is organized as follows. Section 2 reviews definitions of long memory and various statistical tests which allow hypothesis testing. Section 3 briefly reviews the literature on long memory in real and monetary economic variables. Section 4 discusses possible modeling approaches yielding long memory in output variables, including some of our own work on evolutionary selforganization in an endogenous growth setting. Our main empirical results are presented in Sect. 5. Section 6 concludes with a discussion of the 'state of the art.'

2 Long Memory: Definition and Statistical Tests

A good survey of the basic mathematical statistics can be found in [12], while the econometric literature on long memory has been reviewed in [14]. Here we will only recapitulate some of the salient points relevant to our work. There are several related ways of defining long memory of a discrete, real-valued time series y_i formulated either in the time or the frequency domains:

$$\gamma_k = const \cdot k^{2d-1} \text{ask} \longrightarrow \infty, \tag{1}$$

where γ_k is the autocovariance function at lag k;

$$f(\omega) = const \cdot \omega^{-2d} \text{as}\omega \longrightarrow 0^+, \tag{2}$$

where $f(\omega)$ is the spectral density of the process. A short-memory process such as an ARMA process will have exponentially decaying autocovariances and zero spectral density at the origin. Two well-known cases can be singled out: Gaussian white noise, for which $d = 0$, and a random walk, where $d = 1$. Another definition involves the sum of the autocorrelations:

$$\lim_{n \to \infty} \sum_{j=-n}^{n} |\rho_j| = \infty, \tag{3}$$

where ρ_i is the autocorrelation at lag i. Long memory was first related to fractional integration by Granger and Joyeux [15] and Hosking [16]) using the following discrete time stochastic process:

$$(1 - L)^d y_t = \epsilon_t, \tag{4}$$

where L is the lag operator, d is the fractional integration parameter, y_t is the stochastic process, and ϵ_t is iid white noise.

Hurst [11] developed the first estimation procedure for long memory processes, defining the Hurst parameter based on rescaled-range (R/S) analysis, a natural quantity derived from his field of hydrology for determining the appropriate size of a reservoir. While R/S analysis is relatively easy to implement, it suffers from two disadvantages. First, little is known about the statistical properties of the estimator, so that it does not lend itself to hypothesis testing. And second, the results may be biased by the presence of short memory effects in addition to long ones. Lo [17] proposed a modified rescaled range estimator to overcome both of these difficulties, but simulation results have not borne out its robustness.

Another approach is based on the characteristic pattern of the periodogram at low frequencies. Geweke and Porter-Hudak [18] proposed an estimator based on regressing the log of the periodogram on the log of a trigonometric function of frequency. A difficulty with this approach concerns the number of values of the periodogram to use in the regression. They proposed a heuristic based on the length of the time series, but this estimator suffers as well from serious bias in the event of short memory and on small samples [14, p. 33] [19] [20] [21] [22] [23] [24]. Robinson has further refined this approach[25].

An alternative is to use a full or approximate maximum likelihood estimator. A Whittle approximate maximum likelihood estimator is described in detail in [12]. The exact maximum estimator has been derived in [13]. One way of dealing with short memory contamination in the context of maximum likelihood estimators is to simultaneously estimate a full ARFIMA(p,d,q) model of the form

$$A(L)(1-L)^d y_t = B(L)\epsilon_t, \tag{5}$$

where $A(L)$ and $B(L)$ are polynomials with roots outside the unit circle of order p and q, respectively. The basic Fractional Gaussian Noise model is reobtained by setting p and q equal to 0. Computer programs for the Whittle and exact likelihood estimators have been developed by Beran, and Doornik and Ooms, respectively. MLE can also be performed for ARFIMA models within the Ox and Splus packages. With the availability of fast computers, the additional computational complexity of maximum likelihood methods is no longer a problem on PCs.

While ARFIMA methods control for short memory and indeed model it explicitly, they suffer from specification error, i.e., only if the order of the autoregressive and moving average terms p and q are are known a priori is the estimator asymptotically unbiased and consistent. If these terms are unknown, we are confronted with a model selection problem. A standard solution is to apply an information criterion such as the Aikake Information Criterion (AIC) to select the model with the highest value. Even then, however, the estimate may be unstable due to the extreme sensitivity of the MA terms. For this reason, [26] and [27] propose dealing with the short-memory component solely as an autoregressive process. The resultant FAR estimator, they show, is consistent if the BIC or HIC instead of the AIC is used.[1]

While asymptotic results have been derived for the maximum likelihood estimators, their small-sample properties can differ considerably. Recently, [28] has systematically compared the biases, mean square errors, and other measures of the various maximum likelihood estimators on samples of size 100 using Monte Carlo methods. He concludes that the Whittle estimator with tapered data is most reliable. The exact maximum likelihood estimator for demeaned data, in contrast, "should only be used with care for fractionally integrated models due to its potential large negative bias of the fractional integration parameter." [21] are critical of ARFIMA models for estimating persistence in aggregate output.

[1] The definitions of the criteria we use in the following are, for an ARMA(p,q) process with $f=p+q$ degress of freedom: $AIC(f)=2L-2f$, $BIC(f)=2L-f(1+\log n)$, where L is the log likelihood, n is the number of observations, and we maximize the criterion for order selection

They show that overparametrization of an ARMA model may bias the estimates of persistence downwards, an effect which may be reflected in some of our results reported later in this paper. The small-sample properties of the FAR estimator have been examined in a Monte Carlo study in [26]. While the asymptotic results seem to hold for sample sizes of 1000, for 200 observations, the BIC often prefers the ARIMA(0, d, 0) model since it strongly penalizes higher orders.

3 Studies of Long Memory in Real and Financial Economic Variables

One of the main motivations for searching for long memory in economic time series is an observation by Granger [4] about the 'typical shape' of the spectra of economic variables. Considerable interest has focused on long memory in asset price series, a literature which is reviewed in [29]. The evidence is still rather ambiguous, although long memory seems much more likely in asset volatility than in asset returns themselves. Studies of real output measures commenced with [30] and [6]. The former authors examined a number of US output time series using a two-stage process based on the GPH estimator to determine the degree of fractional integration, and an approximate maximum likelihood ARMA on the filtered residuals to handle the short memory aspect. They found significant evidence of long memory in these series. In contrast, Haubich and Lo, using Lo's modified rescaled range estimator, found no evidence for long memory in a time series of quarterly postwar US data, in contrast to using the unmodified R/S statistic. Both of these studies suffer, however, from the estimation problems raised in the previous section.

These estimation problems can be circumvented in some extent by employing a maximum likelihood estimator of a full ARFIMA model. Sowell [20] reexamined postwar US quarterly real GNP with the exact maximum likelihood estimator, nesting deterministic trend and unit root models within the ARFIMA model. He reports a significantly smaller estimate of the fractional integration parameter than [30] due to the bias of the latter method. He concludes that the data are consistent with both trend models. In fact, the best model according to the AIC from a set of estimations with p and q ranging from 0 to 3 is ARFIMA(3, -0.59, 2).

In a recent paper Michelacci and Zaffaroni [2] (referred to as MZ in the following) attempt to reconcile apparently contradictory stylized facts about the presence of unit roots, beta convergence, and long-term exponential trends by invoking the possibility of long memory in the growth process. MZ employ the semiparametric Robinson lpr with joint estimation on the entire sample [25]. Two rule-of-thumb cut-off parameters are employed in this otherwise nonparametric process to determine how many components of the periodogram to use. If short memory is present in the data, Monte Carlo simulations have demonstrated conclusively that this estimator is biased in small samples. Moreover, MZ first subject their raw data to two 'filtering' steps before testing for long memory. The specification of these steps is relegated to an appendix, where their motivation and implications still remain somewhat unclear.

The first step is to subtract an exponential trend (either country specific or common to the whole sample). MZ assert but do not prove that "the memory of the process is entirely reflected in the residuals." The second step consists of fractionally differencing the residual by $\frac{1}{2}$ in order to move the estimates into the region of validity of Robinson's asymptotic test statistic. Two objections can be made here. First, as the authors observe, fractionally differencing by $\frac{1}{2}$ entails an approximation and loss of data since only a finite number of terms of the infinite series representation of the differencing operator can be used, and the series must be initialized with (in their case) 10 observations. We are asked in a footnote to take it on trust that this does not introduce any bias into the test. It is also asserted that the results are 'generally robust' to the choice of the degree of differencing. But if this is the case, why not simply difference by one and avoid the approximation problem entirely (we cannot follow their arguments here concerning the obscuring of the distinction between conditional and unconditional convergence)? Furthermore, if we are free to difference the series as we like, Robinson's restriction of his statistic to $d \leq 1/2$ makes no sense. Thus legitimate questions remain concerning the extent to which the main results are influenced by nontrivial data filtering techniques.

4 Theoretical Basis for Long Memory in Macroeconomic Variables

Although long memory has been established in numerous domains of the natural sciences and is at least an open question in social sciences, until now there has been a dearth of theoretical explanations for its occurrence. Of course one might argue that climatological long memory might spill over into the economic domain without the need for a specifically economic mechanism, much as sunspots had been proposed as the driving force of business cycles by Jevons. While this is not entirely implausible, it seems more reasonable to look for an explanation for long memory intrinsically rooted in the economic process.

An observation due to Granger [5] is useful in this regard. He remarked that if the economy is composed of a large number of individual units each following an AR(1) process, and whose coefficients are drawn at random from a distribution (such as a beta distribution on (0,1)), then the additive aggregate of these variables will display long memory. Thus cross-sectional heterogeneity can be a source. However, this cross-sectional heterogeneity is simply assumed to be an exogenous and invariant given, and does not itself emerge from a more fundamental economic mechanism. [6] also invoke an argument in this vein. [7] and [2] take this argument a step further towards endogenization by relating it to a technological vintage model in which small and large firms position themselves at different relative distances to a moving technological frontier, leading to a modified Gibrat-like stochastic process.

Endogenous cross-sectional heterogeneity, however, is one of the hallmarks of evolutionary models. In these models firms are usually characterized by bounded rationality and rule-of-thumb search behavior, possibly with some form of adaptive learning, basic uncertainty, disequilibrium interactions, and system-emergent selection along the lines of Schumpeterian competition. Cross-sectional heterogeneity will be a natural outcome of such a model, as numerous authors have remarked, but heterogeneity that may be time-varying, and that involves constant reordering of the units in the distribution as a result of their relative performance.

A sequence of related growth models developed by Silverberg and Lehnert and Silverberg and Verspagen [31], [9], [32], [10] will furnish the basis for the following discussion. Consider a 'technology space' consisting of a line of discrete technologies, with technologies to the right being improvements on their immediate predecessors by a fixed factor. At any point in time the firms in this economy will be utilizing some (finite) subset of the nodes in this space. The economy is assumed to progress by a) discovering new nodes to the right, one step at a time, and b) diffusing the known superior technologies into the capital stock by investment and eliminating obsolete technologies by scrapping. The discovery of new technologies will be assumed to be governed by a stochastic process (such as Poisson) whose rate will either be exogenously given or endogenously determined by the profit-seeking activities of the agents themselves.

A common feature of all versions of this underlying structure is that time series of aggregate variables display Granger's 'typical shape' in the frequency domain. At the same time as technologies diffuse into and out of the economy according to the well-known logistic replacement laws, firms grow and contract, and structural breaks can occur in such system-level characteristics as concentration indexes. In other words, a complicated, endogenously generated process of structural change is induced which displays many well-known empirical micro and meso-regularities as well as an aggregate growth process.

[32] take the model a step further into the international domain by allowing for a world of different national economies structured in this way with limited possibilities of technological and behavioral transfer between them. The main result is that such a world economy itself will be characterized by aperiodic cycles of convergence and divergence of the long-memory type. Measures of inequality such as the coefficient of variation of GDP per capita between countries (standard deviation divided by mean) show an irregular pattern of fluctuation at all time scales. What may look like a deterministic pattern of convergence over, say, 100 years, may later be reversed by an equally long period of divergence. Thus over finite windows of observation it becomes impossible to differentiate between apparent trends and segments of superlong 'cycles.'

We believe the exclusive focus on apparent convergence in OECD countries in the postwar period obscures the more complex overall picture, as several authors have also pointed out. First, there is a problem of sample bias due to testing convergence only on a sample of countries with high per capita income in the final period. Looking at a more inclusive sample clearly demonstrates that there is no unconditional tendency to convergence (in the sense, e.g., that countries with lower per capita incomes in the initial period have proportionally higher growth rates over the period – one need only think of the African countries). Second, there is also a temptation on the basis of the Solow growth model to expect universal convergence to a common world frontier without ever asking what were the sources of backwardness in the more distant past (or else there never would have been a problem of convergence in the first place) nor of divergence in the 'present' (as the Twentieth Century history of some of the more advanced Latin American countries such as Argentina, or the most recent experience of the Asian tigers demonstrate). Prima facie one could argue that precisely such an irregularly fluctuating pattern accords better with the historical record than a basic underlying tendency to convergence and that it is more reasonable to posit two opposing forces at work: catch-up due to technological and behavioral imitation on the one hand, and the operation of virtuous and vicious cycles in the technological and institutional domains allowing the rich to grow richer and the poor poorer for extended periods of time. Combined with the operations of chance and threshold effects, these same positive feedbacks can allow some countries to take off and even overtake the leaders while others may fall behind and forfeit long-dominant positions. While at first glance such patterns can appear to be devoid of interpretable regularity and of the simplicity of a convergence hypothesis, the net result may precisely be long-memory time series properties of the type we will investigate in the next section. The convergence issue is discussed in more detail in [33]

5 Long Memory and Economic Growth Revisited

Two problems cloud the literature on long memory and economic growth. First, the reliability of the econometric methods used, and second, the dependence on the dataset employed. In this section we intend to address both of these questions, drawing on an extension of the Maddison [1] data obtained by 'splicing' on more recent data from the GGDC Total Economy Database [34]. The attraction of the Maddison dataset is that it extends over a considerably longer period of time than the post-war (quarterly) US dataset used by the first researchers on the subject and encompasses a large number of countries, ostensibly on a consistent basis. Since it consists of annual observations, the time-aggregation problem of the fractional integration parameter implies that estimates based on annual and quarterly data are not immediately comparable (moreover, seasonality issues are thereby also avoided). While the original Maddison dataset contains 25 countries, including many contemporary developing countries, the series are of varying length and many are too short to permit sensible long memory estimation. Thus we had to reach a compromise between the breadth of the dataset and its length. We have selected a core subset of sixteen countries[2] for which

[2] Australia, Austria, Belgium, Canada, Denmark, Finland, France, Germany, Italy, Japan, The Netherlands, New Zealand, Norway, Sweden, United Kingdom, and United States.

Table 1. Hurst parameter estimated using Beran's FGN estimator for GDP and GDP per capita. One, two and three stars indicate significance at the 10%, 5% and 1% levels, respectively.

	H(GDP)	t	H(GDPpc)	t
Australia	0.531	0.56	0.523	0.42
Austria	0.512	0.22	0.528	0.51
Belgium	0.647	2.59 **	0.641	2.48 **
Canada	0.676	3.08 ***	0.671	3.00 ***
Denmark	0.487	-0.24	0.486	-0.26
Finland	0.664	2.87 ***	0.665	2.89 ***
France	0.665	2.89 ***	0.613	2.01 **
Germany	0.710	3.65 ***	0.697	3.43 ***
Italy	0.665	2.89 ***	0.655	2.73 ***
Japan	0.572	1.21	0.574	1.25
Netherlands	0.546	0.83	0.537	0.67
New Zealand	0.496	-0.07	0.397	-1.97 *
Norway	0.549	0.88	0.550	0.91
Sweden	0.636	2.40 **	0.630	2.29 **
United Kingdom	0.744	4.22 ***	0.742	4.19 ***
United States	0.655	2.73 ***	0.656	2.75 ***

complete data are available for the period 1870 to 2001 (with the exception of the Japanese series, which starts in 1885).

Our methodology differs from that of MZ in the following respects:

1. Instead of subtracting a trend (however selected) and fractionally differencing, we simply first difference the (log) series.
2. To avoid the problems associated with the GPH 'semiparametric' parameters, we employ the nonparametric FGN[3] estimator [12] as well as the Robinson estimator that MZ use and compare the results.
3. We test the sensitivity of the GPH/Robinson estimators with respect to the bandwidth and trimming parameters, for which only heuristic guidelines are available.
4. To tackle the short memory contamination problem, we employ a parametric maximum likelihood ARFIMA estimator as well as FAR estimation procedures.

Table 1 summarizes the results for the FGN estimator for GDP and GDP per capita (pc) time series, expressed as the Hurst parameter H $(= d + 1/2)$. The t-statistic, which Beran shows is asymptotically $N(0,1)$ according to the model assumptions, is calculated for a two-sided test of $H = 1/2$.

The estimates for GDP and GDP per capita are quite similar with respect to both point estimates and significance levels. For seven countries no long memory is evident for either time series. Seven countries display significant long memory

[3] Beran's FGN estimator models the time series as a fractional Gaussian noise process. We make use of the Splus script available in the appendix of his book.

at the 1% level and two at the 5% level. New Zealand shows significant *antipersistence* ($H < 1/2$) at the 10% level, but only for GDP per capita. The effect of the watershed of the Second World War on these results is investigated in [33], where it is shown that the prewar and postwar periods differ significantly with respect to long memory. There is very little evidence for long memory prewar but considerably more postwar, with a general increase in estimated H-values from the prewar to the postwar period.

The picture changes radically when we apply Robinson's LPR estimator[4], the tool used by MZ. Recall that this semiparametric estimator employs two parameters: trimming, which specifies how many of the lowest frequency periodogram components to omit from the regression, and bandwidth or power, which determines a high-frequency cutoff (in the form N^{power}, where N is the number of observations).While it shares the latter parameter with the GPH estimator, the use of the former was proposed by Robinson. However, [24] argue that nonzero values of trimming are not appropriate. Robinson's LPR permits joint estimation of multiple time series as well as univariate analysis. This has no effect on the point estimates. Since the Japanese data are shorter in any case, which means we would have to exclude them from a joint dataset, we present only the univariate estimates here. The results do not differ substantially from the joint estimates. Table 2 summarizes the estimates for three values of the trimming parameter and three values of the power, starting with the value of 0.525 often recommended in the literature.

First of all, there is essentially no robust evidence for long memory, in contrast to the FGN case, in which nine countries had significant results for $H > 1/2$ (corresponding to $d > 0$). Most of the significant results indicate on the contrary *antipersistence* ($d < 0$), but even then, this is never true of more than five countries for given values of the parameters. For nine countries (Australia, Belgium, Canada, Denmark, Italy, New Zealand, Norway, Sweden, and the UK), both hypotheses are rejected consistently across the board. Neither hypothesis can be consistently maintained for any other country. One pattern does seem to emerge. Increasing the power seems to raise the value of the estimates, so that the number of significant negative estimates declines and positive ones increases. Setting the trimming parameter equal to one also seems to increase the total number of significant results of either sign. Of course none of these casual observations is based on any statistical testing, but it is nevertheless interesting to gauge the variability of the estimator on a real dataset. A well-designed Monte Carlo study would obviously be useful here. In any event, MZ's claim (p. 152) "It is nevertheless important to point out that the empirical results are very robust to changes in the choice of the trimming and the bandwidth" is clearly not substantiated in our work.

[4] We employed a Stata script developed by Christopher F. Baum and Vince Wiggins available at http://ideas.uqam.ca/ideas/data/Softwares/bocbocodeS411001.html.

Table 2. Sensitivity of the Robinson estimator to the trimming and power parameters. Significance levels are indicated by one, two and three stars (10%, 5%, 1%, respectively).

trimming	0			1			2		
power	0.525	0.6	0.7	0.525	0.6	0.7	0.525	0.6	0.7
Australia	-0.118	-0.077	0.029	-0.458	-0.269	-0.052	0.134	0.115	0.191
Austria	-0.261	-0.184	-0.099	-0.544 ***	-0.328 **	-0.158	-0.809 ***	-0.418 **	-0.172
Belgium	0.157	0.140	0.120	0.032	0.057	0.068	0.078	0.089	0.086
Canada	-0.193	-0.081	0.068	0.019	0.090	0.208	0.018	0.110	0.248
Denmark	0.143	-0.013	0.040	0.186	-0.046	0.033	0.153	-0.130	0.000
Finland	0.054	-0.136	0.033	-0.201	-0.359 *	-0.049	-0.014	-0.295	0.044
France	-0.087	0.124	0.262 *	-0.139	0.179	0.326 *	-0.147	0.274	0.399 *
Germany	-0.283 *	-0.263 ***	-0.086	-0.370 *	-0.312 **	-0.067	-0.413	-0.323 *	-0.025
Italy	-0.031	0.122	0.046	-0.163	0.093	0.011	-0.509	-0.040	-0.073
Japan	0.062	0.130 **	0.182 **	0.033	0.138 *	0.198 *	-0.054	0.113	0.192
Netherlands	-0.204	-0.187 **	-0.065	-0.360 **	-0.275 **	-0.084	-0.218	-0.170	0.007
New Zealand	0.086	-0.004	-0.208	0.118	-0.012	-0.269	0.157	-0.017	-0.326
Norway	0.022	0.092	0.113	-0.366	-0.112	0.005	-0.607	-0.167	-0.001
Sweden	0.104	0.063	0.190	-0.139	-0.103	0.122	-0.291	-0.182	0.115
United Kingdom	-0.157	-0.161	-0.015	-0.563	-0.405	-0.110	-0.241	-0.175	0.072
United States	-0.471 **	-0.502 ***	-0.040	-0.538 *	-0.555 ***	0.055	-0.217	-0.379 *	0.269
# significant >0	0	1	2	0	1	2	0	0	1
# significant <0	2	3	0	4	5	0	1	3	0

Let us now turn to parametric estimators. Table 3 summarizes the results for these two parametric estimators for GDP per capita time series.[5] Order selection is performed for ARFIMA by calculating all 25 models for p and q between 0 and 4 and maximizing the AIC, and for FAR for p values between 0 and 8 maximizing BIC, and for purposes of comparison, AIC. A caveat is that for some few combinations of parameter values and countries the iterative algorithm does not converge, so that we cannot be sure that the values we are left with really always include the maximum.

ARFIMA/AIC only returns two countries with significant antipersistence, Germany and the United States. FAR/BIC captures these two countries plus three more, Canada, Italy, and New Zealand. Surprisingly, the BIC criterion prefers low or even zero autoregressive orders, while using AIC for the same exercise selects for much higher-order AR terms. In fact, they only agree on the order for four countries. Furthermore, FAR/AIC eliminates Italy and New Zealand from the list of significantly antipersistent countries. The three parametric methods are roughly consistent with respect to Canada, Germany and the United States in claiming antipersistence, and with respect to Australia, Austria, Belgium, Denmark, Finland, France, Japan, the Netherlands, Norway, Sweden, and the UK in rejecting both hypotheses. FAR/BIC is the only para-

[5] We used the Ox package to perform the estimations using maximum likelihood (cf. [35]). Similar results can be obtained using Splus and with Beran's ARFIMA estimator based on Whittle approximate ML.

Table 3. Parametric estimators ARFIMA and FAR with order selection according to the appropriate information criterion, for GDP per capita series. Stars indicate significance levels (10%, 5% and 1%).

	ARFIMA/AIC			FAR/BIC		FAR/AIC	
	p	q	d	p	d	p	d
Australia	3	2	0.017	0	0.016	0	0.016
Austria	1	0	-0.206	0	-0.006	1	-0.206
Belgium	2	3	0.014	0	0.123	2	0.089
Canada	3	4	-0.701	1	-0.523 ***	1	-0.523 ***
Denmark	1	1	-0.107	0	-0.052	3	-0.257
Finland	1	1	0.067	4	0.007	4	0.007
France	4	3	-0.351	0	0.087	6	-0.010
Germany	0	1	-0.222 **	1	-0.376 **	2	-0.394 *
Italy	1	0	-0.108	0	0.141 *	1	-0.108
Japan	0	0	0.067	0	0.067	0	0.067
Netherlands	1	1	-0.106	0	-0.007	1	-0.259
New Zealand	3	2	-0.272	0	-0.169 **	2	-0.033
Norway	2	2	0.069	0	0.039	0	0.039
Sweden	4	4	-0.019	0	0.119	5	0.092
United Kingdom	4	1	0.013	1	-0.222	8	0.115
United States	2	4	-0.707 ***	1	-0.534 ***	2	-0.939 **

metric method which claims (weakly) significant long memory for one country (Italy, where it agrees with FGN). Otherwise, none of the parametric methods provide any evidence for long memory, and only a few cases of antipersistence.

We can obtain some insight into the consistency of N estimators by simply tabulating whether they fail to reject the hypothesis $d = 0$ (when we shall enter a value of 0) or find a significant result $d \lessgtr 0$ at the 10%, 5% or 1% levels (when we shall enter $\pm i$, respectively, where $i = 1, 2, 3$ corresponds to the number of significance 'stars'). If for each country we count the number of significant positive (n_+), negative (n_-), and 'zero' ($n_0 = N - n_+ - n_-$) cases, we can then compute a 'consistency index' CI (in analogy to the well-known Herfindahl index) of the set of estimators:

$$CI = (n_+^2 + n_-^2 + n_0^2)/N^2. \tag{6}$$

This will assume the maximum value of 1 when all results fall into a single category and a minimum value of 1/3 when they are equally divided between the three categories, and thus maximally inconsistent (maximum entropy case). Table 4 summarizes these qualitative results for the nonparametric FGN estimator, three versions of the LPR semiparametric estimator (all with trimming zero), and the three parametric maximum likelihood estimators just discussed.

The only totally consistent estimates are for countries for which the null hypothesis cannot be rejected, i.e., where no evidence is found for either long memory or antipersistence: Australia, Austria, Denmark and Norway. The FGN estimator has a special status in this comparison, since it is known to give mis-

Table 4. Qualitative summary of consistency of nonparametric, semiparametric and ML parametric estimators. The values 1,2,3 correspond to the significance level (10%, 5%, 1%, respectively), the sign corresponds to the sign of the estimated d.

	estimator							with FGN				without FGN			
	1	2	3	4	5	6	7	n_-	n_+	n_0	CI	n_-	n_+	n_0	CI
Australia	0	0	0	0	0	0	0	0	0	7	1	0	0	6	1
Austria	0	0	0	0	0	0	0	0	0	7	1	0	0	6	1
Belgium	2	0	0	0	0	0	0	0	1	6	0.76	0	0	6	1
Canada	3	0	0	0	0	-3	-3	1	2	4	0.43	1	1	4	0.5
Denmark	0	0	0	0	0	0	0	0	0	7	1	0	0	6	1
Finland	3	0	0	0	0	0	0	0	1	6	0.76	0	0	6	1
France	2	0	0	1	0	0	0	0	2	5	0.59	0	1	5	0.72
Germany	3	-1	-3	0	-2	-2	-1	5	1	1	0.55	5	0	1	0.72
Italy	3	0	0	0	0	1	0	0	2	5	0.59	0	1	5	0.72
Japan	0	0	2	2	0	0	0	0	2	5	0.59	0	2	4	0.56
Netherlands	0	0	-2	0	0	0	0	1	0	6	0.76	1	0	5	0.72
New Zealand	-1	0	0	0	0	-2	0	2	0	5	0.59	1	0	5	0.72
Norway	0	0	0	0	0	0	0	0	0	7	1	0	0	6	1
Sweden	2	0	0	0	0	0	0	0	1	6	0.76	0	1	5	0.72
United Kingdom	3	0	0	0	0	0	0	0	1	6	0.76	0	0	6	1
United States	3	-2	-3	0	-3	-3	-2	5	1	1	0.55	5	0	1	0.72
# significant <0	1	2	3	0	2	4	3	14				13			
# significant >0	9	0	1	2	0	1	0		14				6		
# null hyp. (=0)	6	14	12	14	14	11	13			84				77	

Column 1=FGN, 2=lpr(0, 0.525),3=lpr(0,0.6), 4=lpr(0, 0.7)
5=ARFIMA/aic, 6=FAR/bic, 7=FAR/aic

leading results in the presence of short memory, and indeed it is the only estimator to produce a large number of significant long-memory findings (nine, against only one case of antipersistence and six of non-rejection). If we remove it from the set, Belgium and the United Kingdom also join the neutral class in which all remaining estimators cannot reject the null hypothesis ($d = 0$). We consequently calculate the consistency index for both FGN included and excluded. Even then, no long memory or antipersistence result is consistently maintained by the two sets of estimators. The closest we come are the two parallel cases Germany and the US, where FGN produces the only 'positive' result and all other estimates are significant for $d < 0$ except that produced by lpr(0, 0.7). These are the only two countries for which one could reasonably argue that our battery of estimators produces convincing evidence of *antipersistence*. All of the findings of long memory, unfortunately, are always accompanied by even more instances of findings of antipersistence and null-hypothesis non-rejection from the other estimators.

6 Discussion and Conclusions

Our analysis recapitulates to some extent the historical development of long-memory studies in the field of economic growth. Beginning with nonparametric and heuristically motivated semiparametric methods, evidence was adduced, particularly for quarterly postwar US data, for the presence of long memory. Relatively soon, however, objections were raised about short-memory contamination leading to spurious results. This has lead to a cycle of objections and statistical refinements which has made the issue increasingly ambiguous. Refinements of the semiparametric GPH estimator due, for example, to Robinson, have allowed asymptotic properties to be rigorously established. Nevertheless, the need to set parameters by hand means that in small samples these estimators still are poorly understood, and as we have shown, highly sensitive to small changes. The arrival of approximate and full maximum likelihood parametric methods appeared to solve the short-memory problem satisfactorily except that they introduced the equally intractable problem of specification error. Once again, rigorous results are only asymptotic and subject to correct order selection. The use of information criteria, while to some extent justifiable theoretically, is still a matter of black magic in the small-sample case.

Monte Carlo studies on small samples have repeatedly confirmed many of these ambiguities, against theoretical expectations. While useful in a negative sense, they are always performed against a testbed of simple, well-understood artificially generated time series and do not necessarily shed light on larger classes of processes. Confronted with real-world data for which no data-generating process is a priori known, the properties and applicability of the estimators become even more doubtful. In this paper we have therefore proceeded in an empirical spirit and compared estimators from the three classes of models against each other to establish their mutual consistency. We find this consistency to be extremely low, both between classes and internally. One reason of course is that our data may not be stationary. It would not be surprising to find structural breaks in the world economy since the late 19th century, in particular due to the watershed of the Second World War. Thus it may be impossible in principle to find sufficiently long economic growth time series that are stationary, even after first differencing of logs or some other simple transformation. Thus even though by the standards of natural science long memory studies our time series are extremely short, they are perhaps still too long to expect stationarity. Another issue is how the data are handled – for example, whether taking trend residuals and half-differencing, as MZ do, does not lead to spurious results.

Our disappointment is that a study that began several years ago with the intention of elucidating long-term properties of the world economy on the basis of the large multicountry Maddison dataset 'degenerated' rather quickly into a study of the pitfalls of statistical methodology. What can we extract from the different approaches that seems unimpeachably established? Mainly the negative result that four countries show no evidence whatsoever for either long memory or antipersistence. For the countries that do, the results are always contradictory depending on the estimator used. Therefore, no unambiguous conclusions are possible at the moment. Further work by statisticians to establish reliable small-sample properties of these or other estimators on real-world data will be necessary before further progress in this field is possible, we are afraid.

References

1. A. Maddison: *Monitoring the World Economy 1820-1992* (OECD, Paris 1995)
2. C. Michelacci, P. Zaffaroni: Journal of Monetary Economics **45**, 129–153 (2000)
3. R. Nelson, C.I. Plosser: Journal of Monetary Economics **10**, 139–162 (1982)
4. C.W.J. Granger: Econometrica **34(1)**, 150–161 (1966)
5. C.W.J. Granger: Journal of Econometrics **14**, 227–238 (1980)
6. J.G. Haubich, A.W. Lo: (1989), 'The sources and nature of long-term memory in the business cycle' (NBER working paper no. 2951, Cambridge, MA)
7. C. Michelacci: (1997), 'Cross-sectional heterogeneity and the persistence of aggregate fluctuations' (LSE working paper, London)
8. G. Silverberg, D. Lehnert: 'Evolutionary chaos: Growth fluctuations in a Schumpeterian model of creative destruction'. In: *Nonlinear Dynamics in Economics*, ed. by W.A. Barnett, A. Kirman, M. Salmon (Cambridge University Press, Cambridge 1996)
9. G. Silverberg, B. Verspagen: Industrial and Corporate Change **3**, 199–223 (1994)
10. G. Silverberg, B. Verspagen: 'From the artificial to the endogenous: Modelling evolutionary adaptation and economic growth'. In: *Behavioral Norms, Technological Progress and Economic Dynamics: Studies in Schumpeterian Economics*, ed. by E. Helmstädter, M. Perlman (University of Michigan Press, Ann Arbor, MI 1996)
11. H. Hurst: Transactions of the American Society of Civil Engineers **116**, 770–799 (1951)
12. J. Beran: *Statistics for Long-Memory Processes* (Chapman and Hall, New York 1994)
13. F.B. Sowell: Journal of Econometrics **53**, 165–188 (1992)
14. R. Baillie: Journal of Econometrics **73**, 5–59 (1996)
15. C.W.J. Granger, R. Joyeux: Journal of Time Series Analysis **1**, 15–39 (1980)
16. J.R.M. Hosking: Biometrika **68**, 165–176 (1981)
17. A.W. Lo: Econometrica **59(5)**, 1279–1314 (1991)
18. J. Geweke, S. Porter-Hudak: Journal of Time Series Analysis **4**, 221–238 (1983)
19. C. Agiakloglou, P. Newbold, M. Wohar: Journal of Time Series Analysis **14**, 235–246 (1992)
20. F.B. Sowell: Journal of Monetary Economics **29**, 277–302 (1992)
21. M.A. Hauser, B.M. Pötscher, E. Reschenhofer: Empirical Economics **24**, 243–269 (1999)
22. Y.W. Cheung: Journal of Time Series Analysis **14**, 331–345 (1993)
23. C.M. Hurvich, B.K. Beltrao: Journal of Time Series Analysis **15**, 285–302 (1994)
24. C.M. Hurvich, R. Deo, J. Brodsky: Journal of Time Series Analysis **19**, 19–46 (1998)
25. P. Robinson: Annals of Statistics **23**, 1048–1072 (1995)
26. J. Beran, R.J. Bhansali, D. Ocker: Biometrika **85(4)**, 921–934 (1998)
27. R.J. Bhansali, P.S. Kokoszka: 'Estimation of the long memory parameter: a review of recent developments and an extension'. In: *Proceedings of the Symposium on Inference for Stochastic Processes. IMS Lecture Notes in Statistics*, ed. by I. Basawa, C. Heyde, R. Taylor (forthcoming)

28. M.A. Hauser: Journal of Statistical Planning and Inference **80(1-2)**, 229–255 (1999)
29. W. Brock, P. de Lima: 'Nonlinear time series, complexity theory, and finance'. In: *Handbook of Statistics 14: Finance*, ed. by G.S. Maddala, C.R. Rao (North Holland, Amsterdam 1996)
30. F.X. Diebold, G.D. Rudebusch: Journal of Monetary Economics **24**, 189–209 (1989)
31. G. Silverberg, D. Lehnert: Structural Change and Economic Dynamics **4**, 9–37 (1993)
32. G. Silverberg, B. Verspagen: Journal of Evolutionary Economics **5(3)**, 209–227 (1995)
33. G. Silverberg, B. Verspagen: (1999), 'Long memory in time series of economic growth and convergence' (MERIT Research Memorandum 99-016, http://www-edocs.unimaas.nl/files/mer99016.pd, Maastricht)
34. GGDC total economy database 2002 (http://www.eco.rug.nl/ggdc, Groningen and The Conference Board 2002)
35. J.A. Doornik, M. Ooms: (1996), 'A package for estimating, forecasting and simulating arfima models' (mimeo description of Ox ARFIMA package, Nuffield College, Oxford and Erasmus University, Rotterdam)

Correlations and Memory in Neurodynamical Systems

André Longtin, Carlo Laing, and Maurice J. Chacron

University of Ottawa, Physics Department, 150 Louis Pasteur, Ottawa, Ont., Canada K1N 6N5

Abstract. The operation of the nervous system relies in great part on the firing activity of neurons. This activity exhibit significant fluctuations on a variety of time scales as well as a number of memory effects. We discuss examples of stochastic and memory effects that have arisen in our neuronal modeling work on a variety of systems. The time scales of these effects range from milliseconds to seconds. Our discussion focusses on 1) the effect of long-range correlated noise on signal detection in neurons, 2) the correlations and multistability produced by propagation delays, and 3) the long time scale "working memory" effects seen in cortex, and their interaction with intrinsic neuronal noise. We describe both analytical and numerical approaches. We further highlight the challenges of constraining theoretical analyses with experimental data, and of attributing functional significance to correlations produced by noise and memory effects.

1 Introduction

The study of the nonlinear dynamical properties of excitable systems poses great challenges. This is true from the single cell level all the way up to the macroscopic level of electric and magnetic fields recordings from large numbers of cells. The basic element of the nervous system is the nerve cell, or neuron, and its behavior is intrinsically nonlinear. Further, a single neuron is usually stimulated (or "forced", in the nonlinear dynamics context) by time-dependent signals of varying time scales and levels of stationarity, as well as by intrinsic noise sources related to the other neurons and to the intrinsic function of synapses and ionic conductances. The study of such stochastic nonlinear dynamics is at the forefront of research into nonlinear dynamics and how such dynamics are influenced by noise [26]. Further, in humans, there are on the order of 10^{10} nerve cells, and 10^{14} connections between them [27]. Basic groups of cells thought to perform a specific function, such as hypercolumns in the visual cortex usually contain thousands of cells. Yet even the study of two coupled neurons leads to a wealth of new phenomena not found at the single cell level, such as synchronization.

Neural modelers are well aware of this fact, and try to adapt the complexity of their models to the various phenomena they are interested in modeling, as well as to the available data. One common challenge is to decide whether or not to formulate the model in terms of Markovian dynamics such as ordinary differential equations, or in terms of non-Markovian models such as integro-differential models, or delay-differential equations which are a special case of them and arise often in the modeling of neural fee dback. The modeler keen to formulate nonlinear dynamical equations of motion for the assumed state variables of the system must also decide on whether or not to couple stochastic forces to deterministic models.

There are certainly philosophical issues underlying these choices. For example, one can argue that the evolution of the whole dynamical system of the brain relies only on the current state of all its components. And within this purview, activity in a given cell or group of cells may display correlations because it influenced the activity of another group of cells, which in turn later affect the behavior of the first group. In other words, due to the spatially distributed nature of the brain, activity recirc ulates in a more or less modified form. To avoid modeling the system as a whole, one may want to simplify using delayed dynamical systems; this allows one to dispose of spatial dimensions and concentrate on the time lag aspect of the recirculated activity [1,15,10]. Thus, non-Markovian dynamics result as a consequence of a modeling simplification. Another possibility is that by restraining a model to a certain number of state variables, the neglected ones which act on the system may do so in a non-stationary way, i.e. they cause non-stationary fluctuations on the time scale of the chosen state variables.

Given the complexity of excitable cell assemblies such as those in the nervous system of mammals, and even invertebrates, it is not surprising that long range correlations appear in recordings from these systems. Nevertheless, our level of understanding of the origin of these correlations is not well developed, and the functional implications of these correlations are even less clear. One certainly can think of many reasons why the nervous system should have long range correlations in its firing activity, s uch as for the storage and recall of significant experiences. The mechanisms governing this obvious potential use of long range correlations are beginning to be understood, and there is growing consensus that modifications of the strengths of connections between cells are implicated.

There are however many other kinds of nervous system activity that display correlations. For example, the firing of single cells in the auditory system, and other senses as well, have been shown to display long term correlations (see e.g. [47] for a review). Such correlations have been associated with various fractal noises, and it has been suggested that such noises arise through the normal functioning of the synaptic machinery that drives these cells. Also, as we mentioned above, the propagation of neural activity may be deemed slow in comparison to the time scale of the state variables of interest, or may not necessarily be slow, but has to propagate to other parts of t he nervous system before influencing the dynamics of interest. Thus, delay-differential models are commonly seen, especially in the context of neural feedback and control [1,20].

Finally, it is known that nervous activity can be activated in a spatially specific manner, as occurs e.g. when we try to remember the spatial position where we saw a light appear for a second. The memory of this position is important e.g. if one wants to point to this position, either right away or after a minute. Thus, the activation of a spatially specific group of cells can be "remembered" by a spatially distributed model of the area of the brain involved in this so called "working memory" task. Thi s memory will be either strengthened or decay away, and in any case influence the activity of the associated neurons on a long time scale. In this sense, it can be seen as a source of long range correlations. This memory will also be subjected to the effect of neuronal noise, which a priori would be considered as a limiting factor for the performance of this type of memory. We will report below on a paradoxical finding that suggests the opposite.

In this paper we present examples of neural dynamics that display long range correlations, and describe what is known about the functional implications for these correlations. We have a definite *bias towards nonlinear dynamical models*, rather than in developing more sophisticated techniques to characterize the correlations. In fact, the techniques discussed here to establish the presence and basic properties of the correlations are well-known. Further, the phenomena we have considered in our work oper ate at certain time and spatial scales, and do not include for example the important fields of ionic channel dynamics and neurohumoral dynamics (associated with the modulation of the concentration of various neurotransmitters and hormones on the time scale of seconds to days).

Our paper is organized as follows. In Sect. 2, we discuss the measurement, analysis and modeling of long range correlated firing in a class of neurons known as electroreceptors; the results are of general interest for all neurons. We further discuss how the correlations conspire to set up a time scale on which electrosensory system may best perform signal discrimination tasks. Section 3 considers the role of neural conduction delays in producing long range correlated activity, and briefly summarize what is known about the influence of noise on, and theoretical analysis of, noise in systems with delays. Correlations due to short term spatial memory, and their interaction with noise are the subject of Sect. 4. The paper is not meant to be comprehensive of long range correlated effects in neural systems. It does illustrate however how such correlation can arise in systems that range from the single cell level to the large scale spatial level. We have also arranged our presentation so that there is a progression from the small single cell spatial scale with "long range" correlations on the order of milliseconds, up to large spatial scales with longer range correlations on the order of seconds. In all examples, the correlations can be considered "long" because their time scale exceeds those of the dynamical components making up the systems.

2 Correlated Firing in Sensory Neurons

It is known that neural spike trains exhibit long range correlations [27,47] and there has been much speculation about their presence [47]. To this day, their functional role largely remains a mystery. However, many natural signals encoded by sensory neurons exhibit such long range correlations (e.g. music [49]). It has thus been speculated that long range correlations in neurons provides some advantages in terms of matching the detection system to the expected signal [46,47]. Many neural spike trains also exhibit short range correlations [36,42]. It has been shown that the spike train variability of these same neurons displays a minimum as a function of the integration time. However, very little is known about how this can be generated in a biophysical neuron model.

Here we study how such a minimum can be generated in a simplified version of a model that has been used to model weakly electric fish electroreceptors [6–8]. The core of the model resides in the addition of threshold fatigue after an action potential to model refractory effects [19]. It is known that the model under Gaussian white noise stimulation gives rise to negative interspike interval (ISI) correlations that decay over short ranges [9]. We show that the addition of correlated noise gives rise to positive long range ISI correlations in the spike train generated by the model. These positive long range correlations lead to an increase in spike train variability at long time scales similar to that seen experimentally. They further interact with the short range negative ISI correlations to give rise to a minimum in spike train variability as observed experimentally. The functional consequences of both types of correlations is studied by considering their effects on the detectability of signals.

2.1 The Firing Model

A basic model for a neuron is the so-called integrate-and-fire model [27]: the neuron integrates its input in a leaky fashion (due to the RC properties of the membrane), and fires when the voltage reaches a firing threshold. After a firing, the voltage is reset to a value (such as zero) and the process repeats. *The reset destroys all memory of past firing events*, i.e. it makes the dynamics a "renewal process". Here we consider the following model modified integrate-and-fire model:

$$\frac{dv}{dt} = \Theta(t - T_{refrac})\left[-v/\tau_v + I + \xi(t) + \eta(t)\right] \quad \text{if } v(t) < s(t), \tag{1a}$$

$$\frac{dw}{dt} = \frac{w_0 - w}{\tau_w} \quad \text{if } v(t) < w(t), \tag{1b}$$

$$v(t^+) = 0 \quad \text{if } v(t) = w(t) \tag{1c}$$

$$w(t^+) = w(t) + dw \quad \text{if } v(t) = w(t)\ , \tag{1d}$$

where v is the membrane voltage, w is the threshold, and I is the current. $\Theta(\cdot)$ is the Heaviside function ($\Theta(x) = 1$ if $x \geq 0$ and $\Theta(x) = 0$ otherwise). In between action potentials, the threshold w decays exponentially with time constant τ_w towards its equilibrium value w_0. We say that an action potential occurred at time t where $v(t) = w(t)$. Immediately after (i.e. at time t^+), we reset the voltage v to zero while we increment the threshold by a positive constant dw. The voltage v is kept equal to zero for the duration of the absolute refractory period T_{refrac}.

To model the intrinsic variability seen experimentally in neurons, we use two noise sources in the model (for details and justification see [7]): $\xi(t)$ is zero mean Gaussian white noise with variance D^2. $\eta(t)$ is zero mean Ornstein Uhlenbeck (OU) noise [18] with time constant τ_η and variance E^2. We take τ_η to be much larger than the intrinsic neural time scales (τ_v and τ_w).

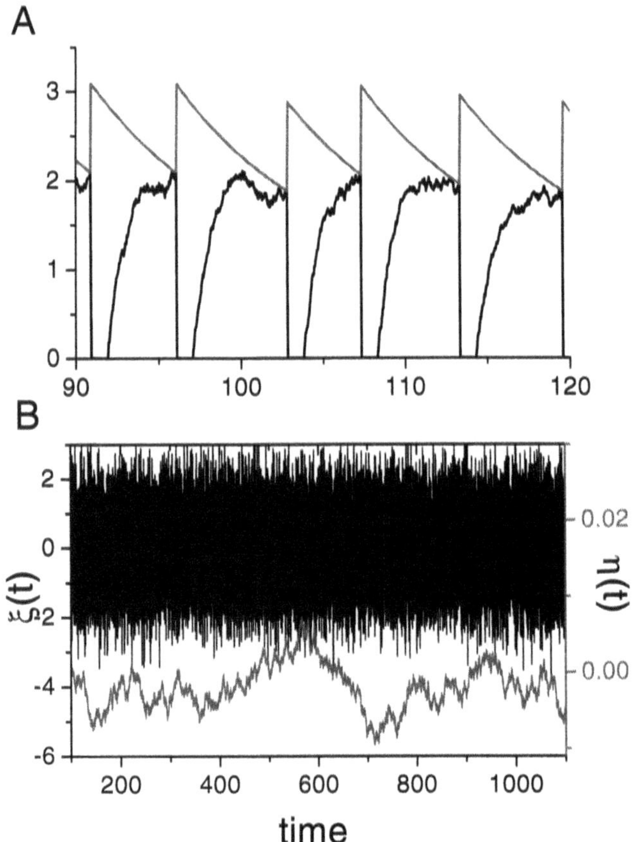

Fig. 1. (**A**) voltage (*black solid line*) and threshold (*gray solid line*) time series obtained with the model. (**B**): The noise sources obtained with the model: $\xi(t)$ (*upper black solid line*) is two orders of magnitude greater than $\eta(t)$ (*lower grey solid line*). Parameter values are: $\tau_v = 1$,$\tau_w = 8.63$,$I = 2$,$T_{refrac} = 1$,$\tau_\eta = 2000$,$D = 0.95$,$E = 0.00306$

Figure 1A illustrates the model dynamics. The noise sources $\xi(t)$ and $\eta(t)$ are shown in Fig. 1B. Note that we chose the noise source $\eta(t)$ to be two orders of magnitude weaker than the noise source $\xi(t)$. Also, $\xi(t)$ varies over short time scales while the noise $\eta(t)$ varies over longer time scales.

2.2 Interspike Interval Correlations

From the spike time sequence $\{t_i\}_{i=1}^{N}$, one can define the ISI sequence $\{I_i\}_{i=1}^{n} = \{t_{i+1} - t_i\}_{i=1}^{N-1}$. The ISI serial correlation coefficients (SCC) ϱ_j are a measure of linear correlations between successive ISIs. They are defined by

$$\varrho_j = \frac{< I_n I_{n+j} > - < I_n >^2}{< I_n^2 > - < I_n >^2} , \tag{2}$$

where < . > denotes an average over the ISI sequence. The SCC ϱ_j is positive if the jth ISI and the current one are both (on average) shorter or longer than average. However, it is negative when the present ISI is shorter (longer) than average while the jth ISI is longer (shorter) than average.

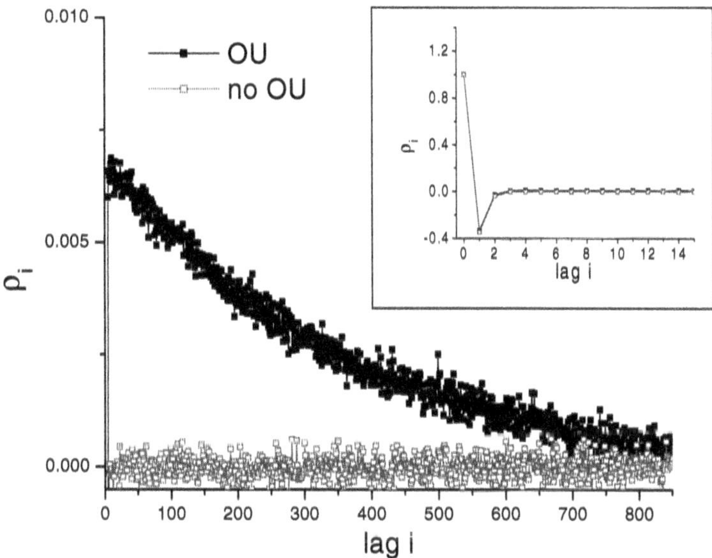

Fig. 2. The ISI correlation coefficients obtained from the model with (*black*) and without (*grey*) OU noise. Weak positive ISI correlations that decay exponentially are present with OU noise but are not without. However, both cases show the presence of a negative SCC at lag one (inset). It was necessary to generate 10^7 action potentials to reveal the presence of the weak positive ISI correlations

Figure 2 shows the SCC sequence obtained with the model (black curve). One observes the presence of long range weak ISI correlations that decay exponentially with increasing lag. These are due to the presence of the OU noise $\eta(t)$ [7]. The SCC sequence obtained in the absence of Ornstein-Uhlenbeck noise (i.e. $E = 0$) does not show long range correlations (gray curve). The presence of long range correlations can be explained intuitively in the following manner [7]. The noise $\eta(t)$ varies on a slower time scale than the neural dynamics: it can thus be considered quasi-constant on the average ISI time scale. Thus, if the noise $\eta(t)$ is positive, it will stay positive for a long time (see Fig. 1B): this will lead to a long sequence of ISIs that will be shorter than average.

Note that in both cases (i.e. with and without OU noise), we have $\varrho_1 < 0$ (see inset). This is due to the deterministic properties of the model [9]. Figure 1A shows that if two spikes are fired during a relatively short time interval, there tends to be a summation effect in the threshold and it becomes higher than average. Consequently, the next ISI will tend to be long since the threshold takes a longer time to decay. A similar argument holds if two spikes are separated by a long time interval. Thus short ISIs will tend to be followed by long ISIs and vice versa, and this will give rise to $\varrho_1 < 0$. This property can be studied by looking at the model's deterministic response to perturbations [9].

We now explore the consequences of these short and long range ISI correlations on spike train variability. We denote by $p(n, T)$ the probability distribution of the number of action potentials obtained during a counting time T. $p(n, T)$ is referred to as the spike count distribution [3,45]. The Fano factor [14] measures the spike train variability on multiple time scales. It is defined by

$$F(T) = \frac{\sigma^2(T)}{\mu(T)} , \tag{3}$$

where $\mu(T)$ and $\sigma^2(T)$ are the respective mean and variance of $p(n, T)$. The asymptotic value of the Fano factor is related to the SCC sequence [11]

$$F(\infty) = CV^2 \left(1 + 2 \sum_{i=1}^{\infty} \varrho_i \right) , \tag{4}$$

where CV is the coefficient of variation: it is given by the ratio of the standard deviation to the mean of the ISI distribution.

We plot the resulting Fano factor curves (Fano factor versus counting time) for the model in Fig. 3A. The Fano factor curve obtained in the absence of OU noise (triangles) decreases monotonically. However, the presence of the weak OU noise affects the spike train variability at long time scales by increasing the Fano factor (squares). Due to the finite correlation time of the OU process, the Fano factor saturates to a finite value. We now explain the phenomenon in more detail. Let us assume the following form for the SCC's with $i > 1$:

$$\varrho_i = -0.34355\delta_{1i} + 0.007 \exp(-i/347.768) , \tag{5}$$

where δ_{ij} is the Kronecker-delta ($\delta_{ij} = 1$ if $i = j$ and 0 otherwise). This was obtained by an exponential fit of the data in Fig. 2 with OU noise as well as adding the value of ϱ_1. Substituting (5) into (4) yields an asymptotic expression (upper black horizontal line) for the Fano factor that is close to that observed numerically. This justifies our assumptions about exponentially decaying ISI correlations.

We compare the results with the Fano factor obtained from the randomly shuffled ISI sequence; the shuffling eliminates all ISI correlations and a renewal process results (circles). The Fano factor $F(T)$ now decreases monotonically towards the asymptotic value given by CV^2 ("CV^2" line in Fig. 3A). This is in accordance with (4). We observe that the presence of short term negative ISI correlations decreases spike train variability at long time scales. However, weak long range ISI correlations will give an increase in spike train variability at long time scales. This is consistent with the predictions from (4).

Figure 3B shows the corresponding mean (open squares) and variances of the spike count distribution obtained under all three conditions. It is observed that the reduction in the Fano factor caused by negative ISI correlations is primarily

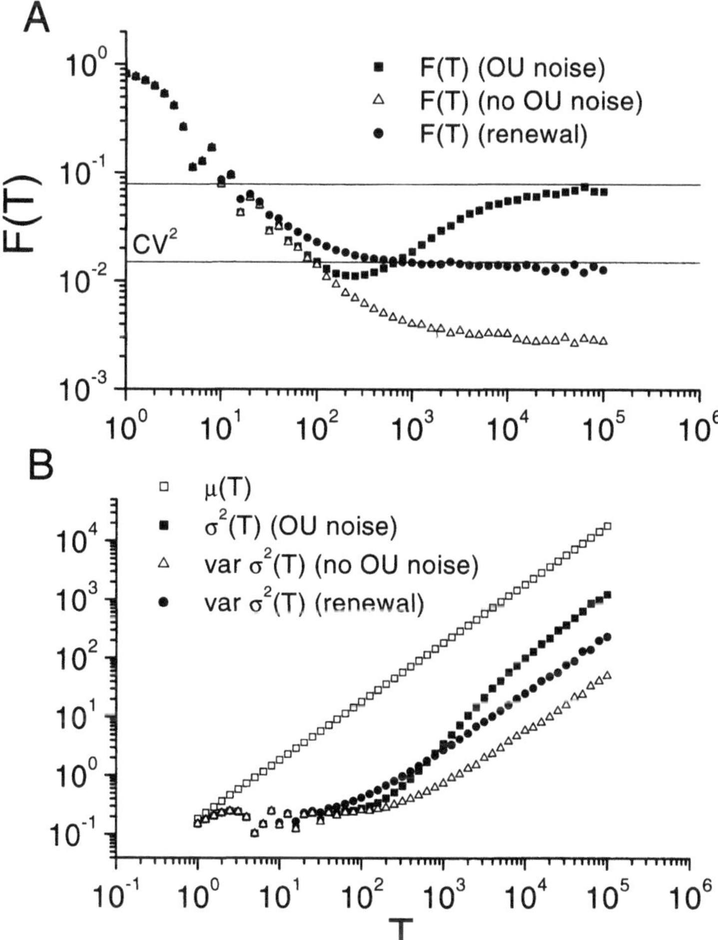

Fig. 3. (**A**) Fano factor obtained with (*filled squares*), and without (*open triangles*) OU noise. Also plotted is the Fano factor obtained without any ISI correlations (*filled circles*). (**B**) The mean (*open squares*) and variance of the spike count distribution obtained with (*filled squares*) and without (*open triangles*) OU noise. Also shown is the variance obtained without any ISI correlations (*filled circles*)

due to the fact that the variance is reduced at long time scales in comparison with a renewal process (compare filled circles and open triangles). The noise $\eta(t)$ is too weak to have any noticeable effect on the mean. However, it significantly increases the variance of the spike count distribution at long time scales (filled squares).

It is thus an interaction between negative and positive ISI correlation coefficients that gives rise to the minimum in the Fano factor curve. This minimum counting time depends upon the strength E and time constant τ_η of the OU process [7]: it is possible to change the counting time at which the minimum occurs by changing the parameter E [7]. Furthermore, the counting time at which the saturation occurs depends on τ_η: the saturation point can be set to arbitrarily large counting times by increasing τ_η.

2.3 Detection of Weak Signals

We now discuss the use of signal detection theory [21] to assess the consequences of ISI correlations on the detectability of signals. This is based on the optimal discrimination of spike count distributions in the absence and presence of a stimulus ($P_0(n,T)$ and $P_1(n,T)$, respectively). Let us suppose that the firing rate is f_0 in the absence of stimulus and that it is f_1 in the presence of a stimulus. If the stimulus is weak and excitatory, it will give an increase in the mean of the spike count distribution without significantly increasing the variance [42,7]. Thus, we assume $\sigma_0^2(T) = \sigma_1^2(T)$ and $\mu_1(T) = (f_1/f_0)\mu_0(T)$. One can then quantify the discriminability between $P_0(n,T)$ and $P_1(n,T)$ by [21]

$$d' = \frac{\mu_0(T)|f_1/f_0 - 1|}{\sqrt{2}\,\sigma_0(T)} \,. \tag{6}$$

The situation is illustrated in Fig. 4A. The discriminability d' is inversely related to the amount of overlap between the distributions $P_0(n,T)$ and $P_1(n,T)$. To quantify the changes in d' caused by both types of ISI correlations, we form the difference between d' obtained from the ISI sequence with ISI correlations and the d' obtained for the corresponding renewal process (i.e. the shuffled ISI sequence); we will use the symbol $\Delta d'$ to denote this difference. Figure 4B shows the measure $\Delta d'$ as a function of counting time T. A maximum can be seen around $T = 200$. Note that this corresponds to the counting time at which the Fano factor $F(T)$ is minimal. Thus, the improvement in signal detectability is maximal when spike train variability (as measured by the Fano factor) is minimal.

2.4 Discussion of Correlated Firing

We have shown that a simple model with correlated noise can give rise to both short and long range ISI correlations. The dynamic threshold could model synaptic plasticity [16], recurrent inhibition [13], or intrinsic ionic conductances that lead to adaptation [32]. The effects discussed here can thus originate from very different physiological and biophysical mechanisms. Deterministic properties of the model have been shown to lead to negative interspike interval correlations at short lags. These negative ISI correlations lead to lower spike train variability at intermediate time scales: they thus regularize the spike train at long time scales.

It was further shown that the addition of a very weak noise with long range correlations could induce exponentially decaying long range positive ISI correlations in the model. These positive ISI correlations were shown to lead to increased spike train variability at long time scales. This increase in spike train variability at long time scales has been observed in many neural spike trains [47,27]. It

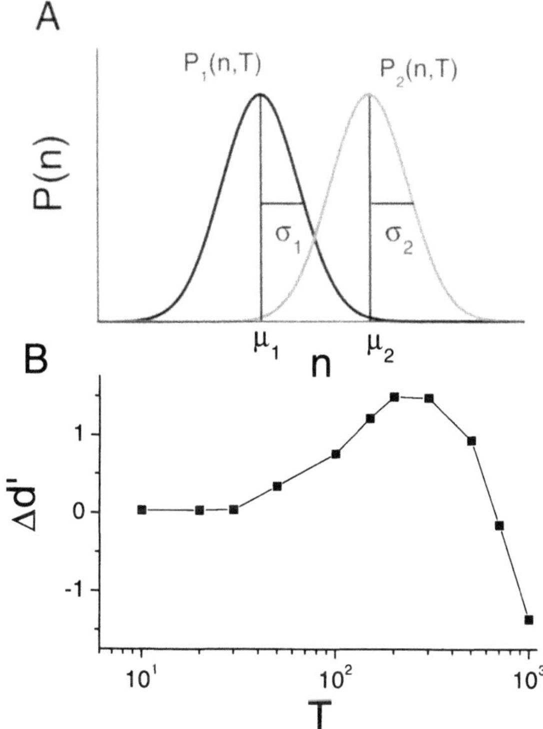

Fig. 4. (**A**) The spike count distributions $P_0(n,T)$ and $P_1(n,T)$ along with their respective means and variances. (**B**) Difference between the d' measure obtained with OU noise and the d' measure obtained with a renewal process $\Delta d'$ as a function of counting time T. A optimal counting time at which signal detectability is maximal is seen. We took $f_1 = 215Hz$ and f_0 was equal to the mean firing rate of the neuron in the absence of stimulus

has been observed that synaptic vesicle release rates display long range correlations [37] and our OU process $\eta(t)$ could model these fluctuations. The noise $\xi(t)$ could then model fluctuations occurring on very short time scales such as channel and/or synaptic noise.

We have shown that both short range negative and long range positive ISI correlations interact to give a minimum in spike train variability at an intermediate counting time. This minimum has been observed experimentally in auditory fibers [36] as well as in weakly electric fish electroreceptors [42]. We have previously reproduced this effect using a detailed electroreceptor model [7]. However, the present study shows that similar effects can be obtained in a simple model with dynamic threshold and correlated noise. The functional consequences of this minimal counting time were assessed using signal detection theory. It was shown that the model under Gaussian white noise stimulation displayed only short range negative ISI correlations [6,9]. That result was reproduced here. It was further shown that negative ISI correlations reduced spike train variability as measured by the Fano factor at long time scales.

Using signal detection theory, we have shown that the improvement in signal detectability was maximal, as compared to a renewal process, when the Fano factor was minimal. It has been shown in weakly electric fish electroreceptors that the counting time at which the Fano factor is minimal corresponds to the behavioral time scale for prey detection [39]. Animals must make decisions in short amounts of time (usually less than a second) in order to survive (i.e. to detect prey or avoid predators). Thus, the presence of short term negative ISI correlations might serve to increase signal detectability while long term positive ISI correlations will give rise to an integration time at which spike train variability is minimal. It could be that the animal does not want to integrate the signal over longer time scales for which the spike train variability increases again.

3 Delayed Neurodynamical Systems

We now turn to the role of delays in inducing long range correlations. Even though we focus on neural systems, given our past work on such systems, the ideas we will be discussing are sufficiently general to be applicable to many other areas where delays are important such as laser physics and economics. We will discuss how delayed dynamics arise, and focus on how they generate correlations that outlast the delay (which is the intrinsic memory of the system) and on how to approach their analysis in the presence of noise, which is of particular importance in the nervous system [27].

3.1 Delay-Differential Equations

We focus on correlations in delay-differential equations (DDEs) of the general form

$$\frac{dx}{dt} = f\left(x(t), x(t-\tau), \boldsymbol{\mu}, \xi(t)\right) . \tag{7}$$

We will restrict ourselves to such first order delay equations, and in particular, to those having only one delay. We will also discuss specific instances of this equation in the context of neuronal modeling, statistical physics of bistable systems and atmospheric physics. The DDE (7) can be seen as a dynamical system that combines aspects of ordinary differential equations (ODEs) and discrete time maps. This equation can also be seen as a special case of the integro-differential system

$$\frac{dx}{dt} = f(z(t), \boldsymbol{\mu}) \tag{8}$$

with

$$z(t) = \int_{-\infty}^{t} K(t-s)x(s)\, ds . \tag{9}$$

In this more general case, the kernel $K(t)$ weighs the past values of $x(t)$ that influence the dynamics of dx/dt in the present. If the function f is linear, the problem is completely solvable, e.g. using Laplace transforms. The solutions are of the same types as for any other linear ordinary differential equation. However, note that since a DDE is an infinite-dimensional dynamical system, even though there appears to be only one dynamical variable, the class of equations above can exhibit oscillatory solutions, which are not possible in one-dimensional ODEs. For the linear case, such solutions are either continually growing or decaying in amplitude, or are marginally stable (pure imaginary eigenvalues).

Below we focus on discrete delays. However, it is important to realize that in most situations, the delay is actually distributed. In laser cavities, it is a good approximation to make the delay fixed, as it corresponds to the travel time of light around some optical path. DDEs are also an approximation to the whole dynamics governing propagation of effects in various media. For example, in physiology, delays are often introduced to take the maturation time of cell populations into account simply, rather th an having extra compartments (and associated dynamical equations) for the maturation process. Likewise in neural systems, delays represent propagation times along axons, as well as synaptic delays. A description of such processes in terms of delays avoids the complexities of partial differential equations for the propagation of action potentials along the nerve, and extra ordinary differential equations associated with synaptic activation and release processes.

It is important to include delays in the description of a physical process when it is of a magnitude comparable to or larger than the other time scales found in the system. For example, if the relaxation of a system to its steady state is dominated by a time scale τ_{sys}, then it is important to take into account any delays on the order of or larger than τ_{sys}.

3.2 Correlations

Delays induce correlations. This can be simply understood by considering the time evolution of the dynamical system: $x(t+dt)$ depends on $x(t)$ and on $x(t-\tau)$. For example, if for the current value of $x(t)$, the derivative $f(x(t), x(t-\tau))$ is positive, $x(t)$ will increase. Since $x(t-\tau)$ influences the sign of f, there can be a positive or negative correlation between $x(t)$ and $x(t-\tau)$. The linear correlation between these two quantities can be measured using the standard autocorrelation function $\langle x(t)x(t-\tau)\rangle$. If the solution is chaotic, it will typically display peaks at multiples of the delay, superimposed on an overall decaying background [38]. Such structure will also be seen if noise is incorporated into the dynamical equations, leading to a "stochastic delay-differential system" (SDDE) in the latter case. Of course, the peaks will be repeated periodically, along with all the rest of the autocorrelation function, if the solution is periodic.

Usually, there is sufficient knowledge about the system under study to conclude the presence of delays. However, sometimes that fact is not clear, or the value of the delay is not known, or there could be more than one delay. There exist time series methods that allow the detection of delays, as well as estimation of these delays [25].

3.3 Delayed Neural Control

Delays arise often in the context of neural control, such as in the control of light flux on the retina in the pupil light reflex (PLR). This reflex is in fact a paradigm of neural control systems [44]. This system is interesting from an experimental point of view because it can be manipulated non-invasively using infrared video-pupillometry coupled with special light stimulation. In past work on the human pupil light reflex [33], we have been able to study the onset of oscillations as the gain of this control system is artificially increased. We have been quickly confronted with noisy fluctuations in the analysis and modeling of this system. This is not surprising given that th is reflex naturally exhibits ongoing fluctuations ("hippus") in healthy humans under constant light stimulation. Such baseline fluctuations are in fact common in neural and physiological control, and their origin and meaning is a subject of great interest and debate [20]. They just happen to be especially prominent in the light reflex. The origin of these fluctuations there is still elusive, but our modeling using a first order DDE strongly suggests that they are the manifestation of neuro nal noise injected into the reflex arc. The delayed negative feedback dynamics of this system can be simply modeled by

$$\frac{dA}{dt} = -\alpha A(t) + f(A(t-\tau)) = -\alpha A(t) + C\frac{\theta^n}{\theta^n + A^n(t-\tau)} + K\,, \tag{10}$$

where A is the pupil area. A supercritical Hopf bifurcation occurs as the parameter n, controlling the slope of the feedback function (i.e. the feedback gain), or the fixed delay τ, are increased. This model correctly predicts the oscillation frequency, as well as the slow variation and robustness of this frequency across the bifurcation. It does not exhibit other bifurcations as the gain is increased from zero (open-loop) or as τ is increased at fixed gain. Also, the fact that hippus occurs in open-loop signifies that its origin is not deterministic chaos arising from the nonlinear delayed feedback, as seen for example in the Mackey-Glass equation [20]. The noisy oscillations may then arise because neural noise perturbs the dynamics of the PLR. This hypothesis has been tested by introducing noise on K (additive noise) or on C (multiplicative noise) in (10). The questions we are ultimately seeking to answer are, is it noise, and if so, what is the origin of this noise, its spectrum before entering the reflex arc, and its coupling to the reflex arc? Any progress along this line would be of great interest given that the pupil reflects brainstem function as well as higher brain function such as attention and wakefulness, and any better discrimination of the associated states using non-invasive monitoring of the pupil size can yield important new diagnostics.

We do not have the space to provide details about our analysis here. In summary, we have found by numerical integration of (10) that pupil area fluctuations can arise from the coupling of the PLR to colored neural noise of Ornstein-Uhlenbeck type with a correlation time of one second. This fact is based on the ability of the model to reproduce key features of the time series (beyond the frequency), such as the behavior, as a function of gain, of the mean and variance of the period and amplitude of the fluctuations [33]. Thus the hypothesis of feedback dynamics coupled to noise as in (10) seems very appropriate. The experimental data are insufficient however to establish the proportions of additive and multiplicative noise. Also, the analysis highlights the difficulty of pinpointing the gain value at which the Hopf bifurcation occurs. The problem is that oscillations are always visible due to the noise, even when the system should exhibit a fixed point if the noise were not present. The power spectra of area time series do not exhibit critical behavior. In fact, the behavior of this system at a Hopf bifurcation under the influence of noise does not differ from that of nonlinear stochastic ODEs near such a bifurcation.

The usual way around this problem of pinpointing a bifurcation when noise is present is to devise an order parameter which does exhibit critical behavior, and that can be calculated theoretically. An order parameter that has been proposed for a noisy Hopf bifurcation [26,33] is based on the invariant density of the state variable. The order parameter is the distance between the two peaks of this density, i.e. it measures its bimodality. The density is unimodal on the fixed point side, and bimodal on the limit cycle side, with the peaks roughly corresponding to the mean maximal and minimal amplitude of a stochastic oscillation. The behavior of this parameter can be compared between simula tions, theory (stationary density of the Fokker-Planck equation) and actual measured data.

Unfortunately, Fokker-Planck analysis is not possible for DDEs such as (10), because they are non-Markovian. Furthermore, such an approach requires many data points to resolve the peaks, and thus compute the order parameter, especially in the vicinity of the deterministic bifurcation. Thus the approach is limited for physiological data. However, numerics reveal the interesting fact that noisy oscillations look qualitatively similar to the data as a function of the gain. Simulations also reveal the fact that additive or multiplicative noise on (10) actually move the onset of the bifurcation point towards the limit cycle side [33,34]. From the point of view of the order parameter, there is a "postponement" of the Hopf bifurcation in the first order DDE (10) (see the general discussion of such effects in [26]). From the time series point of view, even though the order parameter is still zero, the time series is clearly oscillatory, with the mean oscillation amplitude increasing with the gain parameter, while the frequency varies little across the stochastic bifurcation.

Even though it is not possible to compute the invariant density for (10) due to the delay, its solutions are useful to validate hypotheses about noise. One can, for example, compute various statistics about the noisy oscillations and compare them between model and data. This approach allows us to state that additive or multiplicative lowpass Gaussian noise injected in the pupil light reflex can explain the results seen as the gain is varied. This approach is interesting because it allows one to put the noise under the magnifying glass, given that noise dominates the behavior of systems in the vicinity of bifurcations where the usual dominant linear terms are weak or vanish. Wha t the approach does not give is the origin of this noise, although it can be used to test for some of its properties, such as its variance and possibly higher order moments, and what its correlation structure is. In fact, the best results occurred in our comparative study when the noise had a correlation time on the order of 300 msec (lowpass filtered Gaussian white noise, i.e. Ornstein Uhlenbeck noise, was used). Better data would allow to better pinpoint the correlation time, which in turn can give more insight about its origin. For example, it could be fast synaptic noise, slow retinal adaptation noise, or slow neural discharge activity from the reticular activating system that governs wakefulness and which is injected in the PLR at the brainstem level.

There have been recent efforts at analyzing noise in DDEs, and the field is wide open and rife with potential Ph.D. projects. We have done a systematic study of stochastic DDEs in the low noise limit using Taylor expansions of the flow, and indicated the limits of validity of this approach [22]. The Taylor expansions allow one to use the Fokker-Planck formalism. The agreement between analytical approximations and the numerical simulations are good for small delays. However, the agreement decr eases when the underlying dynamics are oscillatory; in particular, this arises when the characteristic equation that arises from the linearization around a fixed point has complex eigenvalues. The Taylor expansions of first order DDEs can only produce first order ODEs, which can not have complex eigenvalues, thus the limitation of this approach.

The noise-induced transitions between the two wells of a delayed bistable system have also been investigated following the same formalism and show the same quality of agreement and intrinsic limitations [23]. That delayed dynamical system reads

$$\frac{dx}{dt} = x(t-\tau) - x^3(t-\tau) + \xi(t) \; , \tag{11}$$

where $\xi(t)$ is Gaussian white noise. There has also been recent work by Ohira [40] that approximates this DDE by a special random walk. From this walk, approximations to the stationary density, correlation functions and even an approximate Fokker-Planck equation have been obtained. There has also been work applying the ideas of two-time scale perturbation theory to stochastic DDEs [28]. Finally, a master equation in which the transition rates are delay-dependent has been recently proposed [48] for a stochastic DDE which is the standard quartic system (particle in a bistable potential) with linear additive feedback and noise

$$\frac{dx}{dt} = x(t) - x^3(t) + \alpha x(t-\tau) + \xi(t) =; \, , \tag{12}$$

where α is a constant. In [48], the authors actually solve only for the case where $x(t-\tau)$ is replaced by its sign, thereby losing information on the precise dynamics within the wells. However, this formalism is very promising as it allows the computation of the mean switching time between the wells as a function of the delay τ and the feedback strength α. It is left to be seen how well the formalism works for a wider range of parameters, since we have recently shown that the deterministic dynamics of (12) are rather complex, and involve a Takens-Bogdanov bifurcation [43]. It should be mentioned that this equation is of special interest in the atmospheric physics literature as well, where it has been used as a toy model of the El Niño Southern Oscillation phenomenon (see a discussion of this in [43] and references therein). Another interesting study of a stochastic delay equation in the context of neurology with comparison to experimen ts can be found in [10].

Finally, let us mention two other aspects where delays and noise are bound to be increasingly under scrutiny in the future. One has to do with neural dynamics at the single neuron level, and involves delays of a few milliseconds. When a neuron fires at its soma, the effect is propagated to other neurons it is connected to, but also to the dendritic tree of the neuron. Depending on the ionic channels in this tree, the spike may propagate back down, influencing and even causing further spiking at the soma. Th is can lead to temporally correlated sequences of spikes, such as bursting patterns [12]. The modeling of such phenomenon is complex, given the spatio-temporal nature of the problem in complex geometry, resulting in partial differential equations that must be solved numerically. However, we have recently shown that the resulting correlation and memory effects on the millisecond time scale that occur in such backpropagation of action potentials from the soma to the dendrites can be modeled with a D DE [31]. We suspect that such dynamics, in combination with noise, can yield long range correlations as well. Also, systems in which the delay is larger than the system response time can often exhibit multistability, i.e. coexistence of deterministic attractors, each with their own basin of attraction. This implies that two different initial functions may yield different steady state behaviors [2]. Further, noise will kick the trajectories from one attractor to another, which may also produce long range correlations. A discussion of this multistability in the neural context can be found in [15].

4 Noise Induced Stabilization of Bumps

4.1 Background

There has been much recent interest in spatially-localized regions of active neurons ("bumps") as models for feature selectivity in the visual system [4,5] and working memory [24,29,50], among others. Working memory involves the holding and processing of information on the time scale of seconds. For these models, which involve one-dimensional arrays of neurons, the position of a bump is thought to encode some relevant aspect of the computational task. Simple versions of these models do reproduce the basic experimental aspects of short term memory [24,29].

One problem, however, is that more realistic models of the neurons involve what are known as adaptation effects. Spike frequency adaptation, in which the firing frequency of a neuron slowly declines when it is subject to a constant stimulus, is ubiquitous in cortical neurons [32]. It is known that including such adaptation in a general model for bump formation destabilizes stationary bumps, causing moving bumps to be stable instead [30]. This was shown in [30], for a particular model, to be due to a supercritical pitchfork bifurcation in bump speed as the strength of adaptation was increased. The bifurcation occurred for a non-zero value of adaptation strength. The implication of this destabilization is that, while the bump may initially be at the position that codes for the spatial stimulus, the movement of the bump will destroy this information, thus degraded the performance of the working memory.

It was also shown in [30] that adding spatiotemporal noise to a network capable of sustaining bumps effectively negated the effect of the adaptation, "restabilizing" the bump. This beneficial aspect of noise is similar in spirit to stochastic resonance [17], in which a moderate amount of noise causes a system to behave in an optimal manner. We now demonstrate this phenomenon, and later summarize its analysis.

4.2 Stochastic Working Memory

The system we study is a network of integrate and fire neurons, in which each neuron is coupled to all other neurons, but with strengths that depend on the distance between neurons. The equations for the network are

$$\frac{dV_i}{dt} = I_i - a_i - V_i + \frac{1}{N}\sum_{j,m} J_{ij}\alpha(t - t_j^m) - \sum_l \delta(t - t_i^l) \tag{13a}$$

$$\tau_a \frac{da_i}{dt} = A\sum_l \delta(t - t_i^l) - a_i \ , \tag{13b}$$

for $i = 1, \ldots, N$, where the subscript i indexes the neurons, t_j^m is the mth firing time of the jth neuron. $\delta(\cdot)$ is the Dirac delta function, used to reset the V_i and increment the a_i. The sums over m and l extend over the entire firing history of the network, and the sum over j extends over the whole network. The post-synaptic current is represented by $\alpha(t) = \beta e^{-\beta t}$ for $0 < t$ and zero otherwise. The variable a_i, representing the adaptation current, is incremented by an amount A/τ_a at each firing time of neuron i, and exponentially decays back to zero with time constant τ_a otherwise. The coupling function we use is

$$J_{ij} = 5.4\sqrt{\frac{28}{\pi}}\exp\left[-28\left(\frac{i-j}{N}\right)^2\right] - 5\sqrt{\frac{20}{\pi}}\exp\left[-20\left(\frac{i-j}{N}\right)^2\right] . \tag{14}$$

This is a "Mexican hat" type of coupling, for which nearby neurons excite one another, but more distant neurons inhibit one another. Periodic boundary conditions are used. Parameters used are $\tau_a = 5$, $\beta = 0.5$, $N = 50$, $A = 0.1$. I_i was set to 0.95 for all i, except for a brief stimulus to initiate a bump, as explained in the caption of Fig. 5. When $I_i = 0.95$ for all i, the quiescent state, $(V_i, a_i) = (0.95, 0)$ for all i, is a solution. However, it is known that with coupling of the form (14), and $A = 0$, stationary patterns of localized activity are also possible solutions [29,30].

Noise was included in (13a)-(13b) by adding or subtracting (with equal probability) current pulses of the form σe^{-t} $(0 < t)$ to each otherwise constant current

I_i. The arrival times of these pulses were chosen from a Poisson distribution. The mean frequency of arrival for both positive and negative pulses was 0.1 per time unit, so the frequency for all pulses was 0.2 per time unit. The arrival times were uncorrelated between neurons. The noise intensity was varied by changing σ.

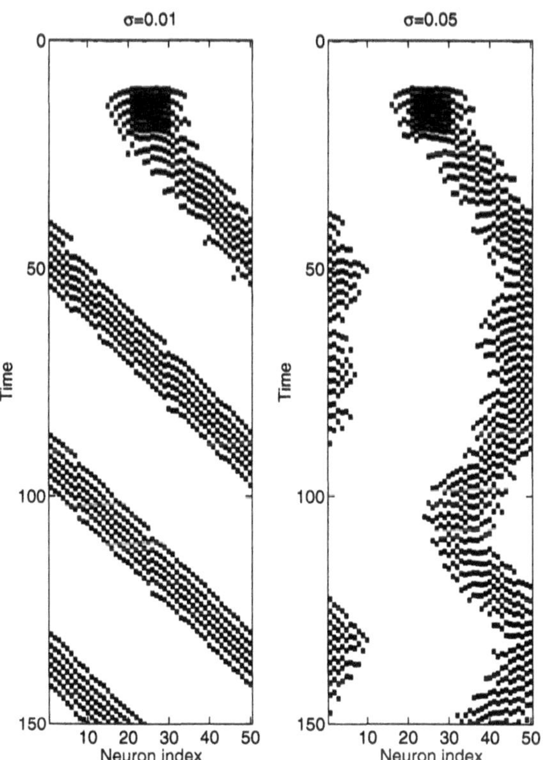

Fig. 5. Typical simulations of (13a)-(13b). (**Left**) $\sigma = 0.01$, (**Right**) $\sigma = 0.05$. A black bar is plotted each time a neuron fires (its voltage reaches 1). I was set to 1.2 for neurons 21 to 30 for $10 < t < 20$, otherwise it was set to 0.95. Other parameters are given in the text. The boundary conditions are periodic.

In Fig. 5 we show typical simulation results for (13a)-(13b) for two different noise levels. A spatially localized current was injected for a short period of time ($10 < t < 20$) to move the system from the "all off" state to a bump state. This could mimic e.g. the briefly illuminated light in an oculomotor delayed response task [24,41]. We see that the activity persists after the stimulus has been removed. This is a model of the high activity seen during the "remembering" phase of a working memory task [41].

The behavior of the system for the two different values of noise is quite different: for low noise values, the bump moves with an almost constant speed. For the example in Fig. 5 (left), the bump moves to the right. This is due to the lack of symmetry in the initial conditions – it could have just as easily traveled to the left. Note that the boundary conditions are periodic. For higher noise intensities (Fig. 5, right) the bump does not have a constant instantaneous speed, rather it moves in a random fashion, often changing direction. As a result of this, the average speed during a finite-time simulation (determined by measuring how far the bump has traveled during this time interval and dividing that distance by the length of the interval) is markedly less than in the low noise case.

This is quantified in Fig. 6 where we plot the absolute value of the average speed during an interval of 200 time units (not including the transient stimulus phase) as a function of σ, averaged over eight simulations. We see that there is a critical value of σ (approximately 0.01) above which the absolute value of the velocity drops significantly – this is the "noise-induced stabilization".

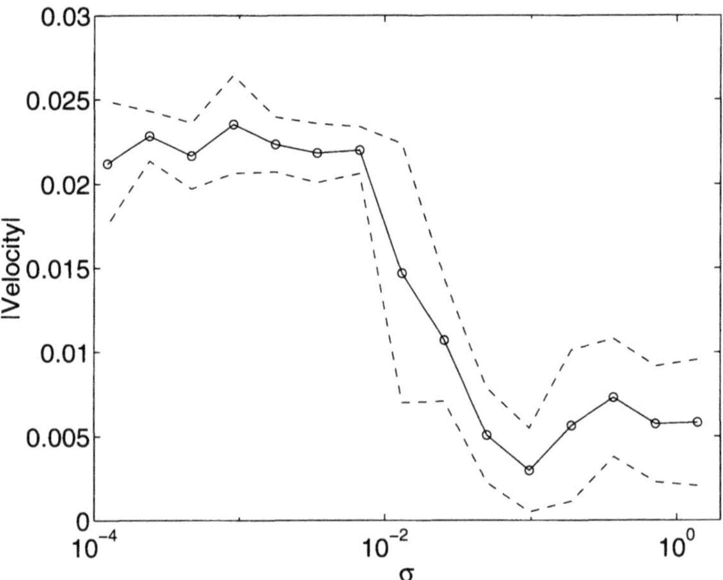

Fig. 6. Absolute value of the velocity of a bump for (13a)-(13b) during a period of 200 time units, as a function of noise intensity σ. The *dashed lines* indicate plus and minus one standard deviation about the mean

4.3 Discussion of Noisy Bumps

The phenomenon discussed above is quite robust with respect to changes in parameters, and has also been observed in a rate model description of bump formation, in which the individual action potentials of each neuron have been temporally averaged so that each neuron is only described by its instantaneous firing rate [30]. Noise-induced stabilization was analyzed in some detail in [30], and we now briefly summarize the results.

As mentioned, the phenomenon was also observed in a spatially-extended rate model. For a particular choice of the coupling function (the spatially continuous version of J_{ij} (14)) it was shown that the bifurcation destabilizing a stationary bump and giving rise to traveling bumps was a supercritical pitchfork bifurcation in bump speed as A increased. Motivated by this, the noisy normal form of such a bifurcation was studied and (when velocity was measured in the same way as for the spiking neuron model above) qualitatively similar slowing down was found as the noise intensity was increased.

Motivated by these results we modeled the dynamics of the noisy normal form of a supercritical pitchfork bifurcation as a persistent random walk in one dimension. The behavior of a particle undergoing such a walk is governed by the stochastic differential equation

$$\frac{dx}{dt} = I(t) \; , \tag{15}$$

where x is the position of the particle (the bump), $I(t) \in \{-v, v\}$, and the probability that $I(t)$ changes from $-v$ to v, or from v to $-v$, in time interval dt is $(\beta/2)dt$. The probability density function of x, $p(x,t)$, satisfies the telegrapher's equation:

$$\frac{\partial^2 p}{\partial t^2} + \beta \frac{\partial p}{\partial t} = v^2 \frac{\partial^2 p}{\partial x^2} \; . \tag{16}$$

This equation can be explicitly solved, and the mean absolute position at time t, and the variance of this quantity, can both be found analytically [30]. They are

$$\langle |x(t)| \rangle = vte^{-\beta t/2}[I_1(\beta t/2) + I_0(\beta t/2)] \tag{17}$$

and

$$\langle |x(t)|^2 \rangle = 2v^2(\beta t - 1 + e^{-\beta t})/\beta^2 \; , \tag{18}$$

where the angled brackets denote averaging over realizations, and $I_{0,1}$ are modified Bessel functions of the first kind of order 0, 1. Once we link β to noise intensity via an Arrhenius-type rate expression, e.g. $\beta = e^{-1/\sigma}$, we can use (17) and (18) to generate a plot like Fig. 6 – see Fig. 7. This agrees qualitatively with the results of the simulations of the full spiking network, (13a)-(13b), and provides an explanation for the phenomenon of noise-induced stabilization.

5 Conclusion

We have presented an overview of various ways in which long range correlations can arise in neurodynamical systems. These range from the millisecond time scale to time scales of many seconds in our work. There are of course much longer time scales (relating e.g. to behavior) on which correlations can be seen, and which are related to changes in the level of expression of various receptors and hormonal concentrations. This is beyond the scope of our own research. Rather we have focussed on how noise can induce short and long range correlations in single cell firing activity, and the potential importance of this effect for information coding. Also, we have considered how correlations arise from the delays in the propagation of neural activity, with particular attention paid to the

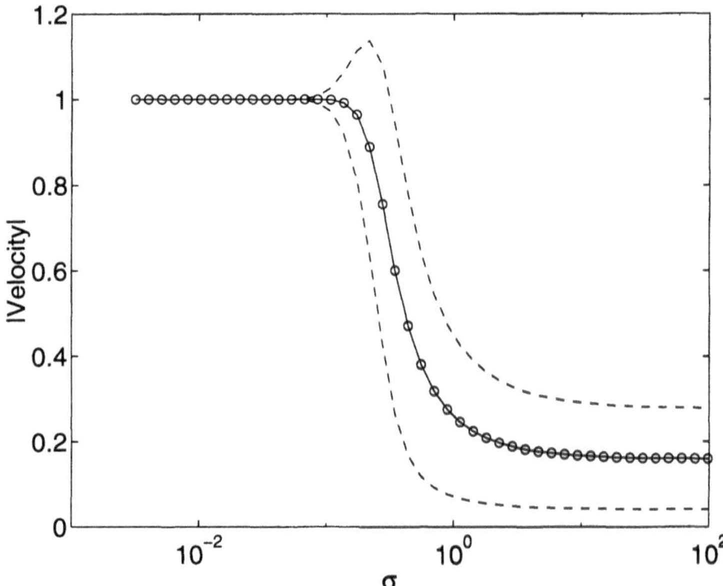

Fig. 7. Absolute value of velocity, $\langle|x(t)|\rangle/t$, from (17), (*circles*), and this quantity plus and minus its standard deviation, from (17) and (18), (*dashed*), as function of noise intensity. β is given by $\beta = e^{-1/\sigma}$, $t = 50$ and $v = 1$

effect of external fluctuations on such processes, and on bifurcations to oscillatory behavior in particular. Finally, we have considered how localized "bumps" of activity arise in realistic neural nets. These are thought to underlie short term memory processes, and the ongoing formation and movement of these bumps will result in correlated firing in single or multi-electrode recordings. Our analysis of such phenomenon with noise has enabled us to reduce a complex spatiotemporal problem to a simple first order differential equation with noise, an approach that may prove useful in other contexts where patterns of activity in excitable systems are under study.

There are still a large number of open problems in these areas, as we have alluded to in the discussion of the three main sections of our paper. Multistability and memory (non-renewability) in first passage time problems, as well as the analysis of noise in PDEs for excitable system, will in our view offer theorists and numerical analysts some of the most challenging problems for decades to come, and their study will certainly forward our understanding of the nervous system.

References

1. U. an der Heiden: J. Math. Biol. **8**, 345 (1979)
2. F.T. Arecchi, A. Politi and L. Ulivi: Il Nuovo Cimento, **71**, 119 (1982)
3. H.B. Barlow HB, W.R. Levick WR: J. Physiol. (Lond) **200**, 11 (1969)

4. R. Ben-Yishai, R. L. Bar-Or and H. Sompolinsky: Proc. Natl. Acad. Sci. USA. **92**, 3844 (1995)
5. P. C. Bressloff, N. W. Bressloff and J. D. Cowan: Neural Comp. **12**, 2473 (2000)
6. M.J. Chacron, A. Longtin, M. St-Hilaire, L. Maler: Phys. Rev. Lett. **85**, 1576 (2000)
7. M.J. Chacron, A. Longtin, L. Maler: J. Neurosci.**21**, 5328 (2001)
8. M.J. Chacron, A. Longtin, L. Maler: Neurocomputing **38**, 129 (2001)
9. M.J. Chacron, K. Pakdaman, A. Longtin: Neural Comput. (2002) (in press)
10. Y. Chen, M. Ding and J.A.S. Kelso, Phys. Rev. Lett. **79**, 4501 (1997)
11. D.R. Cox, P.A.W. Lewis: *The statistical analysis of series of events* (Methuen, London, 1966)
12. B. Doiron, C.R. Laing, A. Longtin and L. Maler: J. Comput. Neurosci. **12**, 5 (2002)
13. B. Ermentrout, M. Pascal, B. Gutkin: Neural Comput. **13**, 1285 (2001)
14. U. Fano: Physiol. Rev. **72**, 26 (1947)
15. J. Foss, A. Longtin, B. Mensour and J.G. Milton: Phys. Rev. Lett. **76**, 708 (1996)
16. G. Fuhrmann, I. Segev, H. Markram, M. Tsodyks: J. Neurophysiol., **87**, 140 (2002)
17. L. Gammaitoni, P. Hänggi, P. Jung and F. Marchesoni: Rev. Mod. Phys. **70**, 223 (1998)
18. C.W. Gardiner: *Handbook of stochastic methods* (Springer, Berlin, 1985)
19. C.D. Geisler, J.M. Goldberg: Biophys. J., **6**, 53 (1966)
20. L. Glass and M.C. Mackey: *From Clocks to Chaos. The Rhythms of Life.* (Princeton U. Press, 1988)
21. D. Green, J. Swets: *Signal Detection Theory and Psychophysics* (Wiley, New York, 1966)
22. S. Guillouzic, I. L'Heureux and A. Longtin: Phys. Rev. E **59**, 3970 (1999)
23. S. Guillouzic, I. L'Heureux and A. Longtin: Phys. Rev. E **61**, 4906 (2000)
24. B. S. Gutkin, C. R. Laing, C. L. Colby, C. C. Chow and G. B. Ermentrout: J. Comput. Neurosci. **11**, 121 (2001)
25. R. Hegger, M.J. Bunner, H. Kantz and A. Giaquinta: Phys. Rev. Lett. **81**, 558 (1998)
26. W. Horsthemke and R. Lefever: *Noise-Induced Transitions. Theory and Applications in Physics, Chemistry and Biology.* (Springer Verlag, New York, 1984)
27. C. Koch: *Biophysics of Computation*, (Oxford UP, New York 1999)
28. R. Kuske, preprint (2002)
29. C. R. Laing and C. C. Chow: Neural Comput. **13** (7), 1473 (2001)
30. C. R. Laing and A. Longtin: Physica D **160**, 149 (2001)
31. C. R. Laing and A. Longtin: Bull. Math. Biol. (2002) (in press)
32. Y.-H. Liu and X.-J. Wang: J. Comput. Neurosci. **10**, 25 (2001)
33. A. Longtin, J.G. Milton, J. Bos and M.C. Mackey: Phys. Rev. A **41**, 6992 (1990)
34. A. Longtin: Phys. Rev. A **44**, 4801 (1991)
35. A. Longtin and J.G. Milton: Math. Biosc. **90**, 183 (1988)
36. S.B. Lowen, M.C. Teich: J. Acoust. Soc. Am. **92**, 803 (1992)
37. S.B. Lowen, S.S. Cash, M. Poo, M.C. Teich: J. Neurosci. **17**, 5666 (1997)
38. B. Mensour and A. Longtin: Physica D **113**, 1 (1998)
39. M.E. Nelson, M.A. MacIver: J. Exp. Biol. **202**, 1105 (1999)
40. T. Ohira: Phys. Rev. E. **55**, R1255 (1997)
41. S. G. Rao, G. V. Williams and P. S. Goldman-Rakic: J. Neurophysiol **81**, 1903 (1999)
42. R. Ratnam, M.E. Nelson: J. Neurosci. **20**, 6672 (2000)
43. B. Redmond, V.G. LeBlanc and A. Longtin: Physica D (2002) (in press)
44. L. Stark: *Neurological Control Systems: Studies in Bioengineering.* (Plenum, New York, 1969)

45. M.C. Teich, S.M. Khanna SM: J. Acoust. Soc. Am. **77**, 1110 (1985)
46. M.C. Teich: IEEE Trans. Biomed. Eng. **36**, 150 (1989)
47. M.C. Teich: 'Fractal neuronal firing patterns'. In: *Single neuron computation.* ed. by T. McKenna, J. Davis, S.F. Zornetzer (Academic, San Diego 1992) pp. 589-622
48. L.S. Tsimring and A. Pikovsky: Phys. Rev. Lett. **87**, 250602 (2001)
49. R.F. Voss, J. Clarke: J. Acoust. Soc. Am. **63**, 258 (1978)
50. X.J. Wang: J. Neuroscience **19**, 9587 (1999)

Long Range Dependence in Human Sensorimotor Coordination

Yanqing Chen[1], Mingzhou Ding[2], and J.A. Scott Kelso[2]

[1] The Neurosciences Institute, San Diego, California 92121, USA
[2] Center for Complex Systems and Brain Sciences, Florida Atlantic University, Boca Raton, Florida 33431-0991, USA

1 Introduction

When humans coordinate their limb movements with a periodic stimulus train (e.g. a rhythmic metronome) by either tapping on the beat (synchronizing) or tapping off the beat (syncopating), their performance is variable and the inevitable timing error from each cycle constitutes a stochastic time series. The analysis of this time series is an important step toward understanding how the human brain controls rhythmic movement. Experimental psychologists have studied this simple task for nearly a century [1]. Early work focused on the mean and variance of the timing errors, but it did not study the possible sequential aspects of the cycle-to-cycle fluctuations. Subsequent researchers attempted to relate correlations among timing errors to the performance strategy [2]. The current prevailing model assumes that the human subject perceives the timing error and makes local corrective adjustments, resulting in short-range correlated timing errors [3,4]. We did a series of experiments to test such hypotheses, and our result revealed a very different picture [5]. In this chapter we summarize our main findings by showing that the error time series exhibits long range correlation and is characterized as a $1/f^{\alpha}$ type of long memory process. Firstly, we describe our analysis of synchronization timing errors using rescaled range (R/S) analysis and frequency domain Maximum Likelihood Estimation (MLE) method. Secondly, we demonstrate that the exponent (α) is different between the synchronization task ($\alpha \approx 0.5$) and the syncopation task ($\alpha \approx 0.8$), and can be altered when human participants employ different coordination strategies. The finding of long range dependency in this simple movement coordination task provides an interesting example of stochastic long range dependence in neuroscience. This phenomenon reveals the complexity of human nervous system, and demonstrates the dynamic interactions between the human brain, its action and the external environment signal.

2 Long Range Dependence of Synchronization Timing Errors

The first experiment required subjects to make finger movements in synchrony with periodic stimuli [6]. Their timing errors, defined as the time differences between certain point of the movement cycle and the corresponding stimulus, were analyzed. Traditional studies of such a task typically only involved less than 100 cycles. Therefore the long range characteristic of timing errors remains unclear. We sought to identify the long term dynamics of the timing fluctuations associated with sensorimotor synchronization. Besides the standard statistical measurements such as mean and variance, we emphasize the correlation and spectral structures among sequential timing fluctuations. Several statistical tools were used, including rescaled range analysis and Maximal Likelihood Estimation. The purpose is to establish the time series characteristics in sensorimotor synchronization.

2.1 Methods

Five normal right handed male subjects, age ranging from 25 to 35, took part in the synchronization experiment. The experiment protocol was approved by local Institutional Review Board (IRB). All subjects were informed of the purpose and procedures of the study and voluntarily participated in the experiment. They were seated in a sound attenuated chamber, instructed to cyclically press their index finger against a computer key in synchrony with a periodic series of auditory beeps, delivered through a headphone. Two frequency conditions, $F_1 = 2$Hz ($T_1 = 500$ms) and $F_2 = 1.25$Hz ($T_2 = 800$ms), were studied. For each frequency the subjects were required to finish 1200 taps without interruption. The 800ms period is the rate that resulted in the smallest percentage variation in synchronization errors according to previous studies [1]. 1200 taps at this frequency require 16 minutes. The other frequency (2Hz) was chosen because subjects could tap with it for a fairly long time (10 minutes) without fatigue. Its period (1/2 sec.) also happens to be the "optimal interval" for synchronization and corresponds to a dominant (preferred) biological rhythm (2Hz) in finger movements. Both of the frequencies are within the range in which humans can synchronize easily. Previous experiments normally only required fewer than 100 taps. We substantially extended the duration of the experiment to obtain longer time series (at least 1024 points) to find out how statistical properties are manifested in the longer time domain.

Each experimental session consisted of 1200 continuous taps for a given frequency. A computer program was used to register the time of a specific point in the tapping cycle in millisecond resolution. The data collected were inter-response-intervals (IRIs), I_i, and synchronization or tapping errors, e_i. As defined in Fig. 1, e_i is the time between the computer recorded response time R_i and the metronome onset S_i, i.e., $e_i = R_i - S_i$, and I_i is the time between successive tapping responses, i.e., $I_i = R_{i+1} - R_i$. From the figure it is clear that I_i and e_i are not independent variables. Specifically,

$$\begin{aligned} I_i &= T + e_{i+1} - e_i \\ e_i &= e_0 + \sum_{k=1}^{i} (I_{k-1} - T) \ . \end{aligned} \qquad (1)$$

This relation determines that synchronization errors may be viewed as the integration (summation in the discrete case) of the IRIs after the mean (stimulus period) is removed. IRIs are the derivative (difference) of the error time series plus the stimulus period. One time series can thus be easily obtained from the other.

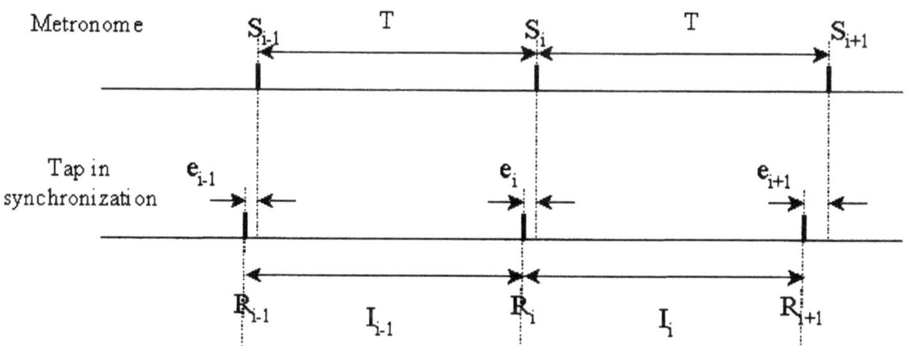

Fig. 1. Definitions for the synchronization error e_i and Inter-Response-Interval I_i.

2.2 Data Analyses and Results

Twenty time series, each consisting of 1200 points, were collected from five subjects, each of whom performed two sessions for a given frequency condition. Each time series was indexed by the order of responses. Figure 2 shows a typical example of an error time series and the corresponding IRIs for $F_1 = 2$Hz. The distribution of the error variable can be well fitted by a Gaussian with mean -16.9ms and standard deviation 20.3ms. From (1) the IRI variable I_i is also Gaussian distributed.

Firstly, static measures of all time series – mean and standard deviation for all the error times series and standard deviation of IRIs were calculated. The total average error at 2Hz is -23.94ms (4.8% of the period) and -32.63ms at 1.25Hz (4% of the period). This result confirms the observation that synchronization at a period of 800ms has lower percentage error [1]. The average SD is 24.15ms at 2Hz and 31.18ms at 1.25Hz. This means the distribution of error becomes wider at longer periods. The negative averaged synchronization error is an interesting phenomenon. It implies that on average subjects actually tap before the onset of external stimuli and anticipate the next one. Studies of this effect and its possible explanation can be found in a list of papers [7,8].

The mean and standard deviation are static measures which cannot provide information about the sequential temporal structure of a time series. The latter can be obtained through power spectrum and/or correlation analysis. Figure 3 (a) and (b) show the power spectra of the time series in Fig. 2 (a) and (b). Both power spectra, plotted on a log-log scale, roughly follow a straight line,

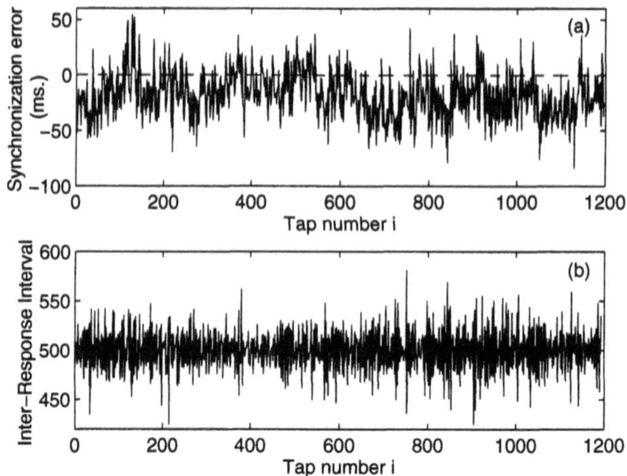

Fig. 2. Example of synchronization error time series at 2Hz (a) and the corresponding Inter-Response-Intervals (b) in millisecond (msec.). (Adapted from [6]).

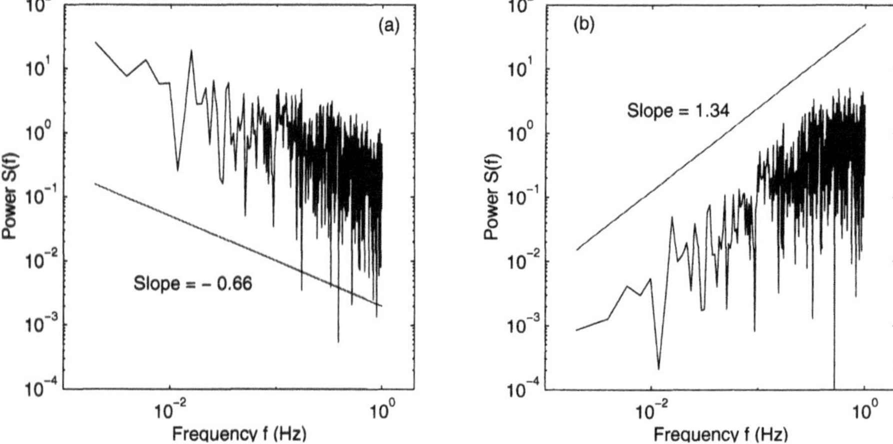

Fig. 3. (a) Spectral density of the error time series in Fig. 2(a); (b) Spectral density of the IRIs time series in Fig.2 (b), calculated from 1024 data points after discarding the first 50 points to eliminate any transient. Unit of frequency has been converted from 1/beat to Hz in the following way. Because the interval between two data points in the error time series is the stimulus period T (0.5 s in this case), the frequency f is defined as $f_k = (k-1)/(mT)$, for $k = 1, 2, ..., m/2 - 1$ where $m(1024)$ is the total length of the time series. (Adapted from [6]).

suggesting that the spectral density $S(f)$ scales with frequency f as a power law, with $S(f) \sim f^{\alpha}$, where $\alpha \approx -0.54$ in the error time series (a) and $\alpha \approx 1.46$ in the IRI series (b).

The power law becomes more evident in the average power spectra. Figure 4 (a) and (b) show the averaged power spectra of the error time series at 2Hz and

1.25Hz. Again the -0.5 power law is clearly observed, extending more than two decades. ¿From the Wiener-Khinchin theorem, the negative slope in the power spectrum implies that the auto-correlation function $C(k)$ of the original time series decays with the time lag k also as a power law

$$C(k) \sim k^{-\beta}, \tag{2}$$

where $\beta = 1 - \alpha$. The presence of inverse power laws in auto-correlations and power spectra means that the temporal structure of the time series has scale invariance. For example, if we calculate the correlations in 2 or 3 lags instead of 1 or scale the frequency to $2f$ or $3f$, the auto-correlation and power spectrum still observe the power law with the same α and β values, and have the same type of curve. In this sense, the time series has self-similarity across many temporal scales with no characteristic frequency or special time duration, and can be identified as a type of fractal time series. Early researchers [2] applied correlation analysis on less than 100 tapping steps and plotted their data on a linear scale. Thus it is not surprising that the power law decay of auto-correlations was not apparent in their results. This type of decay is much slower than the usual exponential decay type of auto-correlation. As a result, its summation goes to infinity: $\sum_{k=0}^{\infty} C(k) = \infty$. This meets the mathematical definition for a long memory process [9,10]. More specifically, since the errors are Gaussian distributed, the synchronization error series may be modeled by fractional Gaussian noise (fGn) [11], which also may be termed long range positive correlated Gaussian noise. The IRI series, because of the relation in (1), must be defined as the derivative of a fractional Gaussian noise process. There are several statistical methods developed to analyze this type of long memory process, in the following section we review two methods we used to rigorously establish the above conclusions.

Rescaled Range (R/S) Analysis. Rescaled range analysis was originally developed by Hurst to analyze the Nile River yearly minimum height [11]. The basic idea is the following. If we construct a random walk from certain stationary random processes, it is observed that the averaged range of this random walk (distance between the highest and lowest point after removing the linear trend – rescaled range) will increase as a power law with respect to the time window length (number of data points). Let $\overline{e}_i = e_i - \overline{e}$ where $\overline{e}$ denotes the sample mean of the given error time series. Consider the cumulative sum, $L(n,s) = \sum_{i=1}^{i=s} \overline{e}_{n+i}$, where $L(n,s)$ can be regarded as the position of a random walk after s steps. Define the range of the random walk as $R(n,s) = \max\{L(n,p), 1 \leq p \leq s\} - \min\{L(n,p), 1 \leq p \leq s\}$. Let $S^2(n,s)$ denote the sample variance of the data set $\{\overline{e}_{n+i}\}_{i=1}^{i=s}$. If the average rescaled statistic $Q(s) =< R(n,s)/S(n,s) >_n$ scales with s as a power law for large s,

$$Q(s) \sim s^{H}, \tag{3}$$

then H is the Hurst exponent. It can be proved that if the random process has no correlation (independent white noise), the exponent is 1/2. But for a large group

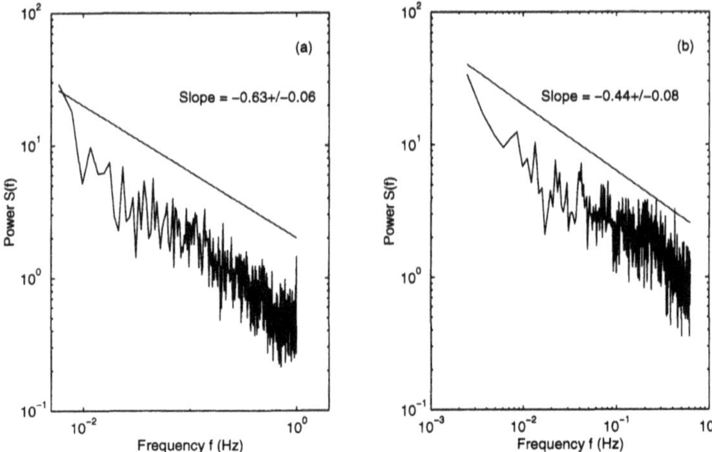

Fig. 4. Average spectral density of the error time series at $2Hz$ (a), $1.25Hz$ (b). Ten time series were averaged in each condition. Slopes and their standard deviations are calculated from MLE method described in the text.

of natural processes (including the yearly water level of Nile river), the exponents are significantly larger than 0.5. This means that the rescaled ranges calculated through the integration of the random process are persistently larger than the integration of uncorrelated noise. These processes are called persistent (long range positive correlated). If the exponents are smaller than 0.5, the processes are called anti-persistent (long range negative correlated). One can further show that if the auto-correlation function decays sufficiently fast such that $C(k)$ sums to a finite number, then generally $H = 1/2$. If $C(k)$ sums to infinity, then $1/2 < H < 1$, and the time series is said to have long persistent memory. It is interesting that this power law in rescaled range is directly related to the power law in its power spectrum [10]. Both are the characteristic properties of a long range correlated process. If the slope in the power spectrum is α, then:

$$H = (1 + \alpha)/2 . \tag{4}$$

Rescaled range analysis gives a direct way to find out the correlation property of a single time series. We applied a trend-corrected rescaled range analysis [12] on all the error and IRI time series we collected. All the R/S plots of the error time series give straight lines on log-log scale with slopes significantly larger than 0.5, while all the IRI plots are curved, which is expected for a derivative time series. Figure 5 shows the R/S plots for the time series in Fig. 2. Only Fig. 5(a) has a linear relationship and is described by a power law. This comparison reveals that only the error time series are long range correlated with well-defined H values. The IRI time series, on the other hand, have no well-defined Hurst exponents from the R/S plot. From the relation in (1), one can readily see that the IRI is the derivative of the long range correlated process. This means that it is not necessary and not correct to fit two slopes on the R/S plot of IRI time series in Fig. 5(b), and to calculate two different H values based upon its parabolic

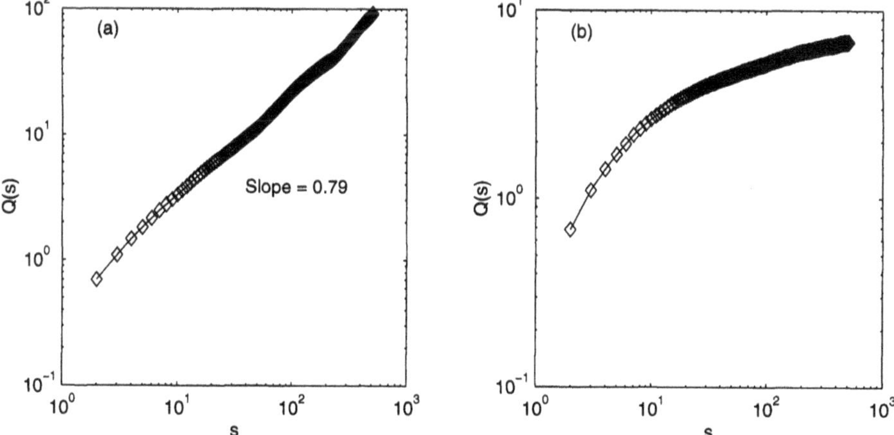

Fig. 5. The rescaled range (R/S) plots for error time series (a) and the IRIs time series (b) from Fig. 2. R/S of errors has a well-defined linear slope on a log-log plot, which gives the Hurst exponent. The R/S of the IRI time series has a parabolic shape which will not give well defined Hurst exponents. Because IRI time series in (b) is the derivative of the error time series in (a), this parabolic curve can not be fitted to two slopes and given to two different H values. (Adapted from [6]).

shape. We can mathematically prove that the curve is a characteristic signature of a differenced time series. The above analysis provides a clear way to identify the two inter-related time series. Only one of them, the synchronization errors, has a single clear positive correlation across all the time scales we studied. The H values from all the experimental error time series based on R/S plots can be found in [6]. All of them are significantly greater than 0.5. Their mean and standard deviation are 0.723 ± 0.071, with no significant difference between 2Hz and 1.25Hz conditions.

Maximal Likelihood Estimation. The Gaussian characteristic established in Fig. 2 further enables us to apply a more systematic statistical method – the frequency domain Maximum Likelihood Estimator [10] – to calculate the Hurst exponent. Statistically, the Maximum Likelihood Estimation (MLE) is a more robust method to evaluate long memory. The frequency domain MLE estimates Hurst exponent from the power spectrum of the experimental data, based upon the fractional Gaussian noise spectral model. Details of the method and the mathematical background can be found in [10], Chapter 6. MLE gives the value of H with a well derived standard deviation of estimation, related only to the number of data points available. The longer the time series, the more accurate the estimation. From our $n = 1024$ data points:

$$SD = \left(\frac{0.6079}{n}\right)^{\frac{1}{2}} = 0.024\,. \tag{5}$$

The overall H value mean and SD is 0.767 ± 0.085. Notice that this value is slightly higher than the H values from R/S analysis. Since it is known that the R/S method underestimates H value when it is larger than 0.75 [15], we believe the H values obtained from MLE provide a better estimate of long memory process. Values of H from both R/S analysis and MLE are consistent with the -0.5 slope in the average power spectra.

Surrogate Data Analysis. The above two methods showed that the synchronization error time series are long range positively correlated. This correlation is manifested in the sequential order of the time series. One way to confirm this result is to destroy the original order of the errors to create a randomly shuffled time series (surrogate set). Figure 6 shows the surrogate time series from Fig. 2(a) and the resulting power spectrum. The power spectrum of the surrogate data obviously changes to a flat white noise power spectrum and its Hurst exponent is close to 0.5 as calculated from MLE. This confirmatory result emphasizes that the sequential variations of the error time series determine the underlying correlation properties. Timing errors in the original order are not independent, but sequentially correlated.

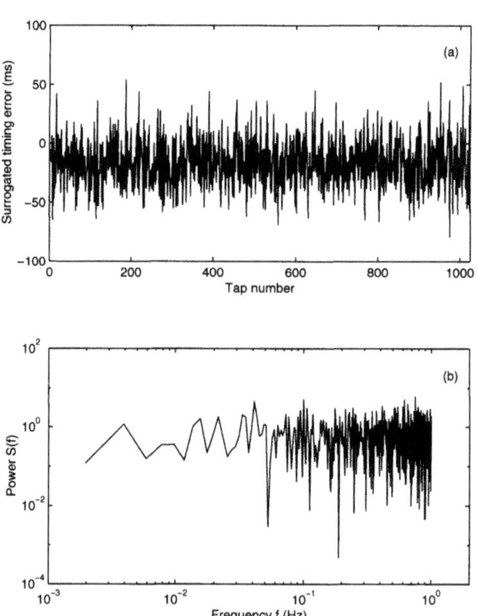

Fig. 6. (a) Randomly shuffled synchronization error time series from Fig. 2(a). (b) Spectral density function of the surrogated data in a log-log scale. The spectrum is basically flat, indicative of a white noise process.

3 Dependence of Scaling Exponent on Task Conditions and Coordination Strategies

Experiment 2 was designed to explore the origins of long memory processes in human timing fluctuations [13]. Similar statistical techniques were used to analyze the timing fluctuations in syncopation (i.e., making movements off the beat) and synchronization under the same stimulus condition. Subjects were also required to use different strategies to syncopate. The objective was to test whether the long range correlations observed in Experiment 1 are related to cognitive factors such as task requirements and coordination strategies. If such correlations originated only from neuromuscular skeletal factors such as muscle force and motor delays, the same fluctuation power spectrum should be observed in syncopation and synchronization experiments, and should not depend upon the specific coordination strategy subjects employ. On the other hand, if long range correlations in timing fluctuations are related to task conditions or coordination strategies, its origins must lie in the higher order, distributed functions of the human brain. It is known that syncopation (e.g. flexion movements off the beat) can be performed at high movement rates if the subject consciously decides to place extension movements on the beat [14]. Might it then be possible to shift the values of the scaling exponents by altering cognitive strategy? Were this so, long range correlation in timing behavior may be said to originate in the higher-level functioning of the human brain. Experiment 2 analyzed the long term temporal structures in synchronization and syncopation tasks, and tested how they change when subjects use different coordination strategies.

3.1 Methods

Eight right-handed male subjects took part in the experiment [13]. Each subject signed the Informed Consent form that had been approved by the local Institutional Review Board (IRB). Subjects were seated in front of a computer inside a sound-attenuated experimental chamber and instructed to rhythmically press their right index finger on a computer key in a specific phase relationship with periodic auditory tunes (duration = 50ms), delivered through the computer. The frequency of the metronome was 1Hz. All eight subjects performed the first two conditions: synchronization and syncopation. In the synchronization task the subject was required to tap in synchrony with the metronome. The syncopation condition required the subject to press the computer key in between adjacent stimuli. Each subject performed two trials for each condition, and a total of 32 time series were collected in the first part of the experiment. The second part of the experiment was designed to produce syncopation by using different coordination strategies. For the "2:1" strategy, subjects (N=4) made two flexion movements during every 1 sec. stimulus interval, one flexion movement synchronized with the stimuli, and a second key depression movement in the middle of the cycle, thus syncopating with the 1Hz stimuli. In a way this is like playing a drum twice as fast as the basic musical rhythm. The second strategy required subjects (N=4) to intentionally control an extension movement (key-release) with the stimulus instead of a downward flexion key-press, so that when the finger relaxed back to the lowest position, the (computer-registered) key-press was in the middle of the cycle (syncopation). At $1Hz$ all these modes of coordination can be maintained, thus providing us a unique opportunity to investigate changes in the coordination dynamics even though the environmental conditions remain the same. A total of 16 time series were collected in the second part of the experiment. To find out how the statistical properties are manifested in the longer time domain, we substantially extended the duration of the experiments to obtain longer continuous time series (550 responses in every single trial). A computer program was used to register the time of a specific point in the movement cycle with 1 ms. resolution. We analyzed the error time series, defined as the time between the computer recorded response time and the metronome onset time, indexed by the number of taps. All data are based on subjects' successful completion of the requirement, i.e., synchronization or syncopation without interruption, without missing stimuli or making extra responses.

Data Analysis. In all time series, the first 10 data points were eliminated to remove transients. The subsequent 512 points were used for further analysis. Spectral analysis was based on standard methods. Each individual spectrum was calculated after removing the mean and normalizing by the standard deviation for the individual time series. To calculate the spectral exponents of each individual time series, the frequency domain Maximal Likelihood Estimation (MLE) [10] was used, which estimates the long range correlations in a time series from its power spectrum based on a fractional Gaussian noise model. The details of the method and the mathematical background can be found in the Chapter 6 of [10]. The second method we used is rescaled range analysis [12], which constructs a random walk from each time series (by integration), and estimates the power law increases of the range of the random walk, as a function of the number of data points. The exponent of this power law is defined as the Hurst exponent. These two methods are the same as that used in Experiment 1. Both MLE and rescaled range analysis give an estimate of the Hurst exponent (H), which we transform to the power spectrum slope through the relation: $\alpha = 2H - 1$. MLE is believed to be the most accurate statistical estimator available, with minimal standard deviation of the estimation limited only by data length (0.07 for 512 data points).

3.2 Results and Discussion

Figure 7 shows the averaged power spectra for the two basic coordination conditions on a log-log scale. Excellent straight line fits mean that the power $S(f)$ scales with the frequency f as a power law $S(f) \sim 1/f^{\alpha}$. By virtue of the Fourier transform, the auto-correlation function C(k) also scales with the time lag k as a power law $C(k) \sim k^{\beta}$, where $\beta = 1 - \alpha$. This slow decrease of the correlation between errors separated by the time interval k is in contrast with the rapid exponential decay predicted by theories that posit local correlation [4]. In fact,

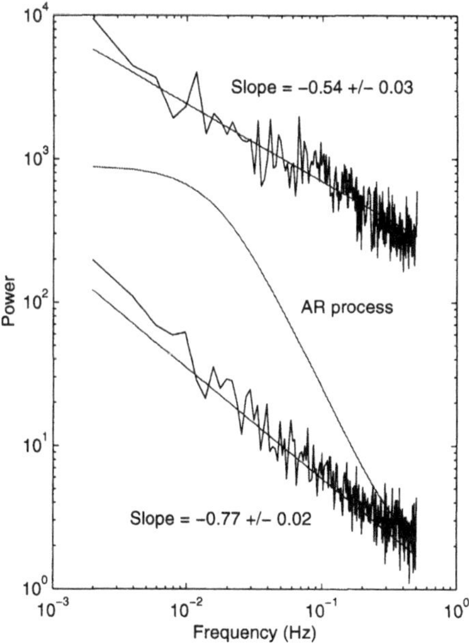

Fig. 7. Averaged spectral density of the error time series for synchronization (top) and syncopation (bottom) from experiment 2. Unit of frequency has been converted from 1/beat to Hz the same way as in Fig. 3. Power spectrum of the synchronization errors has been multiplied by 100 to separate from the syncopation spectrum. The power spectrum of a first-order auto-regressive (AR) process is also plotted as a comparison. Notice that in the low frequency portion the auto-regressive process shows a flattened spectrum. (Adapted from [13]).

for α between 0 and 1, as is the case in the present experiment, the $C(k)$ function decays so slowly with k that the underlying statistical process is said to have long memory. Thus, the value of one synchronization or syncopation error is related to and affects the errors into the distant future. Remarkably, the scaling exponent α allows us to distinguish clearly the two basic modes of coordination. Comparison of the numerical values of $\alpha = 0.77$ for syncopation and $\alpha = 0.54$ for synchronization further reveals that the correlation between syncopation errors decays much more slowly than synchronization. The longer the separation between timing errors, the more pronounced is the difference in the strength of the correlation between the syncopation and synchronization errors. In Fig. (7) we have used the averaged spectra to summarize the results from synchronization and syncopation. To evaluate the statistical significance of the difference between the two conditions, we applied the frequency domain Maximal Likelihood Estimator (MLE) to each of the 32 time series (16 from synchronization and 16 from syncopation) to calculate the individual exponent α. The overall α value for synchronization is 0.45 ± 0.16, while for syncopation, $\alpha = 0.73 \pm 0.17$. The two exponents are significantly different by two-tailed t-test ($p < 0.025$). As a verification step, we applied rescaled range analysis to each of the 32 time series. The estimated α values were again different between the two conditions ($\alpha = 0.45 \pm 0.16$ for synchronization and $\alpha = 0.61 \pm 0.12$ for syncopation), despite the fact that the rescaled range analysis has the tendency to underestimate H when $H > 0.75$. This convergence between the two different methods, one in the spectral domain and the other in the time domain, is important since it is known that each method alone is susceptible to certain artifacts in the data. Together, however, they form a robust integrated approach [15]. Synchronization is a relatively simple task for humans to perform. Syncopation, however, is more demanding, requiring longer training and a higher level of concentration to ensure quality performance. Moreover, it is known that syncopation is a less stable form of coordination, and that spontaneous switching to synchronization can occur at both behavioral and neural levels. The difference between the α values for the two conditions suggests that the scaling exponent is a correlate of task difficulty. It is known that syncopation (e.g. flexion movements off the beat) can be performed at high movement rates if the subject consciously decides to place extension movements on the beat [14]. Is the change of cognitive strategy accompanied by a shift of the scaling exponents? The second part of the experiment was designed to probe this idea. The task required subjects to use different sensorimotor coordination strategies while producing movement responses between stimuli, thus giving rise to variable timing errors mimicking the syncopation task. Using Maximal Likelihood Estimation, the scaling exponent from "2:1" strategy was found to be $\alpha = 0.48 \pm 0.19$ which is nearly the same as that of the original synchronization. The overall averaged spectrum from "extension-on" strategy is -0.83, still close to the original simple syncopation condition. But careful inspection of individual time series and their spectra shows that spectra from subject who well followed the instructions have slope close to -0.45 [13].

The difference between the 2:1 syncopation strategy and the normal 1:1 strategy is that the extra movement is actually doing synchronization, even though objectively the recorded key-tapping movement is syncopating with the metronome. In this case the fluctuation dynamics goes with the "subjective synchronizing" strategy, shifting toward $1/f^{0.5}$ spectrum. The extension-on movement, different from the above two conditions in that: 1) it requires a different muscle group (extensors vs. flexors), and 2) the association point with the metronome occurs when the finger is released from the keyboard, corresponding to the losing of tactile information, instead of gaining touch information with the key during flexion keypress. Because of this, the association between movement and the stimulus is less stable than the flexion movement for some subjects, who may feel the coupling is not as strong as synchronization. But for subjects who can synchronize their extension movements with the metronome better, the fluctuation spectra can be stabilized toward $1/f^{0.5}$ spectrum as the synchronization case. This demonstrates the variability of the human performance; it also well indicates the sensitivity of the dynamics in the timing fluctuations when subjects adopt different coordination strategies, suggesting that the slope in power spectrum could really be a measure of the different motor control strategies.

4 Discussion

Analyses on the sequential temporal structures of timing errors in synchronization and syncopation revealed the existence of long range correlations. In the following we discuss the methodology of our analyses, the meaning of long memory in synchronization behavior, and its implication for existing timing models.

Our results in this chapter were based on spectrum and rescaled range analysis, both of which are important for our conclusions. R/S analysis on timing errors and IRIs revealed that only the error time series could be defined as a long range correlated process whereas the IRIs are the derivatives. This cannot be determined only from the comparison of their power spectra. On the other hand, as pointed out before [10], certain deterministic processes (such as an exponential decaying process) may generate a rescaled range which follows a power law. Only power spectrum analysis can exclude this case and identify whether the error time series truly comes from a $1/f^{\alpha}$ type of long range correlated random process. These two methods complement one another, enabling us to concentrate our analysis and modeling on the synchronization error time series because its long range property determines the dynamical characteristics of the process.

Long range correlations in sensorimotor coordination were manifested in the strong low frequency variations in the time series (see Fig. 2(a)). Specifically, the errors seem to occur in one direction (e.g., below the overall mean) for a while then switch back and keep on occurring in the other direction (e.g., larger than the overall mean). This type of "drifting tendency" has been reported before [1]. We now understood that this is a clear indication of "persistent" behavior which should be described as a characteristic of long memory processes. The long duration of our experimental trials allowed us to examine the large time scale (up to several minutes) dynamics in which this persistence and long memory property is most clearly evident. To test whether this low frequency trend is coming from fatigue or boredom due to the long experimental session, we asked one of the subjects to perform 16 trials of the synchronization task (70 points each) where the trials are separately by at least 3 minutes. We calculated 16 power spectra from 64 synchronization errors from every 70 points. The averaged power spectrum also follows $-1/2$ slope on log-log scale. Comparing it with the average power spectra we obtained in Fig. 4 revealed that they obeyed the same power law. The only difference is that by extending the duration of the experiment, we find that this same $-1/2$ slope can be extended to much lower frequency ranges (corresponding to longer time durations); the variation at this longer interval still follows the same power law. Based on this observation, we argue that the low frequency drift is not caused by fatigue or loss of concentration on the part of the subject. Rather, they come from the same dynamical process that appears to extend from the short range to the time limit as long as the subject is correctly performing the task.

Current prevailing human timing models postulate that the subject makes self-correcting adjustments in response to perceived timing errors [3,4]. They predict that the timing errors exhibit short-range correlation and the autocorrelation function decays exponentially, leading to a Lorentzian type of power spectrum, as shown in Fig. 7. The inconsistency between the local correlation theory and the experimental data shown in Fig. 7 is apparent. Our results indicated that fluctuations in human timing errors are more complex than what can be expected from a simple error correction mechanism. Complex temporal structures of timing errors actually contain deeper information about task difficulties, human cognitive strategies, and possibly higher human brain functions.

It is worth noting that power law relations in the spectral domain can appear quite naturally in systems with many time scales [16]. If we view the error variable measured in the present experiment as the final outcome of neural activities from many structures on multiple time scales, it seems that the 1/f framework is more appropriate than the simple self-correcting models for describing the temporal dynamics of the error series [17]. Our experiments established a functional role for these types of processes by showing that the scaling exponent differentiates basic modes of sensorimotor coordination and can be altered by cognitive manipulations. Relating changes of scaling exponent to different functional states of the human sensorimotor coordination is a first step toward understanding the origin of long range dependence in such human behavior. Research in this direction could lead to a "physical" model of long range dependence in the human brain, not just a phenomenological one.

Acknowledgments

Research was supported by NSF, NIMH and ONR grants. Preparation of this article (Y. C.) was also supported by the Neurosciences Research Foundation.

References

1. K. Dunlap: Psych. Rev. **17**, 399 (1910) H. Woodrow: J. Exp. Psy. **XV(4)**, 357 (1932) N.R. Bartlett, S.C. Bartlett: Psych. Rev. **66**, 20 (1959)
2. D. Hary, G.P. Moore: Biol. Cybern. **56**, 305 (1987) J. Mates: Biol. Cybern. **70**, 463 (1994)
3. D. Vorberg, A.M. Wing: 'Modeling variability and dependence in timing'. In: *Handbook of Perception and Action.* eds. by H. Heuer, S.W. Keele (New York, Academic Press 1996) pp.181-262
4. J. Pressing, G. Jolley-Rogers: Biol. Cybern. **76**, 339 (1997)
5. Y. Chen: Dynamics of Human Sensorimotor Coordination: from Behavior to Brain Activity. Ph.D. Dissertation, Florida Atlantic University, Boca Raton (2000) Online at wwwlib.umi.com/dissertations/fullcit/9981150
6. Y. Chen, M. Ding, J.A.S. Kelso: Phys. Rev. Lett. **79**, 4501 (1997)
7. P. Fraisse: 'Rhythm and tempo'. In: D. Deutsch (Ed.), *The Psychology of Music.* (New York, Academic Press 1982) pp.149-180
8. G. Aschersleben, W. Prinz: Percep. Psychophys. **57**, 305 (1995); J. Paillard: Annee. Psychol. **48**, 28 (1980)
9. G. Samorodnitsky, M.S. Taqqu: *Stable Non-Gaussian Random Processes: Stochastic Models with Infinite Variance.* (Chapman and Hall, New York, 1994)
10. J. Beran: *Statistics for Long-memory Processes.* (Chapman and Hall, New York 1994)

11. H.E. Hurst, Trans. Amer. Soc. Civ. Engrs. **116**, 770 (1951) B.B. Mandelbrot, J.W. Van Ness, SIAM Rev. **10**, 422 (1968) B.B. Mandelbrot, J.R. Wallis, Water Resour. Res. **3**, 967 (1969)
12. J.B. Bassingthwaighte, L.S. Liebovitch, B.J. West, *Fractal Physiology* (Oxford University Press, New York, 1994).
13. Y. Chen, M. Ding, J.A.S. Kelso: J. Mot. Behav. **33**, 3 (2001)
14. J.A.S. Kelso, J.D. DelColle, G. Schöner: (1990). 'Action-perception as a dynamic pattern formation process'. In M. Jeannerod M. (Ed.), *Attention and Performance XIII* ed. by M. Jeannerod (Erlbaum, New Jersey 1990) pp. 139–169
15. H.E. Scepers, J. van Beek, J.B. Bassingthwaighte: IEEE Engr. Med. Biol. Mag. **June**, 57 (1992) G. Rangarajan, M. Ding: Phys. Rev. E **61**, 4991 (2000)
16. C.W.J. Granger: J. Econometrics **14**, 227 (1980)
17. M. Ding, Y. Chen, J.A.S. Kelso: Brain and Cognition **48**, 98 (2002)

Scaling and Criticality in Large-Scale Neuronal Activity

Klaus Linkenkaer-Hansen

BioMag Laboratory, Engineering Centre, Helsinki University Central Hospital, P.O. Box 340, FIN-00029 HUS, Finland

Abstract. The human brain during wakeful rest spontaneously generates large-scale neuronal network oscillations at around 10 and 20 Hz that can be measured non-invasively using magnetoencephalography (MEG) or electroencephalography (EEG). In this chapter, spontaneous oscillations are viewed as the outcome of a self-organizing stochastic process. The aim is to introduce the general prerequisites for stochastic systems to evolve to the critical state and to explain their neurophysiological equivalents. I review the recent evidence that the theory of self-organized criticality (SOC) may provide a unifying explanation for the large variability in amplitude, duration, and recurrence of spontaneous network oscillations, as well as the high susceptibility to perturbations and the long-range power-law temporal correlations in their amplitude envelope.

1 Introduction

This chapter concerns the nature of oscillatory activity in the human brain, in particular at 10 Hz. These oscillations are generated spontaneously in several areas of the cerebral cortex as neuronal networks transiently form assemblies of synchronously firing cells. Sensory input can modulate or induce neuronal activity at around 10 Hz, but a large variability in amplitude, duration, and recurrence of these oscillations is observed even in the absence of such perturbations. This variability has intrigued researchers for more than 70 years [6] [27] [24], yet its mechanistic origin and functional significance have remained poorly understood.

To understand the source of variability in spontaneously generated or stimulus-induced neuronal oscillations, it seems intuitively natural to investigate the detailed biophysical mechanisms underlying the generation of this activity. Nevertheless, there may be an alternative approach. Recent years of research on self-organization and complexity has lead to novel methods and frameworks for analyzing and interpreting variability *per se*, which are now changing the way complex fluctuations are perceived in many areas of science [4] [12] [43] [38]. While previously considered a nuisance or "noise", variability is increasingly being acknowledged as a potentially valuable source of information about the developmental history and spatiotemporal organization of non-linear dynamical systems.

The central hypothesis of this chapter is that theories of self-organization, aiming to explain the often statistically stereotyped variability of structures and dynamics in stochastic multi-unit systems, may also apply to the dynamics of 10-Hz oscillations in the brain. The most fundamental prediction of these theories is a "critical" type of dynamics of these network oscillations, which should be reflected in intermittent amplitude fluctuations with power-law decay of the temporal correlations. Regarding the functional significance, it is conjectured that network oscillations in a critical state may provide a beneficial susceptibility for the brain to detect weak inputs and to reorganize swiftly from one state to another during processing demands.

The chapter is organized as follows. In Sect. 2, I introduce some characteristics of neurons and neuronal networks that are relevant to the spatio-temporal dynamics of large-scale brain activity. The focus of Sect. 3 is on common prerequisites for complex systems to self-organize to a critical state and their neuronal counterparts. Sect. 4 is a review of recent empirical evidence for critical fluctuations in human brain activity followed by a brief discussion in Sect. 5.

2 Self-organization in Neuronal Systems

Contemporary neuroscience primarily studies neuronal activity in relation to specific tasks or the processing of sensory stimuli. Pronounced activity is, however, also generated endogenously because of the intrinsic dynamics of neuronal systems and it plays a critical role in shaping the spatial and temporal complexity of neural systems.

2.1 Brain Ontogeny

The macro-structural organization of the brain, such as the division of the cerebral cortex into left and right hemispheres, is relatively constant across subjects and mainly determined by the genetic code [34]. The information contained in the human genome is, however, vastly insufficient to code for the microstructural wiring of the cortex involving the establishment of around 10^{14} synaptic connections between 10^{11} neurons [9].

Thus, instead of following some master plan, the brain constructs itself in a complex interplay between internal constraints defined by the genetic makeup and a number of activity-dependent physico-chemical processes. While it has long been acknowledged that exogenous activity, i.e., neuronal activity driven by environmental influences through our senses, is crucial for the normal development of neuronal circuits [34], recent years of research have also stressed the importance of *endogenously* generated activity as a strong determining factor in shaping the morphology of individual neurons and the topology of large-scale neuronal networks from early stages of development and throughout life [34]. The closely related terms "endogenous", "intrinsic", and "spontaneous" imply that the activity is generated from within the system, which may be viewed as synonymous with the term "self-organized", which is used frequently in complexity science to denote dynamical changes that result from the interactions among components without explicit involvement from outside the system.

Isolated neurons are generally thought to be able to fire an action potential spontaneously in a probabilistic manner that is random in the sense of being independent of previous activity, and often they are assumed to do so even when being part of a network [35]. There is, however, little justification for this "randomness assumption", since even down to the molecular scale of single Na^+ channels, activity-dependent structural constraints translate into a memory that affects the future neuronal excitability on time scales from milliseconds to minutes [35]. Moreover, electrical activity causes the neurons to release biochemical substances that regulate the morphological development of neurons such as axonal and dendritic arborization [42], and the establishment of synaptic connections with neighboring neurons is also activity-dependent. Altogether, the cellular-level structural development causes electrical activity to propagate and reverberate in increasingly larger and more complex and integrated networks, which in turn exert an influence on the structure and function of individual neurons [42].

The reciprocal relationship between structural changes and temporal dynamics is a hallmark of complex self-organizing systems, in particular those that exhibit SOC [38] (p. 358–360). Although the correspondence between past and present activity is intuitively clear, most self-organizing systems – including the brain – are also subjected to stochastic driving forces, which makes it highly nontrivial to reveal the non-randomness hidden in the complex response patterns of these systems [1] [15]. The best one can do with empirical data at present is to analyze for temporal or spatiotemporal correlations, hoping to detect correlations that deviate from that of a random process, e.g., the tendency of a signal to follow the same path as previously possibly because of underlying structural constraints.

In summary, one may view the brain as a self-organizing non-linear system with a rich variety of emergent structures and dynamics that reflects its developmental history. Insight into the character of self-organization processes in general and the mechanisms that shape their dynamics may therefore help us understand certain aspects of brain ontogeny.

2.2 Oscillations in Neuronal Networks

Neuronal network oscillations are prominent emergent large-scale phenomena generating bursts of neuronal firing at frequencies mainly between 1 and 200 Hz [10] and their biophysical mechanism of generation depends on synaptic interactions between different types of neurons and anatomical structures [40]. Network oscillations at around 10 Hz are particularly pronounced during wakeful rest when they emerge spontaneously in primary sensory areas of the adult brain [27] and propagate as traveling waves in the tangential plane of the cerebral cortex [22]. The "10-Hz oscillations" are absent in babies less than 4 months of age, after which it emerges as a slow 4-Hz oscillation, increasing to 6, 8, and 10 Hz at the age of 1, 3, and 10 years, respectively [27]. The ontogeny of 10-Hz oscillations is therefore a good example of a neuronal process that exhibits changes in dynamics on very long time scales.

Depending on the experimental condition, however, 10-Hz oscillations can be suppressed or increased transiently by various stimuli [27] [32] [28]. To leave open to what extent they are spontaneously or endogenously generated vs. modulated also by externally imposed processing demands, such oscillations are termed "ongoing oscillations" in the remainder of this chapter.

Despite the long tradition of studying 10-Hz oscillations, there is no consensus about their functional significance. The tendency of 10-Hz oscillations to emerge during rest and to be suppressed when the given brain region is activated have earned them the name "idling rhythms" [13]. Although possibly idling for no good reason, it has also been speculated that this activity keeps neuronal networks in a state of readiness to respond quickly, e.g., by setting the mean level of the membrane potential within extensive populations of neurons and thereby affecting their input-output transfer function [23].

2.3 Low-Dimensional Chaos in 10-Hz Oscillations?

The occipito-parietal 10-Hz activity is characterized by conspicuously irregular patterns of waxing and waning oscillations, which have attracted widespread attention in the context of low-dimensional chaos over the past two decades [31] [8] [39].

The conclusions from the many chaos analyses of human EEG signals have, however, been rather bleak: chaos in terms of a low-dimensional process is presumably more a sign of pathology than a general characteristic feature of the normal 10-Hz oscillations [19].

One may think of various explanations for the unsuccessful outcome of the many attempts to pin-point low-dimensional chaos in 10-Hz oscillations. First, the involved networks consist of millions of neurons and are therefore not low-dimensional *per se*, like, e.g., the forced damped pendulum. Secondly, if large-scale synchronized activity would emerge, following an analogous path, e.g., as the formation of ordered flow in Rayleigh-Benard convection [38] (p. 213–216), some mechanism in the brain would have to fine-tune essential parameters governing the dynamics of the large populations of neurons participating in network oscillations at around 10 Hz. It is, however, far from clear even theoretically what such delicate control mechanisms would be.

3 Self-organization and Complexity

The aim of the science of self-organization and complexity is to understand the source and character of structures and temporal dynamics in systems that naturally exhibit large variability. In addition, when applied to biological systems, the aim is to gain insight into the functional implications of spatiotemporal complexity. Finally, one may wish to develop methods that can predict and/or control the evolution of such systems.

A system is said to be *self-organized* when its structure emerges without detailed control from outside the system, i.e., the constraints on the organization result from the interactions among the components and are internal to the system. *Complexity* does not have a strict definition, but a lot of work on complexity centers around statistical power laws, which describe the scaling properties of fractal processes and structures that are common among systems that at least qualitatively are considered complex. The success of describing self-similar fluctuations in time and space with fractal geometry, i.e., in terms of power-law scaling exponents, does, however, not in itself reveal why fractals are ubiquitous in nature. The theory of self-organized criticality (SOC), on the other hand, provides deep insight into this question [4] [38].

3.1 Self-organized Criticality

The concept of self-organized criticality (SOC) was originally introduced by Bak, Tang, and Wiesenfeld as a generic mechanism for producing $1/f$ noise [3] and for explaining the widespread occurrence of fractal structures in nature. The core discovery was that self-similar patterns in time and space are a consequence of each other [26].

Using agent-based modeling, Bak *et al.* simulated the emerging large-scale dynamics of avalanches when grains of sand were added slowly and randomly to a sandpile [3]. The term *self-organized* was coined because a power-law relationship between the size and the frequency of avalanches appeared without any external guidance other than randomly dropping grains of sand. The term *criticality* was chosen on the basis of the analogy with the dynamics of equilibrium systems near the *critical point* of a phase transition, where domains of different phases of a substance (e.g., ordered vs. disordered magnetic spins) appear in all sizes and with long-range power-law correlations in space and time. Although this dynamics is similar to that of the sandpile, there is a dramatic difference in that critical dynamics in equilibrium systems only appears through the fine-tuning of some parameter (e.g., temperature), whereas the sandpile self-organizes naturally to the critical point.

To understand the essential physics of the self-organization process that drives certain nonlinear systems naturally or unavoidably to the critical state, one has to understand how non-randomness emerges from random initial conditions and random input (the driving force). Consider a simple model of forest fires with just two opposing mechanisms: trees are planted at random on a large grid of potential tree locations while lightning strikes also at random [25]. When a tree is hit by lightning, a fire is initiated which propagates to its nearest-neighboring trees. To begin with, the amount of trees that burn down must clearly be random, since neighboring sites are occupied by trees only by chance, but as time goes by regions that have not been ignited by lightning become linked into growing clusters by the newly planted trees. This way, spatial correlations between wide-apart trees emerge, which materializes as a large fire once a tree somewhere in the cluster is set on fire. After such a large fire, trees have vanished which prevents fires in the near future to propagate through that area. As this process goes on for a long time, the amount of and density of trees in larger and larger areas become correlated through the history of fires further and further back in time until the state of every site of the grid (tree or not a tree) depends on in principle past fire activity in every other site. In short, the scale-free spatial and temporal structures emerge hand-in-hand as a natural consequence of the growing correlations via emerging structures (clusters of trees) that in turn bias the path of the temporal dynamics (fires) etc. until the system is fully correlated in a statistical sense.

What counts in the critical state are not complex details, but simple underlying features of geometry that control how influences propagate. This is why such a simple model can actually capture the essential physics and reproduce the empirical power-law statistics of wild-forest fires [25].

Put in general terms, the following features are characteristic of SOC systems or vice versa: systems that meet the criteria below are apt to exhibit SOC [4] [38]:

- Spatially extended
- Units interact locally
- Driven slowly
- Dissipative (or nonlinear)
- Activity propagates in avalanches
- Plastic (i.e., amenable to modifications)
- Insensitive to initial conditions
- Statistics are robust to random perturbations
- "Old" (i.e., evolved for a long time relative to its size)

and the critical state is expressed in terms of:

- Finite-size effects
- Long-range correlations in space
- Long-range correlations in time
- Power-law scaling behavior
- Criticality: high susceptibility to perturbations

How these general prerequisites of SOC systems apply to neuronal networks, is considered in the following Sect. (3.2).

3.2 SOC in Large-Scale Neuronal Activity?

Here the equivalence of the SOC prerequisites listed in the previous Sect. (3.1) and mechanisms of large-scale neuronal activity is established.

The ideal SOC system is spatially extended with locally interacting units. These features are clearly met by the brain with its typical density of neurons reaching approximately 10^5 neurons per mm^2 of cortical tissue and on the order of 10^3–10^4 synaptic connections pr. neuron [29]. Although many neurons project to distant brain regions, most synaptic connections are made within approximately 3 mm of the cell body [29]. The prerequisite of being spatially extended (i.e., having many units) relates to the need of the system to generate spatial structures that can hold a memory of past activity. The concepts of memory in neuroscience and complexity physics differ a bit. In cognitive neuroscience, memory often has the connotation of a neuronal representation of a previous event that is accessible to the consciousness. In physics or physiology, it has the meaning of a temporal correlation: if the realization of some process is not independent of a previous process, some information about the former process must be stored somewhere, i.e., in space (time is not a "medium" for information storage). This is why structural changes must take place also when the brain encodes consciously accessible memories, e.g., by establishing new synapses or adjusting the efficacy of already existing synapses in an activity-dependent manner [18].

SOC systems are driven slowly, which is of course a relative term [38]. In the forest-fire example (Sect. 3.1), it means that there has to be a reasonable trade-off between the speed of planting trees and the frequency of lightning. This is the logic: if trees are planted very quickly relative to the frequency of lightning, every site becomes correlated with every other site and all fires will be large. Conversely, a very high rate of lightning opposes the formation of large clusters of trees and all fires will be small so that long-range spatial correlations are not allowed to form. Critical behavior is found for ratios in between these extremes. In terms of neurons and neuronal networks, a high driving force could, e.g., be such a high rate of afferent signals that synaptic vesicles would deplete and drive the network into a state of refractoriness with very limited possibilities for neuronal signals to propagate.

The dissipation of energy refers to the need for a nonlinear component for energy or stress to build up, which is then occasionally dissipated via a so-called avalanche effect where activity propagates over large areas via nearest-neighbor interactions. The threshold nature of neurons is clearly nonlinear, e.g., a certain amount of spatial and/or temporal summation of postsynaptic potentials is required for a neuron to generate action potentials. Although, not expressed as an avalanche effect, there are sev eral examples that neuronal activity spreads in this way, e.g.:

"The spontaneous appearance of synchronized oscillations results from the initiation by one or a small number of cells followed by the progressive recruitment of large numbers of neighboring neurons into the synchronized network activity." [5].

Or:

"Propagating neural activity in the developing mammalian retina is required for the normal patterning of retinothalamic connections. This activity exhibits a complex spatiotemporal pattern of initiation, propagation, and termination" [7].

The last quote as well as the comments on neuronal ontogeny in Sect. 2.1 indicate that neuronal networks are certainly plastic and thus able to host a memory of past activity through the modification of their circuitry.

An SOC system may develop to the same statistical state irrespective of the initial conditions and a wide range of random perturbations throughout its development, i.e., the critical state is a spatiotemporal attractor of its dynamics [26]. Here the argument for a neuronal equivalence has to be more qualitative, since little is known about the statistics of the brain's spatiotemporal organization. Nevertheless, adult human brains are characterized by a high degree of functional and macro-anatomical similarity, despite large variations in the environments in which these brains develop.

Finally, the system must be old in order reach the self-organized critical state, i.e., it must have evolved for a long time relative to its size because the correlations in space and time develop slowly and the system is only truly critical when fully correlated according to statistical power laws [30].

The above interaction between size and age of a self-organized critical system explains why SOC systems are characterized by finite-size effects, i.e., there are cutoffs where the spatial and temporal power-law scaling behavior breaks down. The spatial correlations cannot exceed the size of the system and the smallness of the system defines an upper limit to how long temporal correlations the system can hold, because eventually all the structures that were created by some event have been "overwritten". SOC systems are not only "critical" in the sense of having power-law correlations in space and time as explained in Sect. 3.1, but also in terms of exhibiting high susceptibility to perturbations, which is also a characteristic of equilibrium systems at the critical point of a phase transition. Qualitatively, neuronal systems may definitely also be considered highly susceptible to perturbations, since quick and reliable responsiveness to incoming signals is a prerequisite for the information processing in the brain.

4 Evidence for SOC in Neuronal Systems

The evidence that neuronal structures may exhibit SOC is currently limited to a couple of reports. Jung *et al.* observed power-law scaling behavior in the spatiotemporal organization of calcium waves in a culture of glial cells [17] and in a computational model of the same system [16]. In the human electroencephalogram, power-law scaling behavior has been reported for the duration of bursts of oscillations [11], which was interpreted as evidence for SOC. The following Sects. (4.1–4.2) review the additional supporting evidence from temporal correlation analysis of spontaneous oscillations in humans [20].

4.1 Materials and Methods

The electrical activity of the brain may be recorded noninvasively either as the electric potentials it produces on the scalp (electroencephalography, EEG) or the associated magnetic fields outside the head (magnetoencephalography, MEG) [14]. The amplitude of an oscillation, as recorded with electroencephalography or magnetoencephalography, is determined by three factors of the underlying neuronal network: the number of neurons that take part in the assembly, the degree of synchrony, and the discharge rate of those neurons [37].

Normal human subjects ($n = 10$) during wakeful rest with eyes closed were measured with multi-channel EEG and MEG (Vectorview^{TM}, 4D-Neuroimaging) for 20 minutes each. Subsets of 4 channels were chosen on the basis of anatomical location and high signal-to-noise ratio for the detailed analysis.

The oscillation-amplitude envelope at around 10 Hz was estimated with a wavelet filter [41] and the time series were analyzed for temporal correlations using three complementary methods: power spectral density, autocorrelation, and detrended fluctuation analysis.

The power spectral density of a signal reveals the contribution of different frequencies to the total power of the signal. White-noise signals contain equal power at all frequencies and are usually considered uncorrelated, while long-range correlated signals often have log-log linear power spectra with a non-zero power-law exponent ($1/f^{\beta}$ signal). Periodic signals have peaks in the spectrum at frequencies corresponding to the inverses of these periods. The important difference between a power spectrum of the recorded broadband signal (e.g., 0.1–100 Hz) and the amplitude envelope of narrow-band neural oscillations (e.g., 7–13 Hz), is shown in Fig. 1.

The autocorrelation function gives a measure of how a signal is correlated with itself over different time lags. A normalized autocorrelation function attains its maximum value of one at zero time lag, decays towards zero with increasing time lag for finite correlated signals, and fluctuates close to zero at time lags free of correlations. Signals that are modulated at a characteristic scale produce autocorrelation functions that are also modulated with the period of the characteristic scale.

The detrended fluctuation analysis has been developed for quantifying correlations in non-stationary signals, e.g., in physiological time series, because long-range correlations – revealed by the autocorrelation function – can arise also as an artifact of the "patchiness" of non-stationary data [33]. The detrended fluctuation analysis measures the variance of linearly detrended signals, $F(\tau)$, as a function of window size, τ. The average fluctuation, $\langle F(\tau)\rangle$, is often of a power-law form:

$$\langle F(\tau)\rangle \approx \tau^{\alpha} . \tag{1}$$

The scaling exponent, α, is extracted with linear regression in double-logarithmic coordinates using a least-squares algorithm. An exponent of $\alpha = 0.5$ characterizes the ideal case of an uncorrelated signal, whereas $0.5 < \alpha < 1.0$ indicate power-law scaling behavior and the presence of temporal correlations over the range of τ where relation (1) is valid. Periodic signals have $\alpha = 0$ for time scales larger than the period of repetition.

The above three complementary tests were used because of recent reports pointing out that one of these methods used alone may indicate the presence of long-range correlations as an artifact of the patchiness of non-stationary data [33], while agreement between independently obtained scaling exponents according to theoretically derived relationships lowers the risk of falsely concluding that a given data set contain long-range correlations [36].

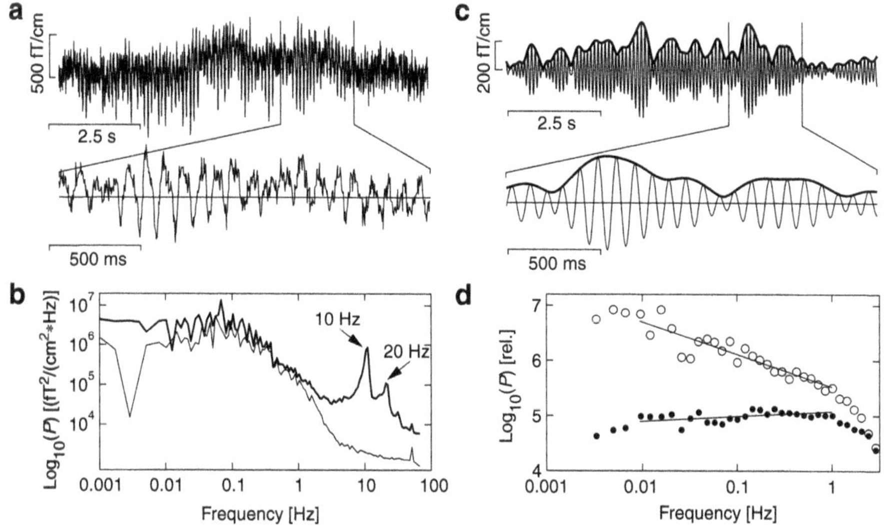

Fig. 1. The frequency content of broadband signals and of the amplitude envelope of narrow-frequency-band neural oscillations. (**a**) A representative epoch of the broadband signal as obtained from a single MEG channel at the acquisition (0.1–100 Hz) is shown at two time scales. Note from the *upper trace* that the high-amplitude 10-Hz oscillations are riding on slow fluctuations (< 1 Hz). The slow fluctuations in the broadband signals are likely to originate mainly from environmental noise as it can be inferred from (**b**) showing the spectral density of the entire 20-min long signal in (**a**) (*thick line*) and reference data (i.e., no subject in the MEG device) from the same channel (*thin line*). The neural signals clearly dominate at frequencies above but not below 1 Hz with prominent peaks at 10 and 20 Hz (see *arrows*). (**c**) The signal shown in (**a**) has been filtered at 10 Hz with a Morlet wavelet (passband 6.7–13.3 Hz). The *thin lines* are the real part and the amplitude envelopes (*thick lines*) are the modulus of the wavelet transform. (**d**) The power spectrum of the amplitude envelope of the neural data at 10 Hz exhibits a $1/f^{\beta}$ power spectrum (*circles*) with $\beta = 0.58$ in the range from 0.01 to 1 Hz, thereby indicating that the fluctuations of the amplitude envelope of these oscillations are correlated at time scales of 1 to 100 s. On the contrary, the 10-Hz amplitude envelope of the reference data at 10 Hz gave rise to a white-noise spectrum, which is characteristic of a temporally uncorrelated process (*dots*). Note that the analyses of temporal correlation reported in the following were performed on the amplitude *envelope* of oscillatory activity (*thick lines* in (**c**))

4.2 Long-Range Temporal Correlations and Scaling Behavior

The human brain spontaneously generates neuronal oscillations with a large variability in frequency, amplitude, duration, and recurrence. Little, however, is known about the long-term spatio-temporal structure of these complexly fluctuating oscillations. We hypothesized that ongoing oscillations may self-organize to a critical state, which is characterized by large variability and long-range power-law temporal correlations, similar to that of other stochastic multi-unit systems (cf. Sects. 3.1 and 3.2) [20].

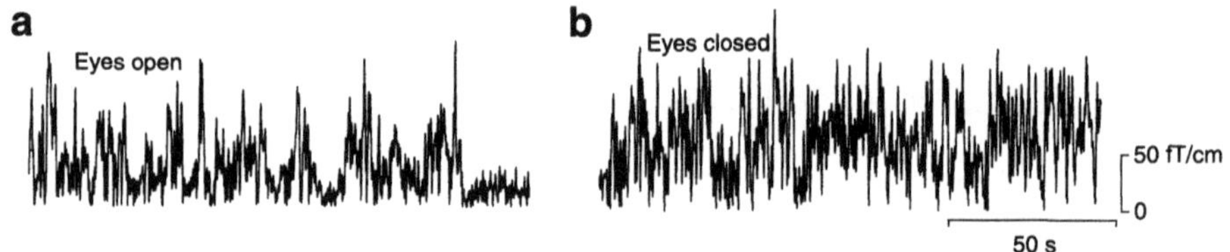

Fig. 2. 10-Hz oscillations fluctuate in amplitude on a wide range of time scales. Intermittent fluctuations in the 10-Hz-oscillation amplitude are seen in 150-s epochs from conditions of eyes open (**a**) and eyes closed (**b**). Modified from Linkenkaer-Hansen *et al.* (2001), [20]

We investigated the temporal structure of 10-Hz oscillations in the normal human brain at time scales ranging from a few seconds to several minutes. Ongoing activity during eyes-open and eyes-closed conditions was recorded from 10 normal subjects with simultaneous magnetoencephalography (MEG) and electroencephalography (EEG).

Highly irregular fluctuations of the amplitude envelope of the oscillations at around 10 Hz were observed in both conditions over the occipito-parietal region (Fig. 2a,b). A linear decay of power with increasing frequency in double-logarithmic coordinates was observed in the range of 0.005–0.5 Hz for both MEG and EEG data and the conditions of eyes open and eyes closed, indicating a $1/f^{\beta}$ type of power spectrum and thus a lack of a characteristic time scale for the duration and recurrence of these oscillations (Fig. 3a). The autocorrelation analysis indicated the presence of statistically significant correlations up to time lags of more than a hundred seconds (Fig. 3b). The decay of the autocorrelation function was slow over two decades and well fitted by a power law, which is in congruence with the $1/f^{\beta}$ power spectra.

To further consolidate the presence of long-range correlations, we implemented the detrended fluctuation analysis [33]. The lack of temporal correlations in the reference recording and surrogate EEG data was confirmed, while the 10-Hz oscillations exhibited robust power-law scaling behavior across conditions (Fig. 3c).

Altogether, these results indicate that the intermittent amplitude fluctuations of occipito-parietal 10-Hz oscillations (Fig. 2) are embedded with correlations at many time scales and that the decrease in correlation with temporal distance is governed by a power law (Fig. 3). Moreover, we did not find differences in scaling exponents between the two conditions or between the MEG and EEG recordings.

Previous studies, aimed at understanding the irregular nature of ongoing 10-Hz oscillations, have usually hypothesized a low-dimensional chaotic process and analyzed short data segments ($< 10\,\mathrm{s}$) [31] [8] [39]. The evidence for chaos has, however, not been robust and those analyses have also left unknown to what extent oscillatory activity is statistically dependent beyond the time scale of about 10 s. The present scaling analyses indicate that successive oscillations indeed are correlated – even over thousands of oscillation cycles – possibly be-

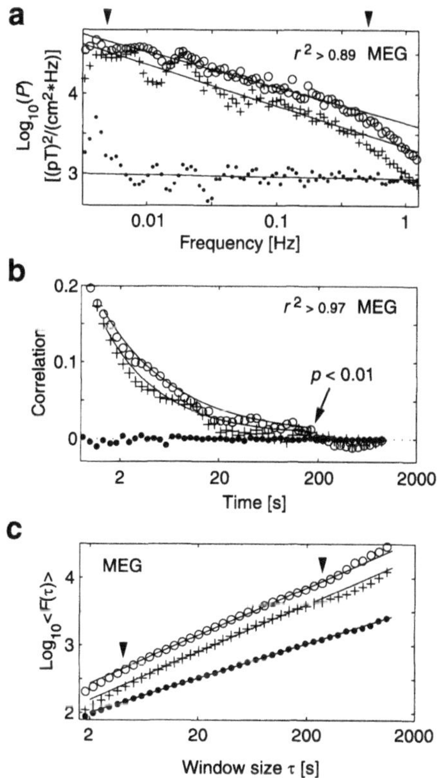

Fig. 3. Occipito-parietal 10-Hz oscillations exhibit long-range temporal correlations and power-law scaling behavior for their amplitude modulation. (**a**) The grand-averaged power spectral density of 10-Hz-oscillation amplitude fluctuations is plotted in double-logarithmic coordinates for the MEG data. *Circles*, eyes closed condition; *Crosses*, eyes open. The *dots* represent an empty-room reference recording. (**b**) The autocorrelation functions exhibit a power-law decrease in correlation with increasing time lag. The abscissas are logarithmic and the solid lines are power-law fits to the data. Significant correlations were obtained at time lags of nearly 200 s for the brain data. (**c**) Double-logarithmic plots of the fluctuation measure, $F(\tau)$, show power-law scaling in the time-window range of 5 to 300 s. Modified from Linkenkaer-Hansen *et al.* (2001), [20]

cause these population oscillations induce changes in the functional connectivity when propagating through the cortical networks and thereby leave behind a physiological memory trace that affects the future recruitment of neurons.

To pursue this idea further, a more recent study investigated the effect of electrical stimulation of the median nerve on the temporal correlations of the amplitude envelope of 10- and 20-Hz oscillations in the sensorimotor region. Preliminary evidence for changes in the long-range temporal correlations and scaling behavior was reported, which suggests that stimulus-induced perturbations of ongoing oscillations can degrade a neuronal network's memory of its past [21].

5 General Discussion

The understanding of the self-organization process that drives many complex systems to the critical state and the properties of the spatio-temporal organization and dynamics in the critical state has conveyed a deep insight, at least conceptually, into the workings of a wide range of complex systems in nature [4] [38].

In this chapter we have seen that there is a close correspondence between the physiological properties of neuronal networks generating ongoing oscillations and the fundamental mechanisms of SOC (cf. Sects. 3.1 and 3.2). I therefore propose that the large variability, the long-range correlations, and the power-law scaling behavior of ongoing oscillations find a unifying explanation within the theory of self-organized criticality.

Power laws in nonlinear dynamical systems may, however, also result from other mechanism, e.g., the fine tuning of a parameter to a value that maintains the system in a point of instability [38]. However, the oscillatory activities that were analyzed in Sect. 4.2 tend to emerge during conditions of rest in brain areas that are "idling", which seems to favor a mechanism of self-organization over some type of imposed control. Computational models of the neuronal network oscillations are likely to prove useful in elucidating the essential mechanisms underlying the observed long-range correlations.

The brain provides an interesting model system for exploring complexity both empirically and theoretically. From an experimental point of view it is convenient that the dynamics of the brain can be modified more easily (e.g., by changing the experimental conditions) than many of the other systems under study in complexity science, e.g., the dynamics of biological evolution, financial markets, earthquakes, or forest-fires. Moreover, there may be interesting functional consequences of self-organized criticality in the brain that have no meaning in inanimate systems.

Here I have established a number of analogies between general characteristic features of SOC systems and neuronal networks. Our data have provided evidence that the ongoing or spontaneous large-scale oscillatory activity of the human brain may be subjected to the same statistical laws as numerous other complex systems in nature. From a physicist's perspective this may not be very surprising, but from a biological point of view this is interesting because the brain, e.g., unlike the crust of the earth, presumably could have avoided this phenomenon during the evolutionary selection of neuronal properties. In other words, if critical dynamics would have adverse effects on the performance of neuronal networks, it probably would not appear in the brain. It is therefore natural to ask whether SOC dynamics could instead offer a beneficial state or context for neuronal computations. Indeed there have been proposals that the SOC state is desirable from a functional perspective because the dynamics in the critical state is highly susceptible to perturbations and therefore may support swift re-organization of neuronal activity when the networks are imposed with processing demands [2] [9] [20].

Acknowledgements

I am grateful to Vadim V. Nikulin, J. Matias Palva, Titia van Zuijen, and Risto J. Ilmoniemi for fruitful discussions and to the Danish Research Agency for financial support.

References

1. M. Abeles, H. Bergman, I. Gat, I. Meilijson, E. Seidemann, N. Tishby, E. Vaadia: Proc. Natl. Acad. Sci. USA **92**, 8616 (1995)
2. P. Alstrøm, D. Stassinopoulos: Phys. Rev. E **51**, 5027 (1995)
3. P. Bak, C. Tang, K. Wiesenfeld: Phys. Rev. Lett. **59**, 381 (1987)
4. P. Bak: *How Nature Works* (Oxford University Press, Oxford 1997)
5. T. Bal, D.A. McCormick: Neuron **17**, 297 (1996)
6. H. Berger: I. Mitteilung Arch. Psychiatr. Nervenkr. **87**, 527 (1929)
7. D.A. Butts, M.B. Feller, C.J. Shatz, D.S. Rokhsa: J. Neurosci. **19**, 3580 (1999)
8. R. Cerf, A. Daoudi, M. Ould Hénoune, E.H. El Ouasdad: Biol. Cybern. **77**, 235 (1997)
9. D.R. Chialvo, P. Bak: Neuroscience **90**, 1137 (1999)
10. B.W. Connors, Y. Amitai: Neuron **18**, 347 (1997)
11. Y. Georgelin, L. Poupard, R. Sarténe, J.C. Wallet: Eur. Phys. J. B **12**, 303 (1999)
12. N. Goldenfeld, L.P. Kadanoff: Science **284**, 87 (1999)
13. R. Hari, R. Salmelin: Trends Neurosci. **20**, 44 (1997)
14. M. Hämäläinen, R. Hari, R. Ilmoniemi, J. Knuutila, O.V. Lounasmaa: Rev. Mod. Phys. **65**, 413 (1993)
15. P.C. Ivanov *et al.*: Nature **383**, 323 (1996)
16. P. Jung: Phys. Rev. Lett. **78**, 1723 (1997)
17. P. Jung, A. Cornell-Bell, K.S. Madden, F. Moss: J. Neurophysiol. **79**, 1098 (1998)
18. C. Koch, G. Laurent: Science **284**, 96 (1999)
19. M. Le van Quyen *et al.*: The Lancet **357**, 183 (2001)
20. K. Linkenkaer-Hansen, V.V. Nikouline, J.M. Palva, R.J. Ilmoniemi: J. Neurosci. **21**, 1370 (2001)
21. K. Linkenkaer-Hansen, V.V. Nikouline, J.M. Palva, K. Kaila, R.J. Ilmoniemi: Soc. Neurosci. Abstr. **21.6** (2001)
22. F.H. Lopes da Silva, W. Storm van Leeuwen: 'The cortical alpha rhythm in dog: the depth and surface profile of phase'. In: *Architectonics of the Cerebral Cortex* (Raven Press, New York, 1998) pp. 319–333
23. F. Lopes da Silva: Electroenceph. Clin. Neurophysiol. **79**, 81 (1991)
24. F.H. Lopes da Silva, J.P. Pijn, D. Velis, P.C.G. Nijssen: Int. J. Psychophysiol. **26**, 237 (1997)
25. B.D. Malamud, G. Morein, D.L. Turcotte: Science **281**, 1840 (1998)
26. S. Maslov, M. Paczuski, P. Bak: Phys. Rev. Lett. **73**, 2162 (1994)
27. E. Niedermeyer, F. Lopes da Silva *Electroencephalography: basic principles, clinical applications, and related fields*, 2nd edn. (Lippencott, Williams & Wilkins, 1992)
28. V.V. Nikouline, K. Linkenkaer-Hansen, H. Wikstrom, M. Kesäniemi, E.V. Antonova, R.J. Ilmoniemi, J. Huttunen: Neurosci. Lett. **294**, 163 (2000)
29. P.L. Nunez: *Neocortical dynamics and human EEG rhythms* (Oxford University Press, Oxford, 1995)
30. M. Paczuski, S. Maslov, P. Bak: Phys. Rev. E **53**, 414 (1996)

31. M. Paluš: Biol. Cybern. **75**, 389 (1996)
32. G. Pfurtscheller: J. Clin. Neurophysiol. **6**, 75 (1989)
33. C.K. Peng, S. Havlin, H.E. Stanley, A.L. Goldberger: Chaos **5**, 82 (1995)
34. A.A. Penn, C.J. Shatz: Pediatr. Res. **45**, 4471 (1999)
35. A. Toib, V. Lyakhov, S. Marom: J. Neurosci. **18**, 1893 (1998)
36. G. Rangarajan, M. Ding: Phys. Rev. E **61**, 4991 (2000)
37. W. Singer: Science **270**, 758 (1995)
38. D. Sornette: *Critical Phenomena in Natural Sciences* (Springer, Berlin, 2000)
39. C.J. Stam, J.P.M. Pijn, P. Suffczynski, F.H. Lopez da Silva: Clin. Neurophysiol. **110**, 1801 (1999)
40. M. Steriade, P. Gloor, R.R. Llinás, F.H. Lopes da Silva, M. Mesulam: Electroenceph. Clin. Neurophysiol. **76**, 481 (1990)
41. C. Torrence, G.P. Compo: Bull. Amer. Meteorol. Soc. **79**, 61 (1998)
42. A. van Ooyen: Activity-dependent neurite outgrowth. Implications for network development. Ph.D. Thesis, University of Amsterdam, The Netherlands (1995)
43. T. Vicsek: Nature **418**, 131 (2002)

Long-Range Dependence in Heartbeat Dynamics

Plamen Ch. Ivanov

[1] Center for Polymer Studies and Department of Physics, Boston University, Boston, USA
[2] Harvard Medical School, Beth Israel Deaconess Medical Center, Boston, USA

1 Introduction

Physiologic signals are generated by complex self-regulating systems that process inputs with a broad range of characteristics [1–3]. Many physiological time series are extremely inhomogeneous and nonstationary, fluctuating in an irregular and complex manner. An important question is whether the "heterogeneous" structure of physiologic time series arises trivially from external and intrinsic perturbations which push the system away from a homeostatic set point. An alternative hypothesis is that the fluctuations are, at least in part, due to the underlying dynamics of the system. The key problem is how to decompose subtle fluctuations (due to intrinsic physiologic control) from other nonstationary trends associated with external stimuli. Till recently, the analysis of the fractal properties of such fluctuations has been restricted to second moment linear characteristics such as the power spectrum and the two-point autocorrelation function. These analyses reveal that the *fractal* behavior of healthy, free-running physiological systems is often characterized by $1/f$-like scaling of the power spectra over a wide range of time scales [4–8]. A signal which exhibits such power-law long-range dependence and is homogeneous (i.e. different parts of the signal have different statistical properties) is called a monofractal signal. Many physiologic time series, however, are inhomogeneous with different parts of the signal characterized by different statistical properties. In addition, there is evidence that physiologic dynamics exhibits nonlinear properties [9–15]. Such features are often associated with multifractal behavior, i.e., presence of long-range power-law dependence in the higher moments which is a nonlinear function of the scaling of the second moment [16]. Up to now, robust demonstration of multifractality for nonstationary time series has been hampered by problems related to a drastic bias in the estimate of the singularity spectrum due to diverging negative moments. Moreover, the classical approaches based on the box-counting technique and structure function formalism fail when a fractal function is composed of a multifractal singular part embedded in regular polynomial behavior [17]. By means of a wavelet-based multifractal formalism, we show that healthy human heartbeat dynamics exhibits even higher complexity (than previously expected from the finding of monofractal 1/f scaling) which is characterized by a broad *multifractal* spectrum [18].

In recent years the study of the statistical properties of heartbeat interval sequences has attracted the attention of researchers from different fields [19–23].

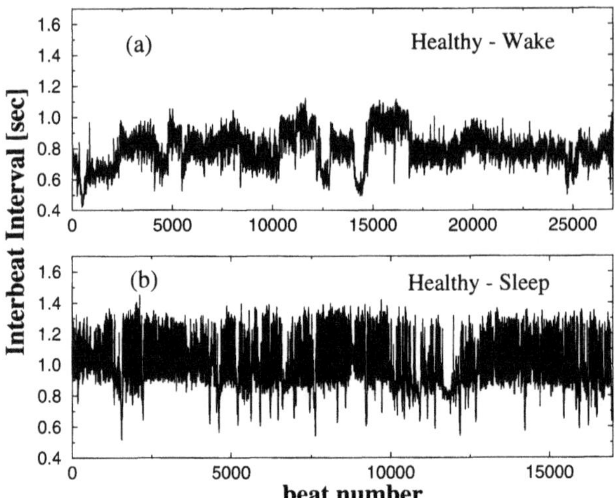

Fig. 1. Consecutive heartbeat intervals are plotted vs beat number for 6 hours recorded from the same healthy subject during: (a) wake period: 12pm to 6pm and (b) sleep period: 12am to 6am. (Note that there are fewer interbeat intervals during sleep due to the larger average of the interbeat intervals, i.e. slower heart rate.)

Analysis has focused extensively on interbeat interval variability as an important quantity to help elucidate possibly nonhomeostatic physiologic variability because: (i) the heart rate is under direct neuroautonomic control, (ii) interbeat interval variability is readily measured by noninvasive means, and (iii) analysis of these heart rate dynamics may provide important practical diagnostic and prognostic information. Figure 1 shows a cardiac interbeat time series – the output of a spatially and temporally integrated neuroautonomic control system. The time series shows "erratic" fluctuations and "patchiness". These fluctuations are usually ignored in conventional studies which focus on averaged quantities. In fact, these fluctuations are often labeled as "noise" to distinguish them from the true "signal" of interest. Generally, in the conventional approach it is assumed that there is no meaningful structure in apparent noise and, therefore, one does not expect to gain any understanding about the underlying system through the study of these fluctuations. However, by adapting and extending methods developed in modern statistical physics and nonlinear dynamics, we find that the physiologic fluctuations shown in Fig. 1 exhibit an unexpected hidden *scaling* structure [6,13,18,24–26]. Furthermore, the dynamical patterns of these fluctuations and the associated scaling features *change* with pathological perturbations. These findings raise the possibility that understanding the origin of such temporal structures and their alterations with disease (*a*) may elucidate certain basic aspects of heart rate control mechanisms, and (*b*) may have potential for clinical monitoring.

2 1/f Fluctuations in Heartbeat Dynamics

A quantity widely used to measure correlations in a time series is the power spectrum, which measures the relative frequency content of a signal. Fourier and related power spectrum analysis have proved particularly useful for recognizing the existence and role of characteristic frequencies (time scales) in cardiac dynamics. The analysis of heart beat fluctuations focused initially on short time oscillations associated with breathing and blood pressure as well as other control [21,22]. Studies of longer heartbeat records revealed 1/f-like scale-free behavior [4,5]. A power spectrum calculation assumes that the signal studied is stationary [27,28], and when applied to nonstationary time series can lead to misleading results. However, time series of beat-to-beat (RR) heart rate intervals obtained from digitized electrocardiograms are typically nonstationary and fluctuate in an irregular manner in healthy subjects, even at rest [Fig. 1b] [29,30]. Because of this property, researchers were faced with the task to consider only portions of the data and to test these portions for stationarity before performing power spectrum analysis.

To illustrate the limitations of the power spectrum analysis for nonstationary time series, we consider 6 hour records ($n \approx 10^4$ beats) of interbeat intervals for a healthy subject during sleep and wake activity. We show that there is *no true* $1/f$ power spectrum for the interbeat intervals in the real heart. Instead, we find that the power spectrum of the interbeat intervals has different regimes with different scaling behavior and that the rounded crossover between the different regimes is the reason why it seems, to first approximation, to scale as $1/f$ [Fig. 2].

Recent analyses of very long time series (up to 24h: $n \approx 10^5$ beats) show that under healthy conditions, interbeat interval increments $I(n)$ exhibit power-law anticorrelations [6]. Since taking increments $I(n)$ reduces to certain extend the effect of nonstationarity, we can apply standard spectral analysis techniques [Fig. 3] and we show that *true* scaling does exist.

The fact that the log-log plot of the power spectrum $S_I(f)$ vs. f is linear implies:

$$S_I(f) \sim f^{-\beta}. \tag{1}$$

The exponent β is related to the mean fluctuation function exponent α by $\beta = 2\alpha - 1$ [32,33] and can serve as an indicator of the presence and type of correlations: (i) If $\beta = 0$, there is no correlation in the time series $I(n)$ ("white noise"); (ii) If $0 < \beta < 1$, then $I(n)$ is correlated such that positive values of I are likely to be close (in time) to each other, and the same is true for negative I values; (iii) If $-1 < \beta < 0$, then $I(n)$ is also correlated. However, the values of I are organized such that positive and negative values are more likely to alternate in time ("anticorrelation") [32].

For interbeat interval increments from records of healthy subjects we obtain $\beta \simeq -1$ (corresponding to $\alpha = 0$, see Fig. 6), suggesting non-trivial power-law long-range *anticorrelations* in the heartbeat increments. Furthermore, the anti-correlation properties of I indicated by the negative β are consistent with a nonlinear feedback system that "kicks" the heart rate away from extremes

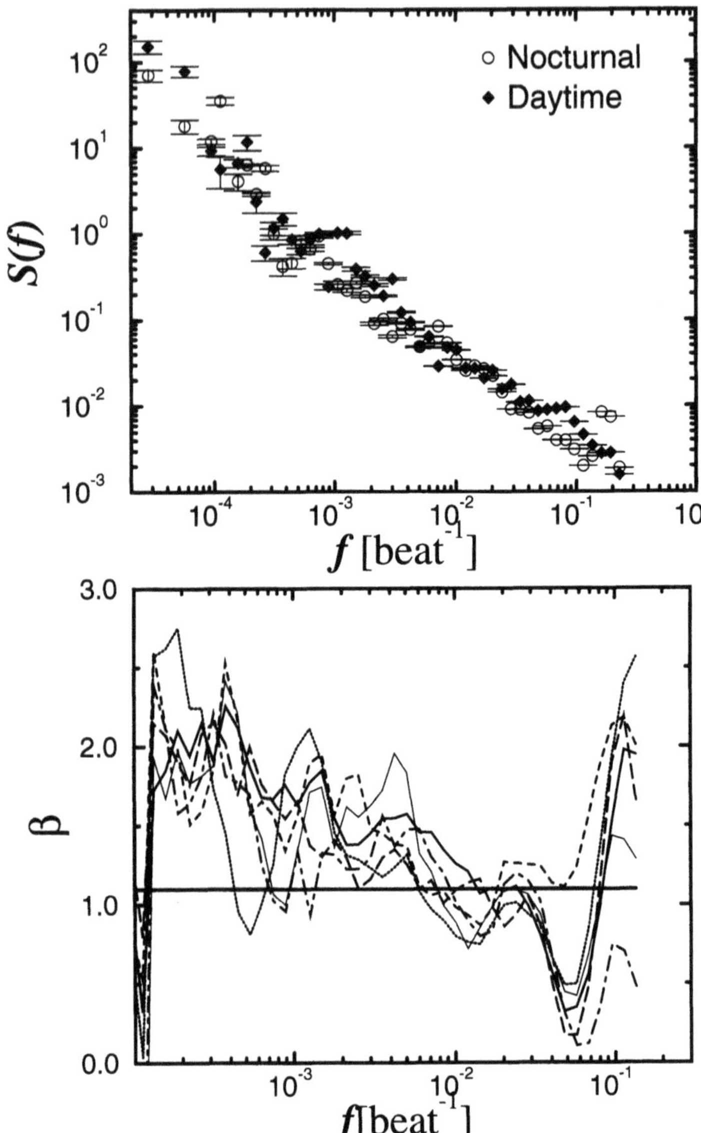

Fig. 2. (top) Power spectrum from 6 hour records of interbeat intervals for a healthy subject during day and night. (bottom) We plot the local exponent β calculated from the power spectrum for 6 healthy subjects. The local value of β shows a persistent drift, so *no true scaling exists*. This is not surprising, having in mind the non-stationarity of the signals. The horizontal line shows the value of the exponent obtained from a least square fit to the data. After [31]

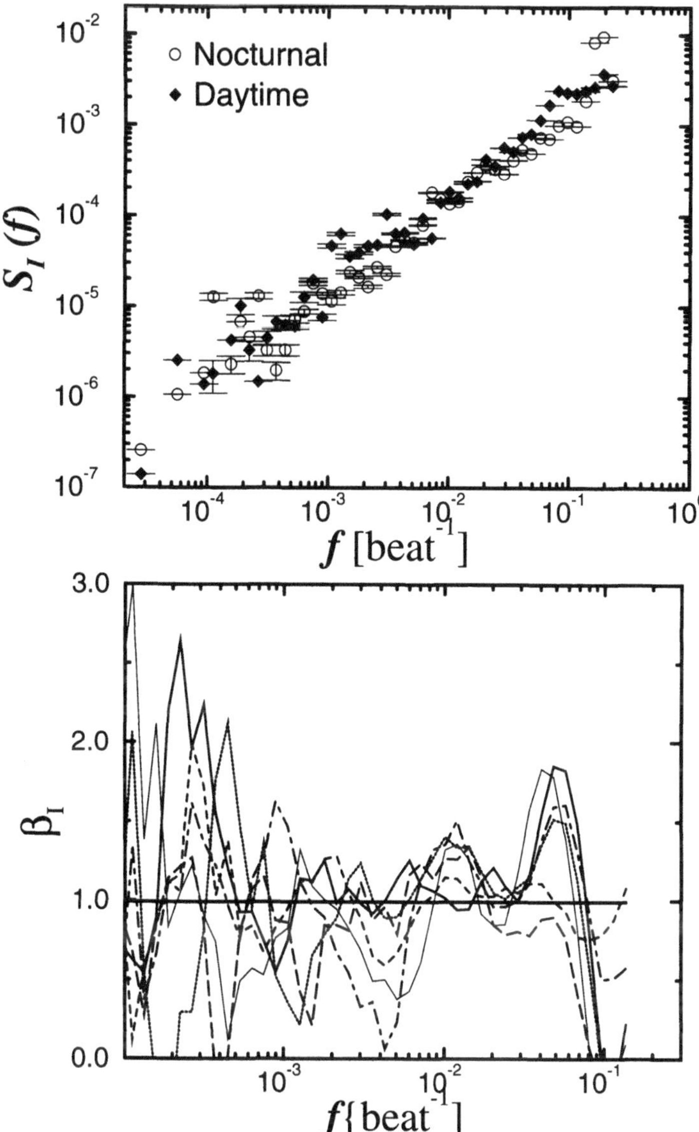

Fig. 3. (top) Power spectrum of the interbeat interval increments from 6h record for the same healthy subject as in Fig. 2. Error bars are calculated as the standard deviation of the power spectrum values for frequencies within the binning interval. (bottom) The local exponent β_I for the power spectrum of the increments for the same 6 healthy subjects as in Fig. 2. Note that the exponent β_I fluctuates around an average value close to one, so *true scaling does exist.* The horizontal line shows the value of β_I obtained from a least square fit. Note however, that the difference between wake and sleep dynamics cannot be observed from the power spectra. After [31]

[34,35]. This tendency, however, does not only operate locally on a beat-to-beat basis, but over a wide range of time scales up to thousands of beats [Fig. 3]. The emergence of such scale-invariant properties in the seemingly "noisy" heart beat fluctuations is believed to be a result of highly complex, nonlinear mechanisms of physiologic control [33,38].

Extracting increments from a time series is only a first step in effectively treating problems related to nonstationarities. Note that the power spectrum of the increments in the heartbeat intervals [Fig. 3] *does not* distinguish between wake and sleep dynamics. One needs to do better – e.g. by taking into account the presence of polynomial trends in the times series. We discuss such an approach in the following Section.

3 Monofractal Analysis: Long-Range Anticorrelations in the Heartbeat Fluctuations

Recently the *detrended fluctuation analysis* (DFA) method [39] was introduced to detect long-range correlations in physiological fluctuations when these are embedded in a seemingly nonstationary time series. The advantage of the DFA method over conventional methods, such as power spectrum analysis, is that it avoids the spurious detection of apparent long-range correlations that are an artifact of nonstationarity related to linear and higher order polynomial trends in the data. The essence of the DFA method is as follows: the average root-mean-square fluctuation function $F(n)$ is obtained after integrating and detrending the data, i.e., subtracting the local polynomial trend in a box of size n data points. The power-law relation between $F(n)$ and the number of data points n in a box indicates the presence of scaling: the fluctuations can be characterized by a scaling exponent α, a self-similarity parameter, defined as $F(n) \sim n^{\alpha}$. The DFA method has been tested on control time series of "built-in" long-range correlations with superposition of a nonstationary external trend [45]. It has also been successfully applied to detect long-range correlations in human gait, ion channel kinetics, and highly heterogeneous DNA sequences [7,8,39–42]. Of note is a recent independent review of fractal fluctuation analysis methods which determined that DFA was one of the most robust methods [43].

It is known that circadian rhythms are associated with periodic changes in key physiological processes [3,38,48]. Typically the differences in the cardiac dynamics during sleep and wake phase are reflected in the average and standard deviation of the interbeat intervals [47,48]. Such differences can be systematically observed from plots of the interbeat intervals recorded from subjects during sleep and wake [Fig. 1]. In recent studies we have reported on sleep-wake differences in the distributions of the amplitudes of the fluctuations in the interbeat intervals – a surprising finding indicating higher probability for larger amplitudes during sleep [13,25,49]. Next, we ask the question if there are characteristic differences in the scaling behavior between sleep and wake cardiac dynamics. We hypothesize that sleep and wake changes in cardiac control may occur on a very broad range of time scales and thus could lead to systematic changes in the scaling properties of the heartbeat dynamics. Elucidating the nature of these sleep-wake rhythms could lead to a better understanding of the neuroautonomic mechanisms of cardiac regulation.

To answer this question we apply the detrended fluctuation analysis (DFA) method. We analyze 30 datasets – each with 24h of interbeat intervals – from 18 healthy subjects and 12 patients with congestive heart failure [50]. We analyze the nocturnal and diurnal fractions of the dataset of each subject, which correspond to the 6h ($n \approx 22,000$ beats) from midnight to 6am and noon to 6pm. These periods incorporate the segments with lowest and highest heart rate in the time series, which we and others found to be the best indirect marker of sleep [47,48]. We find that at scales above $\approx$ 1min ($n > 60$) the data during wake hours display long-range correlations over two decades with average exponents $\alpha_W \approx 1.05$ for the healthy group and $\alpha_W \approx 1.2$ for the heart failure patients. For the sleep data we find a systematic crossover at scale $n \approx 60$ beats followed by a scaling regime extending over two decades characterized by a smaller exponent: $\alpha_S \approx 0.85$ for the healthy and $\alpha_S \approx 0.95$ for the heart failure group [Fig. 4a,c]. Although the values of the sleep and wake exponents vary from subject to subject, we find that for all individuals studied, the heartbeat dynamics during sleep are characterized by a smaller exponent [51].

This analysis suggests that the observed sleep-wake scaling differences are due to intrinsic changes in the cardiac control mechanisms for the following reasons: (i) The DFA method removes the "trends" in the interbeat interval signal which are due, at least in part, to activity, and quantifies the fluctuations along the trends. (ii) Responses to external stimuli should give rise to a different type of fluctuations having characteristic time scales, i.e. frequencies related to the stimuli. However, fluctuations in both diurnal and nocturnal cardiac dynamics exhibit scale-free behavior. (iii) The weaker anticorrelated behavior observed for all wake phase records cannot be simply explained as a superposition of stronger anticorrelated sleep dynamics and random noise of day activity. Such noise would dominate at large scales and should lead to a crossover with an exponent of 1.5. However, such crossover behavior is not observed in any of the wake phase datasets [Fig. 4]. Rather, the wake dynamics are typically characterized by a stable scaling regime up to $n = 5 \times 10^3$ beats.

To test the robustness of our results, we analyze 17 datasets from 6 cosmonauts during long-term orbital flight on the Mir space station under the extreme conditions of zero gravity and high stress activity [52]. Each dataset contains continuous periods of 6h data under both sleep and wake conditions. We find that for all cosmonauts the heartbeat interval series exhibit long-range correlations with scaling exponents consistent with those found for the healthy terrestrial group: $\alpha_W \approx 1.04$ for the wake phase and $\alpha_S \approx 0.82$ for the sleep phase. The values of these exponents indicate that the fluctuations in the interbeat intervals are anticorrelated for the wake phases and even stronger anticorrelated for the sleep phase. This sleep-wake scaling difference is observed not only for the group averaged exponents but for each individual cosmonaut dataset [Fig. 4b]. Moreover, the scaling differences are persistent in time, since records of the same

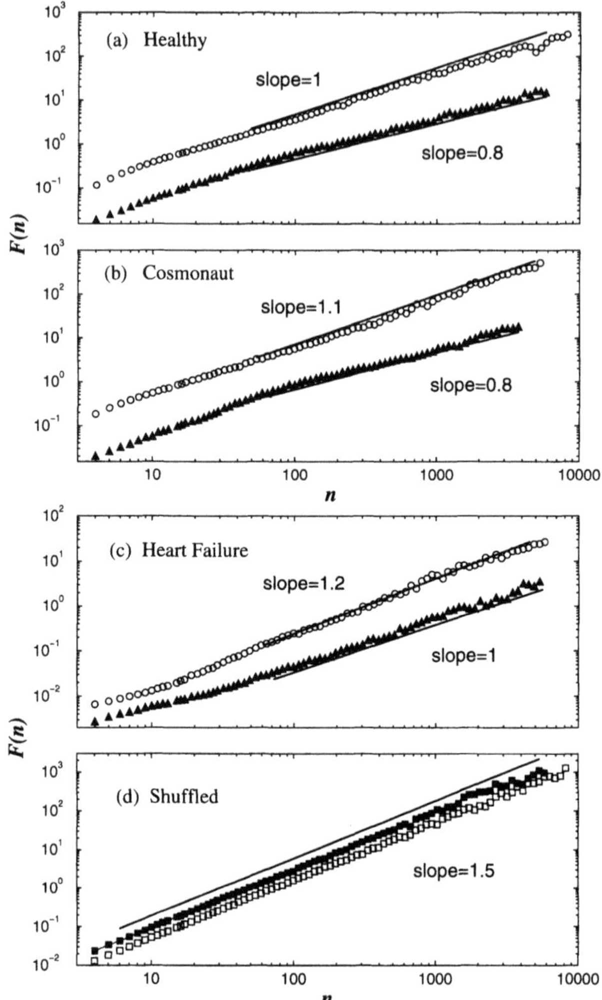

Fig. 4. Plots of $\log F(n)$ vs. $\log n$ for 6h wake (open circles) and sleep records (filled triangles) of (a) one typical healthy subject; (b) one cosmonaut (during orbital flight); and (c) one patient with congestive heart failure. Note the systematic lower exponent for the sleep phase (filled triangles), indicating stronger anticorrelations. (d) As a control, we reshuffle and integrate the interbeat increments from the wake (open squares) and sleep data (solid squares) of the healthy subject presented in (a). We find a Brownian noise scaling over all time scales for both wake and sleep phases with an exponent $\alpha = 1.5$, as one expects for random walk-like fluctuations. After [51]

cosmonaut taken on different days (ranging from the 3rd to the 158th day in orbit), exhibit a higher degree of anticorrelation in sleep.

Thus, the larger values for the wake phase scaling exponents observed for healthy subjects cannot be a trivial artifact of activity. Furthermore, the larger value of the average wake exponent for the heart failure group compared to the other two groups cannot be attributed to external stimuli either, since patients with severe cardiac disease are strongly restricted in their physical activity. We note, however, that the average sleep-wake scaling difference remains the same (≈ 0.2) for all three groups. Such sleep-wake changes in the scaling characteristics ranging from seconds to hours may indicate different regimes of intrinsic neuroautonomic regulation of the cardiac dynamics, which may "switch" on and off in accordance with circadian rhythms. A very recent study confirms our finding of lower value for the scaling exponent during sleep and shows that different stages of sleep (e.g. light sleep, deep sleep, rapid eye movement stages) could be associated with different correlations in the heartbeat fluctuations [53]. The findings of *stronger* anticorrelations [51], as well as higher probability for larger heartbeat fluctuations during sleep [13,25,49] are of interest from a physiological viewpoint, since they suggest that the observed dynamical characteristics in the heartbeat fluctuations during sleep and wake phases are related to intrinsic mechanisms of neuroautonomic control, and support a reassessment of the sleep as a surprisingly *active dynamical* state. The finding of scaling features in the human heartbeat and thier change with disease or sleep-wake transition have motivated new modeling approaches which may lead to better understanding the underlying control mechanisms of heartrate regulation [35].

Before concluding this Section we note that recent work [36] provides evidence of surprising complexity present in the temporal organization of the heterogeneities (e.g. trends) in human heartbeat dynamics. Trends in the interbeat interval signal are traditionally associated with external stimuli. To probe the temporal organization of such heterogeneities we introduce a segmentation algorithm [37] and find that the lengths of segments with different local mean heart rates follow a power-law distribution. This scale-invariant structure is not a simple consequence of the long-range correlations present in the heartbeat fluctuations discussed in this Section. These new findings suggest that relevant physiological information may be hidden in the heterogeneities of the heartbeat time series, the understanding of which remains an open question.

4 Long-Range Correlations in the Magnitudes and Signs of Heartbeat Fluctuations

We consider the time series formed by consecutive cardiac interbeat intervals (Fig. 5a) and focus on the correlations in the time *increments* between consecutive beats. This time series is of general interest, in part because it is the output of a complex integrated control system, including competing stimuli from the neuroautonomic nervous system [56]. These stimuli modulate the rhythmicity of the heart's intrinsic pacemaker, leading to complex fluctuations. Previous reports indicate that these fluctuations exhibit scale-invariant properties [4,13,31], and are anticorrelated over a broad range of time scales (i.e., the power spectrum follows a power-law where the amplitudes of the higher frequencies are dominant) [6,24]. By long-range anticorrelations we also mean that the root mean square

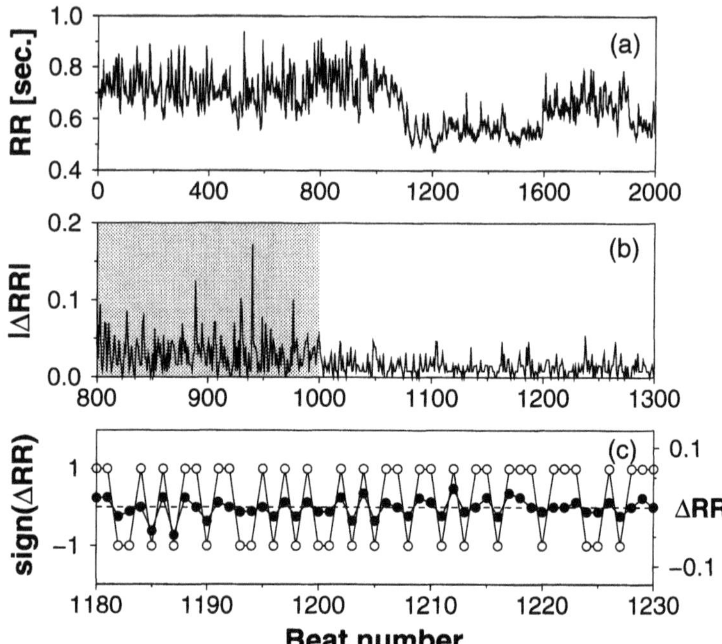

Fig. 5. (a) An example of 2000 time intervals between successive R peaks (RR intervals) taken from electrocardiogram (ECG) recording of a healthy subject during daytime. (b) The magnitude series of a portion of the RR series (beat numbers 800-1300) shown in (a). Patches of more "volatile" increments with large magnitude (beat numbers 800-1000) are followed by patches of less volatile increments with small magnitude (beat numbers 1000-1300), consistent with our quantitative conclusion that there is correlation in the magnitude time series. (c) The sign series (∘), as well as the ΔRR series (•) of a portion of the RR series (beat numbers 1180-1230) shown in (a). The positive sign (+1) represents a positive increment, while the negative sign (−1) represents a negative increment in the RR series of interbeat intervals. The tendency to alternation between +1 and −1 is consistent with our quantitative conclusion that there is (multiscale) anticorrelation in the sign time series. After [26].

fluctuations function of the integrated series is proportional to n^α where n is the window scale and the scaling exponent α is smaller than 0.5. In contrast, for uncorrelated behavior $\alpha = 0.5$, while for correlated behavior $\alpha > 0.5$.

To analyze the ways that heartbeat fluctuations obey scaling laws we "decompose" the time series of the heartbeat intervals into two different time series. We analyze separately the time series formed by the magnitude and the sign of the increments in the time intervals between successive heartbeats (Fig. 5b,c). We use 2nd order detrended fluctuation analysis [39,44–46] and not the conventional power spectrum, since it has the ability to accurately estimate correlations in the heartbeat fluctuations even when they are masked by linear trends. The 1st order detrended fluctuation analysis (DFA) eliminates constant trends from the original series (or, equivalently, linear trends from the integrated series); the 2nd

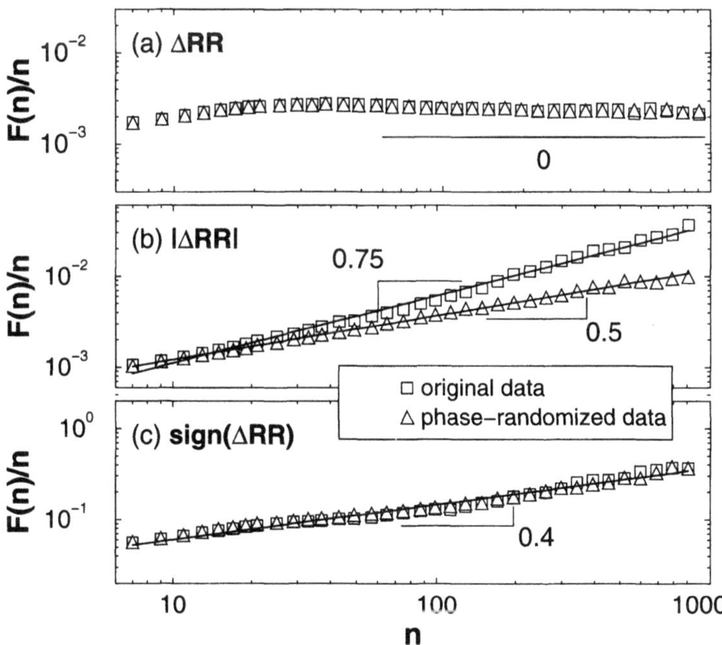

Fig. 6. (a) Root mean square fluctuation, $F(n)$, for ≈ 6 hour record ($\approx 32,000$ data points) for the interbeat interval RR_i series (□) of healthy subject. Here, n indicates the time scale (in beat numbers) over which each measure is calculated. The scaling is obtained using 2nd order detrended fluctuation analysis, and indicates long-range anticorrelations in the heartbeat interval increment series ΔRR_i [24]. As expected, the scaling properties of the heartbeat interval increment series remain unchanged after the Fourier phase randomization (△). (b) The root mean square fluctuation of the integrated magnitude series (□) indicates long-range correlations in the magnitude series $|\Delta RR_i|$ (group average exponent of $\alpha - 1 = 0.74 \pm 0.08$ where $F(n)/n \propto n^{\alpha-1}$). After Fourier phase randomization of the interbeat interval increment series we find random behavior with exponent 0.5 (△). This change in the scaling (after removing the nonlinear features in the time series) suggests that the magnitude series carries information about the nonlinear properties of the heartbeat dynamics. (c) The root mean square fluctuation of the integrated sign series (□) indicates anticorrelated behavior in sign(ΔRR_i) (group average exponent of $\alpha - 1 = 0.42 \pm 0.03$ where $F(n)/n \propto n^{\alpha-1}$). The scaling properties of the sign series remain unchanged after the Fourier phase randomization (△), which suggests that the sign series relates to linear properties of the heartbeat interval time series. We note the apparent crossovers at $n \approx 20$ beats and $n \approx 100$ beats. A gradual loss of anticorrelation in the sign series is observed at time scales larger than $n \approx 100$ beats. The numbers in the figure indicate the scaling exponents before and after the phase-randomization procedure. After [26].

order DFA removes linear trends, and the kth order DFA eliminates polynomial trends of order $k - 1$.

The magnitude/sign decomposition consists of the following steps: (i) given a series of successive interbeat intervals RR_i we create the increment series,

$\Delta RR_i = RR_{i+1} - RR_i$; (ii) we decompose the increment series into a magnitude series ($|\Delta RR|$) and a sign series (sign(ΔRR)); (iii) to avoid artificial trends we subtract from the magnitude and sign series their average; (iv) because of limitations in the accuracy of the detrended fluctuation analysis method for estimating the scaling exponents of anticorrelated signals ($\alpha < 0.5$), we integrate the magnitude and sign series; (v) we perform a scaling analysis using 2nd order detrended fluctuation analysis on the integrated magnitude and sign series; (vi) to obtain the scaling exponents for the magnitude and sign series we measure the slope of $F(n)/n$ on a log-log plot, where $F(n)$ is the root mean square fluctuation function and n is the scale of analysis (in beat numbers).

We find for each subject in a group of 18 healthy individuals [50], that the time series of the magnitudes exhibits correlated behavior (Fig. 6b) (unlike the original heartbeat increment time series, which is anticorrelated, Fig. 6a). The sign series, however, exhibits anticorrelated behavior (Fig. 6c). Correlation in the magnitude series indicates that an increment with large magnitude is more likely to be followed by an increment with large magnitude. Anticorrelation in the sign series indicates that a positive increment is more likely to be followed by a negative increment. Our result for the temporal organization of heartbeat fluctuations thus suggests that, under healthy conditions, a large increment in the positive direction is more likely to be followed by a large increment in the negative direction [26]. We find that this empirical "rule" holds over a broad range of time scales from several up to hundreds of beats (Fig. 6) [57].

To show that fluctuations following an identical scaling law can exhibit different time ordering for the magnitude and sign, we perform a Fourier transform on a heartbeat interval increment time series, preserving the amplitudes of the Fourier transform but randomizing the Fourier phases. Then we perform an inverse Fourier transform to create a surrogate series. This procedure eliminates non-linearities, preserving only the linear features (i.e. two-point correlations) of the original time series [58]. The new surrogate series has the *same* power spectrum as the original heartbeat interval increment time series, with a scaling exponent indicating long-range anticorrelations in the interbeat increments (Fig. 6a). Our analysis of the sign time series derived from this surrogate signal shows scaling behavior almost identical to the one for the sign series from the original data (Fig. 6c). However, the magnitude time series derived from the surrogate (linearized) signal exhibits *uncorrelated* behavior – a significant change from the strongly *correlated* behavior observed for the original magnitude series (Fig. 6b). Thus, the increments in the surrogate series do not follow the empirical "rule" observed for the original heartbeat series, although these increments follow a scaling law identical to the original heartbeat increment series. Moreover, our results raise the interesting possibility that the magnitude series carries information about the nonlinear properties of the heartbeat series, while the sign series relates importantly to linear properties.

We test our analysis on a group of 12 subjects with congestive heart failure [50]. Compared to the healthy subjects, the magnitude exhibits weaker correlations with a scaling exponent closer to the exponent of an uncorrelated series.

Table 1. Comparison of the statistics of the root mean square fluctuation, $F(n)$ (calculated using the 2nd order detrended fluctuation analysis method [39,45,46] where n is the time scale in beat numbers over which each measure is calculated), and the scaling exponents for 18 healthy subjects and 12 subjects with heart failure [50] (obtained from 6-hour records during the day). The scaling features of the magnitude and sign change significantly for the subjects with heart failure, raising the possibility of bedside applications. α is the best fit to the range $6 < n < 1024$. $F(n)$ is estimated at the crossover position ($n = 16$) (Fig. 6b) where the largest separation between the two groups is estimated. Since we observe two apparent crossovers in the scaling behavior of the sign series, we calculate the scaling exponents in three different regions : (i) the short range regime for time scales $6 < n < 16$ with scaling exponent α_1, (ii) the intermediate regime for time scales $16 \leq n \leq 64$ with scaling exponent α_2, (iii) and the long range regime for time scales $64 < n \leq 1024$ with scaling exponent α_3. For each measure, the group average $\pm$ 1 standard deviation is presented. The values which show highly significant differences ($p \leq 0.01$ by Student's t-test) between the healthy and heart failure groups are indicated in boldface. We note, surprisingly, that the short range and the intermediate range scaling exponents α_1 and α_2 of the sign series may provide even more robust separation between healthy and heart failure compared to previous reports [24] based on the scaling exponents of the original heartbeat series.

magnitude			
measure	healthy	heart failure	p value
$\log_{10} F(n)$	$\mathbf{-1.49 \pm 0.16}$	$\mathbf{-1.92 \pm 0.17}$	1×10^{-7}
α	$\mathbf{1.74 \pm 0.08}$	$\mathbf{1.66 \pm 0.06}$	0.01
α_1	1.55 ± 0.08	1.6 ± 0.08	0.13
α_2	1.66 ± 0.08	1.61 ± 0.08	0.14
α_3	$\mathbf{1.82 \pm 0.1}$	$\mathbf{1.71 \pm 0.1}$	4×10^{-3}
sign			
measure	healthy	heart failure	p value
$\log_{10} F(n)$	$\mathbf{0.14 \pm 0.05}$	$\mathbf{0.02 \pm 0.06}$	1×10^{-6}
α	1.42 ± 0.03	1.44 ± 0.02	0.08
α_1	$\mathbf{1.43 \pm 0.12}$	$\mathbf{1.15 \pm 0.12}$	7×10^{-7}
α_2	$\mathbf{1.27 \pm 0.07}$	$\mathbf{1.41 \pm 0.07}$	1×10^{-5}
α_3	1.53 ± 0.065	1.49 ± 0.04	0.04

The change in the magnitude exponent for the heart failure subjects is consistent with a previously reported loss of nonlinearity with disease [12,14,18,59]. The sign time series of heart failure subjects shows scaling behavior similar to the one observed in the original time series, but significantly different from the healthy subjects (Table 1).

Next, we investigate how the heart rhythms of healthy subjects change within the different sleep stages. Typically the differences in cardiac dynamics during wake or sleep state, and during different sleep stages are reflected in the average and standard deviation of the interbeat interval time series [60,61]. Recent studies show that changes in cardiac control due to circadian rhythms or different sleep stages can lead to systematic changes in the correlation (scaling) properties of the heartbeat dynamics. In particular it was found that the long-range

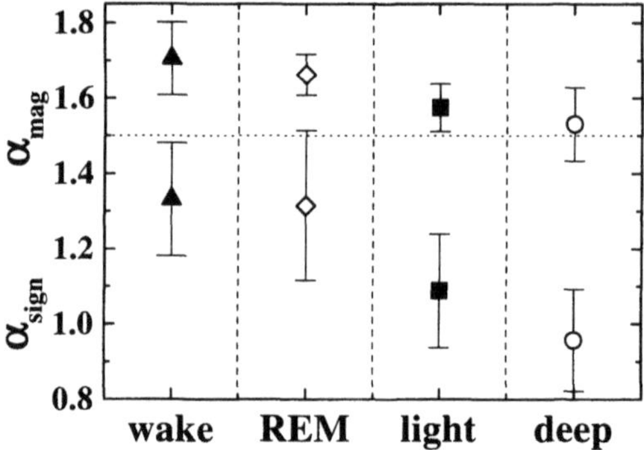

Fig. 7. The average values of the fluctuation exponents α_{mag} for the magnitude series and α_{sign} for the sign series for the different phases (wake state, REM sleep, light sleep, and deep sleep). For each of the 24 records from 12 healthy subjects the corresponding 2nd order DFA fluctuation functions $F(n)$ have been fit in the range of $8 \leq n \leq 13$ and $11 \leq n \leq 150$ heartbeats for α_{sign} and α_{mag}, respectively, where the most significant differences between the sleep stages occur. After [64]

correlations in heartbeat dynamics change during wake and sleep periods [51], indicating different regimes of intrinsic neuroautonomic regulation of the cardiac dynamics, which may switch on and off with the circadian rhythms.

Healthy sleep consists of cycles of approximately 1–2 hours duration. Each cycle is characterized by a sequence of sleep stages usually starting with light sleep, followed by deep sleep, and rapid eye movement (REM) sleep [62]. While the specific functions of the different sleep stages are not yet well understood, many believe that deep sleep is essential for physical rest, while REM sleep is important for memory consolidation [62]. Different sleep stages during nocturnal sleep were found to relate to a specific type of correlations in the heartbeat intervals [53], suggesting a change in the mechanism of cardiac regulation in the process of sleep.

We consider 24 records of interbeat intervals obtained from 12 healthy individuals during sleep [64]. The records have an approximate duration of 7.5 hours. The annotation and duration of the sleep stages were determined based on standard procedures [63]. We apply the detrended fluctuation analysis (DFA) method [39,44–46] on both the sign and the magnitude time series. We find that the sign series exhibits anticorrelated behavior at short time scales which is characterized by a correlation exponent with smallest value for deep sleep, larger value for light sleep, and largest value for REM sleep. The magnitude series, on the other hand, exhibits uncorrelated behavior for deep sleep, and long-range correlations are observed for light and REM sleep, with a larger exponent for REM sleep. The observed increase in the values of both the sign and magnitude correlation exponents from deep through light to REM sleep is systematic and significant. We also find that the values of the sign and magnitude exponents for REM sleep are very close to the values of these exponents for the wake state [64].

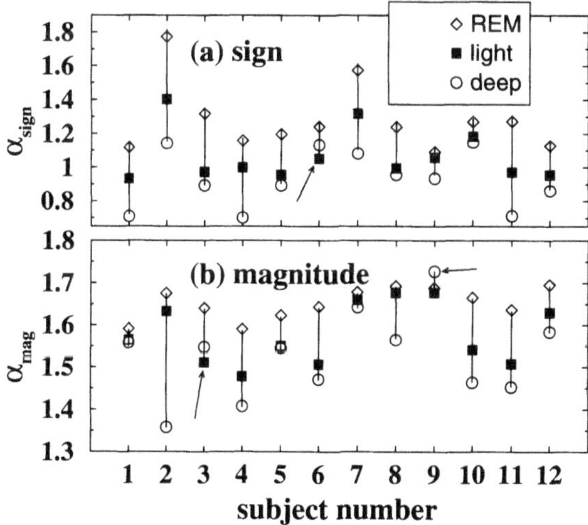

Fig. 8. The values of the effective fluctuation exponents α for the integrated sign series (a) and the integrated magnitude series (b) are shown for all 12 healthy subjects (second night of recording). While the α values are fluctuating, for REM sleep the α is larger than the α for light sleep, which is larger than the α for deep sleep (the 3 arrows indicate the cases which are not ordered in the same way as the majority). The exponent values have been determined over the fitting ranges as described in the caption of Fig. 7. After [64].

The mean values and their standard deviations for the different sleep stages are shown in Fig. 3. We estimate the exponents α from the slopes in the log-log plot of $F(n)$ versus n for all records. Since the most significant differences for the short-range sign correlations occur in the range of $8 \leq n \leq 13$ heartbeats, we use this fitting range for the exponents α_{sign}. For the magnitude exponent α_{mag}, we use the range $11 \leq n \leq 150$, since the long-range correlations occurring in light and REM sleep can be observed best in this region. We find that there is a significant difference in the sign series exponent α_{sign} observed for all three sleep stages (the p-value, obtained by the Student's t-test, is below 0.001). The magnitude correlation exponents for REM sleep and for intermediate wake states are significantly larger than those for the non-REM stages (light and deep sleep). Here also, the p-values are less than 0.001. Note that we do not find a significant difference between the average exponents for REM sleep and for the intermediate wake states. This is not surprising because heartbeat activity during REM sleep is very close to heartbeat activity during the wake state and the heartbeat time series during REM and wake exhibit similar scaling properties [51,53]

More significant than the differences for the average exponents are the differences between the exponents for each individual. Figure 4 shows the α values for REM, light, and deep sleep for all 12 healthy subjects (second night only). In almost all cases the exponent of the REM sleep is the largest, the exponent of the light sleep is intermediate, and the exponent of the deep sleep is smallest (there are three exceptions, indicated by arrows). In our group of 24 records from 12 healthy individuals, we find larger exponents in REM sleep than in deep sleep for 100% of the sign series and for 88% of the magnitude series.

Our results suggest that the empirical "rule" mentioned above also applies to REM sleep, while in deep sleep small and large increments seem to appear in a random fashion. On the other hand, the stronger sign anticorrelations in deep sleep indicate that a positive increment is more likely – even more likely than in REM sleep – to be followed by a negative increment. Thus, the correlation behavior of the heartbeat increments and their signs and magnitudes during daytime activity is similar to the behavior we find in REM sleep, but quite different from the behavior we observe in deep sleep. This is consistent with our finding [Fig. 7] of average exponent values for the wake episodes similar to the exponent values for REM sleep.

We conclude that series with identical correlation properties can have completely different time ordering which can be characterized by different scaling exponents for the magnitude and sign series. Moreover, we show that the magnitude series carries information regarding the nonlinear properties of the original series while the sign series carries information regarding the linear properties of the original series. The significant decrease in the short-range scaling exponent for the sign series in heart failure may be related to perturbed vagal control affecting relatively high frequency fluctuations. The decrease of the long-range scaling exponent for the magnitude series of the heart failure patients indicates weaker correlations and loss of nonlinearity which may be related to impaired feedback mechanisms of neurohormonal cardiac regulation. Further we observe short-range anticorrelations in the sign of the interbeat interval increments which are stronger during deep sleep, weaker during light sleep, and even weaker during REM sleep. In contrast, the magnitude of the increments is long-range correlated with a larger exponent during REM sleep suggesting stronger nonlinear contributions to the heartbeat dynamics in this stage compared with weaker nonlinear contributions in the non-REM stages.

5 Self-similar Cascades in the Heartbeat Fluctuations

Many simple systems in nature have correlation functions that decay with time in an exponential way. For systems comprised of many interacting subsystems, physicists discovered that such exponential decays typically do not occur. Rather, correlation functions were found to decay with a power-law form. The implication of this discovery is that in complex systems, there is no single characteristic time [65–67]. If correlations decay with a power-law form, we say the system is "scale-free" because there is no characteristic scale associated with a power law. Since at large time scales a power law is always larger than an exponential function, correlations described by power laws are termed "long-range" correlations – they are of longer range than exponentially-decaying correlations.

The findings of long-range power-law correlations [24,51] and the recently reported scaling in the distributions of heartbeat fluctuations [13,49] (i.e. "data collapse" of the distributions for different time scales) suggest the absence of a characteristic scale and indicate that the underlying dynamical mechanisms regulating the healthy heartbeat have statistical properties which are *similar* on different time scales. Such statistical self-similarity is an important characteristic of *fractal* objects [68]. However, how can this purported fractal structure be "visualized" in the seemingly erratic and noisy heartbeat fluctuations? The wavelet decomposition of beat-to-beat heart rate signals can be used to provide a visual representation of this fractal structure [Fig. 9]. The brighter shades indicate larger values of the wavelet amplitudes (corresponding to large heartbeat fluctuations) and white tracks represent the wavelet transform maxima lines. The structure of these maxima lines shows the evolution of the heartbeat fluctuations with scale and time. The wavelet analysis performed with the second derivative of the Gaussian (the Mexican hat) as an analyzing wavelet uncovers a hierarchical scale-invariance [Fig. 9 (top panel)], which is characterized by the stability of the scaling form observed for the distributions and the power-law correlations [24,51,13]. The plots reveal a self-affine cascade formed by the maxima lines – a magnification of the central portion of the top panel shows similar branching patterns [Fig. 9 (lower panel)]. Such fractal cascade results from the interaction of many nonlinearly coupled physiological components, operating on different scales (polynomial trends due to daily activity are filtered out).

Thus the wavelet transform, with its ability to remove local trends and to extract interbeat variations on different time scales, enables us to identify fractal patterns (arches) in the heartbeat fluctuations even when the signals change as a result of background interference. Analysis of data from pathologic conditions (e.g. sleep apnea) show a breakdown of these patterns [25]. Fractal characteristics of cardiac dynamics and other biological signals can be usefully studied with the generalized multifractal formalism based on the wavelet transform modulus maxima method which we discuss in the next Section.

6 Multifractality: Nonstationarity in Local Scaling

Monofractal signals are homogeneous in the sense that they have the same scaling properties, characterized locally by a single singularity exponent h_0, throughout the entire signal [67–69,54,70,55]. Therefore monofractal signals can be indexed by a single *global* exponent – the Hurst exponent $H \equiv h_0$ [71] – which suggests that they are *stationary* from viewpoint of their local scaling properties. On the other hand, multifractal signals, can be decomposed into many subsets – possibly infinitely many – characterized by different *local* Hurst exponents h, which quantify the local singular behavior and thus relate to the local scaling of the time series [Fig. 10]. Thus multifractal signals require many exponents to fully characterize their scaling properties [68,54,55] and are intrinsically more complex, and *inhomogeneous*, than monofractals.

The statistical properties of the different subsets characterized by these different exponents h can be quantified by the function $D(h)$, where $D(h_o)$ is the fractal dimension of the subset of the time series characterized by the local Hurst exponent h_o [68,54,55,72–74]. Thus, the multifractal approach for signals, a con-

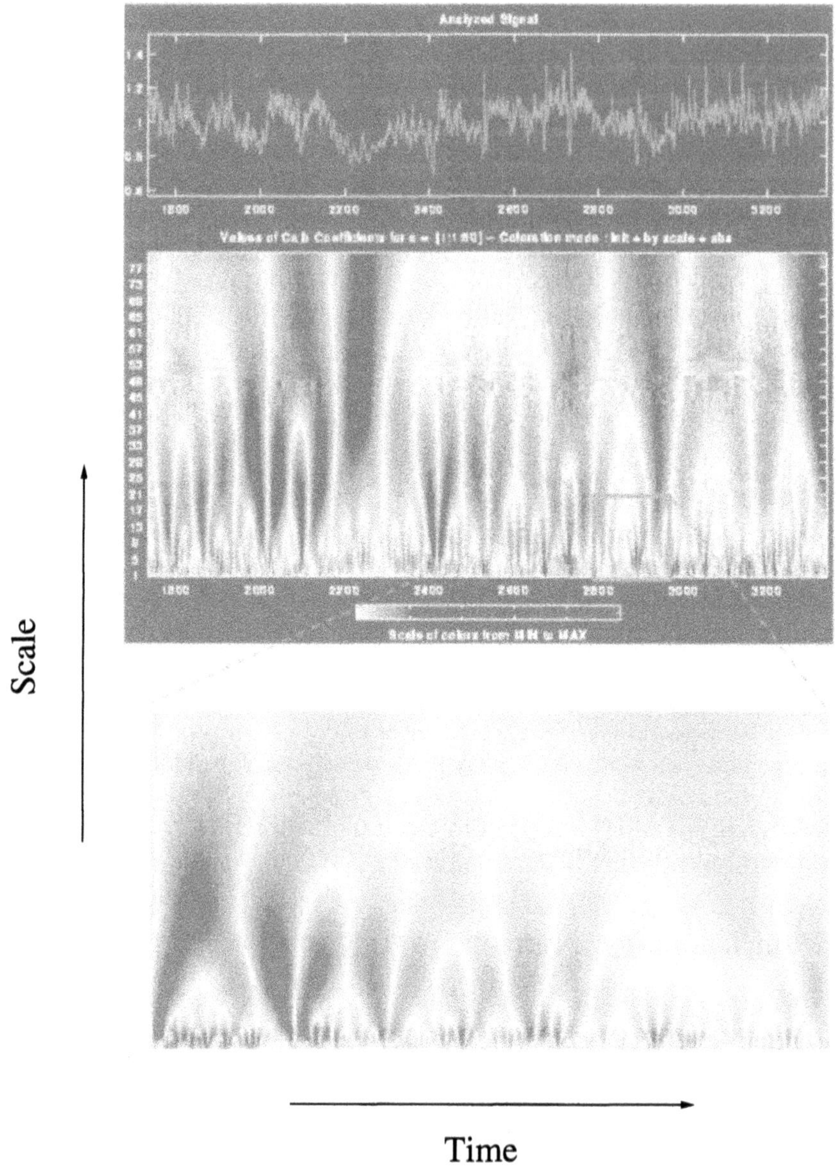

Fig. 9. Visualization of wavelet analysis of a heartbeat interval signal using different shades from white to black. The x-axis represents time ($\approx$ 1700 beats) and the y-axis indicates the scale of the wavelet used ($a = 1, 2, \ldots, 80$; i.e. $\approx$ from 5 seconds to 5 minutes) with large scales at the top. This wavelet decomposition reveals a self-similar fractal structure in the healthy cardiac dynamics – a magnification of the central portion of the top panel with 200 beats on the x-axis and wavelet scale $a = 1, 2, \ldots, 20$ on the y-axis shows similar branching patterns (lower panel). After refs. [25,31]

Fig. 10. Local Hurst exponents h for a multifractal signal (top panel) and the decomposition of this signal into subsets (subsequent panels) with each local Hurst exponent indicated by different shade of gray and each fractal dimension indicated by the density of vertical bars. The x-axis represents time and the vertical bars (y-axis) indicate local Hurst exponents. After [31]

cept introduced in the context of multi-affine functions [75,76], has the potential to describe a wide class of signals that are more complex then those characterized by a single fractal dimension (such as classical 1/f noise).

In a recent study, we establish the relevance of the multifractal formalism for the description of a physiological signal – the human heartbeat [18]. The motivation for our work is not merely looking for yet another example of multifractality, this time in the biological sciences. In fact, if we consider the neuroautonomic control mechanisms responsible for the generation of heartbeats, it is natural to expect the need for multifractal concepts for their description, since the heartbeats are a result of the interaction of many physiological components operating on different time scales. These interactions are nonlinear and self-regulating (through feedback control), leading to the *nonlinear* character of the output signal and to the heterogeneous features of heartbeat time series.

In contrast, the assumption of heartbeat monofractality – which has been the scope of studies in the field so far – is unrealistic because the monofractal hypothesis assumes that the scaling properties of the signal are the same throughout time, and are characterized by the same local Hurst exponent h [Fig. 11c]. However, inspection of heartbeat signals shows them to be heterogeneous and suggests they might require more exponents for their description. Since the power spectrum and the correlation analysis (DFA method) can measure only *one* exponent characterizing a given signal, these methods are more appropriate for the study of monofractal signals. Moreover, the power spectrum and the correlation analysis reflect only the linear characteristics, while the heartbeat dynamics exhibits nonlinear properties. Thus the multifractal analysis may reveal new information on the nature of the nonlinearity encoded in the Fourier phases [Fig. 16].

The first problem, therefore, is to extract the local value of h. To this end we use methods derived from wavelet theory [77]. The properties of the wavelet transform make wavelet methods attractive for the analysis of complex nonstationary time series such as one encounters in physiology [13]. In particular, wavelets can remove polynomial trends that could lead box-counting techniques to fail to quantify the local scaling of the signal [78]. Additionally, the time-frequency localization properties of the wavelets makes them particularly useful for the task of revealing the underlying hierarchy in the cascade of fluctuations [Fig. 9] that governs the temporal distribution of the local Hurst exponents. Hence, the wavelet transform enables a reliable multifractal analysis [78]. As the analyzing wavelet, we use derivatives of the Gaussian function, which allows us to estimate the singular behavior and the corresponding exponent h at a given location in the time series. The higher the order n of the derivative, the higher the order of the polynomial trends removed and the better the detection of the temporal structure of the local scaling exponents in the signal.

The concept of multifractality is exemplified in Fig. 11a,b for a heartbeat intervals record from a healthy subject. The heterogeneity of the healthy heartbeat is represented by the broad range of local Hurst exponents h (colors) present and the complex temporal organization of the different exponents. The middle and bottom panels illustrate the different fractal structure of two subsets of the time series characterized by different local Hurst exponents. The value of the local Hurst exponent for each subset is represented with different gray shade. The two subsets display different temporal structures which can be quantified by different fractal dimension $D(h)$. The healthy signal is represented by a *multishade* plot, reflecting *multifractal* behavior through the variety of values for the local

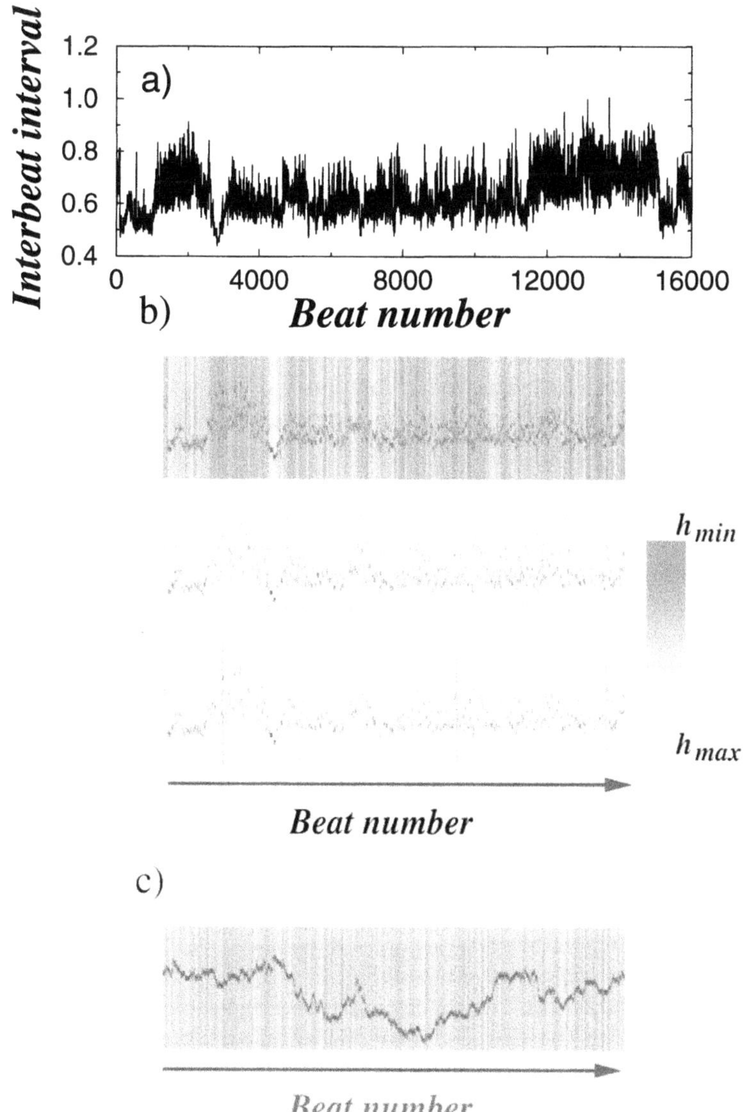

Fig. 11. (a) Consecutive heartbeat intervals measured in seconds are plotted vs beat number from approximately 3 hours record of a representative healthy subject. The time series exhibits very irregular and nonstationary behavior. (b) The top panel displays in differnt shades the local Hurst exponents calculated for the same 3 hours record shown in (a). The other two panels represent two subsets of the heartbeat interval time series in (a) each with a local Hurst exponent (indicated by different shades) and with a different fractal dimension (indicated by the density of the vertical bars). (c) The panel displays in different gray shades the local Hurst exponents calculated for a *monofractal* signal – fractional Brownian motion with $H = 0.6$. The homogeneity of the signal is represented by the nearly monochromatic appearance of the signal which indicates that the local Hurst exponent h is the same throughout the signal and identical to the global Hurst exponent H. After [18]

Hurst exponents. In contrast, fractional Brownian motion (a *monofractal* signal) is essentially *monochromatic* indicating that the local Hurst exponent h is the same throughout the signal [Fig. 11c].

7 Multifractality in Heartbeat Dynamics

We evaluate the local exponent h through the modulus of the maxima values of the wavelet transform at each point in the time series using the wavelet transform modulus maxima method [78]. However, heartbeat time series contain densely packed, *non-isolated* singularities which unavoidably affect each other in the time-frequency decomposition. Therefore, rather than evaluating the distribution of the inherently unstable local singularity exponents (as shown in color in Fig. 11), we estimate the scaling of an appropriately chosen global measure – a partition function $Z_q(a)$, which is defined as the sum of the qth powers of the local maxima of the modulus of the wavelet transform coefficients at scale a. For each scale a these local maxima values are traced along the maxima lines obtained after the wavelet decomposition of the heartbeat signal (maxima lines appear in bright/white color in Fig. 9). As analyzing wavelet we use the 3rd derivative of the Gaussian function. For small scales, we expect

$$Z_q(a) \sim a^{\tau(q)} . \tag{2}$$

For certain values of q, the exponents $\tau(q)$ have familiar meanings. In particular, $\tau(2)$ is related to the scaling exponent of the Fourier power spectra, $S(f) \sim 1/f^{\beta}$, as $\beta = 2+\tau(2)$. For positive q, $Z_q(a)$ reflects the scaling of the large fluctuations and strong singularities, while for negative q, $Z_q(a)$ reflects the scaling of the small fluctuations and weak singularities [54,55]. Thus, the scaling exponents $\tau(q)$ can reveal different aspects of cardiac dynamics [Fig. 12] . Monofractal signals display a linear $\tau(q)$ spectrum, $\tau(q) = qH-1$, where H is the global Hurst exponent. For multifractal signals, $\tau(q)$ is a nonlinear function: $\tau(q) = qh(q)-1$, where $h(q) \equiv d\tau(q)/dq$ is not constant.

A previous obstacle to the determination of the multifractal spectrum of a time series has been the calculation of the negative moments. Until the application of the wavelet modulus maxima method, it was not possible to estimate $Z_q(a)$ for $q < 0$. We calculate $\tau(q)$ for moments $q = -5, 4, \ldots, 0, \ldots, 5$ and scales $a = 2 \times 1.15^i$, $i = 0, \ldots, 41$ from 6 hours records obtained from a healthy subject and a subject with congestive heart failure. In Fig. 12a,b we display the calculated values of $Z_q(a)$ for scales $a > 8$. The top curve corresponds to $q = -5$, the middle curve (shown heavy) to $q = 0$ and the bottom curve to $q = 5$. The exponents $\tau(q)$ are obtained from the slope of the $Z_q(a)$ curves in the region $16 < a < 700$, thus eliminating the influence of any residual small scale random noise due to electrocardiogram signal pre-processing as well as extreme, large scale fluctuations of the signal. A monofractal signal would correspond to a straight line for $\tau(q)$, while for a multifractal signal $\tau(q)$ is nonlinear. Note the clear differences between the $\tau(q)$ curves for healthy and heart failure records [Fig. 12c]. The constantly changing curvature of the $\tau(q)$ curves for the healthy

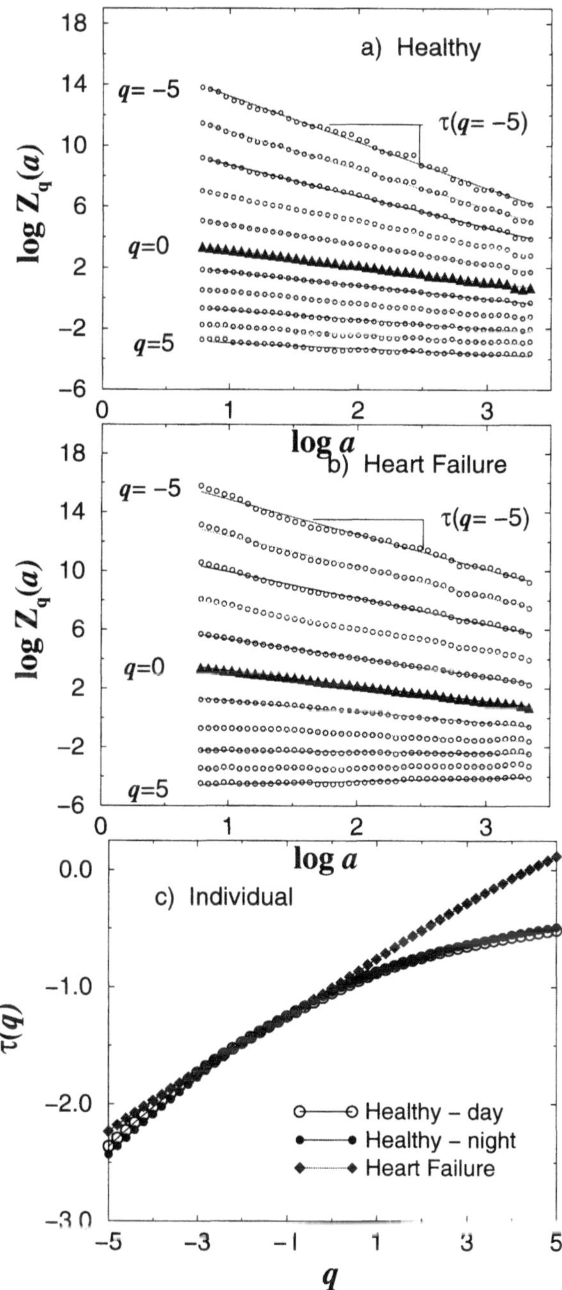

Fig. 12. Scaling of the partition function $Z_q(a)$ with scale a obtained from daytime records consisting of $\approx 25,000$ beats for (a) a healthy subject and (b) a subject with congestive heart failure. (c) Multifractal spectrum $\tau(q)$ for the individual records in (a) and (b). After [18]

records, suggests multifractality. In contrast, $\tau(q)$ is almost linear for the congestive heart failure subject, indicating monofractality.

We analyze both daytime (12:00 to 18:00) and nighttime (0:00 to 6:00) heartbeat time series records of healthy subjects, and the daytime records of patients with congestive heart failure. These data were obtained by Holter monitoring [50].

Next, we obtain the fractal dimension $D(h)$. It is related to $\tau(q)$ through a Legendre transform,

$$D(h) = q\frac{d\tau(q)}{dq} - \tau(q)\,. \tag{3}$$

For all healthy subjects, we find that $\tau(q)$ is a nonlinear function [Fig. 12c and Fig. 13a], which indicates that the heart rate of healthy humans is a multifractal signal. Figure 13b shows that for healthy subjects, $D(h)$ has nonzero values for a broad range of local Hurst exponents h. The multifractality of healthy heartbeat dynamics cannot be explained by activity, as we analyze data from subjects during nocturnal hours. Furthermore, this multifractal behavior cannot be attributed to sleep-stage transitions, as we find multifractal features during daytime hours as well [79]. The range of scaling exponents – $0 < h < 0.3$ – with nonzero fractal dimension $D(h)$, suggests that the fluctuations in the healthy heartbeat dynamics exhibit anti-correlated behavior ($h = 1/2$ corresponds to uncorrelated behavior while $h > 1/2$ corresponds to correlated behavior).

In contrast, we find that heart rate data from subjects with a pathological condition – congestive heart failure – show a clear *loss of multifractality* [Figs. 13a,b]. For the heart failure subjects, $\tau(q)$ is close to linear and $D(h)$ is non-zero only over a very narrow range of exponents h indicating monofractal behaviour [Fig. 13].

Our results show that, for healthy subjects, local Hurst exponents in the range $0.07 < h < 0.17$ are associated with fractal dimensions close to one. This means that the subsets characterized by these local exponents are statistically dominant. On the other hand, for the heart failure subjects, we find that the statistically dominant exponents are confined to a narrow range of local Hurst exponents centered at $h \approx 0.22$. These results suggest that for heart failure the fluctuations are less anti-correlated than for healthy dynamics since the dominant scaling exponents h are closer to 1/2. Thus, our findings support previous reports of long-range anti-correlations in healthy heartbeat fluctuations [see caption to Fig. 13] [24].

We present panels with the local Hurst h exponent for 6 healthy individuals [Fig. 14] and 6 subjects with congestive heart failure [Fig. 15]. Each panel represents 6h long record. The shade code for these panels is the following: with increasing value of h, the spectrum goes from light to dark gray shade. A wider range of shades indicate a higher degree of multifractality. For this reason, records from healthy individuals should be more polychromatic. On the other hand, records from heart failure patients should be more monochromatic (with a single gray shade predominating), indicating loss of multifractality. In addition, the shade spectrum for the healthy individuals is shifted to the light gray and

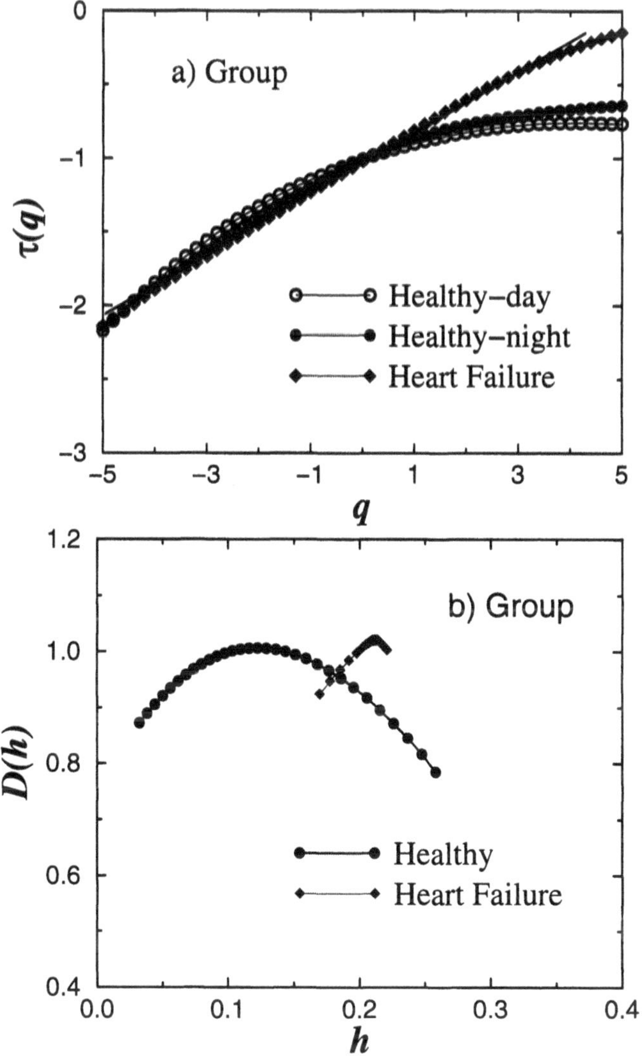

Fig. 13. (a) Multifractal spectrum $\tau(q)$ of the group averages for daytime and nighttime records for 18 healthy subjects and for 12 patients with congestive heart failure. The results show multifractal behavior for the healthy group and distinct change in this behavior for the heart failure group. (b) Fractal dimensions $D(h)$ obtained through a Legendre transform from the group averaged $\tau(q)$ spectra of (a). The shape of $D(h)$ for the individual records and for the group average is broad ($\Delta h \approx 0.25$), indicating multifractal behavior. On the other hand, $D(h)$ for the heart failure group is very narrow ($\Delta h \approx 0.05$), indicating loss of multifractality. The different form of $D(h)$ for the heart failure group may reflect perturbation of the cardiac neuroautonomic control mechanisms associated with this pathology. Note, that for $q = 2$ the heartbeat fluctuations of healthy subjects are characterized by $h \approx 0.1$, which corresponds to $\alpha \approx 1.1$ for the interbeat interval series obtained from DFA analysis (Sect. III). After [18]

for the heart failure patients is shifted to the dark gray. This is in agreement with the results in Fig. 13 where the peak of the multifractal spectrum $D(h)$ is centered at smaller values of h for the healthy group and at larger values of h for the heart failure group. These findings may have a potential for diagnosis [80].

8 Multifractality and Nonlinearity

The multifractality of heart beat time series also enables us to quantify the greater complexity of the healthy dynamics compared to pathological conditions. Power spectrum and detrended fluctuation analysis define the complexity of heart beat dynamics through its scale-free behavior, identifying a *single* scaling exponent as an index of healthy or pathologic behavior. Hence, the power spectrum is not able to quantify the greater level of complexity of the healthy dynamics, reflected in the heterogeneity of the signal. On the other hand, the multifractal analysis reveals this new level of complexity by the *broad* range of exponents necessary to characterize the healthy dynamics [Fig. 13]. Moreover, the change in shape of the $D(h)$ curve for the heart failure group may provide insights into the alteration of the cardiac control mechanisms due to this pathology.

To further study the complexity of the healthy dynamics, we perform two tests with surrogate time series. First, we generate a surrogate time series by shuffling the interbeat interval increments of a record from a healthy subject. The new signal preserves the distribution of interbeat interval increments but destroys the long-range correlations among them. Hence, the signal is a simple random walk, which is characterized by a single Hurst exponent $H = 1/2$ and exhibits monofractal behavior [Fig. 16a]. Second, we generate a surrogate time series by performing a Fourier transform on a record from a healthy subject, preserving the amplitudes of the Fourier transform but randomizing the phases, and then performing an inverse Fourier transform. This procedure eliminates nonlinearities, preserving only the linear features of the original time series. The new surrogate signal has the *same* $1/f$ behavior in the power spectrum as the original heart beat time series; however it exhibits monofractal behavior [Fig. 16a]. We repeat this test on a record of a heart failure subject. In this case, we find a smaller change in the multifractal spectrum [Fig. 16b]. The results suggest that the healthy heartbeat time series contains important phase correlations canceled in the surrogate signal by the randomization of the Fourier phases, and that these correlations are weaker in heart failure subjects. Furthermore, the tests indicate that the observed multifractality is related to nonlinear features of the healthy heartbeat dynamics. A number of recent studies have tested for nonlinear and deterministic properties in recordings of interbeat intervals [9–11,14,15]. Our results suggest an explicit relation between the nonlinear features (represented by the Fourier phase interactions) and the multifractality of healthy cardiac dynamics [Fig. 16].

Fig. 14. Panels obtained from healthy individuals illustrating how the local Hurst exponent h (vertical shaded bars) changes with time (x-axis). Each panel represents a 6h record. A broad range of colors indicates broad multifractal spectrum. After [31]

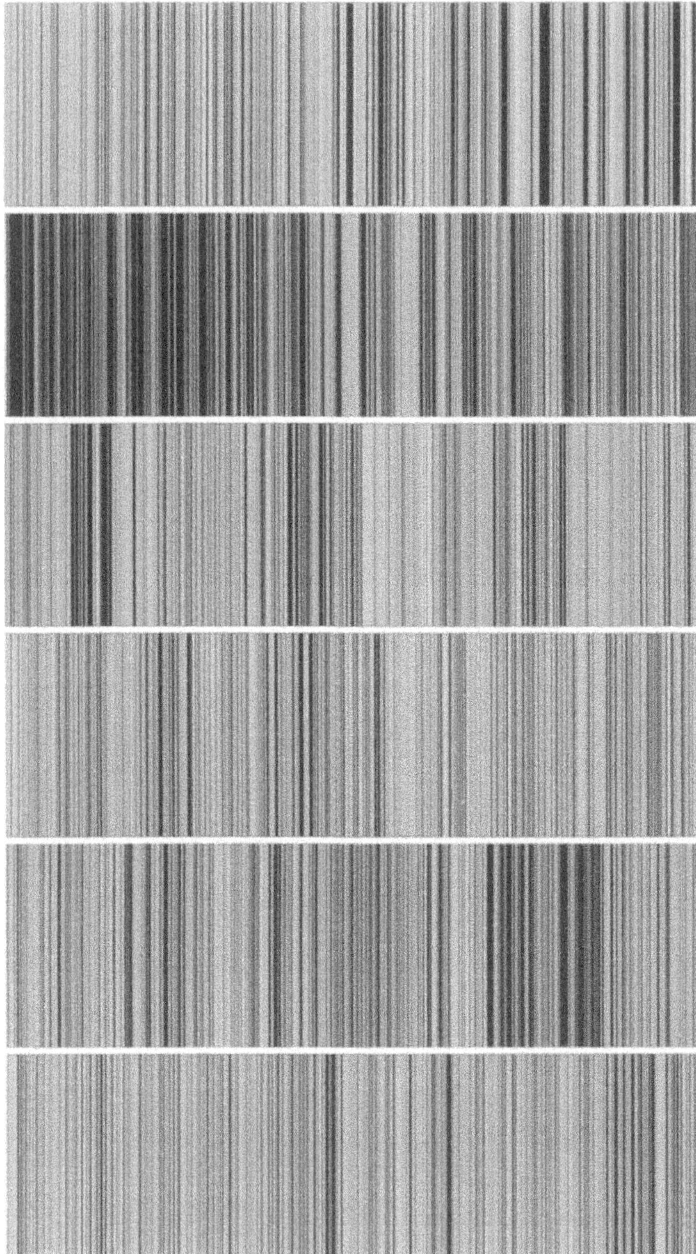

Fig. 15. Panels obtained from subjects with congestive heart failure illustrating how the local Hurst exponent h (vertical shaded bars) changes with time (x-axis). Each panel represents a 6h record. An almost monochromatic appearance indicates narrow multifractal spectrum, i.e. loss of multifractality. After [31]

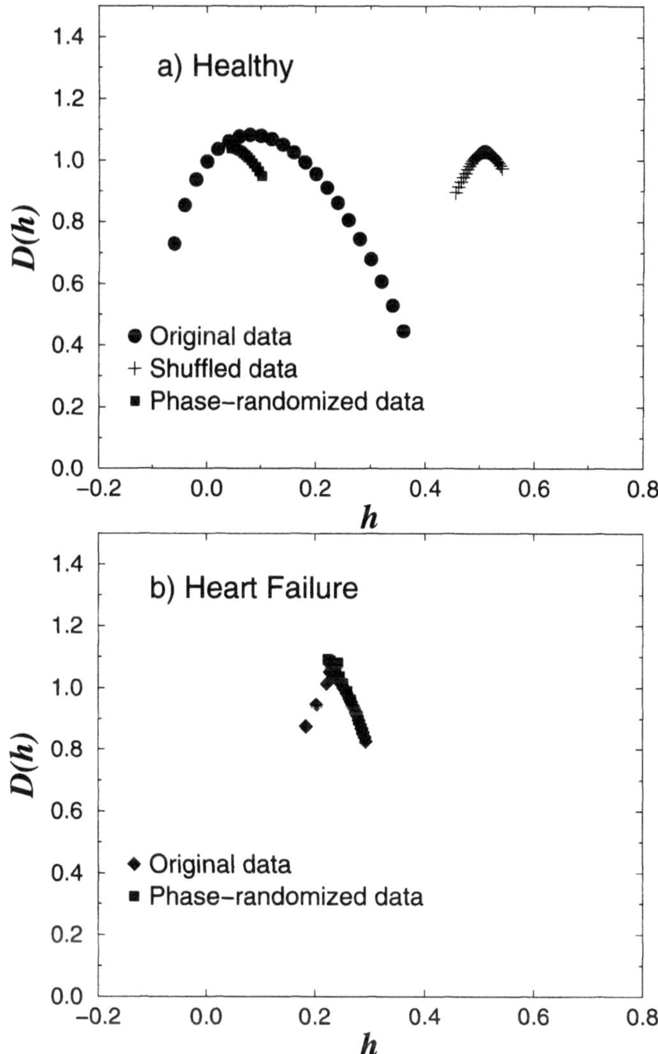

Fig. 16. (a) The fractal dimensions $D(h)$ for a 6h daytime record of a healthy subject. After reshuffling and integrating the increments in this interbeat interval time series, so that all correlations are lost but the distribution is preserved, we obtain monofractal behavior – a very narrow point-like spectrum centered at $h = H - 1/2$. Such behavior corresponds to a simple random walk. A different test, in which the $1/f$-scaling of the heart beat signal is preserved but the Fourier phases are randomized (i.e., nonlinearities are eliminated) leads again to a monofractal spectrum centered at $h \approx 0.07$, since the linear correlations were preserved. These tests indicate that the observed multifractality is related to nonlinear features of the healthy heart beat dynamics rather than to the ordering or the distribution of the interbeat intervals in the time series. (b) The fractal dimensions $D(h)$ for a 6h daytime record of a heart failure subject. The narrow multifractal spectrum indicates loss of multifractal complexity and reduction of nonlinearities with pathology. After [18]

9 Summary

The discovery of multifractality in a physiological time series and its breakdown with pathology is significant from a number of perspectives.

First, contemporary analysis of heartbeat fluctuations, and the study of physiological time series in general, have emphasized two important, but apparently unconnected properties: (i) the presence of nonlinearities and (ii) 1/f-behavior (monofractality). The monofractal hypothesis assumes that the scaling properties of the signal are the same throughout. Yet the heterogeneous nature of the heartbeat interval time series clearly indicates nonlinear features. The finding of a multifractal mechanism for heart rate control provides a unifying connection between nonlinear and fractal properties and, indeed indicates that they are aspects of a more fundamental type of mechanism. In particular, we show that both the multifractal character and the nonlinear properties of the signal are encoded in the Fourier phases [Fig. 16]. The origin and nature of these Fourier phase interactions is an open question.

Second, our analysis indicates that the healthy heartbeat is described by a broad range of scaling exponents h with a well-defined set of bounding parameters, $h_{\min}$ and $h_{\max}$. Furthermore, certain exponents appear to be "forbidden" ($h < h_{\min}$ and $h > h_{\max}$) and the exponents present occur with a given structure characterized by the function $D(h)$.

Third, our findings may lead to new diagnostic applications. Further detailed studies on a larger number of datasets are needed to establish the advantages of given methods compared to others and to find optimal combinations of methods for diagnostic and prognostic purposes.

Fourth, our analysis is based on a "microscopic" approach which can identify the statistical properties of the self-affine cascade of heartbeat fluctuations at different scales [Fig. 9]. Our finding of multifractality quantifies the complex dynamics of this cascade and suggests that a *multiplicative* mechanism might be the origin of this phenomena. The detailed features of the cascades and how they relate to other processes with cascades (e.g., turbulence) remain to be addressed.

On a more general level, our approach provides a way of testing a broad range of 1/f-type signals to see if they represent multifractal or monofractal processes. As such, these findings should be of interest to a very wide audience given the historic interest in elucidating the nature of different types of 1/f noise.

Finally, from a physiological perspective, the detection of robust multifractal scaling in the heart rate dynamics is of interest because our findings raise the intriguing possibility that the control mechanisms regulating the heartbeat interact as part of a coupled cascade of feedback loops in a system operating far from equilibrium – an extraordinarily complex behavior which in physical systems has been connected with turbulence and related multiscale phenomena [81–83]. Furthermore, the present results indicate that the healthy heartbeat is even more complex than previously suspected, posing a challenge to ongoing efforts to develop realistic models of the control of heart rate and other processes under neuroautonomic regulation [19,35,84–86].

Acknowledgments

This work was supported by NIH/National Center for Research Resources (P41 RR13622).

References

1. M. F. Shlesinger: Ann. NY Acad. Sci. **504**, 214 (1987)
2. J.B. Bassingthwaighte, L.S. Liebovitch, and B.J. West: *Fractal Physiology* (Oxford Univ. Press, New York 1994)
3. M. Malik and A.J. Camm, eds.: *Heart Rate Variability* (Futura, Armonk NY 1995).
4. M. Kobayashi and T. Musha: IEEE Trans. Biomed. Eng. **29**, 456 (1982)
5. J.P. Saul, P. Albrecht, D. Berger, and R.J. Cohen: Computers in Cardiology (IEEE Computer Society Press, Washington DC), 419 (1987)
6. C.-K. Peng, J. Mietus, J.M. Hausdorff, S. Havlin, H. Eugene Stanley, and A.L. Goldberger: Phys. Rev. Lett. **70**, 1343 (1993)
7. J.M. Hausdorff, P.L. Purdon, C.-K. Peng, Z. Ladin, J. Y. Wei, and A.L. Goldberger: J. Appl. Physiol. **80**, 1448 (1996)
8. L.S. Liebovitch: Adv. Chem. Ser. **235**, 357 (1994); S.B. Lowen, L.S. Liebovitch, J.A. White: Phys. Rev. E **59**, 5970 (1999)
9. J. Lefebvre, D.A. Goodings, M.V. Kamath, and E.L. Fallen: Chaos **3**, 267 (1993)
10. Y. Yamamoto, R.L. Hughson, J.R. Sutton, C.S. Houston, A. Cymerman, E.L. Fallen, and M.V. Kamath: Biol. Cybern. **69**, 205 (1993)
11. J.K. Kanters, N.H. Holstein-Rathlou and E. Agner: J. Cardiovasc. Electrophysiol. **5**, 128 (1994)
12. J. Kurths, A. Voss, P. Saparin, A. Witt, H.J. Kleiner, and N. Wessel: Chaos **5**, 88 (1995)
13. P. Ch. Ivanov, M. G. Rosenblum, C.-K. Peng, J. Mietus, S. Havlin, H. E. Stanley and A. L. Goldberger: Nature **383**, 323 (1996)
14. G. Sugihara, W. Allan, D. Sobel and K. D. Allan: Proc. Natl. Acad. Sci. USA **93**, 2608 (1996)
15. C-S. Poon, and C.K. Merrill: "Decrease of Cardiac Chaos in Congestive Heart Failure", Nature **389**, 492 (1997).
16. J. Feder: *Fractals* (Plenum, NY 1988)
17. J. F. Muzy, E. Bacry, and A. Arneodo: Phys. Rev. E **47**, 875 (1993)
18. P. Ch. Ivanov, L. A. N. Amaral, A. L. Goldberger, S. Havlin, M. G. Rosenblum, Z. Struzik, and H. E. Stanley: Nature **399** 461 (1999)
19. M. Mackey and L. Glass: Science **197**, 287 (1977)
20. M.M. Wolf, G.A. Varigos, D. Hunt, and J.G. Sloman: Med. J. Australia **2**, 52 (1978)
21. R.I. Kitney and O. Rompelman: *The Study of Heart-Rate Variability* (Oxford Univ. Press, London 1980)
22. S. Akselrod, D. Gordon, F. A. Ubel, D. C. Shannon, A. C. Barger, and R. J. Cohen: Science **213**, 220 (1981)
23. L. Glass, P. Hunter, and A. McCulloch, eds.: *Theory of Heart* (Springer Verlag, New York 1991)
24. C.-K. Peng, S. Havlin, H. E. Stanley, and A. L. Goldberger: in *Proc. NATO dynamical disease conference*, edited by Glass L. Chaos **5**, 82 (1995)
25. P.Ch. Ivanov, M.G. Rosenblum, C.-K. Peng, J. Mietus, S. Havlin, H.E. Stanley, and A.L. Goldberger: Physica A **249**, 587 (1998).

26. Y. Ashkenazy, P.Ch. Ivanov, S. Havlin, C.-K. Peng, A.L. Goldberger, and H.E. Stanley: Phys. Rev. Lett. **86**, 1900 (2001)
27. D. Panter: *Modulation, noise and spectral analysis* (McGraw-Hill, New York 1965)
28. R. L. Stratonovich: *Topics in the theory of random noise, vol. I* (Gordon and Breach, New York 1981)
29. R.I. Kitney, D. Linkens, A.C. Selman, and A.A. McDonald: Automedica **4**, 141 (1982)
30. A.L. Goldberger: Lancet **347**, 1312 (1996)
31. P.Ch. Ivanov, L.A.N. Amaral, A.L. Goldberger, S. Havlin, M.G. Rosenblum, H.E. Stanley, Z. Struzik: *Chaos* **11**, 641 (2001)
32. S. Havlin et al.: Phys. Rev. Lett. **61**, 1438 (1988)
33. M.F. Shlesinger and B.J. West: *Random Fluctuations and Pattern Growth: Experiments and Models* (Kluwer Academic Publishers, Boston 1988)
34. M. N. Levy: Circ. Res. **29**, 437 (1971)
35. P. Ch. Ivanov, L.A.N. Amaral, A.L. Goldberger, and H.E. Stanley: Europhys. Lett. **43**, 363 (1998)
36. P. Bernaola-Galvan, P.Ch. Ivanov, L.A.N. Amaral, and H.E. Stanley: Phys. Rev. Lett. **87**, 168105 (2001)
37. P. Bernaola-Galvan, I. Grosse, P. Carpena, J.L. Oliver, R. Roman-Roldan, H.E. Stanley: Phys. Rev. Lett. **85**, 1342 (2000)
38. R. M. Berne and M. N. Levy: *Cardiovascular Physiology* 6th ed. (C.V. Mosby, St. Louis 1996)
39. C.-K. Peng, S. V. Buldyrev, S. Havlin, M. Simons, H. E. Stanley, and A. L. Goldberger: Phys. Rev. E **49**, 1691 (1994)
40. S. V. Buldyrev, A. L. Goldberger, S. Havlin, R. N. Mantegna, M. E. Matsa, C.-K. Peng, M. Simons, and H. E. Stanley: Phys. Rev. E **51**, 5084 (1995)
41. S. V. Buldyrev, A. L. Goldberger, S. Havlin, C.-K. Peng, H. E. Stanley, and M. Simons: Biophys. J. **65**, 2673 (1993)
42. S. M. Ossadnik, S. V. Buldyrev, A. L. Goldberger, S. Havlin, R. N. Mantegna, C.-K. Peng, M. Simons, and H. E. Stanley: Biophys. J. **67**, 64 (1994)
43. M. S. Taqqu, V. Teverovksy, and W. Willinger: Fractals **3**, 185 (1996)
44. J.W. Kantelhardt, E. Koscielny-Bunde, H.H.A. Rego, S. Havlin, and A. Bunde: Physica A **294**, 441 (2001)
45. K. Hu, P.Ch. Ivanov, C. Zhi, P. Carpena, and H.E. Stanley: Phys. Rev. E **64**, 011114 (2001)
46. Z. Chen, P.Ch. Ivanov, K. Hu, and H.E. Stanley: Phys. Rev. E **65**, 041107 (2002)
47. H.V. Huikuri, K.M. Kessler, E. Terracall, A. Castellanos, M.K. Linnaluoto, R.J. Myerburg: Am. J. Cardiol. **65**, 391 (1990)
48. H. Moelgaard, K.E. Soerensen, and P. Bjerregaard: Am. J. Cardiol. **68**, 777 (1991)
49. P.Ch. Ivanov M. G. Rosenblum, C.-K. Peng, S. Havlin, H.E. Stanley, and A.L. Goldberger, 'Wavelets in Medicine and Physiology'. In *Wavelets in Physics*, ed. H. van der Berg (Cambridge University Press, Cambridge 1998)
50. *Heart Failure Database* (Beth Israel Deaconess Medical Center, Boston, MA). The database now includes 18 healthy subjects (13 female and 5 male, with ages between 20 and 50, average 34.3 years), and 12 congestive heart failure subjects (3 female and 9 male, with ages between 22 and 71, average 60.8 year) in sinus rhythm. Data is freely available at the following URL: http://www.physionet.org
51. P.Ch. Ivanov, A. Bunde, L.A.N. Amaral, S. Havlin, J. Fritsch-Yelle, R.M. Baevsky, H.E. Stanley and A.L. Goldberger: Europhys. Lett. **48**, 594 (1999)
52. A.L. Goldberger, M.W. Bungo, R.M. Baevsky, B.S. Bennett, D.R. Rigney, J.E. Mietus, G.A. Nikulina, and J.B. Charles: Am. Heart J. **128**, 202 (1994)

53. A. Bunde, S. Havlin, J.W. Kantelhardt, T. Penzel, J.H. Peter, K. Voigt, Phys. Rev. Lett. **85**, 3736 (2000)
54. T. Vicsek: *Fractal Growth Phenomenon* 2nd edn (World Scientific, Singapore 1993)
55. H. Takayasu: *Fractals in the Physical Sciences* (Manchester Univ. Press, Manchester, UK, 1997)
56. R.M. Berne and M.N. Levy: *Cardiovascular Physiology*, 7th edn (Mosby, St. Louis 1996)
57. Heartbeat increment series were investigated by A. Babloyantz and P. Maurer, Phys. Lett. A **221**, 43 (1996) and P. Maurer, H.-D. Wang, and A. Babloyantz, Phys. Rev. E **56**, 1188 (1997). These studies differ from ours because we investigate, quantitatively, normal heartbeats by evaluating the scaling properties of the magnitude and sign series. In addition, our calculations are based on window scales larger than 6 and up to one thousand heartbeats.
58. P.F. Panter, *Modulation, Noise, and Spectral Analysis Applied to Information Transmission* (New York, New York 1965). We also applied a test for nonlinearity using the phase randomization procedure described in J. Theiler, S. Eubank, A. Longtin, B. Galdrikian, and D.J. Farmer, Physica D **58**, 77 (1992), and find that the magnitude scaling exponent drops to 0.5 while the sign scaling exponent remains unchanged.
59. L.A.N. Amaral, P.Ch. Ivanov, N. Aoyagi, I. Hidaka, S. Tomono, A.L. Goldberger, H.E. Stanley, and Y. Yamamoto. *Physical Review Letters* **86**, 6026 (2001)
60. H. Moelgaard, K. E. Soerensen, and P. Bjerregaard, *Am. J. Cardiology* **68**, 77 (1991)
61. H. V. Huikuri, K. M. Kessler, E. Terracall, A. Castellanos, M. K. Linnaluoto, and R. J. Myerburg, *Am. J. Cardiology* **65**, 391 (1990)
62. M. A. Carskadon and W. C. Dement: In *Principles and Practice of Sleep Medicine*, edited by M. H. Kryger, T. Roth, and W. C. Dement (W. B. Saunders, Philadelphia 1994), pp. 16-25
63. A. Rechtschaffen and A. Kales, *A Manual of Standardized Terminology, Techniques, and Scoring System for Sleep Stages of Human Subjects* (U.S. Government Printing Office, Washington 1968)
64. J.W. Kantelhardt, Y. Ashkenazy, P.Ch. Ivanov, A. Bunde, S. Havlin, T. Penzel, J.-H. Peter, and H.E. Stanley: *Physical Review E* **65**, 051908 (2002)
65. F. Mallamace and H. E. Stanley: Eds. *Physics of Complex Systems: Proc. Enrico Fermi School on Physics, Course CXXXIV* (IOS Press, Amsterdam 1997)
66. P. Meakin: *Fractals, Scaling and Growth Far from Equilibrium* (Cambridge University Press, Cambridge 1997)
67. H.E. Stanley: Nature **378**, 554 (1995)
68. A. Bunde and S. Havlin: *Fractals in science* (Springer-Verlag, Berlin 1994)
69. A. Bunde and S. Havlin: Eds. *Fractals and Disordered Systems, 2nd Edition* (Springer-Verlag, Berlin 1996)
70. A.-L. Barabasi and H. E. Stanley: *Fractal Concepts in Surface Growth* (Cambridge University Press, Cambridge 1995)
71. H.E. Hurst: Trans. Am. Soc. Civ. Eng. **116**, 770 (1951)
72. T.G. Dewey: *Fractals in Molecular Biophysics* (Oxford University Press, Oxford 1997)
73. Z. R. Struzik: Fractals, **8**, 163 (2000)
74. Z. R. Struzik: Fractals, **9**, 77 (2001)
75. T. Vicsek and A.L. Barabási. J. Phys. A: Math. Gen., **24**, L845 (1991)
76. A.-L. Barabasi and H.E. Stanley: *Fractal Concepts in Surface Growth* (Cambridge University Press, Cambridge 1995), Chapter 24

77. I. Daubechies: *Ten Lectures on Wavelets* (S.I.A.M., Philadelphia 1992)
78. J.F. Muzy, E. Bacry and A. Arneodo: Int. J. Bifurc. Chaos. **4**, 245 (1994)
79. L.A.N. Amaral, P.Ch. Ivanov, N. Aoyagi, I. Hidaka, S. Tomono, A.L. Goldberger, H.E. Stanley, and Y. Yamamoto: Phys. Rev. Lett. **86**, 6026 (2001)
80. L. A. N. Amaral, A. L. Goldberger, P. Ch. Ivanov, and H. E. Stanley: Phys. Rev. Lett. **81**, 2388 (1998)
81. C. Meneveau and K. R. Sreenivasan: Phys. Rev. Lett. **59**, 1424 (1987)
82. H.E. Stanley and P. Meakin: Nature **335**, 405 (1988)
83. U. Frisch: *Turbulence* (Cambridge University Press, Cambridge UK 1995)
84. L. Glass and C.P. Malta: J. Theor. Biol. **145**, 217 (1990)
85. H. Seidel and H. Herzel: Physica D, **115**, 145 (1998)
86. D.C. Lin and R.L. Hughson: Phys. Rev. Lett. **86**, 1650 (2001)

Multifractals: From Modeling to Control of Broadband Network Traffic

P. Murali Krishna, Vikram M. Gadre, and Uday B. Desai

SPANN Lab, Department of Electrical Engineering, IIT – Bombay, INDIA

Abstract. Broadband network traffic has been observed to possess complex scaling behavior which cannot be modeled using the traditional tele traffic models. In this paper, we provide insights into modeling of tele traffic data using multifractal cascade processes which were first proposed for modeling turbulence in fluid dynamics. This model is also applied to explain the phenomena of increased burstiness in the multiplexed tele traffic. Since the phenomenon of traffic multiplexing is non linear in nature, we use non traditional methods like the analysis of the entropy of the multiplexed data in order to arrive at conclusive results.

1 Introduction

Integrated broadband networks are expected to support various traffic types such as data, voice, image and video. The complexity of broadband traffic is such that it requires modeling and analysis which can be quite unconventional in the engineering sense. Traditional modeling tools and techniques, both theoretical and empirical, fail to characterize and understand the behavior of broadband traffic. However the discovery of scaling in the measured tele traffic has led to modeling solutions that can approximate the complexity of the data. Analysis of complex systems and data can be made easier if a suitable analogy of the same can be obtained which is more easily tractable and which offers new insights into the characteristics of data. Of late based on multifractal analysis of the traffic data, broadband tele traffic data is analogously looked upon as turbulent flow through the network [1].

2 Modeling Broadband Tele Traffic: Shifting Paradigms

The traditional tool used for the modeling and analysis of network traffic was the classical Poisson traffic model. This model was proposed initially for the analysis of telephone switching systems by Erlang in the 1920s and with the advent of computer networks, it was modified and adapted for the analysis of queuing systems [2]. This model had the advantage that it was analytically tractable in nature and closed form expressions could be derived for the performance characteristics that were of interest. But with the advent of broadband networks that supported various services, the nature of the traffic flowing through the network changed drastically. The measured traffic began to exhibit characteristics that were quite different from what was predicted by the Poisson/Markov models.

This phenomena was first observed in the early years of the last decade by the group led by Leland [3]. The data which was measured at Bell Core exhibited statistical characteristics which were statistically self similar. A random process $X(t)$ is said to be self similar if it's statistics remain invariant with respect to a change in the time scale at which it is observed. Mathematically, it can be stated as

$$X(t) \stackrel{d}{=} a^{-H} X(at) \tag{1}$$

where the equality is in distribution. The parameter H is called the Hurst parameter $(0 < H < 1)$ which controls the extent of self similarity in the process. An example of a self similar process is the fractional Brownian motion (fBm) introduced by Mandelbrot and Van Ness [4]. The fractional Brownian motion process is an important model used in various modeling applications including traffic modeling [5],[6] The presence of self similarity led to a total change in the view in the analysis and modeling of tele traffic data. The traditional times series models were found insufficient to model this phenomena and the analysis of such processes also called for new techniques and methodologies. The phenomenon of scale invariance in traffic expresses itself as increased burstiness in traffic. The nature of congestion produced by self-similar network traffic models differs drastically from that predicted by standard formal models [7]. It was observed that contrary to commonly held beliefs that multiplexing traffic streams tends to produce smoothed out aggregate traffic with reduced burstiness, aggregating self-similar traffic streams can actually intensify burstiness rather than diminish it. From the view point of performance analysis of networks, this is a very crucial phenomenon as increased burstiness leads to lower resource utilization. As a result, quality of service parameters like available bandwidth, data transfer delay and loss probability of the system get adversely affected. A whole host of problems arise as a result pertaining to management of resources in networks like call admission and congestion control.

2.1 Multiple Scaling: From Self Similarity to Multifractals

In recent years, various researchers have highlighted the existence of multiple scaling phenomena in broadband traffic data. It was Jacques Levy Vehel and Rudolf Riedi who in 1997 first reported multiple scaling in TCP-IP traffic data [8]. This was followed by experiments performed by Willinger and his team at AT&T Bell Labs who confirmed the existence of multiple scaling regimes in traffic data [9],[10], [11],[12]. The concept of multiple scaling can be understood as follows. The wavelet based estimator of scaling parameter in self similar data proposed by Abry and Veitch makes use of the scaling relationship of the energy of the wavelet coefficients at each scale [13]. If $X(t)$ is a self similar process with Hurst parameter $H \in (0.5, 1)$, then the expectation of the energy E_j that lies within a given bandwidth 2^{-j} around frequency $2^{-j}\lambda$ can be expressed as

$$E[E_j] = E\left[\frac{1}{N_j}\sum_k \mid d_{j,k} \mid^2\right] = c \mid 2^{-j}\lambda \mid^{1-2H} \tag{2}$$

where c is a factor that does not depend on j. N_j denotes the number of wavelet coefficients at scale j. It can be seen from the above relation that if the data is monofractal (self similarity parameter is stationary) in nature , then the energy of the wavelet coefficients at various scales follow a linear scaling in the logarithmic scale. For an exactly self similar process, a single slope will appear in the log – log plot and any sort of a deviation from this behavior can be considered as an deviation from the exact self similar nature of the data. Experimental analysis with measured data showed the presence of multiple scaling domains. This kind of a behavior has made it necessary to model broadband traffic in a different manner than using mono fractal models. These multiple scaling behaviors are caused due to variations in the traffic at small intervals. A process $X(t)$ is said to have local scaling properties with a local scaling exponent $\alpha(t)$ if the process behaves like $X(\Delta t) \sim (\Delta t)^{\alpha(t)}$ as $\Delta t \to 0$. For a mono fractal process, the scaling exponent $\alpha(t) = H$ for all time while we use the term multifractal to denote the processes that show non constant scaling parameter $\alpha(t)$. The use of multiplicative cascades to model traces with multifractal behavior also have the additional advantage of capturing the heavy tailed nature of the data (non Gaussianity) in confirmation with measured traffic behavior.

3 Multiplicative Multifractal Cascades

Multiplicative cascades were first proposed by Kolmogorov in the modeling of turbulence [14]. Turbulence flows are characterized by a scaling relation which controls the transfer of energy from from scale to another in a cascade structure [15], [16], [17], [18]. In the recent years the multiplicative cascade model has found applications in the modeling of a wide range of phenomena in various fields as diverse as broadband traffic modeling [19], [20] DNA modeling [21], volatility of market exchange rates [22], [23], mineral distribution [24], texture modeling [25], modeling of geophysical phenomena [26] etc. The relative simplicity of the model along with the flexibility that it provides makes it a very attractive tool to model nonlinear phenomena which show multiplicative structure. They are a paradigm shift from the traditional linear time invariant system based models used for time series analysis. The analysis of these processes is difficult to carry out using traditional techniques like spectral estimation. The distribution of the Holder exponents of the intervals of the cascade is a characteristic used for their study. This distribution is also termed multifractal spectrum or $f(\alpha)$ curve. The multifractal spectrum can be interpreted by using the definition of the fractal dimension or by using large deviation theory.

On one hand the multiplicative cascade processes are simple to visualize and easy to synthesize and on the other hand, they possess very complicated mathematical properties and structure. The synthesis technique is quite flexible as various multiplier families can be chosen to model the situation of interest. The analysis techniques for these processes are also non-traditional in nature since they rely on methods adopted from thermodynamics and statistical physics rather than conventional signal processing or statistical signal analysis.

An added attraction in the analysis of these processes is the inter disciplinary nature by which concepts and techniques like Large Deviation theory, Legendre Transforms, generalized dimensions, etc that are not frequently encountered in electrical and communication engineering, are used. This leads to a better understanding of the various analogies that exist between seemingly different areas of both theoretical and applied science. The modeling of time series using multiplicative cascade processes is a relatively new area. For broadband traffic processes, multiplicative cascades have been proposed as the measured traffic has shown properties like multiple scaling, heavy tailed behavior etc, which can be explained in terms of them. More information on the multiplicative cascade processes and it's properties can be obtained from the tutorial papers by Riedi [27] and the article by Mandelbrot and Evertsz [28].

4 Broadband Traffic Inter Arrival Time Modeling Using V.V.G.M Model

We propose a model for modeling the inter arrival times in broadband LAN traffic. We have named the model as Variable Variance Gaussian Multiplier (V.V.G.M) model taking into account the nature of the multiplier distributions used to generate the cascade. This model is based on the binomial multiplicative multifractal cascade processes. In the case of multiplicative cascade process modeling, the main task is in the estimation of the multiplier distribution from which the multipliers are sampled in order to generate the cascade. The method of estimation of the distribution of multipliers is to generate the histogram of the multipliers at various levels in the cascade generation process and try to parameterize the distribution. A good model is attributed with a few parameters which makes analysis more lucid and simple. Once the model has been parameterized, the statistical and performance measures have to be obtained. In order to accomplish this, the data synthesized by the model has to be compared with the original measured data. For the performance measures, we undertake queuing simulations with the synthesized and original measured data and compare the complimentary distribution of the queue length. The similarity in the manner in which the synthesized data can capture the characteristics of real world traffic will be obtained as a result of these simulations.

4.1 Development of V.V.G.M Multifractal Model

In this section, we discuss the development of the V.V.G.M model for modeling inter arrival times in broadband network traffic. As mentioned earlier, the binomial cascade structure is adapted due to it's relative simplicity. The main features that need to be considered while modeling are as follows

- Data must always be positive.
- Multiple scaling must be present (self similarity is non stationary).
- Data must exhibit non Gaussian distribution (heavy tailedness).

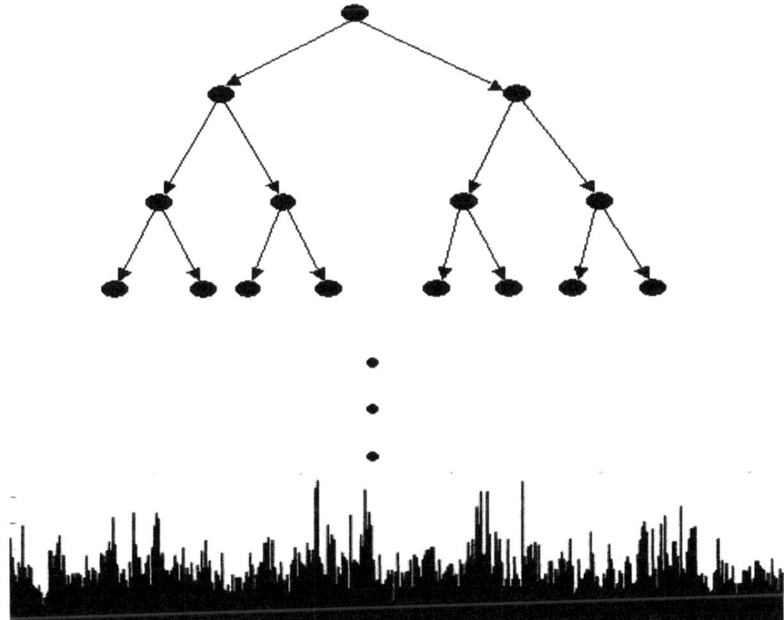

Fig. 1. Construction of cascade process

The first condition is evident from the fact that the data that we are trying to model (inter arrival times) cannot assume negative values. The second and third conditions are based on the measured data that show multiple scaling and also non Gaussian probability distribution with heavy tailed behavior. The model must take into consideration all the above factors. We model the broadband network inter arrival trace as the result of a binomial multiplicative cascade construction process. The data trace that we obtain is a result of the distribution of an initial measure. Fig. 1 illustrates this concept. The original measure is distributed by multiplying with ratios r and $1-r$. The multiplier r is a random variable chosen from a probability distribution $f_{R_j}(r)$, $0 \leq r \leq 1$ where j indicate the stage of the cascade. This process is iterated at each stage resulting in a multiplicative cascade. The main parameters to be estimated are the multiplier distributions $f_{R_j}(r)$. We assume $f_{R_j}(r)$ is symmetric about $r = \frac{1}{2}$ so that both r and $(1-r)$ have the same probability distribution. Let X_i^N, $(i = 1, \ldots, 2^N)$ denote the process obtained as a result of the above construction at stage N. Every point in the sequence X_i^N can be written as the product of several random variables $u_1 u_2 \ldots u_N$, where u_j, $(j = 1, \ldots, N)$ is either r or $(1-r)$ which is the multiplier at stage j.

4.2 Estimation of Multiplier Distributions

Given the data at stage N, X_i^N $(i = 1, \ldots, 2^N)$ (with time resolution of 2^{-N}), the data at stage $(N-1)$ is obtained by aggregating the consecutive values at stage N over non overlapping blocks of size two. Similarly, given the data at a coarser scale $(N-j)$, X_i^{N-j} $(i = 1, \ldots, 2^{N-j})$, we obtain the data at stage $(N-j-1)$ (lesser resolution) by adding consecutive values at stage j over non overlapping blocks of size two. i.e.

$$X_i^{N-j-1} = X_{2i-1}^{N-j} + X_{2i}^{N-j} \tag{3}$$

for $i = 1, \ldots, 2^{N-j-1}$. The procedure ends when the aggregates form one point at the coarsest scale. An estimate of the multipliers for proceeding from stage j to stage $j+1$ can be obtained as

$$r_j^{(i)} = \frac{X_i^{N-j}}{X_{2i-1}^{N-j-1}} \tag{4}$$

for $i = 1, \ldots, 2^{N-j-1}$. We view $\left\{r_j^{(i)}, i = 1, \ldots, 2^{N-j-1}\right\}$ as samples of the multiplier distribution $f_{R_j}(r)$ at stage j. The multiplier distribution at scale j, can now be obtained from the histogram of $r_j^{(i)}$. Our initial data set (Bellcore Aug89) contains 2^{18} points from which we aggregate according to the earlier mentioned rule to produce data at coarser resolutions. From the aggregated data, the multiplier distributions are obtained for the various scales. The histograms for the multipliers are plotted for the scales 2 to 5 in Fig. 2. It can be seen from the histograms that the multipliers can be modeled by using a Gaussian distribution of the appropriate parameter. We considered the case where the multiplier distributions $f_R(r)$ are Gaussian, centered at $r = 0.5$ with variances changing at each stage in the cascade. From the distributions obtained at each stage, we estimated the variance at each stage of the cascade. The change of variation of the multiplier distribution was parameterized by using curve fitting techniques. The parametric equation governing the variation of the variance was obtained as

$$\sigma(i) = \exp\left(-i\, 0.1285 - 1.3378\right) \tag{5}$$

where i denotes the ith cascade stage.

4.3 Synthesis Algorithm

For the synthesis purpose, we start from the coarsest value of the aggregate and multiply it using multiplier values chosen from the Gaussian distributions with variances that we estimated in the estimation phase. The algorithm for synthesis of the traces is as follows:

- Begin with starting value of the aggregate obtained during the estimation phase.
- At stage i, generate random numbers from $N(0.5, \sigma_i^2)$ where σ_i^2 is the variance at stage i.
- Multiply the starting aggregate value by multipliers generated at each stage from the distributions mentioned earlier to obtain the multiplicative cascade.

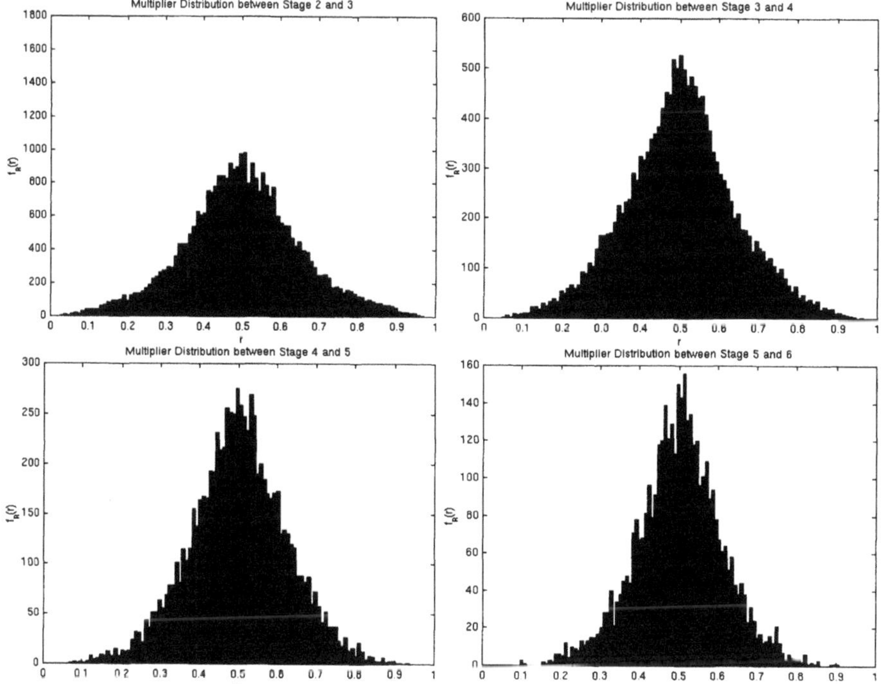

Fig. 2. Histogram of multipliers at various stages in the cascade

5 Comparison of Queuing Performance Analysis

We also present the results of the queuing simulations. We have considered a single server queue of infinite buffer capacity to which the inter arrival traces and corresponding packet lengths are fed in. The simulation tests were done for different link capacities. We have used 2^{18} points for all the simulation experiments. Fig. 3(a) and Fig. 3(b) illustrates the complimentary distribution function of the queue length for a link capacity of 400 bytes per 0.001s for original and fixed packet lengths. It can be seen that the queue length distribution in both the cases are identical in nature. The queuing simulations also have indicated that the tail queue probability of the traffic decays at a slower rate than expected by the Markov models. We had proven the existence of a global scaling exponent for the VVGM multifractal process [29]. This makes it possible to arrive at analytical expressions similar to the self similar models for multifractal processes. This is a crude approximation that does not reveal the complex scaling structure that is inherently present in a multifractal process, but it can be used to evaluate closed form expressions for the various QoS requirements like buffer requirements, rate of service and queue length distribution. The comparison of other performance parameters like packet loss probability and variation of the delay with buffer utilization factor also gave similar matching results. The results of these simulation experiments have shown that the V.V.G.M multiplicative cascade process can

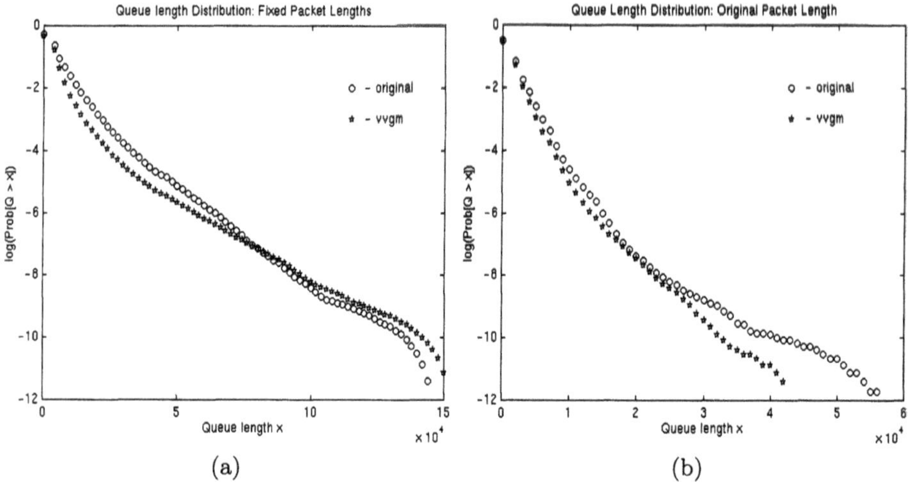

Fig. 3. Comparison of queue length distribution

approximate the statistical and performance characteristics of the original tele traffic data.

6 Analysis of Multiplexing and Aggregation of Multifractal Traffic

The analysis of multiplexing and aggregation of traffic is very important as it can give insight into design of network parameters. An interesting problem to analyze is the analysis of multiplexing and aggregation of traces that show multifractal nature. The analysis of effect of multiplexing of these processes is more complex due to the lack of closed form analytical expressions in comparison with more traditional Markovian models. Typically in a network scenario there will occur multiplexing at various points (routers, switches etc). The aim of the analysis is to show that burstiness in an aggregated traffic stream will increase as with the individual streams. Typically in the case of the inter arrival processes that we address in this study, burstiness implies that the inter arrival times of packets are less. From the data we have at hand (BellCore for LAN) and (LBL Traffic for WAN), this can be ascertained at hand. Figure 4 compares the captured LAN/ WAN traces. The traces in black is the LAN trace and the one in gray is the WAN trace. From observation alone, it can be seen that the LAN traffic inter arrival times are more spread out than that of the WAN. On a comparison note, the Table 1 gives the statistical parameters of both the traces.

6.1 Analysis Using VVGM Model

We can use the VVGM model to explain the increased burstiness of WAN traffic as compared to LAN in light of the above observations. Consider the multiplier

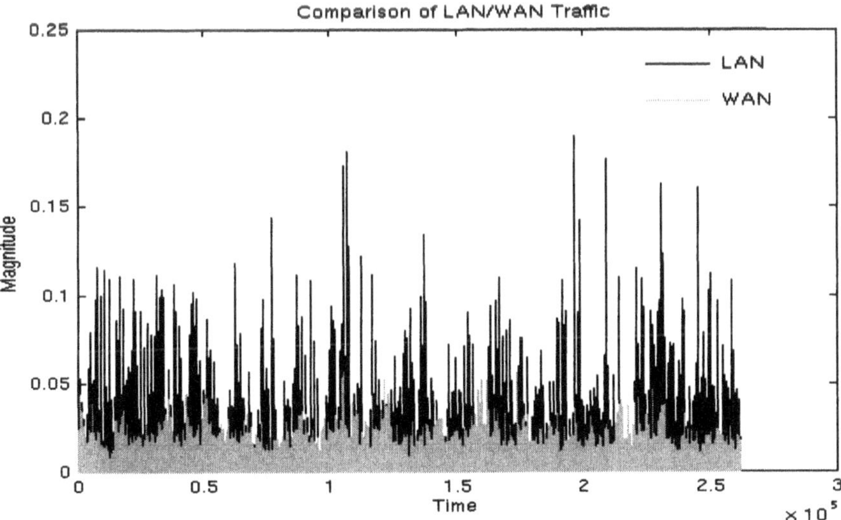

Fig. 4. LAN / WAN interarrival traces

Table 1. Comparison of Interarrival Times

Trace	LAN	WAN
Mean	0.0027	0.0020
Maximum	0.1900	0.00566
Variance	2.2044e-05	8.3156e-06

distribution estimation process that we adopt during the estimation phase of the cascade generation. The traces $X_{N+1}(k)$ at stage $N+1$ are aggregated in steps of two each to obtain the process at stage N (cascade at the earlier stage). This is given by

$$X_N(k) = X_{N+1}(2k-1) + X_{N+1}(2k) \tag{6}$$

The multipliers are estimated as

$$r_N(k) = \frac{X_{N+1}(k)}{X_N(2k)} \tag{7}$$

The pdf of the multipliers $f_N(r)$ is obtained from the histogram of the multiplier values. The variance of the multipliers at each stage indicate the burstiness at that stage. The multipliers are having a mean of 0.5 at each stage. If the given trace is less bursty (the inter arrival times have large peaks), then the range of variation of the multipliers as obtained from the (7) will be large say from 0.2–0.8. If on the other hand, the trace is very bursty in that the inter arrival times are very small and do not exhibit large peaks, then the range of variation of the multipliers will be confined to a smaller region say from 0.4–0.6. Since we do not have a strict VVGM model for all the stages for WAN, we have compared the variances between the two traces at those stages where

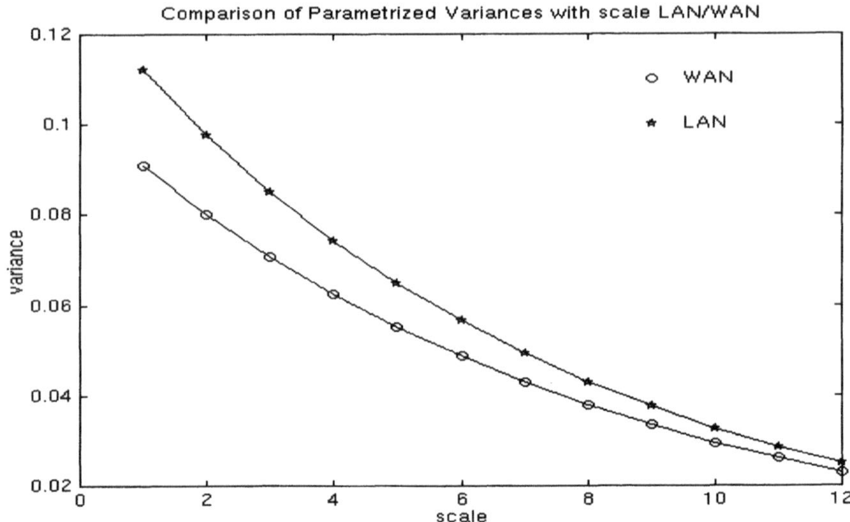

Fig. 5. Comparison of variance per stage between LAN/WAN traces

the Gaussianity assumption for the multipliers are seen to be valid. Figure 5 compares the measured variances at each stage in the cascade generation process. It can be seen from the graph that the variances at each stage are less for WAN than the LAN traffic. This observation has vindicated our understanding of the aggregation/multiplexing process that takes place in a networking environment. In the analysis on the statistics of self similar cascades, we had derived the expression for the effective Hurst parameter of the cascade as [29]

$$H_{eff} = 1 - \frac{1}{2}\log_2(4\mu_2) \tag{8}$$

where μ_2 is the second moment of the multipliers. It can be seen from the above expression that as the variance decreases, the effective Hurst parameter increases. Since the value of the Hurst parameter is an indication of the burstiness, it can be surmised that the WAN traces will be more bursty. We also conducted simulations where a set of LAN traces synthesized by the VVGM model were multiplexed. The multiplexed output trace showed an interesting trend with respect to the variance. The change of variance with the number of multiplexed traces is given in Fig. 6. It can be seen that the variance of the multipliers decrease monotonically with respect to increase in multiplexing action.

6.2 Analysis Using the $f(\alpha)$ Curve

The results of the multifractal analysis done on the multiplexed traces also have provided encouraging results which support the claims we made earlier. The multifractal spectrum $f(\alpha)$ showed increased burstiness with greater level of multiplexing. The multifractal spectrum gets shifted to the left ($\alpha < 1$) with

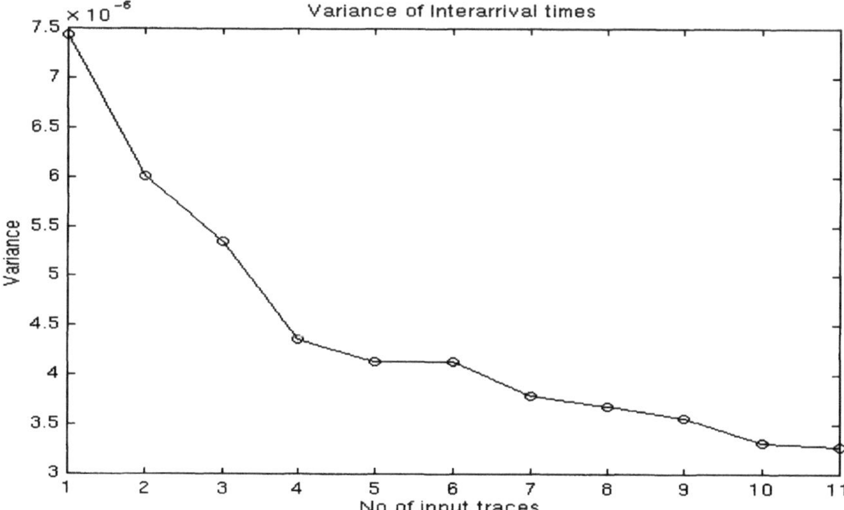

Fig. 6. Variance change of interarrival times

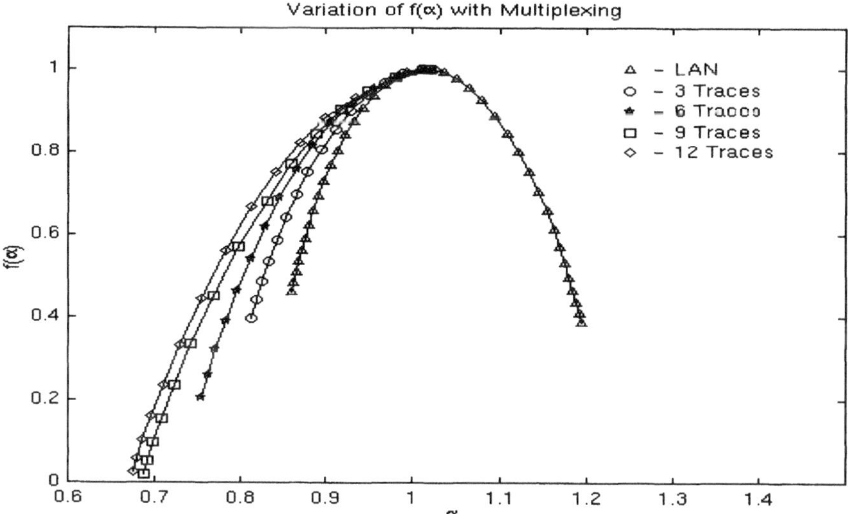

Fig. 7. Variation in $f(\alpha)$ with multiplexing

increasing level of multiplexing. Figure 7 shows the $f(\alpha)$ curve for various degrees of multiplexing. The change in α_{min} for the various levels of multiplexing are given in Table 2. From the table, it can be seen that there is a monotonic decline in the value of α_{min} as the degree of multiplexing increases. This has given conclusive evidence that with respect to the phenomenon of inter arrival times, the multiplexing leads to increased burstiness. In a later section, we will compare the results obtained on the analysis of multiplexing with self similar traffic and arrive at conclusion whether packet lengths of inter arrival times is the parameter of importance for broadband traffic control.

6.3 Analysis of Multiplexing Bursty Traces Using Entropy

The problem of multiplexing or aggregation of bursty traffic which possess multifractal nature is similar to mixing of a turbulent fluid flows. A strict analysis of the process is extremely difficult to perform. However analysis of the aggregated process can be done with techniques borrowed from Information theory and statistical mechanics – namely 'entropy' [30],[31]. The definition of entropy in the classical sense is as follows. Let p_i be the probability of occupancy of the ith state by a system. Let the total number of states be N. The entropy of the system is defined as

$$H = -\sum_{i=1}^{N} p_i \log p_i \tag{9}$$

In the present situation, where we are interested in the study of the interarrival times of traffic flow, we characterize the state space as a two dimensional one composed of the present and the past interarrival time of traffic. The time series $X_i, i = 1, \ldots, N$ is transformed into a two dimensional phase space (X_i, X_{i-1}). On obtaining the phase space of the system, we divide it into cells of equal size ϵ. The phase space thus obtained is used to characterize the trajectory of evolution of the system. The probability that the trajectory passes through some point in the phase space is obtained by counting the number of points in the cell encompassing the given point. A very important observation we made during this experiment was that irrespective of the starting point of the trace, the phase space obtained was identical in the case of all the traces. This implies that the probability of occupancy of the cells constitute an invariant measure or distribution and that the system is ergodic in nature. An example of the phase space plot obtained is as indicated in Fig. 8. The entropy was calculated as according to (9). We also changed the size of cell to increase the number of cells covering the phase space and computed the entropy for each case. For the experiment the data was obtained from multiplexing 3, 6, 9 and 12 multifractal traces (VVGM synthesized). The results obtained has shown that the entropy of the multiplexed trace decreases with increase in the number of traces being multiplexed. When the number of multiplexed traces increases, the perceived interarrival time in the resultant trace becomes very small. Thus the phase space of such a system occupies a relatively smaller area. The probability that the interarrival traces will show significant deviation will decrease as the number of traces that are multiplexed increases. Another interesting aspect was the increase of entropy with

Table 2. Change in α_{min} with multiplexing

Trace	LAN	No of Multiplexed traces			
		3	6	9	12
α_{min}	0.8602	0.8118	0.7537	0.6865	0.6749

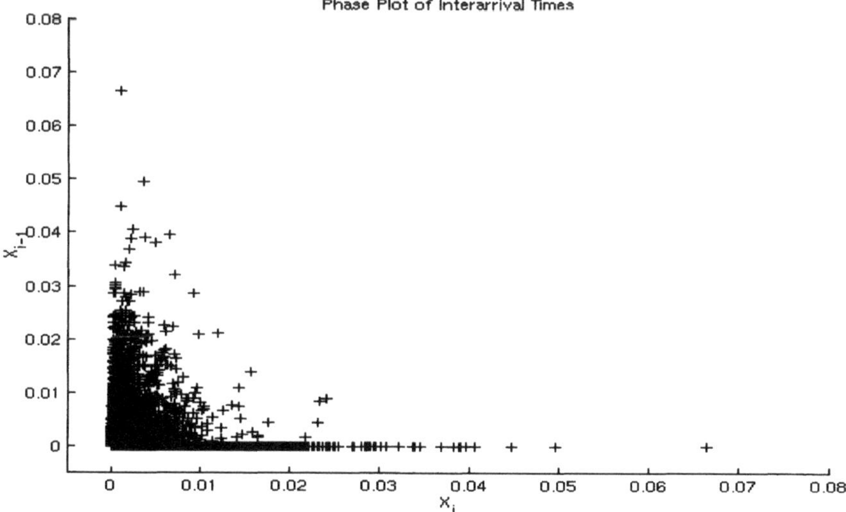

Fig. 8. Phase plot of interarrival times

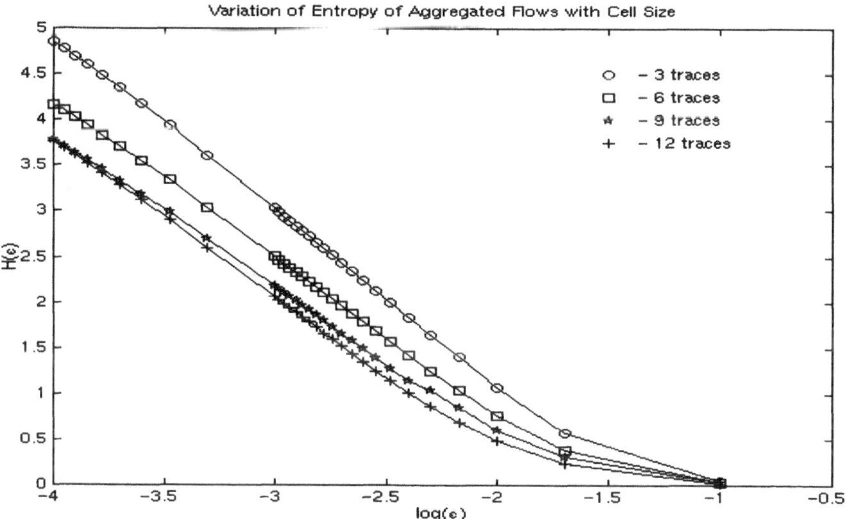

Fig. 9. Variation of entropy with multiplexing

the decrease in cell size. The traces with different levels of multiplexing showed identical behavior in the increase of entropy. Figure 9 illustrates the increase on entropy with the decrease in cell size. It can be seen that there is a scaling relation of the form

$$H(\epsilon) = -K_1 \; \log \epsilon + K_2 \tag{10}$$

where $H(\epsilon)$ is the entropy measured with cell size ϵ. Another interesting feature that we found during the experiment was that the increase in the number of cells

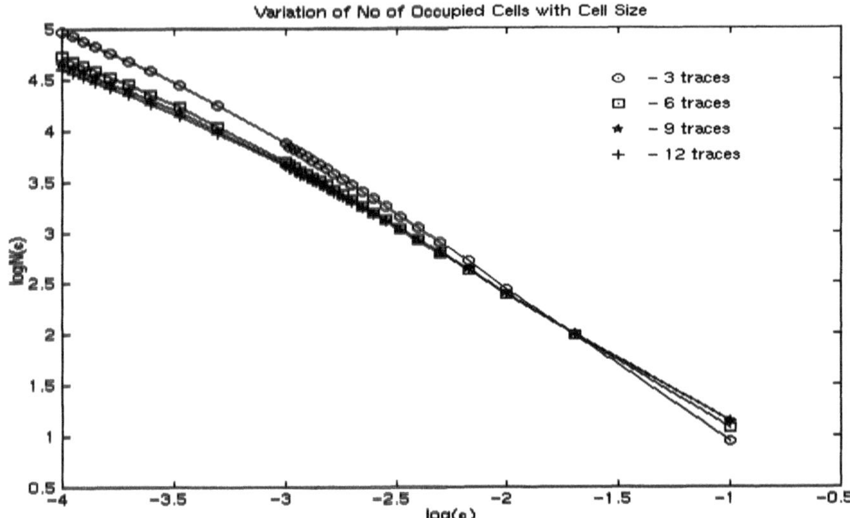

Fig. 10. Variation of number of occupied cells with multiplexing

with non zero occupation probability follows a scaling law. The scaling present is almost independent of the level of multiplexing in the resultant trace. The scaling law for the number of cells with non zero occupation probability is found to be as

$$\log N(\epsilon) = -\beta \ \log(\epsilon) + K \tag{11}$$

where $N(\epsilon)$ is the no of cells with non zero occupation probability, ϵ is the cell size and β is the slope of the log – log regression curve. Figure 10 illustrates the scaling present in the number of cells with non zero probability in the phase space of the system. It can be seen that the slope β is approximately same for all the cases. The value of the slope for the best fit curve was found to be $\beta = -1.2012$. This implies that the nature of the scaling law for the cells with non zero occupation probability is almost the same irrespective of the level of multiplexing present.

7 Information Theoretic Analysis of Multifractal Systems

The traditional linear models used in engineering have limitations in the analysis and study of various physical phenomena that exhibits non linearity. The emergence of chaos theory have helped in the detailed study of many non linear systems. However these methods suffer from the disadvantage that some priori model of the system has to be known which is quite difficult to arrive at in many cases. In this context, the use of information theory for system analysis assumes greater importance. The objective in such a case is to arrive at conclusions regarding the behavior of the system from the trajectories of it's phase space than get explicit closed form relationships between them input and output. A chaotic system is characterized by a phase space which is bounded but at the same time with trajectories that exhibit complex nature. The analysis of the phase space of such systems can be attempted by using entropy as a criterion. We investigate the mathematical relations that govern the entropy of a system that display multifractal properties. The variation of the entropy is analyzed and a mathematical relationship is obtained for the same. The relation with the concept of "Generalized Dimensions" proposed by Renyi [32] is also brought out to supplement the result.

7.1 Variation of Entropy with Scale

The strict definition of a multifractal system is quite hard to arrive at as multifractality is more often seen as a property of a process than that of a system. However one can define the concept of a multifractal system as follows. We assume that the system has a d dimensional phase space which is finitely bounded. We assume a probability measure over the entire phase space. In order to characterize the trajectory of the system, we look at the probability that the trajectory occupies a particular region in the phase space. Divide the phase space into d dimensional cells of size ϵ which can be varied. For a multifractal system, the number of occupied cells in the phase space shows a scaling relation of the form [33]

$$N(\epsilon) \sim \epsilon^{-\beta} \tag{12}$$

This imply that the probability of finding a cell with non zero probability of trajectory passing through scales as

$$p(\epsilon) = \frac{N(\epsilon)}{N} \sim \frac{1}{N}\epsilon^{-\beta} \tag{13}$$

where N is the total number of cells in the phase space. The entropy variation of the system can now be studied. We are considering the Shannon entropy measure given by

$$H = -\sum_i p_i \log p_i \tag{14}$$

For the multifractal system, we get on substituting (13),

$$\begin{aligned} H(\epsilon) &\sim -\sum_{i=1}^{N} \frac{\epsilon_i^{-\beta}}{N} \log \left[\frac{\epsilon_i^{-\beta}}{N}\right] \\ &-\sum_{i=1}^{N} \frac{\epsilon_i^{-\beta}}{N} \left[-\beta \log \epsilon_i - \log \frac{1}{N}\right] \\ &-\sum_{i=1}^{N} \frac{\epsilon_i^{-\beta}}{N}\, \beta\, \log \epsilon_i - \sum_{i=1}^{N} \frac{\epsilon_i^{-\beta}}{N} \log \frac{1}{N} \end{aligned} \tag{15}$$

where the index i runs through each cell ϵ_i. Since the size of each cell is the same, the above relation can be simplified as

$$\begin{aligned} H(\epsilon) &= -\epsilon^{-\beta}\, \beta\, \log \epsilon - \epsilon^{-\beta} \log \frac{1}{N} \\ &= -K_1 \log \epsilon + K_2 \end{aligned} \tag{16}$$

7.2 Relation with Generalized Dimensions

Alfred Renyi defined generalized dimensions as [32]

$$D_q = \lim_{\epsilon \to 0} \frac{1}{q-1} \frac{\log \sum_{i=1}^{N} p_i^q}{\log \epsilon} \tag{17}$$

The expression for the parameter D_1 can be evaluated as

$$D_1 = \lim_{q \to 1} D_q = \lim_{q \to 1} \left\{ \lim_{\epsilon \to 0} \frac{1}{q-1} \frac{\log \sum_{i=1}^{N} p_i^q}{\log \epsilon} \right\} \tag{18}$$

Using L'Hospital's rule, the above expression can be simplified as

$$\begin{aligned} D_1 &= \lim_{\epsilon \to 0} \frac{1}{\log \epsilon} \lim_{q \to 1} \left\{ \frac{\sum_{i=1}^{N} p_i^q \log p_i}{\sum_{i=1}^{N} p_i^q} \right\} \\ &= \lim_{\epsilon \to 0} \frac{\sum_{i=1}^{N} p_i \log p_i}{\log \epsilon} = \lim_{\epsilon \to 0} \frac{H(\epsilon)}{\log \epsilon} \end{aligned} \tag{19}$$

From (19) we can write,

$$H(\epsilon) \sim D_1 \ \log \epsilon \tag{20}$$

Comparing (16) and (19), we can arrive at the conclusion that $D_1 = -K_1$. D_1 is termed as the *Information Dimension* in dynamical systems literature. It gives the change of entropy when the measurement scale (ϵ) is changed. The validity of the above relation can be tested with data having multifractal nature. In the present case, we have shown the results obtained after the analysis of multiplexing of traffic inter arrival times that exhibit multifractal nature. Even though these models do capture the multifractality observed by the traces, the analytical treatment of their multiplexing of is complicated and closed form expression for the system parameters are hard to obtain.

8 Control of Broadband Traffic: Packet Count or Interarrival Times

We had conducted empirical tests on data obtained through simulation with OPNET for investigating the multifractal character of the data traffic. In this experiment, the data that we obtained was the packet counts per unit time (per second). The underlying assumption in this analysis is that since the inter arrival times in a network showed multifractal scaling, the packet traces also could reflect this behavior. From the results, it was concluded that the traces show multifractal character [34]. This result has made it possible for the analytical treatment of aggregation of multifractal data traces. Consider a situation in which traffic from various paths are aggregating to a single flow. The analytical expression for the combination of interarrival times of the packets is not easy to come with. An alternative is to look at the aggregation of the packet traces. In the realistic scenario, the packet lengths get added up on aggregation. This can be used to model the aggregation of multifractal processes since we have found evidence of multifractal behavior in packet lengths. Let there be M multifractal packet traces $X_i(n); i = 1 \ldots M$. Let the traces be aggregated to produce the resultant trace $Y(n)$ as

$$Y(n) = \sum_{i=1}^{M} X_i(n) \tag{21}$$

The traces $X_i(n); i = 1 \ldots M$ are all assumed to be a cascade process at stage N with the identical multiplier distributions with mean 0.5. The aggregated trace can be expressed as

$$Y(n) = \sum_{i=1}^{M} u_1^{(i)} \ldots u_N^{(i)} \tag{22}$$

where $u_k^{(i)}$ is the multiplier for the ith input trace at the kth stage in the cascade. We further assume that the multipliers are independent and proceed to derive closed form expression for the effective Hurst exponent of the aggregated trace. For this analysis, we use the result that we proved earlier where we have shown that multiplicative multifractal cascades possess a global scaling exponent that is analogous to Hurst parameter for a self similar process [29]. The relationship between the second moment of the multipliers and the effective Hurst exponent was shown to be

$$H_{eff} = 1 - \frac{1}{2}\log_2(4\mu_2) \tag{23}$$

Define $W^{(m)}$ as

$$\begin{aligned} W^{(m)} &= \frac{1}{m} \sum_{k=(i-1)m+1}^{im} Y(k) \\ &= \frac{1}{m} \sum_{k=(i-1)m+1}^{im} \left\{ \sum_{j=1}^{M} X_j(k) \right\} \end{aligned} \tag{24}$$

Let $m = 2^l$ so that $W^{(m)}$ is the rescaled cascade at stage $(N-l)$. i.e $W^{(m)} = 2^{-l} Y_{N-l}(k)$.

$$\begin{aligned} var[W^{(m)}] &= var\left[2^{-l} \sum_{j-1}^{M} X_{N-l}^{(j)}(k) \right] \\ &= 2^{-2l} \sum_{j=1}^{M} var\left(X_{N-l}^{(j)}(k) \right) \end{aligned} \tag{25}$$

Substituting the value of the expression of the variance of multifractal cascades as obtained earlier and taking the limiting case where $N \to \infty$, the expression for variance can be simplified as

$$var[W^{(m)}] = 2^{-2l} \left[\sum_{j=1}^{M} \left(\mu_2^{(j)}\right)^N \left\{4\mu_2^{(j)}\right\}^{-log_2 m} \right]$$
$$= K_l \sum_{j=1}^{M} \left(\mu_2^{(j)}\right)^{N-\log_2 m} \tag{26}$$

where K_l is a constant and μ_2^j is the second moment of the multiplier distribution for the jth trace. It can be seen that the above sum of exponentials will be influenced by the term which will have the maximum of μ_2. The effective Hurst parameter or the global scaling parameter will be thus influenced by the term $\mu_2' = \max\left(\mu_2^{(j)}\right); j = 1 \ldots M$

$$H_{eff}' = 1 - \frac{1}{2}\log_2 \mu_2' \tag{27}$$

An interesting factor to notice is that as $\mu_2' = \max\left(\mu_2^{(j)}\right); j = 1 \ldots M, H_{eff}$ will be the minimum of the Hurst parameters of the traces that are aggregated. The aggregation process smoothen the packet count traces while it increases the burstiness of the inter arrival process. This is important in network flow control problems. The interarrival timings are to be monitored closely and the information of whether the traffic is bursty has to be obtained from them rather than the packet count information. This result has made the estimation problem of the burstiness parameter α in a multifractal trace that corresponds to interarrival traces more important.

9 Conclusion

The interdependence of various research disciplines as vast as turbulence analysis, information theory and broadband tele traffic modeling is the main focus of discussion in this article. The measured broadband tele traffic data possess properties that are seen in cascade processes. The range and nature of the scaling present in the data calls for a multiplicative multifractal model to model it's statistical properties. With this observation, we have proposed a new model, namely the V.V.G.M multiplicative process in order to model the traffic interarrival times. The results of the comparison of the statistical and performance tests have confirmed that usefulness of the model in approximating the properties of the measured traffic data. We have not investigated the reason for the multifractal nature of broadband traffic but believe that it is due to the hierarchy inherent to telecommunication protocols where data transfers involve successive sub division among entities. The knowledge of the multifractal scaling will allow more precise statistics than the knowledge of a single Hurst exponent H as in the case of monofractal models. There is a wide scope of work to be done in the area of control of broadband networks incorporating traffic data information [35], [36]. The advent of new techniques and modeling tools like multiplicative multifractal models will help towards this goal. We have derived expressions for the variation of entropy with scale for a system that exhibit multifractal character. We also brought out the relationship of the entropy scaling behavior with information dimension which is a particular instance of generalized dimensions. In order to verify the validity of the expressions obtained, we conducted simulation experiments in in multiplexing of multifractal processes. For the analysis of multiplexing of interarrival times, an information theoretic approach is attempted where we study the variation of entropy of the system with increase in the degree of multiplexing. The experimental results obtained corroborate with the expressions that we derived. The analysis of entropy variation across scale can be used to identify the presence of multifractal nature in time series obtained from systems that are hard to parameterize and analyze using the traditional methods.

References

1. P. Abry and R. Baranuik and P. Flandrin and R. Riedi and D. Veitch: IEEE Signal Processing Magazine **19**, 28 (2002)
2. L. Kleinrock: *Queuing Systems: Vol. I: Theory.* (John Wiley & Sons, Inc, New York 1975)
3. W. E. Leland, M. S. Taqqu, W. Willinger and D. V. Wilson: IEEE/ACM Transactions on Networking **2**, 1 (1994)
4. B. B. Mandelbrot, J. W. Van Ness: SIAM Review **10**, 422 (1968)
5. W. Willinger, M. S. Taqqu, R. Sherman and D. V. Wilson: IEEE/ACM Transactions on Networking **5**, 71 (1997)
6. I. Norros: Queuing Systems **16**, 387 (1994)
7. A. Erramilli and O. Narayan and W. Willinger: IEEE/ ACM Trans. on Networking **4**, 209 (1996)
8. J. L. Vehel, R. Riedi: Multifractal Properties of TCP Traffic, a Numerical Study. INRIA Technical Report No:RR-3129, INRIA Rocenquort (1997)
9. A. Feldmann, A. C. Gilbert and W. Willinger, "Data Networks as Cascades: Investigating the Multifractal Nature of Internet WAN Traffic", In *Proc. of SIGCOMM - 1998*, pp 25-38
10. A. Feldmann, A. C. Gilbert, W. Willinger and T. Kurtz: Computer Communication Review **28**, 5 (1998)
11. A. Feldmann, A. C. Gilbert, P. Huang and W. Willinger: "Dynamics of IP Traffic: A Study of the Role of Variability and the Impact of Control". In *Proc. of SIGCOMM - 1999*, pp. 301 313
12. A. C. Gilbert: Applied and Computational Harmonic Analysis **10**, 185 (2001)
13. P. Abry, D. Veitch: IEEE Transactions on Information Theory, **44**, 2 (1998)
14. A. N. Kolmogorov: J. Fluid Mech. **13**, 82 (1962)
15. C. Meneveau, K. Sreenivasan: Phys. Rev. Lett. **59**, 1424 (1987)
16. U. Frisch: *Turbulence, the legacy of A. N. Kolmogorov.* (Cambridge University Press, Cambridge 1985)
17. B B. Mandelbrot: Journal of Fluid Mechanics **62**, 331 (1974)
18. R. Benzi, G. Paladin, G. Parisi and A. Vulpiani: J. Phys. **18**, (1984)

19. R. H. Riedi, M. S. Crouse, V. J. Ribeiro and R. G. Baraniuk: IEEE Transactions on Information Theory **45**, 992 (1999)
20. P. Mannersalo, I. Norros: Multifractal analysis of real ATM traffic: a first look. COST 257TD(97) Report VTT Information Technology Finland (1997)
21. D. R. Bickel, B. J. West: Fractals **6** 211 (1998)
22. S. Ghashghaie: Nature **381** 767 (1996)
23. A. Fisher, P. Calvet and B. B. Mandelbrot: Multifractal analysis of USD/DM exchange rates. Yale University Working Paper (1998)
24. F. P. Agterberg: Aspects of Multifractal Modeling. Publication of Geological Survey of Canada (2001)
25. A. Turiel, N. Parga: Neural Computation **12**, 763 (2000)
26. V. Gupta, E. Waymire: Journal of Applied Meteorology **32**, 251 (1993)
27. R. H. Riedi: Introduction to Multifractals. Rice University Technical Report 99-06 (1999)
28. C. J. G. Evertsz and B. B. Mandelbrot: In *H. O. Petgen, H. Jurgens, and D. Saupe, editors, Chaos and Fractals: New Frontiers in Science.* (Springer Verlag, New York 1992)
29. M. Krishna, V. M. Gadre, U. B. Desai: 'Global Scaling Exponent For Variable Variance Gaussian Multiplicative (VVGM) Multifractal Cascades' In *Proc. of SPCOM' - 2001* pp. 19-25
30. F. Rief: *Fundamentals of Statistical and Thermal Physics.* (Mc Graw Hill, New York 1985)
31. C. Beck, F. Schlogl: *Thermodynamics and Chaos.* (Cambridge University Press, Cambridge 1993)
32. Alfred Renyi, *Probability theory.* (North-Holland, Amsterdam 1970)
33. Robert C. Hilborn, *Chaos and nonlinear dynamics : An introduction for scientists and engineers.* (Oxford University Press, New York 1994)
34. Murali Krishna. P, Vikram M. Gadre, Uday B. Desai, "Multifractal Analysis of LAN Traffic Generated Using OPNET" *Presented at OPNETWORK - 2001*, Washington D.C, August 2001
35. A. Ephremides and B. Hajek: IEEE Trans. on Information Theory **44**, 2416 (1998)
36. A. Coates,A. O. Hero III, R. Nowak and B. Yu: IEEE Signal Processing Magazine **19**, 47 (2002)

Lecture Notes in Physics

For information about Vols. 1–577
please contact your bookseller or Springer-Verlag
LNP Online archive: http://www.springerlink.com/series/lnp/

Vol. 578: D. Blaschke, N. K. Glendenning, A. Sedrakian (Eds.), Physics of Neutron Star Interiors.

Vol. 579: R. Haug, H. Schoeller (Eds.), Interacting Electrons in Nanostructures.

Vol. 580: K. Baberschke, M. Donath, W. Nolting (Eds.), Band-Ferromagnetism: Ground-State and Finite-Temperature Phenomena.

Vol.581: J. M. Arias, M. Lozano (Eds.), An Advanced Course in Modern Nuclear Physics.

Vol.582: N. J. Balmforth, A. Provenzale (Eds.), Geomorphological Fluid Mechanics.

Vol.583: W. Plessas, L. Mathelitsch (Eds.), Lectures on Quark Matter.

Vol.584: W. Köhler, S. Wiegand (Eds.), Thermal Nonequilibrium Phenomena in Fluid Mixtures.

Vol.585: M. Lässig, A. Valleriani (Eds.), Biological Evolution and Statistical Physics.

Vol.586: Y. Auregan, A. Maurel, V. Pagneux, J.-F. Pinton (Eds.), Sound–Flow Interactions.

Vol.587: D. Heiss (Ed.), Fundamentals of Quantum Information. Quantum Computation, Communication, Decoherence and All That.

Vol.588: Y. Watanabe, S. Heun, G. Salviati, N. Yamamoto (Eds.), Nanoscale Spectroscopy and Its Applications to Semiconductor Research.

Vol.589: A. W. Guthmann, M. Georganopoulos, A. Marcowith, K. Manolakou (Eds.), Relativistic Flows in Astrophysics.

Vol.590: D. Benest, C. Froeschlé (Eds.), Singularities in Gravitational Systems. Applications to Chaotic Transport in the Solar System.

Vol.591: M. Beyer (Ed.), CP Violation in Particle, Nuclear and Astrophysics.

Vol.592: S. Cotsakis, L. Papantonopoulos (Eds.), Cosmological Crossroads. An Advanced Course in Mathematical, Physical and String Cosmology.

Vol.593: D. Shi, B. Aktaş, L. Pust, F. Mikhailov (Eds.), Nanostructured Magnetic Materials and Their Applications.

Vol.594: S. Odenbach (Ed.),Ferrofluids. Magnetical Controllable Fluids and Their Applications.

Vol.595: C. Berthier, L. P. Lévy, G. Martinez (Eds.), High Magnetic Fields. Applications in Condensed Matter Physics and Spectroscopy.

Vol.596: F. Scheck, H. Upmeier, W. Werner (Eds.), Noncommutative Geometry and the Standard Model of Elememtary Particle Physics.

Vol.597: P. Garbaczewski, R. Olkiewicz (Eds.), Dynamics of Dissipation.

Vol.598: K. Weiler (Ed.), Supernovae and Gamma-Ray Bursters.

Vol.599: J.P. Rozelot (Ed.), The Sun's Surface and Subsurface. Investigating Shape and Irradiance.

Vol.600: K. Mecke, D. Stoyan (Eds.), Morphology of Condensed Matter. Physcis and Geometry of Spatial Complex Systems.

Vol.601: F. Mezei, C. Pappas, T. Gutberlet (Eds.), Neutron Spin Echo Spectroscopy. Basics, Trends and Applications.

Vol.602: T. Dauxois, S. Ruffo, E. Arimondo (Eds.), Dynamics and Thermodynamics of Systems with Long Range Interactions.

Vol.603: C. Noce, A. Vecchione, M. Cuoco, A. Romano (Eds.), Ruthenate and Rutheno-Cuprate Materials. Superconductivity, Magnetism and Quantum Phase.

Vol.604: J. Frauendiener, H. Friedrich (Eds.), The Conformal Structure of Space-Time: Geometry, Analysis, Numerics.

Vol.605: G. Ciccotti, M. Mareschal, P. Nielaba (Eds.), Bridging Time Scales: Molecular Simulations for the Next Decade.

Vol.606: J.-U. Sommer, G. Reiter (Eds.), Polymer Crystallization. Obervations, Concepts and Interpretations.

Vol.607: R. Guzzi (Ed.), Exploring the Atmosphere by Remote Sensing Techniques.

Vol.608: F. Courbin, D. Minniti (Eds.), Gravitational Lensing:An Astrophysical Tool.

Vol.609: T. Henning (Ed.), Astromineralogy.

Vol.610: M. Ristig, K. Gernoth (Eds.), Particle Scattering, X-Ray Diffraction, and Microstructure of Solids and Liquids.

Vol.611: A. Buchleitner, K. Hornberger (Eds.), Coherent Evolution in Noisy Environments.

Vol.612 L. Klein, (Ed.), Energy Conversion and Particle Acceleration in the Solar Corona.

Vol.613 K. Porsezian, V.C. Kuriakose, (Eds.), Optical Solitons. Theoretical and Experimental Challenges.

Vol.614 E. Falgarone, T. Passot (Eds.), Turbulence and Magnetic Fields in Astrophysics.

Vol.615 J. Büchner, C.T. Dum, M. Scholer (Eds.), Space Plasma Simulation.

Vol.616 J. Trampetic, J. Wess (Eds.), Particle Physics in the New Millenium.

Vol.617 L. Fernández-Jambrina, L. M. González-Romero (Eds.), Current Trends in Relativistic Astrophysics, Theoretical, Numerical, Observational

Vol.618 M. Degli Esposti, S. Graffi (Eds.), The Mathematical Aspects of Quantum Maps

Vol.619 H.M. Antia, A. Bhatnagar, P. Ulmschneider (Eds.), Lectures on Solar Physics

Vol.620 C. Fiolhais, F. Nogueira, M. Marques (Eds.), A Primer in Density Functional Theory

Vol.621 G. Rangarajan, M. Ding (Eds.), Processes with Long-Range Correlations

GPSR Compliance
The European Union's (EU) General Product Safety Regulation (GPSR) is a set of rules that requires consumer products to be safe and our obligations to ensure this.

If you have any concerns about our products, you can contact us on

ProductSafety@springernature.com

In case Publisher is established outside the EU, the EU authorized representative is:

Springer Nature Customer Service Center GmbH
Europaplatz 3
69115 Heidelberg, Germany

www.ingramcontent.com/pod-product-compliance
Ingram Content Group UK Ltd.
Pitfield, Milton Keynes, MK11 3LW, UK
UKHW021858190726
13853UKWH00003B/1328

* 9 7 8 3 6 6 2 1 4 3 8 4 1 *